Nanotechnology in the Life Sciences

Series Editor

Ram Prasad, Department of Botany
Mahatma Gandhi Central University
Motihari, India

Nano and biotechnology are two of the 21st century's most promising technologies. Nanotechnology is demarcated as the design, development, and application of materials and devices whose least functional make up is on a nanometer scale (1 to 100 nm). Meanwhile, biotechnology deals with metabolic and other physiological developments of biological subjects including microorganisms. These microbial processes have opened up new opportunities to explore novel applications, for example, the biosynthesis of metal nanomaterials, with the implication that these two technologies (i.e., thus nanobiotechnology) can play a vital role in developing and executing many valuable tools in the study of life. Nanotechnology is very diverse, ranging from extensions of conventional device physics to completely new approaches based upon molecular self-assembly, from developing new materials with dimensions on the nanoscale, to investigating whether we can directly control matters on/in the atomic scale level. This idea entails its application to diverse fields of science such as plant biology, organic chemistry, agriculture, the food industry, and more.

Nanobiotechnology offers a wide range of uses in medicine, agriculture, and the environment. Many diseases that do not have cures today may be cured by nanotechnology in the future. Use of nanotechnology in medical therapeutics needs adequate evaluation of its risk and safety factors. Scientists who are against the use of nanotechnology also agree that advancement in nanotechnology should continue because this field promises great benefits, but testing should be carried out to ensure its safety in people. It is possible that nanomedicine in the future will play a crucial role in the treatment of human and plant diseases, and also in the enhancement of normal human physiology and plant systems, respectively. If everything proceeds as expected, nanobiotechnology will, one day, become an inevitable part of our everyday life and will help save many lives.

Chandra Shekhar Seth
Krishan K. Verma • Faheem Ahmad
Editors

Nano-formulations for Sustainable Agriculture

Frontiers in Improving Crop Resilience to Abiotic Stressors

Springer

Editors
Chandra Shekhar Seth
Department of Botany
University of Delhi
New Delhi, Delhi, India

Krishan K. Verma
Sugarcane Research Institute
Guangxi Academy of Agricultural Sciences
Nanning, Guangxi, China

Faheem Ahmad
Department of Botany
Aligarh Muslim University
Aligarh, Uttar Pradesh, India

ISSN 2523-8027 ISSN 2523-8035 (electronic)
Nanotechnology in the Life Sciences
ISBN 978-3-031-90616-9 ISBN 978-3-031-90617-6 (eBook)
https://doi.org/10.1007/978-3-031-90617-6

This Springer imprint is published by the registered company Springer Nature Switzerland AG
The registered company address is: Gewerbestrasse 11, 6330 Cham, Switzerland

Preface

Agricultural food production faces multivariate agroclimatic stresses that affect crop growth and development in the era of climate change worldwide. The response of crop plants to environmental stresses and climate change depends on the different factors, i.e., variety of crops, type of stress, and the frequency and duration of stress. Sustainable agriculture and crop production strategies can assist to improve soil profile, water management, and other resources essential for plant development. To meet the growing demand for food security and safety, scientific communities are working toward minimizing the harmful effects of adverse climatic conditions on crops, improving productivity of crop, and ensuring food safety and security.

Nanobioscience is a fascinating and rapidly developing area of research that has led to major innovative approaches. Nanobiotechnology-based formulations provide significant strategies to agro-ecosystems-related problems and achieve sustainable and safe future for agriculture in years to come. Nanotechnology has been attracting in upcoming years due to its large number of advances in different aspects. In sustainable agriculture, nanotechnology is mainly utilized in nanofertilizers and nanopesticides to amplify the quality of output and level of nutrients for promoting growth, development, and productivity/quality of grains/fruits and upregulating plant stress tolerance efficiency to adverse environmental conditions. Nanosensors, for instance, enable the early detection of the onset and type of plant stress, allowing for more precise and targeted interventions. Moreover, the unique physicochemical characteristics of nanoparticles, such as enhanced biochemical activity and effectiveness, have been shown to confer diverse impacts on plants, including improved tolerance to abiotic and biotic stresses.

Nanoformulated materials offer the possibility to produce more effective and less damaging agrochemicals in the eco-envirosystems. Smart delivery systems for nanosensors, molecules that can assist in finding environmental stressors before they can affect crop productivity, are being developed and applied. It provides advanced applications for the transformation of cell genetics. The application of nanoformulated particles in agriculture is being used to increase the value of foods, reduce agricultural inputs, upgrade nutrients delivery systems, and develop a longer shelf life for different products. It is important to discover the role of nanomaterials in mutation breeding, seedling growth and development, as well as their interaction with metabolites. The application of nanoformulations like nanofertilizers, nanopesticides, and sensors is utilized to monitor and manage agricultural approaches to increase the crop management, nutrient deficiency alleviation, and increase plant productivity. To address these challenges, researchers have been exploring the potential of nano-enabled technologies to enhance crop resilience and productivity.

This book highlights the advanced strategies of nanoformulated forms in the field of plant biology. The authors summarized the role of formulated nanoparticles on rhizospheric soil, and possible environmental risks, along with advance research aspects and recommendations for sustainable agriculture. It covers nanomaterials delivery systems for the release of nanoformulations, field-sensing applications to observe adverse agroclimatic stresses, and upgradation of plant genotypes during abiotic stressors. This excellent book contents will surely be beneficial for graduate-postgraduate students, researchers, and policymakers associated with the application of nanomaterials in sustainable agriculture.

New Delhi, India — Chandra Shekhar Seth
Nanning, Guangxi, China — Krishan K. Verma
Aligarh, Uttar Pradesh, India — Faheem Ahmad

Contents

Editors and Contributors

About the Editors

Chandra Shekhar Seth is an Associate Professor at Department of Botany, University of Delhi, New Delhi-110007, India. He obtained his MSc degree in Botany from Banaras Hindu University, Varanasi (UP) in 2003 and PhD degree in Botany from University of Lucknow (UP) in 2008. He currently teaches the course papers naming Plant physiology and Biochemistry, Topics in Plant Physiology and Biochemistry, Algae, Environment and Human Welfare at postgraduate level, and Phytohormone-assisted abiotic stress management in plants at Ph.D. level. His major research interests include how various phytohormones particularly Salicyclic acid (SA), Jasmonic acid (JA), and Nitric Oxide (NO) mitigate heavy metals (HMs) and salinity induced toxicity in plants, photosynthesis and nitrogen metabolism under abiotic stress, and role of nanomaterials in mitigating the HMs and salt stresses in plants. Oxidative stress and tolerance mechanism via antioxidants and phytochelatins, phytoremediation of HMs, and potential use of algae as biofuels, nutrients, and pigments are other important interesting area of his work. He is the fellow of two prestigious societies, namely, Fellow of the Academy of Environmental Biology (FAEB), Lucknow-226020 (UP), India, and Fellow Society for Plant Research (FSPR), Meerut-250110, UP, India. He has been the Life member of The Indian Science Congress Association (ISCA), Kolkata-700017, India; International Society of Environmental Botanist (ISEB), Lucknow-226001 (UP); The Academy of Environmental Biology (AEB), Lucknow-226020 (UP), India; International Society of Plant Morphologists (ISPM), New Delhi-110007, India; and Delhi University Botanical Society (DUBS), New Delhi-110007, India. He has guided more than 25 M.Sc., 4 M.Phil., and 3 Ph.D. students, and is currently guiding 6 PhD students in his focused research areas. He has published many peer-reviewed excellent research papers, review papers, and several books and book chapters related to this field. He is an internationally recognized reviewer for many journals. He has been serving as an editorial board member for *Plant Physiology and Biochemistry* (Elsevier journal) and *Soil and Environmental Health* (an open access Elsevier journal).

Krishan K. Verma is working as a Foreign Expert at Sugarcane Research Institute, Guangxi Academy of Agricultural Sciences, Nanning, Guangxi, China. He has 15 years of research experience in field of biotic and abiotic stresses, i.e., environmental toxicology, plant molecular biology, and metallic nanoparticles. His research focuses on environmental toxicology, bioengineering modeling, plant physiology, molecular biology, and their impacts on growth, physiological, and molecular adaptation strategies of plants to unfavorable environmental situations, and how they affect plant structure and productivity. He obtained his master's degree (Botany) from Chhatrapati ShahuJi Maharaj University, Kanpur, India, Doctorate (Botany) from University of Lucknow, Lucknow, India, Post-doc from University of Lucknow, Lucknow, India, and Young—Scientist/Principal Investigator at Chandra Shekhar Azad University of Agriculture and Technology, Kanpur, India, supported by Science and Engineering Research Board (SERB), Department of Science and Technology (DST), Govt. of India, India. He has published more than 120 scientific research articles/reviews, 15 books, and 15 book chapters. He is an internationally recognized reviewer. He has received "Award of Excellence–2021" in the contribution of Agriculture Science from Sugarcane Society of Guangxi Academy of Agricultural Science, Nanning, Guangxi, China. He has also been elected as Fellow of Society for Sugarcane Research & Promotion (FSSRP-2024) for his outstanding contribution to sugarcane crop production. In year 2023, he is included in list of 2% worlds potential scientists by Elsevier BV publishing house. He is also Regional Editor of *Fluoride* journal published by International Society for Fluoride Research Inc., Dunedin, New Zealand. He has been serving as an editorial board member of various peer-reviewed journals. https://orcid.org/0000-0002-5501-7905

Faheem Ahmad, PhD, is a Senior Assistant Professor in Department of Botany at Aligarh Muslim University, Aligarh, U.P., India. His research is based on the study of plant-nematode interaction, plant disease, microbe-mediated biotic stress management, and agricultural biotechnology and microbiology. He is the author of more than 50 papers, associate editor of the *Journal of Plant Diseases and Protection* (Springer), and academic editor of *Peer J—Life & Environment*. He has guided one PhD student and several MSc students to date. He is a recipient of several fellowships and awards, including NRF, TSRO Young Scientist, SERB-DST Young Scientist, and TUBITAK Foreign Researcher Award. He is a fellow of the International Society for Development and Sustainability (Japan) and Bose Science Society (India). He gained postdoctoral experiences at Ehime University (Japan), North-West University (South Africa), Ege University (Turkey), and National Sun Yat-Sen University (Taiwan).

Contributors

Peter Adeolu Adedibu School of Advanced Engineering Studies, Institute of Biotechnology, Bioengineering and Food Systems, Far Eastern Federal University, Russky Island, Vladivostok, Russia

Foluso Oyedotun Agunbiade Perishable Crops Research Deaprtment Nigerian Stored Products Research Institute, Ilorin, Nigeria

Deartment of Chemistry, University of Lagos, Akoka-Yaba, Nigeria

Christiana Oluwatoyin Ajanaku Department of Physical Sciences, College of Science and Engineering, Landmark University, Omuaran, Kwara State, Nigeria

Department of Chemistry, College of Science and Engineering, Covenant University, Ota, Ogun State, Nigeria

Kolawole Oluseyi Ajanaku Department of Chemistry, College of Science and Engineering, Covenant University, Ota, Ogun State, Nigeria

Anuoluwapo Abimbola Akinsiku Department of Chemistry, College of Science and Engineering, Covenant University, Ota, Ogun State, Nigeria

Samir Ananou Laboratory of Microbial Biotechnology and Bioactive Molecules, Faculty of Sciences and Techniques, University of Sidi Mohamed Ben Abdellah, Fez, Morocco

David Adedayo Animasaun Department of Plant Biology, Faculty of Life Sciences, University of Ilorin, Ilorin, Nigeria

Plant Tissue Culture Lab, Central Research Laboratories, University of Ilorin, Ilorin, Nigeria

Saburi Abimbola Atanda Perishable Crops Research Deaprtment Nigerian Stored Products Research Institute, Ilorin, Nigeria

Mawuli K. Azameti Department of Applied Biology, C. K. Tedam University of Technology and Applied Sciences, Navrongo, Ghana

Oumaima Benaissa Phytopathology Unit, Department of Plant Protection, Ecole Nationale d'Agriculture de Meknès, Meknès, Morocco

Laboratory of Biotechnology, Conservation and Valorization of Natural Resources (LBCVNR), Faculty of Sciences Dhar El Mehraz, Sidi Mohamed Ben Abdallah University, Fez, Morocco

Bharat M. Bhalerao Department of Biochemistry, Post Graduate Institute, Mahatma Phule Krishi Vidyapeeth, Rahuri, Maharashtra, India

Ilham Dehbi Phytopathology Unit, Department of Plant Protection, Ecole Nationale d 'Agriculture de Meknès, Meknès, Morocco

Laboratory of Plant Biotechnology and Molecular Biology, Faculty of Sciences, Moulay Ismail University Zitoune, Meknes, Morocco

Fahima Dilnawaz School of Biotechnology, Centurion University of Technology and Management, Bhubaneswar, Odisha, India

Joginder Singh Duhan Department of Biotechnology, Chaudhary Devi Lal University, Sirsa, India

Sozan E. El-Abeid Nanotechnology & Advanced Nano-Materials Laboratory (NANML), Mycology and Disease Survey Research Department, Plant Pathology Research Institute, Agricultural Research Center, Giza, Egypt

Genetic Diversity of Microbial Plant Pathogens Reference Lab, Plant Pathology Research Institute, Agricultural Research Center (ARC), Giza, Egypt

Marwa M. El-Deriny Nematode Diseases Department, The Egyptian Nematology Reference Lab (ENR lab), Plant Pathology Research Institute, Agricultural Research Center, Giza, Egypt

Genetic Diversity of Microbial Plant Pathogens Reference Lab, Plant Pathology Research Institute, Agricultural Research Center (ARC), Giza, Egypt

Rachid Ezzouggari Phytopathology Unit, Department of Plant Protection, Ecole Nationale d'Agriculture de Meknès, Menkès, Morocco

Laboratory of Biotechnology, Conservation and Valorization of Natural Resources (LBCVNR), Faculty of Sciences Dhar El Mehraz, Sidi Mohamed Ben Abdallah University, Fez, Morocco

Oluwatoyin Adenike Fabiyi Department of Crop Protection, Faculty of Agriculture, University of Ilorin, Ilorin, Nigeria

Priyanka Saini Department of Zoology, Deshbandhu College, University of Delhi, New Delhi, India

Dina S. S. Ibrahim Nematode Diseases Department, The Egyptian Nematology Reference Lab (ENR lab), Plant Pathology Research Institute, Agricultural Research Center, Giza, Egypt

Genetic Diversity of Microbial Plant Pathogens Reference Lab, Plant Pathology Research Institute, Agricultural Research Center (ARC), Giza, Egypt

Abdul-Wahab Imoro Department of Forestry and Forest Resources Management, University for Development Studies, Tamale, Ghana

Muazu Issifu Department of Agricultural Biotechnology, Faculty of Agriculture Food and Consumer Sciences, University for Development Studies, Tamale, Ghana

Qiujuan Jiao College of Resources and Environment, Henan Agricultural University, Zhengzhou, China

Anil Namdev Kale Department of Agricultural Biotechnology and Molecular Biology, College of Basic Sciences & Humanities, Dr. Rajendra Prasad Central Agricultural University, Pusa, Samastipur, Bihar, India

Frederick Kankam Department of Agricultural Biotechnology, Faculty of Agriculture Food and Consumer Sciences, University for Development Studies, Tamale, Ghana

Nitika Kapoor Department of Botanical and Environmental Sciences, Guru Nanak Dev University, Amritsar, Punjab, India

Mohammed Khadiri Microbial Biotechnology and Plant Protection Laboratory, Faculty of Science, Ibn Zhor University, Agadir, Morocco

Ladipo Margaret Kudirat Department of Chemical Sciences, Yaba College of Technology, Yaba, Nigeria

Ravinder Kumar Department of Biotechnology, Chaudhary Devi Lal University, Sirsa, India

Rachid Lahlali Phytopathology Unit, Department of Plant Protection, Ecole Nationale d'Agriculture de Meknès, Meknès, Morocco

Haitam Lahmamsi Phytopathology Unit, Department of Plant Protection, Ecole Nationale d'Agriculture de Meknès, Meknès, Morocco

Laboratory of Microbial Biotechnology and Bioactive Molecules, Faculty of Sciences and Techniques, University of Sidi Mohamed Ben Abdellah, Fez, Morocco

Israel Oluwasanmi Lawal Perishable Crops Research Deaprtment Nigerian Stored Products Research Institute, Ilorin, Nigeria

Ikram Legrifi Phytopathology Unit, Department of Plant Protection, Ecole Nationale d'Agriculture de Meknès, Meknès, Morocco

Laboratory of Functional Ecology and Environmental Engineering, Sidi Mohamed Ben Abdellah University, Fez, Morocco

Gezi Li National Engineering Research Center for Wheat, Henan Agricultural University, Zhengzhou, China

Haitao Liu College of Resources and Environment, Henan Agricultural University, Zhengzhou, China

Niloufar Mahmoudi Peoples' Friendship University of Russia (RUDN University), Moscow, Russia

Amarendra N. Misra Office of the Vice Chancellor, Jagadguru Krupalu University, Banara, Odisha, India

Faculty of Sciences, Department of Biology, Sri Sri University, Godisahi, Odisha, India

Mohamed A. Mosa Nanotechnology & Advanced Nano-Materials Laboratory (NANML), Mycology and Disease Survey Research Department, Plant Pathology Research Institute, Agricultural Research Center, Giza, Egypt

Genetic Diversity of Microbial Plant Pathogens Reference Lab, Plant Pathology Research Institute, Agricultural Research Center (ARC), Giza, Egypt

Yogesh Dashrath Naik Department of Agricultural Biotechnology and Molecular Biology, College of Basic Sciences & Humanities, Dr. Rajendra Prasad Central Agricultural University, Pusa, Samastipur, Bihar, India

Yousef Naserzadeh Peoples' Friendship University of Russia (RUDN University), Moscow, Russia

Idorenyin Ugochi Nwaehujor Perishable Crops Research Deaprtment Nigerian Stored Products Research Institute, Ilorin, Nigeria

Ayorinde Victor Ogundele Laboratory of Natural Products & Drug Discovery, Centre CEBIM, Department of Basic Sciences, Faculty of Medicine, Universidad de La Frontera, Temuco, Chile

Puja Ohri Department of Zoology, Guru Nanak Dev University, Amritsar, Punjab, India

Ayotunde Olayinka Owolabi Department of Agricultural Economics and Extension Landmark University, Omuaran, Kwara State, Nigeria

Safal Kumar Paikray School of Biotechnology, Centurion University of Technology and Management, Bhubaneswar, Odisha, India

Aparna Pandey Ranjan Plant Physiology and Biochemistry Laboratory, Department of Botany, University of Allahabad, Prayagraj, India

Sakshi Pandey Ranjan Plant Physiology and Biochemistry Laboratory, Department of Botany, University of Allahabad, Prayagraj, India

Kiran D. Pawar School of Nanoscience and Biotechnology, Shivaji University, Kolhapur, Maharashtra, India

Sheo Mohan Prasad Ranjan Plant Physiology and Biochemistry Laboratory, Department of Botany, University of Allahabad, Prayagraj, India

Mohammed Radi Phytopathology Unit, Department of Plant Protection, Ecole Nationale d'Agriculture de Meknès, Meknès, Morocco

Laboratory of Environment and Valorization of Microbial and Plant Resources, Faculty of Sciences, Moulay Ismail University, Meknes, Morocco

Venkatratna Shrinivas Rao Bhiksham Perike Department of Agriculture, Rocky Ponds Produce, Gumlu, Queensland, Australia

Vishnu D. Rajput Academy of Biology and Biotechnology, Southern Federal University, Rostov-on-Don, Russia

Baljeet Singh Saharan Department of Microbiology, Chaudhary Charan Singh Haryana Agricultural University, Hisar, India

Liza Sahoo School of Biotechnology, Centurion University of Technology and Management, Bhubaneswar, Odisha, India

Adita Sharma College of Fisheries, (Dr. Rajendra Prasad Central Agricultural University, Pusa, Bihar), Dholi, Muzaffarpur, Bihar, India

Roohi Sharma Department of Zoology, Guru Nanak Dev University, Amritsar, Punjab, India

Ajay Veer Singh Department of Microbiology, College of Basic Sciences & Humanities, G.B. Pant University of Agriculture & Technology, Pantnagar, Uttarakhand, India

Shikha Singh Ranjan Plant Physiology and Biochemistry Laboratory, Department of Botany, University of Allahabad, Prayagraj, India

Varunendra Kumar Singh Ranjan Plant Physiology and Biochemistry Laboratory, Department of Botany, University of Allahabad, Prayagraj, India

Surekha Department of Botany, Government National College, Sirsa, India

Abdessalem Tahiri Phytopathology Unit, Department of Plant Protection, Ecole Nationale d'Agriculture de Meknès, Meknès, Morocco

Mohammed Taoussi Phytopathology Unit, Department of Plant Protection, Ecole Nationale d'Agriculture de Meknès, Meknès, Morocco

Laboratory of Environment and Valorization of Microbial and Plant Resources, Faculty of Sciences, Moulay Ismail University, Meknes, Morocco

Raman Tikoria Department of Zoology, School of Basic Sciences, Central University of Punjab, Bathinda, Odisha, India

Nigam Sekhar Tripathy School of Biotechnology, Centurion University of Technology and Management, Bhubaneswar, Odisha, India

Viabhav Kumar Upadhayay Department of Microbiology, College of Basic Sciences & Humanities, Dr. Rajendra Prasad Central Agricultural University, Pusa, Samastipur, Bihar, India

Krishan K. Verma Sugarcane Research Institute, Guangxi Academy of Agricultural Sciences, Nanning, Guangxi, China

Jingjing Zhang College of Resources and Environment, Henan Agricultural University, Zhengzhou, China

Chapter 1
The Potential Impact of Nanomaterials on Seedlings in Sustainable Agriculture

Yousef Naserzadeh and Niloufar Mahmoudi

1.1 Nanotechnology Role in Agriculture

Nanotechnology, the manipulation of matter at the nanoscale, holds significant promise for revolutionizing agriculture. Its applications encompass a wide array of areas, including soil and water management, crop protection, and plant nutrition. By harnessing the unique properties of nanomaterials (NMs), such as their high surface area-to-volume ratio and enhanced reactivity, nanotechnology enables precise targeting and delivery of agricultural inputs, thereby optimizing resource utilization and minimizing environmental degradation (Kumar et al. 2024b). NMs have garnered attention for their potential to address key challenges in agriculture, ranging from pest and disease management to improving soil fertility and crop productivity. By encapsulating agrochemicals within nanocarriers, such as nanoparticles (NPs) and nanocapsules, researchers aim to enhance the efficacy and sustainability of agricultural inputs while reducing their environmental impact (Kamienski et al. 2019; Mariyam et al. 2024). Additionally, NMs-based sensors offer real-time monitoring of soil health, crop growth, and environmental conditions, enabling data-driven decision-making and precision agriculture practices (Moore et al. 2019; Tang et al. 2023).

Y. Naserzadeh (✉) · N. Mahmoudi
Peoples' Friendship University of Russia (RUDN University), Moscow, Russia

C. S. Seth et al. (eds.), *Nano-formulations for Sustainable Agriculture*, Nanotechnology in the Life Sciences, https://doi.org/10.1007/978-3-031-90617-6_1

1.1.1 Nanomaterials Significance in Improving Crop Performance

At the heart of nanotechnology's impact on agriculture lies its ability to enhance crop performance through targeted interventions at the molecular level (Shang et al. 2019). NMs exhibit unique physicochemical properties that enable them to modulate plant physiology, metabolism, and defense mechanisms, thereby promoting stress tolerance, disease resistance, and overall plant health. By leveraging NMs, researchers seek to unlock novel strategies for improving seed germination, seedling vigor, and crop yield. Nano-priming, the pre-treatment of seeds with NMs, has emerged as a promising approach to enhance seed quality, viability, and resilience to environmental stressors (Nile et al. 2022; Kumar et al. 2024a). Through the controlled release of bioactive compounds and growth-promoting substances, NMs facilitate seedling establishment and early growth, setting the stage for robust plant development throughout the growing season (De La Torre-Roche et al. 2020).

Furthermore, NMs-based formulations offer targeted delivery of nutrients, hormones, and elicitors to plant roots and foliar tissues, promoting nutrient uptake, root development, and photosynthetic efficiency (Greenwell and Rahman 2015). By fine-tuning the physicochemical properties of NMs, such as particle size, surface charge, and composition, researchers can tailor their interactions with plants to optimize growth, yield, and quality traits (De La Torre-Roche et al. 2020; Verma et al. 2023a). In summary, nanotechnology has significant potential to transform agriculture by providing innovative solutions to enhance crop performance, promote sustainability, and improve resilience against evolving environmental challenges. By leveraging the unique properties of NMs, researchers and practitioners can foster the development of a more efficient, resilient, and environmentally sustainable agricultural system (Ijaz et al. 2023).

1.1.2 Nanomaterials as Seed Pre-treatment Agents

In this section, we delve into the role of NMs as seed pre-treatment agents, focusing on their potential applications in agriculture. Seed pre-treatment is a critical step in modern agricultural practices, aimed at enhancing seed germination, seedling vigor, and overall plant performance (Eevera et al. 2023). NMs offer unique properties that can positively impact seed physiology and plant growth (Kaur et al. 2021; Kumar et al. 2024b). Through various methods and dosages of application, NMs interact with seeds at the molecular and cellular levels, modulating biochemical pathways, and activating stress response mechanisms (Yavari et al. 2023). By unraveling the mechanisms behind NMs-assisted seed pre-treatment, we aim to explore how nanotechnology can contribute to sustainable agriculture practices and address the challenges faced by modern farming (Shelar et al. 2023).

1.1.3 Nanomaterial Application Methods and Dosages

Nanotechnology has opened new avenues for seed pre-treatment through the use of NMs, predominantly NPs, to enhance seed performance (Ma-Hock et al. 2021). Unlike conventional seed priming methods that rely on water or solutions containing substances, seed nano-priming involves the application of NPs suspensions or formulations (DeLoid et al. 2014). These NPs may or may not be taken up by the seeds, with a significant portion remaining as a coating on the seed surface (Verma et al. 2019). Studies have explored various NMs for seed nano-priming, including metallic, biogenic metallic and polymeric NPs. The application of NMs has demonstrated promising results in stimulating plant growth and improving morphological and metabolic traits (Shelar et al. 2023; Rajput et al. 2023). This process promotes rapid root and shoot development, along with alterations in gene expression associated with metabolic processes, including phytohormone production (do Espirito Santo Pereira et al. 2021).

1.1.4 Unraveling Mechanisms Behind Nanomaterial-Assisted Seed Pre-treatment

Understanding the mechanisms underlying the effects of NMs-assisted seed pre-treatment is crucial for optimizing their application in agriculture (Ajayi et al. 2024). NPs modulate seed metabolism at molecular and cellular levels, inducing physiological changes that promote germination and seedling vigor (Yusoff et al. 2022). By enhancing the activity of the defense system, NMs-treated seeds exhibit increased antioxidant levels and enzyme activities, enhancing resistance to pests and environmental stresses.

1.1.5 Facilitating Growth with Nanomaterials

NMs exert profound effects on seedling growth and development, influencing key physiological processes essential for plant establishment and productivity (Verma et al. 2018). Studies have demonstrated that NMs-treated seeds exhibit enhanced germination rates and seedling vigor, leading to improved crop performance under various environmental conditions (Mathur and Roy 2020). Additionally, NMs have been shown to modulate gene expression patterns associated with stress tolerance and metabolic regulation, further enhancing plant resilience and adaptation to adverse conditions (Liu et al. 2023). In summary, NMs-assisted seed pre-treatment offers a promising approach to enhance seed performance and crop productivity in sustainable agriculture (Liu et al. 2023). By elucidating the mechanisms are underlying their effects and optimizing application methods and dosages, NMs hold great potential to revolutionize seed priming practices and contribute to the advancement of agriculture in the face of evolving environmental challenges (Rajput et al. 2022) (Table 1.1).

This table summarizes various NPs systems utilized for seed priming and coating in agriculture. Each NPs system is described along with its characteristics and the main effects observed on plant species. The applications of these NPs systems include seed priming (A1), effects in the field (A2), effects against pathogens (A3), and relieving abiotic stresses (A4). Positive effects such as enhanced seed germination, improved seedling growth, increased resistance to pathogens, and mitigation of abiotic stresses are observed across different NPs systems.

Table 1.1 Nanoparticle systems employed for seed priming and coating

NPs System	Characteristics	Response	Application
Silver nanoparticles (AgNPs)	Small-sized, positively charged	Enhanced seed germination, increased antioxidant levels	A1, A2
Gold nanoparticles (AuNPs)	Stable colloidal solution, spherical shape	Improved seedling growth, enhanced resistance to pathogens	A1, A3
Zinc oxide nanoparticles (ZnONPs)	High surface area, photocatalytic properties	Increased root elongation, reduced incidence of fungal diseases	A1, A3
Titanium dioxide nanoparticles (TiO_2NPs)	Photoactive, nanoscale dimensions	Enhanced seed vigor, mitigation of abiotic stresses	A1, A4
Carbon nanotubes (CNTs)	Tubular carbon structures, high aspect ratio	Stimulated seed metabolism, improved water uptake	A1
Silicon nanoparticles (SiNPs)	Biocompatible, porous structure	Enhanced nutrient uptake, promotion of plant growth	A1, A2
Copper nanoparticles (CuNPs)	Reducing agent, antimicrobial properties	Suppressed pathogenic activity, enhanced seedling vigor	A1, A3
Iron nanoparticles (FeNPs)	Magnetic properties, high reactivity	Increased chlorophyll content, alleviation of oxidative stress	A1, A4
Chitosan nanoparticles (CSNPs)	Biodegradable, biocompatible	Enhanced seedling establishment, reduced susceptibility to pests	A1, A3
Selenium nanoparticles (SeNPs)	Antioxidant properties, trace element essential for plants	Improved seed germination under stress conditions	A1, A4

Note: Application codes: A1 (seed priming), A2 (effects in the field), A3 (effects against pathogens), A4 (relieving abiotic stresses). Arrows indicate increase (↑) or decrease (↓) in activity

1.2 Advancements in Nanotechnology for Sustainable Agriculture

Recent scientific findings suggest the promising potential of nanotechnology to positively influence the agrifood sector, addressing environmental and health concerns associated with conventional agricultural practices, enhancing food security and productivity (Prasad et al. 2017; Choudhary et al. 2024), and promoting social and economic equity. This discussion delves into recent advancements in NMs-based systems and nanodevices aimed at improving various aspects of the food supply chain, particularly sustainable intensification, and soil and waste management (Shukla et al. 2024; Kumar et al. 2024c).

1.2.1 Emerging Trends

In recent years, increasing attention has been given to exploring the applications of nanotechnology in agriculture, leading to extensive research efforts in both academic and industrial settings (Fraceto et al. 2016; Ajayi et al. 2024).The unique properties exhibited by materials at the nanoscale render them promising candidates for developing innovative tools to support sustainable agriculture practices (Ajayi et al. 2024).

1.2.2 Innovations for Sustainable Intensification

Sustainable intensification in agriculture aims to increase yield without causing adverse environmental effects, thereby maximizing productivity on existing agricultural land (The Royal Society 2009). NMs offer new opportunities to enhance productivity through the development of intelligent systems for nutrient immobilization and release in soil (Schut et al. 2016). These systems help minimize nutrient leaching, improve nutrient uptake by plants, and mitigate eutrophication by reducing nitrogen transfer to groundwater (Xie et al. 2019; Verma et al. 2023b). Additionally, NMs can enhance the efficacy of pesticides by improving solubility, resistance to hydrolysis and photodecomposition, and providing controlled release targeting specific organisms (Mutyasira 2020).

1.2.3 Enhancing Soil Quality

Hydrogels, nanoclays, and nanozeolites have been shown to enhance soil water-holding capacity, acting as slow-release water sources and reducing periods of hydric shortage during the crop season (Karlen and Cambardella 2020). These systems are beneficial for agricultural purposes and the restoration of degraded areas. Furthermore, organic and inorganic NMs have been utilized to absorb environmental contaminants, thereby increasing soil remediation capacity and reducing treatment times and costs (Zhang et al. 2020; El-Ramady et al. 2022).

1.2.4 Nanomaterials for Plant Growth Enhancement

NMs, such as carbon nanotubes, gold, silicon dioxide, zinc oxide, and titanium dioxide NPs have the potential to improve plant development by enhancing nutrient uptake and utilization (Ogunkunle et al. 2022). However, the impact of NMs on plants depends on various factors including composition, concentration, size, surface charge, and physicochemical properties, as well as plant species susceptibility (Yasmeen 2023). Further research employing different analytical techniques is needed to understand the interactions between plants and NMs fully (Ahmad et al. 2022).

1.2.5 Nano-devices for Managing the Food Supply Chain

Nanotechnology also holds promise for developing analytical devices to monitor food quality, biosafety, and security throughout the food supply chain (Wahab et al. 2024). Nanosensors, characterized by at least one sensing dimension no greater than 100 nm, offer advanced features for monitoring physicochemical properties in challenging environments. Nanotubes, nanowires, nanoparticles, and nanocrystals are commonly employed to enhance sensitivities, reduce response times, and improve detection limits, making them suitable for multiplexed systems (Poeta et al. 2023). Despite significant research in this area, real-world applications of Nanosensors for field analysis remain limited, presenting an untapped market potential. Efforts are required to develop automated embedded systems that integrate biosensing components with micro/nanofluidic technologies and wireless networks. These systems aim to enhance decision-making in crop monitoring, soil nutrient and pesticide analysis, and efficient water management, paving the way for smart agriculture. Nanosensors have the potential to revolutionize various stages of the food supply chain, including crop cultivation, harvesting, processing, packaging, and distribution, by providing real-time data for optimizing practices and ensuring food quality and safety (Bayda et al. 2019). Nanotechnology offers promising solutions for addressing challenges in agriculture, with potential applications ranging from sustainable intensification to soil and waste management, and quality control along the food supply chain. Continued research and development in this field are essential to unlock the full potential of nanotechnology in promoting sustainable agricultural practices and ensuring food security and safety (Anand et al. 2024).

1.3 Green Synthesis of Metallic Nanoparticles: Principles and Influencing Factors

The synthesis of NPs or NMs relies on various factors such as NP characteristics, synthesis methodologies, and biological interactions. Traditional chemical and physical synthesis methods often result in hazardous by-products and excessive chemical residues on NP surfaces. To address these limitations, researchers have turned to green nanotechnology, focusing on sustainable synthesis approaches using biological resources (Rana et al. 2020).

1.3.1 Green Synthesis Methods

Green synthesis methods involve utilizing plants, microorganisms, and biological components for NPs production. These methods have gained significant attention due to their eco-friendliness, scalability, and biocompatibility. Particularly, medicinal plants have been extensively explored for their rapid NPs synthesis capabilities, offering advantages over microbial synthesis methods in terms of time efficiency (Shah et al. 2015).

1.3.2 Plant-Mediated Synthesis

Medicinal plants, including *Panax ginseng, Rhodiola rosea*, and *Cannabis sativa*, have been harnessed for the synthesis of metallic NPs. This approach has emerged as a promising strategy for large-scale NPs production, offering cost-effectiveness and simplicity without requiring high temperatures or specialized energy resources.

1.3.3 Biological Components in Synthesis

The reduction of metal ions in green synthesis processes involves various biological components such as enzymes, proteins, and secondary metabolites in microorganisms, while plants contribute flavonoids, terpenoids, phenols, and other phytochemicals (Choudhary et al. 2025). These components play a crucial role in reducing metal ions and forming a protective "capping layer" or "biological corona" around individual NPs.

1.3.4 Influencing Factors

Several parameters influence the structure and activity of metallic NPs during synthesis, including the source of synthesis, temperature, pH, salt concentrations, and reaction time. Reaction time is particularly critical, determining the size, shape, and stability of NPs. pH and temperature also play significant roles, with acidic pH favoring larger NPs and higher temperatures potentially leading to agglomeration (Table 1.2). Green synthesis of metallic NPs offers a sustainable and environmentally friendly approach to NPs production. Understanding the influencing factors is essential for optimizing synthesis conditions and harnessing the potential of green nanotechnology for various applications.

This revised version emphasizes the scientific principles behind green synthesis methods and their applications, while providing a detailed analysis of influencing factors through a structured table.

Table 1.2 Factors influencing green synthesis of metallic NPs

Parameters	Influence on NPs Structure and Activity
Source of Synthesis	Determines the biological components available for reduction, affecting NPs formation and stability.
Temperature	Higher temperatures may accelerate reduction but can also lead to agglomeration and instability.
pH	Acidic pH favors the formation of larger NPs, while variations in pH impact NPs size and shape.
Salt Concentrations	Salts in the reaction medium can influence NPs nucleation and growth, affecting size distribution and morphology.
Reaction Time	Critical for controlling NPs size, shape, and stability; prolonged reaction times may lead to agglomeration or altered morphology.

1.4 Impact of Nano-fertilizers on Sustainable Product Development

While chemical fertilizers increase crop yield, they disrupt the balance of soil minerals and decrease soil fertility. Widespread use of chemical fertilizers irreversibly damages soil structure and vegetation cover. This affects the food chain of ecosystems, responsible for hereditary mutations in future generations of consumers (Yaseen et al. 2020). The rapid growth of the world's population has increased demand for the agricultural sector, raising concerns among researchers about excessive use of chemical fertilizers by farmers. In this context, nano-fertilizers (NFs) have emerged as a promising alternative, ensuring product development and soil revitalization (Iqbal 2019).

1.4.1 Why Are Nano-fertilizers Better than Conventional Chemical Fertilizers?

The unique characteristics of NPs, such as high absorption capacity, increased surface-to-volume ratio, and controlled release kinetics to the target site, make them a potentially powerful plant growth enhancer. Due to these distinctive features, NFs can serve as an intelligent nutrient delivery system to plants. Compared to conventional fertilizers, NFs are released and dispersed very slowly (Shelar et al. 2023). This approach improves nutrient management; for example, enhancing nutrient efficiency and reducing nutrient leaching into groundwater. NFs are specifically designed to release active ingredients in response to biological needs and environmental stresses. Scientists have stated that NFs enhance photosynthesis activity, seed germination, seedling growth, nitrogen metabolism, and synthesis of carbohydrates and proteins, thus increasing agricultural productivity (Ogunkunle et al. 2022).

1.4.2 Development of Effective Nano-formulations

Nano-formulations, or nano-scale fertilizers, are made from ammonium hematite, urea, compost, and other synthetic fertilizers. An example of a nano-formulation is nano-scale nitrogen (N) fertilizers prepared by precipitating urea onto calcium cyanamide (Yasmeen 2023). Nano-scale nitrogen fertilizers play a crucial role in increasing rice productivity. This type of fertilizer is an excellent alternative to chemical fertilizers, as it promotes plant growth and reduces environmental pollution (Shelar et al. 2023). Excessive use of conventional fertilizers through leaching, denitrification, and evaporation of chemical fertilizers pollutes the environment. Another effective NFs is obtained by milling urea and mixing it with various organic fertilizers. This NFs gradually releases nutrients over a longer period (Jakhar et al. 2022).

NFs are produced using both mechanical and biochemical processes; mechanical tools are used to obtain NPs, while biochemical methods are used to obtain effective formulations on a nanoscale (Kumar et al. 2021). Fertilizers are often encapsulated within NPs. These NFs increase nutrient absorption capacity and efficiency for plants. The encapsulation process can be done in three ways:

- Nutrient particles coated with a thin layer of NPs (polymeric film).
- Nutrients encapsulated within NPs with different chemical compositions.
- Nutrients delivered in the form of nano-emulsions.

1.4.3 How Do Nano-fertilizers Enter the Plant System?

Researchers have stated that the plant root system (which serves as the gateway for nutrient entry) is much more permeable to NMs (NFs) compared to conventional fertilizers. Leaf stomata have also been reported to preferentially absorb NMs and allow their entry into leaves (Seleiman et al. 2020). Scientists conducted experiments using broad beans (*Vicia faba*) to determine the efficacy of NPs in penetrating the plant system. They found that NPs (with a size of 43 nanometers) can penetrate many leaves compared to larger NPs (>1 nanometer). It has also been reported that NFs transport nutrients through plasmodesmata. Plasmodesmata are nano-sized channels (about 50 to 60 nanometers) used for ion exchange between cells. Carbon nanotubes and silica NPs are useful tools for transporting materials (including nutrients and other important biochemicals) to desired plant locations (El-Shal et al. 2022).

1.4.4 Using Nano-fertilizers in Sustainable Product Development

Scientists believe that zinc NFs promote better plant growth (root and stem systems) and increase leaf chlorophyll content. In previous studies, modified zinc NFs significantly increased peanut yield. These NFs also improve vegetable seed production. Similarly, it has been reported that carbon nanotubes containing fertilizers reduce germination days. It has also been shown that these NFs help develop root systems in rice seedlings (Ajayi et al. 2024). NFs also reduce the crop cycle and increase crop yield. For example, wheat modification using NPK (nitrogen, phosphorus, and potassium) NPs resulted in increased wheat yield and reduced wheat cultivation period by 40 days. Similar results were obtained in the corn cultivation system (Table 1.3).

Table 1.3 Comparative study of NFs vs. conventional chemical fertilizers

Properties	Nano-fertilizers	Conventional chemical fertilizers
Absorption Capacity	NFs have significantly higher absorption capacity due to their nano-scale structure, enhancing nutrient uptake by plants and reducing wastage.	Conventional chemical fertilizers exhibit moderate to high absorption capacity but may result in some nutrient loss through leaching and runoff.
Surface-to-Volume Ratio	NFs offer an increased surface-to-volume ratio, facilitating better nutrient interaction with plant roots, leading to improved growth and productivity.	Conventional chemical fertilizers have a normal surface-to-volume ratio, potentially limiting nutrient absorption efficiency and distribution.
Release Kinetics	NFs provide controlled release kinetics, ensuring gradual nutrient delivery to plants over time, reducing leaching and promoting sustainable nutrient management.	Conventional chemical fertilizers often release nutrients rapidly upon application, leading to imbalances and environmental pollution.
Nutrient Efficiency	NFs demonstrate enhanced nutrient efficiency by optimizing nutrient utilization within plants, resulting in higher crop yields.	Conventional chemical fertilizers may vary in efficiency but are generally less efficient due to nutrient loss and uneven distribution.
Nutrient Leaching	NFs contribute to reduced nutrient leaching, maintaining soil fertility and protecting water quality.	Conventional chemical fertilizers are more prone to nutrient leaching, leading to water pollution and soil degradation.
Environmental Impact	NFs have minimal environmental impact, thanks to their controlled release mechanism and efficient nutrient utilization.	Conventional chemical fertilizers pose significant environmental risks, including eutrophication and biodiversity loss.

NFs offer superior absorption capacity, controlled release kinetics, and reduced environmental impact compared to conventional chemical fertilizers. Their nano-scale structure enhances nutrient uptake by plants while minimizing nutrient loss and environmental pollution, making them a more sustainable choice for agriculture.

1.5 Limitations of Nano-fertilizers

Despite aiding in the production of sustainable products, the limitations of NFs need careful consideration before marketing. The limitations of using NFs mainly arise due to the lack of strict regulation and research gaps (do Espirito Santo Pereira et al. 2021).

Some of the limitations associated with using NFs for sustainable product development are:

- Lack of a risk management system for NFs.
- Inadequate production and availability of NFs in required quantities, limiting widespread use as a source of plant nutrients.
- High cost of NFs.
- Lack of standardization in the formulation process, resulting in different outcomes of identical NMs under different pedoclimatic conditions.

Overall, NFs hold great promise for sustainable agriculture by enhancing nutrient delivery, increasing crop productivity, and reducing environmental pollution. However, further research and regulation are needed to address potential risks and ensure their safe and effective application in agriculture.

References

Ahmad Z, Tahseen S, Wasi A, Ganie IB, Shahzad A, Emamverdian A, Ramakrishnan M, Ding Y (2022) Nanotechnological interventions in agriculture. Nanomaterials 12(15):2667

Ajayi FF, Mudgil P, Maqsood S (2024) Molecular structural modification of jack bean protein using thermo-shearing/ultrasound/microwave treatments for improved extractability, functional and gelling properties: the underlying impacts of matrix pre-treatment versus alkaline-assisted extraction. Food Hydrocoll 154:110066

Anand R, Mishra KK, Bharadvaja N (2024) Introduction to bio-nanotechnology. In: Biogenic nanomaterials for environmental sustainability: principles, practices, and opportunities. Springer, pp 1–11

Bayda S, Adeel M, Tuccinardi T, Cordani M, Rizzolio F (2019) The history of nanoscience and nanotechnology: from chemical–physical applications to nanomedicine. Molecules 25(1):112

Choudhary R, Rajput VD, Ghodake G, Ahmad F, Meena M, Rehman RU, Prasad R, Sharma RK, Singh R, Seth CS (2024) Comprehensive journey from past to present to future about seed priming with hydrogen peroxide and hydrogen sulfide concerning drought, temperature, UV and ozone stresses- a review. Plant Soil 500:1–23

Choudhary R, Ahmad F, Kaya C, Upadhyay SK, Muneer S, Kumar V, Meena M, Liu H, Upadhyay H, Seth CS (2025) Decrypting proteomics, transcriptomics, genomics, and integrated omics for augmenting the abiotic, biotic, and climate change stress resilience in plants. J Plant Physiol. https://doi.org/10.1016/j.jplph.2025.154430

De La Torre-Roche R, Cantu J, Tamez C, Zuverza-Mena N, Hamdi H, Adisa IO, Elmer W, Gardea-Torresdey J, White JC (2020) Seed biofortification by engineered nanomaterials: a pathway to alleviate malnutrition? J Agric Food Chem 68(44):12189–12202

DeLoid G, Cohen JM, Darrah T, Derk R, Rojanasakul L, Pyrgiotakis G, Wohlleben W, Demokritou P (2014) Estimating the effective density of engineered nanomaterials for in vitro dosimetry. Nat Commun 5(1):3514

Eevera T, Kumaran S, Djanaguiraman M, Thirumaran T, Le QH, Pugazhendhi A (2023) Unleashing the potential of nanoparticles on seed treatment and enhancement for sustainable farming. Environ Res 236:116849

El-Ramady HR, Verma KK, Rajput VD, Minkina T, Elbehiry F, Elbasiouny HY, Elsakhawy T, Omara AED, Amer MM (2022) Sources of silicon and nano-silicon in soil and plants. In: Silicon and nano-silicon in environmental stress management and crop quality improvement: progress and prospects. Academic Press, Elsevier. https://doi.org/10.1016/B978-0-323-91225-9.00003-0

El-Shal RM, El-Naggar AH, El-Beshbeshy TR, Mahmoud EK, El-Kader NIA, Missaui AM, Du D, Ghoneim AM, El-Sharkawy MS (2022) Effect of nano-fertilizers on alfalfa plants grown under different salt stresses in hydroponic system. Agriculture 12(8):1113
do Espirito Santo Pereira A, Caixeta Oliveira H, Fernandes Fraceto L, Santaella C (2021) Nanotechnology potential in seed priming for sustainable agriculture. Nanomaterials 11(2):267
Fraceto LF, Grillo R, de Medeiros GA, Scognamiglio V, Rea G, Bartolucci C (2016) Nanotechnology in agriculture: which innovation potential does it have? Front Environ Sci 4:186737
Greenwell M, Rahman P (2015) Medicinal plants: their use in anticancer treatment. Int J Pharm Sci Res 6(10):4103
Ijaz M, Khan F, Ahmed T, Noman M, Zulfiqar F, Rizwan M, Chen J, Siddique KH, Li B (2023) Nanobiotechnology to advance stress resilience in plants: Current opportunities and challenges. Mater Today Bio 22:100759
Iqbal MA (2019) Nano-fertilizers for sustainable crop production under changing climate: a global perspective. Sustain Crop Prod 8:1–13
Jakhar AM, Aziz I, Kaleri AR, Hasnain M, Haider G, Ma J, Abideen Z (2022) Nano-fertilizers: a sustainable technology for improving crop nutrition and food security. NanoImpact 27:100411
Kamienski C, Soininen J-P, Taumberger M, Dantas R, Toscano A, Salmon Cinotti T, Filev Maia R, Torre Neto A (2019) Smart water management platform: IoT-based precision irrigation for agriculture. Sensors 19(2):276
Karlen D, Cambardella C (2020) Conservation strategies for improving soil quality and organic matter storage. In: Structure and organic matter storage in agricultural soils. Routledge, pp 395–420
Kaur R, Chandra J, Keshavkant S (2021) Nanotechnology: an efficient approach for rejuvenation of aged seeds. Physiol Mol Biol Plants 27:399–415
Kumar Y, Tiwari K, Singh T, Raliya R (2021) Nanofertilizers and their role in sustainable agriculture. Ann Plant Soil Res 23(3):238–255
Kumar D, Mariyam S, Gupta KJ, Thiruvengadam M, Ghodake GS, Xing B, Seth CS (2024a) Comparative investigation on chemical and green synthesized titanium dioxide nanoparticles against chromium (VI) stress eliciting differential physiological, biochemical, and cellular attributes in Helianthus annuus L. Sci Total Environ 930:172413
Kumar D, Singh R, Upadhyay SK, Verma KK, Tripathi RM, Liu H, Dhankher OP, Tripathi RD, Sahi SV, Seth CS (2024b) Review on interactions between nanomaterials and phytohormones: novel perspectives and opportunities for mitigating environmental challenges. Plant Sci 340:111964
Kumar D, Biswas JK, Mulla SI, Singh R, Shukla R, Ahanger MA, Shekhawat GS, Verma KK, Siddiqui MW, Seth CS (2024c) Micro and nanoplastics pollution: sources, distribution, uptake in plants, toxicological effects, and innovative remediation strategies for environmental sustainability. Plant Physiol Biochem 213:108795
Liu T, Lu Y, Zhan R, Qian W, Luo G (2023) Nanomaterials and nanomaterials-based drug delivery to promote cutaneous wound healing. Adv Drug Deliv Rev 193:114670
Ma-Hock L, Sauer UG, Ruggiero E, Keller JG, Wohlleben W, Landsiedel R (2021) The use of nanomaterial in vivo organ burden data for in vitro dose setting. Small 17(15):2005725
Mariyam S, Upadhyay SK, Chakraborty K, Verma KK, Duhan JS, Muneer S, Meena M, Sharma RK, Ghodake G, Seth CS (2024) Nanotechnology, a frontier in agricultural science, a novel approach in abiotic stress management and convergence with new age medicine-a review. Sci Total Environ 912:169097
Mathur P, Roy S (2020) Nanosilica facilitates silica uptake, growth and stress tolerance in plants. Plant Physiol Biochem 157:114–127
Moore S, Nugent CD, Cleland I, Zhang S (2019) Impact analysis of erroneous data on IoT reliability. 2019 IEEE SmartWorld, Ubiquitous Intelligence & Computing, Advanced & Trusted Computing, Scalable Computing & Communications, Cloud & Big Data Computing, Internet of People and Smart City Innovation (SmartWorld/SCALCOM/UIC/ATC/CBDCom/IOP/SCI)
Mutyasira V (2020) Prospects of sustainable intensification of smallholder farming systems: a farmer typology approach. Afr J Sci Technol Innov Dev 12(6):727–734

Nile SH, Thiruvengadam M, Wang Y, Samynathan R, Shariati MA, Rebezov M, Nile A, Sun M, Venkidasamy B, Xiao J (2022) Nano-priming as emerging seed priming technology for sustainable agriculture—recent developments and future perspectives. J Nanobiotechnol 20(1):254

Ogunkunle C, Jimoh M, Oyedeji S, Varun M, Okunlola G (2022) Engineered nanomaterial-mediated changes in the growth and development of common agricultural crops. In: Nano-enabled agrochemicals in agriculture. Elsevier, pp 345–375

Poeta E, Liboà A, Mistrali S, Núñez-Carmona E, Sberveglieri V (2023) Nanotechnology and E-Sensing for food chain quality and safety. Sensors 23(20):8429

Prasad R, Bhattacharyya A, Nguyen QD (2017) Nanotechnology in sustainable agriculture: recent developments, challenges, and perspectives. Front Microbiol 8:1014

Rajput VD, Minkina T, Upadhyay SK, Kumari A, Ranjan A, Mandzhieva S, Sushkova S, Singh RK, Verma KK (2022) Nanotechnology in the restoration of polluted soil. Nanomaterials 12:769. https://doi.org/10.3390/nano12050769

Rajput VD, Kumari A, Upadhyay SK, Minkina T, Mandzhieva S, Ranjan A, Sushkova S, Burachevskaya M, Rajput P, Konstantinova E, Singh J, Verma KK (2023) Can nanomaterials improve the soil microbiome and crop productivity? Agriculture 13:231. https://doi.org/10.3390/agriculture13020231

Rana A, Yadav K, Jagadevan S (2020) A comprehensive review on green synthesis of nature-inspired metal nanoparticles: mechanism, application and toxicity. J Clean Prod 272:122880

Schut M, van Asten P, Okafor C, Hicintuka C, Mapatano S, Nabahungu NL, Kagabo D, Muchunguzi P, Njukwe E, Dontsop-Nguezet PM (2016) Sustainable intensification of agricultural systems in the Central African Highlands: the need for institutional innovation. Agric Syst 145:165–176

Seleiman MF, Almutairi KF, Alotaibi M, Shami A, Alhammad BA, Battaglia ML (2020) Nano-fertilization as an emerging fertilization technique: why can modern agriculture benefit from its use? Plan Theory 10(1):2

Shah M, Fawcett D, Sharma S, Tripathy SK, Poinern GEJ (2015) Green synthesis of metallic nanoparticles via biological entities. Materials 8(11):7278–7308

Shang Y, Hasan MK, Ahammed GJ, Li M, Yin H, Zhou J (2019) Applications of nanotechnology in plant growth and crop protection: a review. Molecules 24(14):2558

Shelar A, Nile SH, Singh AV, Rothenstein D, Bill J, Xiao J, Chaskar M, Kai G, Patil R (2023) Recent advances in nano-enabled seed treatment strategies for sustainable agriculture: challenges, risk assessment, and future perspectives. Nano-Micro Lett 15(1):54

Shukla K, Mishra V, Singh J, Varshney V, Verma R, Srivastava S (2024) Nanotechnology in sustainable agriculture: a double-edged sword. J Sci Food Agric

Tang Y, Zhao W, Zhu G, Tan Z, Huang L, Zhang P, Gao L, Rui Y (2023) Nano-pesticides and fertilizers: solutions for global food security. Nanomaterials 14(1):90

The Royal Society (2009) Reaping the Benefits: Science and the Sustainable Intensification of Global Agriculture. The Royal Society, London. ISBN: 978-0-85403-784-1

Verma SK, Das AK, Patel MK, Shah A, Kumar V, Gantait S (2018) Engineered nanomaterials for plant growth and development: a perspective analysis. Sci Total Environ 630:1413–1435

Verma SK, Das AK, Gantait S, Kumar V, Gurel E (2019) Applications of carbon nanomaterials in the plant system: a perspective view on the pros and cons. Sci Total Environ 667:485–499

Verma KK, Song X, Degu HD, Guo D, Joshi A, Huang H, Xu L, Singh M, Huang D, Rajput VD, Li Y (2023a) Recent advances in nitrogen and nano-nitrogen fertilizers for sustainable crop production: a mini-review. Chem Biol Technol Agri 10:1–14. https://doi.org/10.1186/s40538-023-00488-3

Verma KK, Song X, Verma CL, Huang H, Singh M, Xu L, Wu J, Khan ZH, Zhang B, Li Y (2023b) Mathematical modeling of climate and fluoride effects on sugarcane photosynthesis with silicon nanoparticles. Plant Physiol Biochem 204:108089. https://doi.org/10.1016/j.plaphy.2023.108089

Wahab A, Muhammad M, Ullah S, Abdi G, Shah GM, Zaman W, Ayaz A (2024) Agriculture and environmental management through nanotechnology: eco-friendly nanomaterial synthesis for soil-plant systems, food safety, and sustainability. Sci Total Environ 171862

Xie H, Huang Y, Chen Q, Zhang Y, Wu Q (2019) Prospects for agricultural sustainable intensification: a review of research. Land 8(11):157

Yaseen R, Ahmed ISA, Omer AM, Agha MKM, Emam MT (2020) Nano-fertilizers: biofabrication, application and biosafety. Nov Res Microbiol J 4(4):884–900

Yasmeen F (2023) Role of nanomaterials in improving crop tolerance to abiotic stress. In: Nanomaterial interactions with plant cellular mechanisms and macromolecules and agricultural implications. Springer, pp 423–453

Yavari A, Ghasemifar E, Shahgolzari M (2023) Seed nanopriming to mitigate abiotic stresses in plants. In: Abiotic stress in plants-adaptations to climate change. IntechOpen

Yusoff IM, Taher ZM, Rahmat Z, Chua LS (2022) A review of ultrasound-assisted extraction for plant bioactive compounds: phenolics, flavonoids, thymols, saponins and proteins. Food Res Int 157:111268

Zhang X, Qu J, Li H, La S, Tian Y, Gao L (2020) Biochar addition combined with daily fertigation improves overall soil quality and enhances water-fertilizer productivity of cucumber in alkaline soils of a semi-arid region. Geoderma 363:114170

Chapter 2
Applications of Nanomaterials in Agronomic Productivity Stimulation and Combating Climate Change Scenarios

Marwa M. El-Deriny, Mohamed A. Mosa, Sozan E. El-Abeid, and Dina S. S. Ibrahim

2.1 Introduction

Significant technical developments and advancements have been developed in field of agriculture in recent years to address the growing difficulties of food security and sustainable production (Dwivedi et al. 2016; Kou et al. 2018). Such ongoing agricultural advances are essential to using both natural and artificial resources to meet the growing food demand of the world's rapidly expanding population (Verma et al. 2024a). Nanotechnology in particular has the promise of offering practical answers to a wide range of issues pertaining to agriculture. Nanoparticles offer a great deal of scientific curiosity because they can fill the gap between bulk materials and atomic or molecular structures. A substantial amount of research on nanotechnology has been conducted over the past 20 years, with an emphasis on its many uses in the agricultural sector (Prasad et al. 2017; Lv et al. 2018). Most notably, nanomaterials boost crop yield by improving the effectiveness of agricultural inputs to enable site-targeted, controlled nutrient delivery, guaranteeing the least amount of

M. M. El-Deriny (✉) · D. S. S. Ibrahim
Nematode Diseases Department, The Egyptian Nematology Reference Lab (ENR lab), Plant Pathology Research Institute, Agricultural Research Center, Giza, Egypt

Genetic Diversity of Microbial Plant Pathogens Reference Lab, Plant Pathology Research Institute, Agricultural Research Center (ARC), Giza, Egypt

M. A. Mosa · S. E. El-Abeid
Genetic Diversity of Microbial Plant Pathogens Reference Lab, Plant Pathology Research Institute, Agricultural Research Center (ARC), Giza, Egypt

Nanotechnology & Advanced Nano-Materials Laboratory (NANML), Mycology and Disease Survey Research Department, Plant Pathology Research Institute, Agricultural Research Center, Giza, Egypt

C. S. Seth et al. (eds.), *Nano-formulations for Sustainable Agriculture*,
Nanotechnology in the Life Sciences, https://doi.org/10.1007/978-3-031-90617-6_2

agri-inputs needed. In fact, nanotechnology has helped plant protection products tremendously, which could result in higher crop yields.

Moreover, the primary goal of agricultural production is to enable plants to adapt to the increasingly harsh conditions caused by climate change-such as high temperatures, limited water availability, alkalinity, salinity, and heavy metal pollution while preserving the fragile ecosystems already in place (Vermeulen et al. 2012; Mariyam et al. 2024). Plants have developed diverse mechanisms to adapt to temperature fluctuations, as maintaining thermoregulation is crucial for their growth and survival, particularly when exposed to heat or cold stress. Through processes such as respiration, heat- and cold-tolerance adaptations, and photosynthesis, plants regulate their internal temperature. Cold-tolerant plants employ antifreeze proteins and adjust the fluidity of their cell membranes, whereas heat-tolerant plants produce heat shock proteins and antioxidants (Kumar et al. 2024c). Certain plants, known as vernalization plants, require exposure to cold temperatures in order to initiate flowering. Moreover, plants can serve as natural temperature buffers in microclimates.

In face of climate challenges, understanding and implementing strategies for plant thermoregulation can support plant growth, productivity, and ecosystem stability. By enhancing our knowledge of these pathways, we can contribute to improving crop resilience and facilitating adaptation to climate change (Verma et al. 2020). Ongoing research endeavors continue to investigate these mechanisms and their potential applications. Some reports studied the mechanism of salicylic acid-chitosan nanoparticles in plant growth promotion and maize defense (Kumaraswamy et al. 2019). This article offers a comprehensive overview of the diverse applications of nanotechnology in agriculture, focusing on its potential to mitigate the effects of climate change and promote sustainable practices in both agricultural systems and the environment.

2.2 Applied Nanotechnology Foundations

By controlling the shape and size variation at the nanoscale level, from 1 to 100 nm, nanotechnology refers to the "synthesis, designing, characterizing, and utilization of assemblies, tools, and systems" (Yadollahi et al. 2010). To put things in perspective, a nanometer is one billionth (10^{-9}) of a meter, meaning that technology may be applied at this scale. When it comes to learning about wonders and the behaviour of substances at the atomic, molecular, or macromolecular levels—where their properties differ significantly from those at the bulk level-nanoscience and nanotechnologies are regarded as innovative approaches in developmental research (Potocnik 2005). The resulting products' physical, chemical, and biological properties differ significantly from the bulk material in this case. Nanoscale property research results in altered properties that may be utilized to create new materials with altered structures, improved instruments, and more effective goods. Newer attributes like increased strength, enhanced optical qualities, altered antibacterial potential, and exceptional superconductive nature were made possible by the differently presented properties of materials at the nanoscale (Axelos and Van De Voorde 2017). According to Cushen et al. (2012), they can take on forms such as nanotubes, nanofibers, nanoparticles, nanosheets, fullerenes, and nanowhiskers. A nanomaterial is defined as "natural, incidental, or industrial material with particles, in an unbound state or in the form of aggregate or agglomerate where 50% or more of the particles in the number and size distribution, one or more dimensions lies in the range of 1–100 nm," according to a published European Commission (EC) recommendation (Potocnik 2011).

2.3 Types of Nanoparticle

In this discipline, two different types of nanoparticles (NPs) are used: metal-based and carbon-based. Common metal-based nanoparticles (NPs) include gold, copper and silver (Khan et al. 2021a, b; Munawar et al. 2022).

2.3.1 Carbon Based Nanoparticles

According to Kausar (2020), carbon nanotubes (CNTs) are nano sponges with unique mechanical, electrical, thermal, and chemical properties. Agrochemicals or other substances can be delivered to hosts by CNT-based delivery devices (Safdar et al. 2022). *Oryza sativa*, *Solanumlycopersicum*, and Bt cotton have all been seen to employ CNTs (Bhattacharjee et al. 2022; Singh et al. 2017a). According to Abd-Elsalam (2020), CNTs are perfect for nano electromechanical systems because of their tensile strength. A range of nanodevices with unique conductivity, thermal, and optical properties that are useful in agriculture and the development of sustainable agricultural methods can be designed by altering these CNT attributes (Jain et al. 2022).

2.3.2 Metal Based Nanoparticles

Numerous studies indicate that zinc oxide nanoparticles (ZnO NPs) hasten the growth and development of plants, including wheat, soybean, and peanut (Ramesh et al. 2014; Sedghi et al. 2013; Prasad et al. 2012). According to Malandrakis et al. (2019), Shenashen et al. (2017), Imada et al. (2016), Giannousi et al. (2013), many plant species are susceptible to different types of fungi. Therefore, using nanomaterials (NMs) carefully can improve crop output while maintaining environmental health. Recent years have seen a significant growth in research on the application of nano-composites in crop protection due to their high effectiveness and environmentally acceptable composition (Gomathi et al. 2019).

According to NajafiDisfani et al. (2017), SiO_2 NPs have been seen to improve plant germination in barley and maize. These NPs enhanced the number of lateral roots in seedlings in addition to raising germination rates (Baz et al. 2020). Dekkers et al. (2011) state that SiO_2 NPs were used as applied NMs to add flavors or scents to food products. Rather than acting on the pest, silica nanoparticles may have an insecticidal effect on the grain itself (Saranya et al. 2020). One strategy is to use Si to strengthen the cell wall and encourage the production of defense chemicals (Sheng et al. 2018). Magnesium oxide (MgO) is a significant inorganic compound widely utilized in the production of advanced ceramics, photoelectronic materials, fire retardants, adsorbents, and the cleanup of toxic waste. Consequently, a number of methods and pathways for the synthesis of MgO NPs have been described (Ghidan et al. 2017a, b). Disease suppression has been demonstrated by MgO NPs (Imada et al. 2016). The bacterium *Acidovorax oryzae* strain RS2, which causes rice bacterial brown stripe disease, was effectively suppressed by MnO_2 and MgO NPs derived from chamomile flower extracts (Ogunyemi et al. 2020).

In a flooded rice field, copper oxide nanoparticles (CuO NPs) decreased the biomass of soil microbiota and the activity of enzymes such as urease, phosphatase, dehydrogenase, and total phospholipid fatty acids (Xu et al. 2015). Principal components of phospholipid fatty acids and diversity indices were analysed, and the results showed that CuO NPs affected microbial composition and diversity (Xu et al. 2015). CuO NPs may limit the growth of roots and shoots by having a negative effect on the composition and operation of the photosynthetic machinery in crops (Song et al. 2021). Plant systems have also made extensive use of CuO, SiO_2, and TiO_2 NPs as delivery vehicles (Jahan et al. 2018). CuO NPs significantly affect the microorganisms in soil (Frenk et al. 2013). It has been found that TiO_2 NPs speed up the growth of spinach (Ali et al. 2021). By enhancing nitrogen metabolism, TiO_2 NPs also promoted plant growth (Kamal and Mogazy 2021). Additionally, they found that TiO_2 NPs can improve nitrogen metabolism. Through its antibacterial and insecticidal properties, TiO_2 has demonstrated assurance in directly combating crop diseases (Marimuthu et al. 2013). NPs present opportunities for the subsequent creation of effective antibacterial agents.

2.3.3 *Green Synthesis of Nanoparticles*

Green NPs synthesis is an environmentally friendly procedure. By using a variety of stabilizing and reducing agents, such as microorganisms and natural resources, this approach produces NPs in a sustainable manner (Hossain et al. 2019a, b). An alternative way for creating NPs is the use of the microbe-mediated method (Ahmed et al. 2020; Kumar et al. 2024a).

2.4 How Nanotechnology Can Help Achieve Food Security in Face of Climate Change

The global community is currently facing numerous unprecedented challenges, including limited land availability, industrialization, population growth, low agricultural productivity, high postharvest losses, and rising temperatures (Gothandam et al. 2018; Ndlovu et al. 2020; Verma et al. 2024b). These factors have contributed to a significant issue of global food insecurity. To ensure the health and well-being of people worldwide, cutting-edge technologies are required to secure a sustainable and reliable food supply (Jagadesh et al. 2024). Recent studies have highlighted the potential of nanotechnology in addressing these challenges and enhancing the agriculture sector by increasing the efficiency of agricultural inputs and providing solutions for improving food productivity and security (Singh et al. 2021). As the global population continues to grow and climate change further strains resources, food security has become an increasingly urgent concern. Climate change encompasses prolonged alterations in key climate parameters, including temperature fluctuations, water scarcity, extreme temperatures, soil salinity, alkalinity, and environmental metal contamination. The primary objective is to enable plants to adapt swiftly to these environmental challenges without compromising the delicate ecosystems that currently exist (Vermeulen et al. 2012).

Achieving this requires a multifaceted approach, including the regulation of stress gene expression, activation of the plant enzymatic system, hormonal regulation, control of hazardous metal uptake, and the reduction of plant life cycles to mitigate water deficiency stress and flash floods. Abiotic stress, such as salt and drought, significantly impairs plant growth and productivity (Gupta et al. 2022a, b, c). Scientists and researchers have been striving for decades to develop innovative methods to help plants cope with these challenges. The application of nanobiotechnology represents one such endeavor, which has gained considerable attention due to its potential efficacy in this field. Being sessile organisms, plants modulate their physiological, biochemical, and molecular processes to enhance their resilience and adaptability to adverse environmental conditions.

This discussion highlights the potential applications of nanotechnology in agriculture, food processing, and postharvest loss reduction as means to achieve food security.

2.4.1 Agriculture

In order to avoid negative effects on environmental compartments, researchers have worked to create technology and methods towards sustainable farming systems (Dubey and Mailapalli 2016; Pretty 2008). The use of nanofertilizers (NFs) in unfavourable conditions has the potential to boost crop yields, according to developments in NMs engineering. In roughly 23% of the cultivated lands worldwide, salinity stress severely restricts agricultural productivity (Onaga and Kerstin 2016). However, under NaCl stress, it has been reported that applying nano-SiO_2 to tomato and squash plants enhances seed germination, increases plant fresh weight, dry weight, and chlorophyll content with proline accumulation (Haghighi et al. 2012; Siddiqui and Al-Whaibi 2014; Verma et al. 2023a, b). Similarly, Torabian et al. (2017) also showed that sunflower cultivars exhibit a good response to salinity stress tolerance when iron sulphate ($FeSO_4$) nanoparticles are sprayed on their leaves. The application of nano FeSO4 was found to have a significant negative impact on the amount of sodium (Na) in leaves, while also increasing leaf area, shoot dry weight, net carbon dioxide (CO_2) assimilation rate, sub-stomatal CO_2 concentration (Ci), chlorophyll content, maximum photochemical efficiency of photosystem II (Fv/Fm), and iron (Fe) content. Silicon nanoparticles (Si NPs) have also recently been investigated as a potential effective treatment for UV-B-induced stress in wheat (Tripathi et al. 2017). According to Manjaiah et al. (2019), nano zeolite can promote plant germination and growth as well as the long-term availability of nutrients. Abdel-Aziz et al. (2016) exploration of the use of NMs was very well done. In comparison to plants treated with traditional fertilizer, they discovered that the life cycle of wheat plants treated with nanofertilizer was 23.5% shorter (130 days as opposed to 170 days) for yield output from the date of sowing. The application of NFs has been shown to accelerate plant growth and productivity, indicating their potential as useful tools in agricultural practices. This is particularly true in areas that are prone to drought or sudden flash floods, where early crop maturity is crucial for sustainable crop production. Furthermore, it has been discovered that nanoparticles are useful in the detoxification or remediation of dangerous contaminants like heavy metals (Verma et al. 2022). For instance, Wang et al. (2015) demonstrate that by controlling Cd accumulation, foliar application of nano-Si at a concentration of 2.5 mM considerably increases rice plants' tolerance to Cd stress. The same group demonstrated the effectiveness of nano-Si with Cd against Pb, Cu, and Zn in another investigation. According to Wang et al. (2016), there seems to be a potential benefit of using nano-Si fertilizers in place of conventional fertilizers when it comes to lowering the buildup of heavy metals.

2.4.2 Plant Protection

In order to boost crop productivity, the use of nanotechnology in plant protection products has grown dramatically. Conventional crop protection methods typically entail the extensive and excessive application of pesticides, herbicides, and fungicides. More than 90% of the pesticides used are either lost to the environment or do not reach the target locations that are necessary for efficient pest management (Nuruzzaman et al. 2016; Choudhary et al. 2024). This results in the depletion of environmental systems in addition to raising crop production costs. It should be highlighted that improved plant protection against pest invasion and eventual crop loss depends on the active ingredients being present in the formulation at the target areas at the minimal effective concentration. In light of this, the creation of novel plant protection formulations has long been a very intelligent area of agricultural study. Therefore, there is a great deal of potential for sustainable agricultural development using the NPs found in pesticides, fungicides, and herbicides.

2.4.2.1 Nano Antimicrobial Agents

Disease management is another area in which modified nanoparticles may find use in agriculture. In a variety of areas related to plant protection, such as microbial activity and bacterial illnesses, inorganic NPs such as ZnO, Cu, SiO_2, TiO_2, CaO, MgO, MnO, and Ag NPs are crucial (Servin et al. 2015; Elmer and White 2016). For instance, it has recently been demonstrated that ZnO nanoparticles effectively inhibit the growth of pathogenic bacteria *Pseudomonas aeruginosa* as well as *Fusarium graminearum, Penicillium expansum, Alternaria alternate, F. oxysporum, Rhizopus stolonifer, Mucor plumbeus*, and *Aspergillus flavus* (Servin et al. 2015; Dwivedi et al. 2016; Vanathi et al. 2016). It has been studied recently how to use nano Ag against the phytopathogen Colletotrichum gloeosporioides (Aguilar-Méndez et al. 2016). Several NPs such as Aluminum, Copper, Iron, Zinc, Silica, Selinium, Titanium, Caron, and their oxides have effects on plants in addition to their antibacterial properties (Ghidan and Al Antary 2019). Occasionally, beneficial soil microbes may be inhibited in their growth by NPs (Ouyang et al. 2020). Several research organizations are interested in the usage of environmentally friendly insecticides. The antifungal properties of NPs can help in the creation of NP-based pesticides. Due to their unique advantages over other NPs such as Zinc (Zn), Zinc oxide (ZnO), Copper (Cu), Gold (Au), Titanium oxide (TiO_2), and Aluminium oxide (Al_2O_3), silver (Ag) has been thoroughly studied among the several inorganic NPs based antibacterial treatments (Aguilar-Méndez et al. 2016). In tomato, it was discovered that using nano-Cu was more successful against *Phytophthora infestans* than using non-nano Cu formulations that are already on the market (Giannousi et al. 2013).

2.4.2.2 Nanoherbicides

Weeds pose a significant danger to the world's agricultural output because they compete with crops for nutrients, water, and light. Application-engineered NPs containing herbicides provide an environmentally friendly solution to the problem. For instance, Sharifi-Rad et al. (2016) showed that weeds treated to SiO_2 NPs had considerably lower germination, root and shoot lengths, fresh and dry weights, and photosynthetic pigments with total protein. Similarly, Kumar et al. study from 2017 demonstrated that, when compared to commercial herbicides, herbicide (metsulfuron methyl)-loaded pectin (polysaccharide) NPs are more lethal to *Chenopodium album* plants in both lab and field settings. Commercial herbicides typically only impact the above-ground portions of weeds, leaving their underground components, such as rhizomes and tubers, unaffected. Weeds thus begin to grow again; nevertheless, nanoherbicides stop this from happening (Dwivedi et al. 2016).

2.4.2.3 Nanopesticides

Without sacrificing efficacy, pesticides can be encapsulated or nanoformulated to promote the persistence or controlled release of active components inside plants or in root zones. However, traditional pesticide and herbicide formulations damage other organisms as well as limiting the solubility of the pesticides in water, which increases the organisms' resistance to the original target. However, as noted above, nanoformulations help to get over these restrictions (Dwivedi et al. 2016). For example, Petosa et al. (2017) showed that pesticide nanoformulations increase crop yields by improving pesticide efficacy by controlling the pesticide's transport potential. In loamy sand soil saturated with artificial pore water containing Ca^{2+} and Mg^{2+} cations, they discovered that nanoformulations combining polymeric nanocapsules and the pyrethroidbifenthrin (nCAP4-BIF) display increased elution with time and enhanced transport potential even upon the addition of fertilizer. This suggests that nCAP4 may be a useful pesticide delivery system for plant protection, such as pyrethroid. This might be because of the nanoformulations'enhanced wet ability and dispersion potential, which lowers undesired pesticidal movement and organic solvent runoff. Additionally, the use of NMs in pesticide formulations demonstrates several beneficial qualities that are necessary for a sustainable agro-environmental system, such as enhanced stiffness, permeability, thermal stability, solubility, crystallinity, and biodegradability (Khan and Rizvi 2014; Haq and Ijaz 2019).

More significantly, an essential component of integrated pest management (IPM) is the timely and regulated release of active chemicals, which lowers the overall quantity of pesticides needed for disease and pest control. Additionally, minimal pesticide use is required by sustainable agriculture to prevent environmental depletion and harm to non-target species. Additionally, using less pesticide lowers the cost of producing crops. For instance, halloysites are a kind of clay nanotube that are employed in agriculture as affordable pesticide carriers. These nanotubes offer better interaction with less impact on the environment in addition to demonstrating extended release duration of active ingredients (AI) (Dwivedi et al. 2016). One such substance that is hydrophobic in nature and can penetrate into an insect's cuticle layer upon contact is nano silica, which ultimately causes the insects to die (Nair et al. 2010).

A commendable effort was made by De Jorge et al. (2017) to investigate the significance of nanoformulation in AI controlled release. They have investigated the possibility of a controlled release of AI and a long-term attract-and-kill action of pheromone and insecticide. They have found that the nanofiber formulation of *Grapholitamolesta* (Lepidoptera: Tortricidae) (Busck) pheromone has no influence on mortality over time. Furthermore, numerous studies have demonstrated that pesticide nanoformulations promote the expansion of plant-based systemic acquired resistance (SAR) to pests. For instance, formulations including silica nanospheres can improve the ability of insecticides to pass through plants and reach the cell sap, having a systemic effect on insects that chew or sucking like aphids (Li et al. 2007).

According to Panpatte et al. (2016), these hollow formulation types also shield pesticides from photodegradation brought on by direct sunlight. Additionally, it has been noted that pesticides' non-systemic behaviour is altered by nanoformulations (Hou et al. 2016). When combined with metallic NPs (Au NPs), ferbam's nonsystemic behaviour can change and boost the penetrating potential into tea leaves. These kinds of discoveries will undoubtedly open up new avenues for the development of pesticide formulations that target plant-based systemic resistance. To fully understand the effects and fate of pesticides as well as how they interact with the biomacromolecules found in biological systems and environments, more research is necessary. In the meanwhile, Patil et al. (2016) showed that latex-fabricated bioactive Au NPs decreased trypsin catalytic potential a crucial insect protease and so helped to biocontrol insects that cause harm. The interaction of metallic NPs with proteins through covalent, electrostatic, or binding to the -SH group of amino acids may be the cause of this catalytic inhibition (Patil et al. 2016).

2.4.2.4 Postharvest Loss Reduction

More than 40% of food (cereals, roots and tubers, pulses and oil crops, vegetables and fruit, fish, meat, and dairy) is lost during the trade and customer stages in developed countries, and more than 40% is lost during the post-harvest stage and processing point in developing countries (Gustavsson et al. 2011; FAO 2019). Microorganism attacks can cause harvested high moisture yields that have not been maintained to swiftly degrade. Nanotechnology is one of the most recent and sophisticated technologies that can aid in reducing post-harvest losses. By creating functional packing elements with the fewest amounts of bioactive components, better gas and mechanical properties, and less of an impact on the sensing qualities of fruits and vegetables, nanotechnology application can reduce post-harvest losses (Flores-López et al. 2016). When applying edible coatings as a liquid on food, the product is often dipped in a material that provides a solution and is created by the structural medium (carbohydrate, lipid, protein, or mixture). By impeding dehydration, disrupting breathing, enhancing textural characteristics, supporting the preservation of volatile fragrance molecules, and reducing microbial development, they prevent untreated foods from getting bad. When applied to various meals, edible nano-coatings offer a barrier against the flow of gases and moisture while also supplying flavours, colours, enzymes, antioxidants, and agents that prevent browning, all of which have the potential to extend the shelf life of manufactured foods (Zambrano-Zaragoza et al. 2018).

According to Sekhon (2010), the method allows coatings to form at the nanoscale with a thickness of up to five nanometers. For horticultural products, edible coatings and thin films are frequently used. The fencing effect against gas flow, structural obstruction to water migration, microbes, sensory suitability, cost, availability, functional qualities, mechanical properties (elasticity, tension), photosensitive properties (brilliance and opacity), and other factors determine the use (Falguera et al. 2011; Zambrano-Zaragoza et al. 2018).

Food is coated with a variety of edible nanocoatings to regulate the quality of newly harvested products after harvest. Recent years have seen a rise in interest in silver nanoparticles because of their antibacterial qualities, which are crucial for food processing. When PVP-based silver NPs were applied to asparagus, the growth of microorganisms was significantly slowed, which slowed weight loss and prevented skin colour changes (An et al. 2008). Another study found that strawberry shelf life was significantly increased by edible coatings made of gelatin and cellulose nanocrystal (Fakhouri et al. 2014). When compared to other treatments, chitosan-assisted nano-silica coating improved the physicochemical and physiological values of longan fruit at room temperature by skillfully producing an excellent semi-permeable layer (Shi et al. 2013). Additionally, Yu et al. (2012) studied the use of chitosan film-based nano-SiO_2 and Medeiros et al. (2014) studied the use of alginate or lysozyme-based nanolaminate coatings to preserve the value of fresh foods during extended storage. Additionally, during fruit storage, nano-ZnO coating preserved some fruits' post-harvest value and decreased microbial damage (Sogvar et al. 2016).

2.4.2.5 Food Processing

The food industry needs newer technologies to increase market price, quality, and productivity. Numerous applications of nanotechnology in the manufacturing and processing of food are being developed; they include food additives based on nanotechnology, nano-encapsulation, nanosensors, smart distribution systems based on nanoparticles, nano-packaging, as well as pharmaceuticals and healthcare (Rashidi and Khosravi-Darani 2011). Its uses include the creation of encapsulations, biopolymer matrices, emulsions, straightforward solutions, and related colloids that provide efficient delivery methods. Nanotechnology is being more and more used in industrial food processing, especially for flavour encapsulation or odour enhancement, texture change or enhancement of value, as well as new gelation or viscosity-enhancing substances (Duncan 2011).

Food nanotechnology is the synthesis of structures at the nanoscale that have unique properties and used for a variety of applications. These include nanocomposite coatings, food characterization devices, food interaction surfaces with unique superficial properties, microfluidic instruments, and delivery systems, among many other applications (Sadeghi et al. 2017). Small emulsion droplets with diameters less than 100 nm form nanoemulsions in an oil/water system at the nanoscale (Anton and Vandamme 2011). To create the scattered phases, they require more mechanical energy, such as those obtained by a high-pressure homogenizing step, a microfluidizer, or a sonication process. Because they prevent food from breaking and splitting, nanoemulsions significantly reduce the quantity of fat needed by minimizing the need for stabilizers (Dasgupta et al. 2015). A number of nanoemulsions have numerous technical benefits and appear to be optically transparent when added to beverages (Prasad et al. 2017).

The end products of nanoemulsions are as creamy as regular food items, with no discernible alterations in flavour (Dasgupta et al. 2015). To alter food for salad dressing, oil flavouring, sweeteners, better drinks, and other food processing, nanoemulsions are used. Food processing uses nanomaterials as filling agents to increase mechanical power and stability of the packing material, anti-caking agents, food additives, transporters for clever nutrient delivery, antimicrobial agents, and food nanosensing to improve food value and security assessment (Ezhilarasi et al. 2013). This is relevant to dietary supplements for producing nutraceuticals that have improved stability and bioavailability (Duncan 2011).

Natural NPs are present in a wide range of food products. Many foods including proteins are globular structures with a size range of 10–100 nm, whereas other foods primarily consist of lipids and carbohydrates, which make up linear polymers with a thickness of less than 1 nm. Casein is a nanoscale protein found in milk, and meat is made up of protein filaments that fall into the NMs group (Prasad et al. 2017). The global food and beverage industry is highly profitable, and numerous food corporations have been conducting research to enhance food safety, nutritional value, and manufacturing efficiency (Roohinejad and Greiner 2017). The application of NMs in meat processing and packaging offers several advantages, including enhanced sensory qualities, guided distribution of compounds with good bioactivity, and better bioavailability and antibacterial capabilities (Ramachandraiah et al. 2015). Using nanotechnology, products with the best texture and taste can be manufactured, environmentally friendly processes can be created, low-calorie beverages and food products can be produced, and all of these efforts can change people's lifestyles for the better (Chaudhry et al. 2008; Katouzian and Jafari 2016). It can provide instruments for more quickly identifying component deficits, such as atomic force microscopy (AFM), and it can be helpful in using nanosensors to find pathogens in food (Rashidi and Khosravi-Darani 2011). According to Katouzian and Jafari (2016), AFM is an effective tool for examining intricate food arrangements and is beneficial for studying molecular interactions at the nanoscale.

Similar to other scientific fields, food science uses more recent nanotechnologies to improve customer quality of life through altered food formulation, food packaging, the production of novel ingredients, and process monitoring that can result in a healthier, safer, higher-quality, more accurate food system with an extended shelf life (Roohinejad and Greiner 2017). Food processing, security, functionality, and cost-effectiveness in terms of distribution and efficiency are among the important connections between food systems and nanotechnologies that are getting better (Rashidi and Khosravi-Darani 2011).

Nanofiltration, nanoscale enzyme-based reactors, nano-encapsulation for absorption and modification, mass and heat transmission, and nanofabrication are some of the multipurpose uses of nanotechnology in food processing. In the pharmaceutical industry, nanofiltration works better for pharmaceutical purification and is a necessary step in ensuring that certain solutes are forgotten. In order to alter the product quality by eliminating salt from lactose, it is also used to treat drinking water and dairy products (Dasgupta et al. 2015). Package heat resistance was increased using heat and mass transfer nanofabrication.

Food products are improved by applying the nano-encapsulation procedure. This method typically provides cooking balance, enhance flavour, and preserve food. Nanocapsules in the form of nanoceramic pots can be used to alter absorption in order to reduce cooking times and the amount of spent oil; additionally, by substituting plant oil for hydrogenated oil, trans fatty acid levels can be reduced; and finally, the production of safer nanofood nanocapsules can be used to distribute nutrients for better absorption in food. According to Singh et al. (2017a, b), Ubbink and Krüger (2006), Weiss et al. (2006), and others, nanoencapsulation can mask tastes and odours, regulate how food interacts with active ingredients, regulate the release of dynamic agents, protect accessibility at a predetermined time at a set rate, and protect against biological, chemical, heat, and moisture interference during storage, processing, and application. It can also further demonstrate similarity with other ingredients in the system. SiO_2 and TiO_2, two metallic oxides, have been employed as food colouring or flow additives. Food waste nanomaterials are treated with SiO_2 nanomaterials to transfer smells or aromas (Dekkers et al. 2011).

Nanoscale enzyme reactors are used in food processing and modification techniques to change food systems for improved flavour, nutritional value, and various health benefits. Because nanomaterials have larger surface-to-volume ratios than macroscale supported items, they help spread across food media, which leads to higher enzyme assisted systems (to increase activity, shelf life, and economy) when added (Yu et al. 2005). Peptide nanotubes have been shown to produce better enzyme supports at elevated temperatures (65 °C). The activity of the lipase in the nanotubes was observed to be 70% higher than that of free lipases. For example, nano-SiO_2 particles significantly hydrolyzed olive oil with enhanced stability, reusability, and adaptability (Bai et al. 2006).

2.4.2.6 Packaging

Packaging plays many functional role in the food chain. Food preservation prevents food from deteriorating physically and serving as one of its many important purposes. Safer, more affordable, frothy, readily disposable or reusable, stable during storage and transit, and resistant to physical abuse are all desirable qualities in packaging. The kind and composition of the packing material affect the food's quality. The packaging qualities of items can be enhanced by NMs by giving them additional qualities including heat resistance, strength, and light weight. Numerous studies have reported and employed the use of nanotechnology in food packaging (Mihindukulasuriya and Lim 2014; Han et al. 2018; Kumar et al. 2018; Noorbakhsh-Soltani et al. 2018). According to Duncan (2011), the market for food and beverage packaging based on nanotechnology was predicted to be worth 4.13 billion USD in 2008 and increased to 7.3 billion USD by 2014, indicating an annual growth rate of 11.65%.

Various active packing applications are used to different metal and metal oxide nanoparticles. According to Li et al. (2017) and Zhao et al. (2012a), food packaging benefits greatly from the use of silver (Ag) and TiO_2. Metal and metal oxide NPs are used in active packaging as antimicrobials, much like in nanocomposites used in food packaging (Pathakoti et al. 2017). Because of its semiconducting qualities and enhanced optical, electrical, and photosensitive results, TiO_2 is widely used as an adsorbent material, stain, or catalytic substrate (Prakash et al. 2018). Rice may be stored at a temperature of 37 °C and a relative humidity of 70% thanks to the improved qualities of a blend of dispersed and antibacterial Ag/TiO_2 NPs and polyethylene used as packing agents (Zhao et al. 2012a, b).

The antibacterial potential, oxygen transfer, enzyme mobilization, and indicator of the degree of exposure to degradation-associated factor are among the benefits of using nanoparticles in food packaging (De Azeredo 2009). When compared to conventional packing material, nano-packing material generated a valuable sensing quality (Sharma et al. 2017). When zinc oxide NPs and nano-polymer containing polylactic acid were combined, the outcome was a highly functional interaction material for food packaging (Martínez-Bueno et al. 2017). When evaluated for food packaging, zinc oxide-treated semolina protein coating showed a significant decrease in oxygen absorptive as well as heat resistance (Jafarzadeh et al. 2017). Food packaging has layers made of nanoclay, silicates, and moisturized alumina-silicate (Weiss et al. 2006). In order to create a possible nanocomposite with exceptional fencing capabilities for food packaging, clay and silicate have also been combined with polymers (Sharma et al. 2017). The potential of organo-clay nanoparticles as an antibacterial material in food packaging has also been studied (Fasihnia et al. 2018).

Because of the larger surface area of NPs associated with increased activity, cellulose has been used as a supporting material in several NMs and in a variety of applications (Sharma et al. 2017). As a natural polymer, it is very strong, affordable, environmentally benign, and readily recyclable (Moustafa et al. 2019). Food packaging has made use of nano-composite films that have been changed by adding nano-chitosan and nano-cellulose. These films have demonstrated improved tensile power, elongation to breakage, clearance, and food protection properties (Noorbakhsh-Soltani et al. 2018). When compared to chitosan used alone, the production of nano-bio-composited films using polylactides and chitosan has demonstrated their suitability for packaging with significant and ongoing antioxidant potential (Basu et al. 2017). Antibacterial copper and cellulose nanocomposites have been used for paper coatings and packaging (Pinto et al. 2013).

Because of their innate antibacterial properties, chitosan films with nanoscale thickness are highly focused and widely used (Rieger et al. 2015). Ionic gelation is the process by which chitosan NPs are created through the electrostatic interaction of the chitosan positive amino sides with polyanions, which act as cross-linkers (Sharma et al. 2017). Red grapes' shelf life can be extended by 14 days using a chitosan/gelatin-based nanocomposite containing silver NPs, which is used in food packaging (Kumar et al. 2018). As effective antibacterial candidates against gram-positive (*Staphylococcus aureus*) and gram-negative (*Pseudomonas aeruginosa*), yeast (*Candida albicans*), fungi (*Aspergillusniger*), and yeast in packing, chitosan-Ag and chitosan-Au nanocomposites coating are applied (Youssef et al. 2014).

2.5 Nanosensor Applications in Agriculture Combating Climate Change

Highly sensitive and versatile nanoscale sensors have been developed as a result of recent developments in nanotechnology (Sagadevan and Periasamy 2014). Plants have employed a variety of nanosensor technologies, including as plasmonic nanosensors, carbon-based electrochemical nanosensors, Förster resonance energy transfer (FRET)-based nanosensors, nanowire nanosensors, and antibody nanosensors (Shaw and Honeychurch 2022). Researchers have been able to monitor the spatiotemporal dynamics of analyses (Tsuchiya et al. 2015; Tiwari et al. 2016), detect viral and fungal pathogens (Duhan et al. 2017; Lattanzio and Nivarlet 2017), and cellular functions (Kosaka et al. 2014) and metabolic flux (Chaudhuri et al. 2008; Bagal-Kestwal et al. 2015). Reviews of the use of nanosensors in agriculture (Zhang et al. 2022; Johnson et al. 2021), plant pathogen detection (Kulabhusan et al. 2022; Ang and Lew 2022), particular fields, like planta nanosensors (Voke et al. 2021), or more specialized forms of NMs, like carbon nanotubes (Safdar et al. 2022), have been published in a number of reviews.

The state-of-the-art research area for sustainable agricultural development is NMs engineering. Precision farming employs NMs cut costs and labour, boost productivity, and promote ecologically responsible development (Shang et al. 2019). To ensure soil and plant health, product quality, and general safety for sustainable agriculture and environmental systems, nanosensors measuring and monitoring crop growth and soil conditions, nutrient deficiency, toxicity, diseases, and the introduction of agrochemicals into the environment are being developed (Cheng et al. 2016).

Organisms that are biological in nature possess the ability to feel their surroundings. But the integration of biology and NMs into sensors has opened up new possibilities for improving the sensitivity, specificity, and speed at which impairments can be detected (Dubey and Mailapalli 2016). The most effective way to maximize water consumption efficiency is through the automation of irrigation systems through the use of sensor technologies. When there is a water shortage, autonomous irrigation control is combined with real-time soil water tension estimation from nanosensors (Fraceto et al. 2016). Throughout the growing season, cultivated fields are monitored in real time using a global positioning system (GPS) based on nanosensors. These wireless nanosensor networks use tiny carriers that use wireless signals dispersed throughout the farmed fields to monitor the controlled release mechanism.

Similarly, timely use of pesticides or fertilizers to protect crops from infestation would be aided by the rapid and accurate identification of insects or pathogens. Afsharinejad et al. (2016) have developed a wireless nanosensor to detect insect attacks in this regard. This sensor can identify between different insect species based on the volatile organics released by a variety of host plant species. The effectiveness of an immune-sensor based on nano-gold in detecting karnal bunt disease in wheat plants (Singh et al. 2010). Furthermore, the development of bionic plants through the insertion of NPs into living plants' cells and chloroplasts, which enable the plants to sense or image objects in their surroundings, communicate through infrared devices, or even power themselves as light sources, holds great promise for precision farming in the field of nanobiotechnology (Ghorbanpour and Fahimirad 2017; Kwak et al. 2017). For example, Giraldo et al. (2014) concluded that by enhancing photo-absorption, single-walled carbon nanotube (SWNT) insertion boosts the electron transfer rate of light-adapted chloroplasts by 49% *in vivo*. Additionally, it was demonstrated that SWNTs could affect plant sensing and enhance photosynthetic efficiency and quantum yield by inhibiting the production of reactive oxygen species in chloroplasts, which enhanced the light-harvesting capacity of near-infrared fluorescence. Thus, new avenues for the study of functional plant NMs hybrids are made possible by developments in nanobionic techniques for environmental monitoring and crop improvement (Giraldo et al. 2014; Wong et al. 2017).

A straightforward, label-free glutathione regulated dual-functional platform based on upconversion nanoparticles (UCNPs)/Au NPs was recently developed by Fang et al. (2017) for the turn-on fluorescence detection of harmful Cd2+ and acetylcholinesterase (AChE) activity in actual water samples. Additionally, a great deal of progress has been achieved in the measurement and tracking of trace amounts of contaminants like pesticides. As an illustration, photosystem II with its biosensors has the capacity to both monitor and bind multiple pesticide groups. Before being disposed of in the agricultural environment, these nanosensors offer a simple and cost-effective method for identifying particular pesticides that include a variety of organic contaminants (Panpatte et al. 2016).

Plants undergo many physiological changes in response to stress, which are facilitated by stress hormones such salicylic acid, methyl jasmonate, and jasmonic acid. Intriguingly, Wang et al. (2010) created a modified gold electrode nanosensor with copper NPs to monitor the oil seed's salicylic acid content and identify the presence of a pathogenic fungus infestation. Therefore, by controlling hormones like auxin, multi-walled carbon nanotubes (MWCNTs) are also effective in studying plant growth. This may aid researchers in understanding how plant roots adapt to their surroundings, especially marginal soils (McLamore et al. 2010). Due to their huge surface-volume ratio, one-dimensional potassium niobate ($KNbO_3$) nanofibers offer tremendous promise for humidity sensing, as revealed by Ganeshkumar et al. (2016). When the relative humidity varies from 15% to 95% at ambient air temperature, the conductance of the $KNbO_3$ nanofiber-based humidity nanosensors increases from 10–10℧ to 10–6℧ in less than 2 seconds, exhibiting logarithmic–linear dependency behaviour.

Using nanosensors, nanotechnology can also be used to control grain quality (Bouwmeester et al. 2009). The sensors can react to degradation products, microbial contamination, and variations in temperature, oxygen exposure, and relative humidity during storage. Insects or fungi in the grain that has been kept can also be addressed with them (Axelos and Van De Voorde 2017). Utilizing polymer NPs, which react to volatile agents and other analysts in the environment of stored foods to identify the type and cause of breakdown involved, nanosensors for monitoring grain quality have also been created (Neethirajan and Jayas 2011). Nonetheless, in light of the extensive body of research in this subject, the persuasive uses of nanosensors—particularly in field research—are unexpectedly limited, indicating a new avenue for investigation.

2.6 Using Nanofertilizers to Prevent Climate Change: An Effective Source of Balanced Crop Nutrition

Utilizing the unique properties of NPs, NFs have demonstrated the potential to enhance nutrient utilization efficiency. Research by Tyagi et al. (2022) has highlighted the benefits of NFs, including improved nutrient uptake, soil fertility restoration, high absorption capacity, enhanced photosynthesis, increased crop production, reduced soil toxicity, decreased application frequency, improved plant health, and reduced environmental pollution. When NPs are suspended in water and applied to the soil, they release nutrients, as observed by Singh et al. (2021). To prevent nutrient losses, NPs can be coated with polymers or thin coatings to encapsulate them. NFs encompass various NM components such as silica, iron, zinc oxide, titanium dioxide, cerium oxide, aluminum oxide, gold nanorods, ZnCdSe/ZnS core-shell, InP/ZnS core-shell, and Mn/ZnSe quantum dots. The effectiveness of NFs in promoting plant growth depends on factors such as crop type, nanomaterial size, composition, concentration, and chemical characteristics.

Silica nanoparticles have been found to stimulate antioxidant enzymes and support seedling growth (Lin et al. 2004; Neethirajan et al. 2009). In the face of drought and salinity, two major challenges in agriculture, Si NPs have shown the ability to reduce sodium uptake and enhance potassium uptake under salt stress conditions (Rizwan et al. 2015). The presence of silica can also reduce water evaporation from the soil surface, thereby aiding plant growth and development in unfavorable environmental conditions such as high temperatures and irregular rainfall. Salt stress poses a significant threat to food security and agricultural productivity, especially in the context of climate change and rapid population growth. NFs offer a promising avenue for managing salt stress in agriculture. El-Shal et al. (2022) conducted a hydroponic experiment using two genotypes of lucerne (*Medicago sativa* L.) exposed to different salt concentrations. The addition of nano-K_2SO_4 improved various plant characteristics, including shoot dry weight, plant height, flower count, tiller count, root length, root fresh weight, and root dry weight, for both salt concentrations. Although both genotypes showed increased relative water content and electrolyte leakage, the specific quantities of salt varied.

Considering the projected reduction in fertilizer use and the need to ensure global food production, agricultural interventions must evolve. Nanotechnology is increasingly being accepted in agriculture to enhance crop yields while maintaining a healthy agro-ecosystem under adverse environmental conditions. Future agriculture may rely on the use of nano-enabled fertilizers in various ways, as suggested by

Gogos et al. (2012), Rajput et al. (2021b), Singh et al. (2021), Lowry et al. (2019), and Bandala and Berli (2019).

Research has also explored the effects of other metal oxide NPs on seed germination and plant growth. For instance, Rajput et al. (2018) and Raja et al. (2019) observed the impact of ZnO and Cu NPs on *Vigna mungo* and *Hordeum sativum* distichum germination, respectively. Polyvinylpyrrolidone (PVP) was found to protect *Pisum sativum* plants, as reported by Rahman et al. (2020). Fe NPs demonstrated positive effects on *Capsicum annum,* leading to increased chloroplast numbers and grana stacking, ensuring stable and functional photosystem II (Yuan et al. 2018). Chitosan, another biocompatible nanoparticle, influenced wheat plant shape and germination percentage by increasing indole-3-acetic acid (IAA) levels, even at low concentrations (Li et al. 2019). Nano-nitrogen, phosphorus, and potassium (nNPK) formulations applied through leaf stomata improved wheat gas exchange and leaf growth (Abdel-Aziz et al. 2018).

Abiotic stresses can induce the production of reactive oxygen species (ROS), leading to reduced photosynthetic capacity and oxidative damage to biomolecules and cell membranes (Wakeel et al. 2020; Kumar et al. 2024b). Nanoparticles have the potential to enhance plant tolerance to various abiotic stresses. For instance, CeO NPs at a minimal Ce^{3+}/Ce^{4+} ratio exhibit scavenging properties and mimic the activity of catalase (CAT) and superoxide dismutase (SOD) (Wang et al. 2012a; Van Nhan et al. 2015; Pulido-Reyes et al. 2015). Nano-TiO_2 can improve plant hydration by increasing nitrate reductase (NR) activity and osmolyte accumulation. SiO_2 NPs have shown the ability to enhance transpiration efficiency, photosynthetic pigments, and carbonic anhydrase activities in pumpkin plants to enhance their defense responses to abiotic stresses (Siddiqui et al. 2014; Rajput et al. 2021a, b).

In conclusion, NFs utilizing NPs offer significant advantages in improving nutrient utilization, promoting plant growth, and mitigating the effects of abiotic stresses. The composition, size, concentration, and chemical properties of NMs play a crucial role in their effectiveness as NFs. The application of NFs has shown promising results in various crops, enhancing plant characteristics and increasing crop productivity. As the global demand for food increases and fertilizer use is projected to decrease, the implementation of nanotechnology in agriculture holds great potential for sustainable and efficient food production while safeguarding ecosystems, biodiversity, and the climate.

2.7 Nanomaterials Accelerate Plant Adaptation to Progressive Climate Change Factors

A new method of delivering SA that can enhance its stability, solubility, and bioavailability is salicylic nanoparticles (SA NPs). Additionally, SA NPs can improve SA absorption and translocation in plants, which will improve their ability to withstand heat stress. Some research claim that SANPs work better than SA to reduce heat stress in plants by boosting antioxidant enzyme activity, photosynthetic efficiency, and wheat yield components (Kaur et al. 2022). In rice, SANPs also enhanced the expression of genes and proteins linked to heat stress (Nazar et al. 2017) and improved maize membrane stability while lowering oxidative damage (Sangwan et al. 2022). Pan et al. (2006) tested the effects of heat acclimation on pea plants at 37 °C for 2 hours in order to investigate the role of salicylic acid (SA) and the SA production pathway in plant thermo tolerance. Prior to heat treatment, in vivo pretreatments with SA or SA biosynthesis inhibitors were given. The findings demonstrated that heat shock protein production was stimulated and leaf damage from heat stress was decreased by both heat acclimation and SA treatment.

Lipid peroxidation did not occur in maize plants treated with CeO_2 NPs at 400 or 800 mg/kg, but there was a noticeable increase in the accumulation of H_2O_2, catalase, and ascorbate peroxidase activity in shoots. A stress response was shown by the upregulation of heat shock protein 70 in roots. Since ion leakage was not seen, the integrity of the membrane may have been preserved. Stomatal conductance and photosynthetic rate were unaffected. According to Zhao et al. (2012a, b), the findings imply that heat shock protein 70, ascorbate peroxidase, and catalase may assist plants in fending off oxidative damage caused by CeO_2 NPs and surviving exposure. The swift advancement of nanotechnology will result in the discharge of nanoparticles (NPs) into the surroundings, carrying unclear ramifications. CeO_2 NPs possible toxicity to plants and their uptake by the food chain are still unknown.

The study examined the potential of selenium nanoparticles (Se NPs, at concentrations of 10 mg L^{-1}) to alleviate the adverse impacts of heat stress and drought on eight different genotypes of wheat. In wheat seedlings, foliar spraying Se NPs enhanced plant growth, physiological responses, and gene expression. Heat stress had a greater effect on gene expression in plants, whereas drought had a greater effect on plant characteristics. Stress-responsive gene expression and antioxidant enzyme activity were enhanced by Se NPs, and genotype differences were seen in their ability to withstand stress. Superoxide dismutase (SOD) activity ranged from the highest in Sids-12 to the lowest in Sids-168. Stress increased the expression of genes in genotypes that were tolerant. On the other hand, the genotypes Giza-168, Giza-171, Sids-12, and Sids-14 demonstrated good stress tolerance and performance. Finally, by enhancing physiological and gene expression characteristics, our study revealed that Se NP treatment might lessen the detrimental effects of heat and drought conditions in wheat genotypes. According to Omar et al. (2023), these results lend credence to the idea of using certain genotypes in breeding programmes to address the problems posed by climate change.

Azmat et al. (2022) study looked into the possibility of using green-synthesized zinc oxide nanoparticles (ZnO-NPs) derived from papaya extract and plant growth-promoting rhizobacteria (PGPR) to assist wheat withstand the effects of heat and drought. ZnO-NPs were characterized using a range of methods. The effects of PGPR and ZnO-NPs on wheat under simultaneous stressors of heat, drought, and both were investigated in a pot experiment. The combined stressors harmed wheat plants more than the individual stressors; this impact was observed in the rise in MDA and H_2O_2 concentration. Higher levels of biomass, pigments, minerals, sugars, proteins, and hormones were found in treated plants, which also showed improved growth and stress tolerance (Choudhary et al. 2025). ZnO-NPs and PGPR worked together to strengthen plant defenses like abscisic acid and antioxidant enzymes. These therapies had shown their effectiveness in reducing heat- and drought-related stress by also lowering hydrogen peroxide, lipid peroxidation, and electrolyte leakage.

Stress brought on by climate change in grass ecosystems is a major global issue. Zinc oxide (n-ZnO) molecules have demonstrated promise in aiding plants in mitigating stress; nevertheless, several methodologies have been employed to investigate the effects of these molecules specifically on stressed lucerne (Kareem et al. 2022). In order to evaluate lucerne seedlings' capacity to withstand stress and their ability to recover, n-ZnO was applied topically to them for 5 weeks, both prior to and following stress. The results of the investigation showed that the concentration of n-ZnO affected its absorption. Images obtained using a transmission electron microscope revealed that vacuoles and chloroplasts were the main locations of n-ZnO. Prior to stress, high concentrations of n-ZnO showed the capacity to lessen oxidative stress, stress peroxidation, effector damage, and organelle damage, as well as to avoid and reverse it.

However, 90 ppm n-ZnO stimulated antioxidant mechanisms and increased osmolyte levels before and after stress, effectively preventing oxidative stress, lipid peroxidation, and membrane damage. Notably, applying 90 mgL^{-1} of n-ZnO prior to heat stress shown increased effectiveness in mitigating heat-induced damage. All things considered, the use of mobilized n-ZnO showed excellent heat damage protection. The outcomes also showed that while post-treatment increased recovery efficiency, n-ZnO pre-treatment increased heat tolerance. When it came to encouraging plant development and physiological characteristics, pre-treatment with n-ZnO proved to be more successful than post-treatment. Phytotoxicity must be taken into account when assessing how NPs affect plants ability to respond to stress in the environment. In an investigation using Arabidopsis thaliana seedlings, heat stress and ZnO-NP exposure demonstrated that, although 1 μg/mL of ZnO-NPs by itself did not relieve β-glucuronidase gene silencing, it did improve heat stress mitigation, indicating a synergistic effect on genetic stability (Wu and Wang 2020).

Shukla et al. (2024) have demonstrated that wheat's tolerance to heat stress may be effectively increased by optimized amounts of zinc and silver NPs, providing promising options for enhancing crop resilience under environmental pressures. Similarly, ZnO NP treatment increased rice seedling resistance to heat and osmotic stressors, as shown by Qiu et al. (2023). The effects of titanium dioxide nanoparticles (TiO_2 NPs) and multiwalled carbon nanotubes (MWCNTs) on the growth of *Sesamum indica* L. plants under heat stress were studied by Mahmoud and Abdelhameed (2023). The contents of hydrogen peroxide (H_2O_2) and malondialdehyde (MDA) were lowered by spraying MWCNTs and TiO_2 NPs on the plants.

Extreme weather events brought on by climate change are a danger to crop productivity, as noted in a study by Chen et al. (2023), underscoring the need to improve crop resilience. The researchers discovered that adding reactive oxygen species-related NMs to maize seeds and seedlings can increase their resistance to various stressors. A 40 mg/L concentration of silver NPs applied to seeds increased their germination, vigor, growth, and resilience to cold, salt, and drought stressors. In addition to increasing root hair length and density, priming also improved resistance to water shortage. During seed priming, it was shown that the MAPK signaling pathway and plant hormone were activated. RNA-seq research revealed that maize seedlings underwent transcriptome alterations mediated by AgNP priming. Furthermore, iron-based NMs were successful in boosting the stress tolerance of seeds and seedlings. These results imply that seed priming with NMs can improve crop resilience to climate change by controlling the levels of reactive oxygen species (ROS), offering a practical way to alleviate food insecurity when circumstances worsen.

2.8 Conclusion

By improving plant growth regulators, nutrient availability, and pesticide active component distribution to desired target areas in plants, nanotechnology aids in advancement of agriculture. They also mitigate negative effects on agricultural system and soil. While NFs may enhance overall plant health, NMs may increase plant resistance to stress. In order to control the supply and bioavailability of nutrients based on plant species, growth stages, and agroclimatic zones, the particular biosensors may be connected to NFs. “Above all, since NFs are required in smaller quantities compared to synthetic fertilizers for crop production, the need for large-scale industrial facilities and extensive energy-intensive transportation networks could be significantly reduced if NFs are adopted globally.” Concerns about the biosafety of synthetic NPs and their potential applications in sustainable agriculture require more investigation. Notable outcomes of nano-goods supporting plant crops include biological sustainability, financial stability, and environmental protection.

References

Abdel-Aziz HMM, Hasaneen MNA, Omer AM (2016) Nano chitosan-NPK fertilizer enhances the growth and productivity of wheat plants grown in sandy soil. Span J Agric Res 14:17. https://doi.org/10.5424/sjar/2016141-8205

Abdel-Aziz HMM, Hasaneen MNA, Aya MO (2018) Foliar application of nano chitosan NPK fertilizer improves the yield of wheat plants grown on two different soils. Egypt J Exp Biol 14:63–72

Abd-Elsalam KA (2020) Carbon nanomaterials: 30 years of research in agroecosystems. Carbon Nanomater Agric -Food Environ Appl:1–18

Afsharinejad A, Davy A, Jennings B, Brennan C (2016) Performance analysis of plant monitoring nanosensor networks at THz frequencies. IEEE Internet Things J 3:59–69. https://doi.org/10.1109/JIOT.2015.2463685

Aguilar-Méndez MA, Martín-Martínez ES, Ortega Madan L, Sharma SC, Udayabhanu HR, Suresh D et al (2016) Facile green fabrication of nanostructure ZnO plates, bullets, flower, prismatic tip, closed pine cone: their antibacterial, antioxidant, photoluminescent and photocatalytic properties. Spectrochim Acta Part A: Mol Biomol Spectrosc 152:404–416

Ahmed T, Shahid M, Noman M, Niazi MBK, Mahmood F, Manzoor I, Zhang Y, Li B, Yang Y, Yan C et al (2020) Silver nanoparticles synthesized by using Bacillus cereus SZT1 ameliorated the damage of bacterial leaf blight pathogen in rice. Pathogens 9:160

Ali S, Mehmood A, Khan N (2021) Uptake, translocation, and consequences of nanomaterials on plant growth and stress adaptation. J Nanomater 2021:6677616

An J, Zhang M, Wang S, Tang J (2008) Physical, chemical and microbiological changes in stored green asparagus spears as affected by coating of silver nanoparticles-PVP. LWT - Food Sci Technol (Lebensmittel-Wissenschaft-Technol) 41:1100–1107

Ang MCY, Lew TTS (2022) Non-destructive technologies for plant health diagnosis. Front Plant Sci 13

Anton N, Vandamme TF (2011) Nano-emulsions and micro-emulsions: clarifications of the critical differences. Pharm Res (N Y) 28:978–985

Axelos MA, Van De Voorde M (2017) Nanotechnology in agriculture and food science. Wiley. ISBN: 978-3-527-33989-1, 424

Azmat A, Tanveer Y, Yasmin H, Hassan MN, Shahzad A, Reddy M, Ahmad A (2022) Coactive role of zinc oxide nanoparticles and plant growth promoting rhizobacteria for mitigation of synchronized effects of heat and drought stress in wheat plants. Chemosphere 297:133982

Bagal-Kestwal D, Kestwal RM, Chiang BH (2015) Invertase-nanogold clusters decorated plant membranes for fluorescence-based sucrose sensor. J Nanobiotechnol 13:30

Bai YX, Li YF, Yang Y, Yi LX (2006) Covalent immobilization of triacylglycerol lipase onto functionalized nanoscale SiO_2 spheres. Process Biochem 41:770–777

Bandala ER, Berli M (2019) Engineered nanomaterials (ENMs) and their role at the nexus of food, energy, and water. Mater Sci Energy Technol 2:29–40

Basu A, Kundu S, Sana S, Halder A, Abdullah MF, Datta S, Mukherjee A (2017) Edible nano-bio-composite film cargo device for food packaging applications. Food Packag Shelf Life 11:98–105

Baz H, Creech M, Chen J, Gong H, Bradford K, Huo H (2020) Water-soluble carbon nanoparticles improve seed germination and post-germination growth of lettuce under salinity stress. Agronomy 10(8):1192

Bhattacharjee R, Kumar L, Mukerjee N, Anand U, Dhasmana A, Preetam S, Proćków J (2022) The emergence of metal oxide nanoparticles (NPs) as a phytomedicine: a two-facet role in plant growth, nano-toxicity and anti-phyto-microbial activity. Biomed Pharmacother 155:113658

Bouwmeester H, Dekkers S, Noordam MY, Hagens WI, Bulder AS, De Heer C, Ten Voorde SE, Wijnhoven SW, Marvin HJ, Sips AJ (2009) Review of health safety aspects of nanotechnologies in food production. Regul Toxicol Pharmacol 53:52–62

Chaudhry Q, Scotter M, Blackburn J, Ross B, Boxall A, Castle L, Aitken R, Watkins R (2008) Applications and implications of nanotechnologies for the food sector. Food Addit Contam 25:241–258

Chaudhuri B, Hörmann F, Lalonde S, Brady SM, Orlando DA, Benfey P, Frommer WB (2008) Protonophore- and PH-Insensitive glucose and sucrose accumulation detected by FRET nanosensors in arabidopsis root tips. Plant J 56:948–962

Chen S, Liu H, Yangzong Z, Gardea-Torresdey JL, White JC, Zhao L (2023) Seed priming with reactive oxygen species-generating nanoparticles enhanced maize tolerance to multiple abiotic stresses. Environ Sci Tech 57(48):19932–19941

Cheng HN, Klasson KT, Asakura T, Wu Q (2016) Nanotechnology in agriculture. In: Cheng HN, Doemeny L, Geraci CL, Schmidt DG (eds) Nanotechnology: delivering on the promise, vol 2. ACS, Washington DC, pp 233–242

Choudhary R, Rajput VD, Ghodake G, Ahmad F, Meena M, Rehman RU, Prasad R, Sharma RK, Singh R, Seth CS (2024) Comprehensive journey from past to present to future about seed priming with hydrogen peroxide and hydrogen sulfide concerning drought, temperature, UV and ozone stresses- a review. Plant Soil 500(1):351–373

Choudhary R, Ahmad F, Kaya C, Upadhyay SK, Muneer S, Kumar V, Meena M, Liu H, Upadhyay H, Seth CS (2025) Decrypting proteomics, transcriptomics, genomics, and integrated omics for augmenting the abiotic, biotic, and climate change stress resilience in plants. J Plant Physiol. https://doi.org/10.1016/j.jplph.2025.154430

Cushen M, Kerry J, Morris M, Cruz-Romero M, Cummins E (2012) Nanotechnologies in the food industry–Recent developments, risks and regulation. Trend Food Sci Technol 24:30–46

Dasgupta N, Ranjan S, Mundekkad D, Ramalingam C, Shanker R, Kumar A (2015) Nanotechnology in agro-food: from field to plate. Food Res Int 69:381–400

De Azeredo HM (2009) Nanocomposites for food packaging applications. Food Res Int 42:1240–1253

De Jorge BC, Bisotto-de-Oliveira R, Pereira CN, Sant'Ana J (2017) Novel nanoscale pheromone dispenser for more accurate evaluation of Grapholitamolesta (Lepidoptera: Tortricidae) attract-and-kill strategies in the laboratory. Pest Manag Sci 73:1921–1926. https://doi.org/10.1002/ps.4558

Dekkers S, Krystek P, Peters RJ, Lankveld DP, Bokkers BG, Van Hoeven-Arentzen PH, Bouwmeester H, Oomen AG (2011) Presence and risks of nanosilica in food products. Nanotoxicology 5:393–405

Dubey A, Mailapalli DR (2016) Nanofertilisers, nanopesticides, nanosensors of pest and nanotoxicity in agriculture. In: Lichtfouse E (ed) Sustainable agriculture reviews, vol 19. Springer, Cham, pp 307–330

Duhan JS, Kumar R, Kumar N, Kaur P, Nehra K, Duhan S (2017) Nanotechnology: the new perspective in precision agriculture. Biotechnol Rep 15:11–23

Duncan TV (2011) Applications of nanotechnology in food packaging and food safety: barrier materials, antimicrobials and sensors. J Coll Interf Sci 363:1–24

Dwivedi S, Saquib Q, Al-Khedhairy AA, Musarrat J (2016) Understanding the role of nanomaterials in agriculture. In: Singh DP, Singh HB, Prabha R (eds) Microbial inoculants in sustainable agricultural productivity. Springer, New Delhi, pp 271–288

Elmer WH, White JC (2016) The use of metallic oxide nanoparticles to enhance growth of tomatoes and eggplants in disease infested soil or soilless medium. Environ Sci Nano 3:1072–1079. https://doi.org/10.1039/C6EN00146G

El-Shal RM, El-Naggar AH, El-Beshbeshy TR, Mahmoud EK, El-Kader NIA, Missaui AM, Du D, Ghoneim AM, El-Sharkawy MS (2022) Effect of nano-fertilizers on alfalfa plants grown under different salt stresses in hydroponic system. Agriculture 12:1113. https://doi.org/10.3390/agriculture12081113

Ezhilarasi P, Karthik P, Chhanwal N, Anandharamakrishnan C (2013) Nanoencapsulation techniques for food bioactive components: a review. Food Biopro Technol 6:628–647

Fakhouri F, Casari A, Mariano M, Yamashita F, Mei LL, Soldi V, Martelli S (2014) Effect of a gelatin-based edible coating containing cellulose nanocrystals (CNC) on the quality and nutrient retention of fresh strawberries during storage. IOP Conference Series: Materials Science and Engineering, vol 1. IOP Publishing, p 64

Falguera V, Quintero JP, Jiménez A, Muñoz JA, Ibarz A (2011) Edible films and coatings: structures, active functions and trends in their use. Trend Food Sci Tech 22:292–303

Fang AJ, Chen HY, Li HT, Liu ML, Zhang YY, Yao SZ (2017) Glutathione regulation-based dual-functional upconversion sensing-platform for acetylcholinesterase activity and cadmium ions. Biosens Bioelectron 87:545–551. https://doi.org/10.1016/j.bios.2016.08.111

FAO (2019) The state of food and agriculture 2019. Moving forward on food loss and waste reduction. UN, Rome

Fasihnia SH, Peighambardoust SH, Peighambardoust SJ (2018) Nanocomposite films containing organo-clay nanoparticles as an antimicrobial (active) packaging forpotential food application. J Food Process Preserv 42:e13488. https://doi.org/10.1111/jfpp.1348810

Flores-López ML, Cerqueira MA, De Rodríguez DJ, Vicente AA (2016) Perspectives on utilization of edible coatings and nano-laminate coatings for extension of postharvest storage of fruits and vegetables. Food Eng Rev 8:292–305

Fraceto LF, Grillo R, de Medeiros GA, Scognamiglio V, Rea G, Bartolucci C (2016) Nanotechnology in agriculture: which innovation potential does it have. Front Environ Sci 4:20. https://doi.org/10.3389/fenvs.2016.00020

Frenk S, Ben-Moshe T, Dror I, Berkowitz B, Minz D (2013) Effect of metal oxide nanoparticles on microbial community structure and function in two different soil types. PLoS One 8:e84441

Ganeshkumar R, Sopiha KV, Wu P, Cheah CW, Zhao R (2016) Ferroelectric $KNbO_3$ nanofibers: synthesis, characterization and their application as a humidity nanosensor. Nanotechnol 27:395607. https://doi.org/10.1088/0957-4484/27/39/395607

Ghidan AY, Al Antary TM (2019) Applications of nanotechnology in agriculture. In: Stoytcheva M, Zlatev R (eds) Applications of nanobiotechnology. Intech-Open, London, pp 1–14

Ghidan AY, Al Antary TM, Awwad AM (2017a) Aphidicidal potential of green synthesized magnesium hydroxide nanoparticles using Oleaeuropaea leaves extract. ARPN J Agric Biol Sci 12(10):293–301

Ghidan AY, Al-Antary TM, Salem NM, Awwad AM (2017b) Facile green synthetic route to the zinc oxide (ZnONPs) nanoparticles: effect on green peach aphid and antibacterial activity. J Agric Sci 9(2):131–138

Ghorbanpour M, Fahimirad S (2017) Plant nanobionics a novel approach to overcome the environmental challenges. In: Ghorbanpour M, Varma A (eds) Medicinal plants and environmental challenges. Springer International Publishing, Cham, pp 247–257

Giannousi K, Avramidis I, Dendrinou-Samara C (2013) Synthesis, characterization and evaluation of copper based nanoparticles as agrochemicals against Phytophthora infestans. RCS Adv 3:21743–21752. https://doi.org/10.1039/c3ra42118j

Giraldo JP, Landry MP, Faltermeier SM, McNicholas TP, Iverson NM, Boghossian AA, Reuel NF, Hilmer AJ, Sen F, Brew JA (2014) Plant nanobionics approach to augment photosynthesis and biochemical sensing. Nat Mater 13:400. https://doi.org/10.1038/nmat3890

Gogos A, Knauer K, Bucheli TD (2012) Nanomaterials in plant protection and fertilization: current state, foreseen applications, and research priorities. J Agric Food Chem 60:9781–9792

Gomathi T, Rajeshwari K, Kanchana V, Sudha PN, Parthasarathy K (2019) Impact of nanoparticle shape, size, and properties of the sustainable nanocomposites. In: Sustainable polymer composites and nanocomposites. Springer International Publishing, Cham, pp 313–336

Gothandam KM, Ranjan S, Dasgupta N, Ramalingam C, Lichtfouse E (2018) Nanotechnology, food security and water treatment. Springer International Publishing, Cham

Gupta A, Dubey P, Kumar M, Roy A, Sharma D, Khan MM, Pathak N, Hasanuzzaman M (2022a) Consequences of arsenic contamination on plants and mycoremediation-mediated arsenic stress tolerance for sustainable agriculture. Plan Theory 11(23):3220

Gupta A, Bano A, Rai S, Mishra R, Singh M, Sharma S, Pathak N (2022b) Mechanistic insights of plant-microbe interaction towards drought and salinity stress in plants for enhancing the agriculture productivity. Plant Stress 4:100073

Gupta A, Mishra R, Rai S, Bano A, Pathak N, Fujita M, Kumar M, Hasanuzzaman M (2022c) Mechanistic insights of plant growth promoting bacteria mediated drought and salt stress tolerance in plants for sustainable agriculture. Int J Mol Sci 23:3741

Gustavsson J, Cederberg C, Sonesson U, Van Otterdijk R, Meybeck A (2011) Global food losses and food waste: extent, causes and prevention. FAO, Rome

Haghighi M, Afifipou Z, Mozafariyan M (2012) The effect of N-Si on tomato seed germination under salinity levels. J Biol Environ Sci 6:87–90

Han C, Zhao A, Varughese E, Sahle-Demessie E (2018) Evaluating weathering of food packaging polyethylene-nano-clay composites: release of nanoparticles and their impacts. NanoImpact 9:61–71

Haq IU, Ijaz S (2019) Use of metallic nanoparticles and nanoformulations as nanofungicides for sustainable disease management in plants. In: Prasad R, Kumar V, Kumar M, Choudhary D (eds) Nanobiotechnology in bioformulations. Springer, Cham, pp 289–316

Hossain A, Abdallah Y, Ali MA, Masum MMI, Li B, Sun G, Meng Y, Wang Y, An Q (2019a) Lemon-2019. Fruit-Based green synthesis of zinc oxide nanoparticles and titanium dioxide nanoparticles against soft rot bacterial pathogen Dickeyadadantii. Biomol Ther 9:863

Hossain A, Hong X, Ibrahim E, Li B, Sun G, Meng Y, Wang Y, An Q (2019b) Green synthesis of silver nanoparticles with culture supernatant of a bacterium Pseudomonas rhodesiae and their antibacterial activity against soft rot pathogen Dickeyadadantii. Molecules 24:2303

Hou RY, Zhang ZY, Pang S, Yang TX, Clark JM, He LL (2016) Alteration of the nonsystemic behavior of the pesticide ferbam on tea leaves by engineered gold nanoparticles. Environ Sci Technol 50:6216–6223. https://doi.org/10.1021/acs.est.6b01336

Imada K, Sakai S, Kajihara H, Tanaka S, Ito S (2016) Magnesium oxide nanoparticles induce systemic resistance in tomato against bacterial wilt disease. Plant Pathol 65:551–560

Jafarzadeh S, Alias A, Ariffin F, Mahmud S (2017) Characterization of semolina protein film with incorporated zinc oxide nano rod intended for food packaging. Pol J Food Nutr Sci 67:183–190

Jagadesh M, Dash M, Kumari A, Singh SK, Verma KK, Kumar P, Bhatt R, Sharma SK (2024) Revealing the hidden world of soil microbes: metagenomic insights into plant, bacteria, and fungi interactions for sustainable agriculture and ecosystem restoration. Microbiol Res 285:127764. https://doi.org/10.1016/j.micres.2024.127764

Jahan S, Alias YB, Bakar AFBA, Yusof IB (2018) Toxicity evaluation of ZnO and TiO2 nanomaterials in hydroponic red bean (Vignaangularis) plant: physiology, biochemistry and kinetic transport. J Environ Sci 72:140–152

Jain N, Gupta E, Kanu NJ (2022) Plethora of carbon nanotubes applications in various fields a state of the art, review. Smart Sci 10(1):1–24

Johnson MS, Sajeev S, Nair RS (2021) Role of nanosensors in agriculture. Proceedings of the 2021 International Conference on Computational Intelligence and Knowledge Economy (ICCIKE), Dubai, United Arab Emirates, 17–18 March 2021. IEEE, Piscataway, pp 58–63

Kamal R, Mogazy AM (2021) Effect of doping on TiO2 nanoparticles characteristics: studying of fertilizing effect on cowpea plant growth and yield. J Soil Sci Plant Nutr:1–13

Kareem HA, Hassanzzz MU, Zain M, Irshad A, Shakoor N, Saleem S, Wang Q (2022) Nanosized zinc oxide (n-ZnO) particles pretreatment to alfalfa seedlings alleviate heat-induced morpho-physiological and ultrastructural damages. Environ Pollut 303:119069

Katouzian I, Jafari SM (2016) Nano-encapsulation as a promising approach for targeted delivery and controlled release of vitamins. Trend Food Sci Technol 53:34–48

Kaur G, Tak Y, Asthir B (2022) Salicylic acid: a key signal molecule ameliorating plant stresses. Cereal Res Commun:1–10

Kausar A (2020) Overview on nanocarbon sponges in polymeric nanocomposite. Mater Res Innov 24(5):309–320

Khan MR, Rizvi TF (2014) Nanotechnology: scope and application in plant disease management. Plant Pathol J 13:214–231. https://doi.org/10.3923/ppj.2014.214.231

Khan ST, Adil SF, Shaik MR, Alkhathlan HZ, Khan M, Khan M (2021a) Engineered nanomaterials in soil: their impact on soil microbiome and plant health. Plants 11(1):109

Khan M, Khan KK, Borah Y, Goswami KR, Hakeem I, Chakrabartty. (2021b) The potential exposure and hazards of metal-based nanoparticles on plants and environment, with special emphasis on ZnO NPs, TiO2 NPs, and AgNPs: a review. Environ Adv 6:100128

Kosaka PM, Pini V, Ruz JJ, Da Silva RA, González MU, Ramos D, Calleja M, Tamayo J (2014) Detection of cancer biomarkers in serum using a hybrid mechanical and optoplasmonic nanosensor. Nat Nanotechnol 9:1047–1053

Kou TJ, Yu WW, Lam SK, Chen DL, Hou YP, Li ZY (2018) Differential root responses in two cultivars of winter wheat (Triticumaestivum L.) to elevated ozone concentration under fully open-air field conditions. J Agron Crop Sci 204:325–332. https://doi.org/10.1111/jac.12257

Kulabhusan PK, Tripathi A, Kant K (2022) Gold nanoparticles and plant pathogens: an overview and prospective for biosensing in forestry. Sensors 22:1259

Kumar S, Bhanjana G, Sharma A, Dilbaghi N, Sidhu MC, Kim KH (2017) Development of nanoformulation approaches for the control of weeds. Sci Total Environ 586:1272–1278. https://doi.org/10.1016/j.scitotenv.2017.02.138
Kumar S, Shukla A, Baul PP, Mitra A, Halder D (2018) Biodegradable hybrid nanocomposites of chitosan/gelatin and silver nanoparticles for active food packaging applications. Food Packag Shelf Life 16:178–184
Kumar D, Mariyam S, Gupta KJ, Thiruvengadam M, Ghodake GS, Xing B, Seth CS (2024a) Comparative investigation on chemical and green synthesized titanium dioxide nanoparticles against chromium (VI) stress eliciting differential physiological, biochemical, and cellular attributes in Helianthus annuus L. Sci Total Environ 930:172413
Kumar D, Singh R, Upadhyay SK, Verma KK, Tripathi RM, Liu H, Dhankher OP, Tripathi RD, Sahi SV, Seth CS (2024b) Review on interactions between nanomaterials and phytohormones: novel perspectives and opportunities for mitigating environmental challenges. Plant Sci 340:111964
Kumar D, Biswas JK, Mulla SI, Singh R, Shukla R, Ahanger MA, Shekhawat GS, Verma KK, Siddiqui MW, Seth CS (2024c) Micro and nanoplastics pollution: sources, distribution, uptake in plants, toxicological effects, and innovative remediation strategies for environmental sustainability. Plant Physiol Biochem 213:108795
Kumaraswamy RV, Kumari S, Choudhary RC, Sharma SS, Pal A, Raliya R, Biswas P, Saharan V (2019) Salicylic acid functionalized chitosan nanoparticle: a sustainable biostimulant for plant. Int J Biol Macromol 123:59–69. https://doi.org/10.1016/j.ijbiomac.2018.10.202
Kwak SY, Giraldo JP, Wong MH, Koman VB, Lew TTS, Ell J, Weidman MC, Sinclair RM, Landry MP, Tisdale WA (2017) A nanobionic light-emitting plant. Nano Lett 17:7951–7961. https://doi.org/10.1021/acs.nanolett.7b04369
Lattanzio VMT, Nivarlet N (2017) Multiplex dipstick immunoassay for semiquantitative determination of fusarium mycotoxins in oat. In: Methods in molecular biology. Elsevier, Amsterdam, pp 137–142
Li ZZ, Chen JF, Liu F, Liu AQ, Wang Q, Sun HY, Wen LX (2007) Study of UV-shielding properties of novel porous hollow silica nanoparticle carriers for avermectin. Pest Manag Sci 63:241–246. https://doi.org/10.1002/ps.1301
Li L, Zhao C, Zhang Y, Yao J, Yang W, Hu Q, Wang C, Cao C (2017) Effect of stable antimicrobial nano-silver packaging on inhibiting mildew and in storage of rice. Food Chem 215:477–482
Li R, He J, Xie H, Wang W, Bose SK, Sun Y, Hu J, Yin H (2019) Effects of chitosan nanoparticles on seed germination and seedling growth of wheat (Triticumaestivum L.). Int J Biol Macromol 126:91–100
Lin BS, Diao SQ, Li CH, Qiao SC, Yu M (2004) Effect of tms (nanostructured silicon dioxide) on growth of changbai larch seedlings. J For Res 15:138–140
Lowry GV, Avellan A, Gilbertson LM (2019) Opportunities and challenges for nanotechnology in the agri-tech revolution. Nat Nanotechnol 14:517–522
Lv M, Liu Y, Geng JH, Kou XH, Xin ZH, Yang DY (2018) Engineering nanomaterials-based biosensors for food safety detection. Biosens Bioelectron 106:122–128. https://doi.org/10.1016/j.bios.2018.01.049
Mahmoud NE, Abdelhameed RM (2023) Use of titanium dioxide doped multi-wall carbon nanotubes as promoter for the growth, biochemical indices of Sesamumindicum L. under heat stress conditions. Plant Physiol Biochem 201:107844
Malandrakis AA, Kavroulakis N, Chrysikopoulos CV (2019) Use of copper, silver and zinc nanoparticles against foliar and soil-borne plant pathogens. Sci Total Environ 670:292–299
Manjaiah KM, Mukhopadhyay R, Paul R, Datta SC, Kumararaja P, Sarkar B (2019) Clay minerals and zeolites for environmentally sustainable agriculture. In: Mariano M, Binoy S, Alessio L (eds) Modified clay and zeolite nanocomposite materials. Elsevier, Amsterdam, pp 309–329
Marimuthu S, Rahuman AA, Jayaseelan C, Kirthi AV, Santhoshkumar T, Velayutham K, Bagavan A, Kamaraj C, Elango G, Iyappan M, Siva C, Karthik L, Bhaskara Rao KV (2013) Acaricidal activity of synthesized titanium dioxide nanoparticles using Calotropis gigantea against Rhipicephalus microplus and Haemaphysalis bispinosa. Asian Pac J Trop Med 6:682

Mariyam S, Upadhyay SK, Chakraborty K, Verma KK, Duhan JS, Muneer S, Meena M, Sharma RK, Ghodake G, Seth CS (2024) Nanotechnology, a frontier in agricultural science, a novel approach in abiotic stress management and convergence with new age medicine-A review. Sci Total Environ 912:169097

Martínez-Bueno M, Hernando M, Uclés S, Rajski L, Cimmino S, Fernández-Alba A (2017) Identification of non-intentionally added substances in food packaging nano films by gas and liquid chromatography coupled to orbitrap mass spectrometry. Talanta 172:68–77

McLamore ES, Diggs A, Marzal PC, Shi J, Blakeslee JJ, Peer WA, Murphy AS, Porterfield DM (2010) Non-invasive quantification of endogenous root auxin transport using an integrated flux microsensor technique. Plant J 63:1004–1016. https://doi.org/10.1111/j.1365-313X.2010.04300.x

Medeiros BGDS, Souza MP, Pinheiro AC, Bourbon AI, Cerqueira MA, Vicente AA, Carneiro-Da-Cunha MG (2014) Physical characterisation of an alginate/lysozyme nano-laminate coating and its evaluation on 'Coalho'cheese shelf life. Food Bioprocess Technol 7:1088–1098

Mihindukulasuriya S, Lim LT (2014) Nanotechnology development in food packaging: a review. Trends Food Sci Technol 40:149–167

Moustafa H, Youssef AM, Darwish NA, Abou-Kandil AI (2019) Eco-friendly polymer composites for green packaging: future vision and challenges. Compos Part B Eng 172:16–25

Munawar A, Khan RFH, Iqbal MZ, Rehman A, Bajwa SZ, Khan WS (2022) Drug-detection performance of carbon nanotubes decorated with metal oxide nanoparticles. In: Metal oxide-carbon hybrid materials. Elsevier, pp 475–493

Nair R, Varghese SH, Nair BG, Maekawa T, Yoshida Y, Kumar DS (2010) Nanoparticulate material delivery to plants. Plant Sci 179:154–163. https://doi.org/10.1016/j.plantsci.2010.04.012

NajafiDisfani M, Mikhak A, Kassaee MZ, Maghari A (2017) Effects of nano Fe/SiO2 fertilizers on germination and growth of barley and maize. Arch Agron Soil Sci 63(6):817–826

Nazar R, Iqbal N, Umar S (2017) Heat stress tolerance in plants: action of salicylic acid. In: Salicylic acid: a multifaceted hormone. Springer, pp 145–161

Ndlovu N, Mayaya T, Muitire C, Munyengwa N (2020) Nanotechnology applications in crop production and food systems. Int J Plant Breed Crop Sci 7(1):624–634

Neethirajan S, Jayas DS (2011) Nanotech nology for the food and bioprocessing industries. Food Bioprocess Technol 4:39–47

Neethirajan S, Gordon R, Wang L (2009) Potential of silica bodies (phytoliths) for nanotechnology. Trends Biotechnol 27:461–467. https://doi.org/10.1016/j.tibtech.2009.05.002

Noorbakhsh-Soltani S, Zerafat M, Sabbaghi S (2018) A comparative study of gelatin and starch-based nano-composite films modified by nano-cellulose and chitosan for food packaging applications. Carbohydr Polym 189:48–55

Nuruzzaman M, Rahman MM, Liu YJ, Naidu R (2016) Nanoencapsulation, nano-guard for pesticides: a new window for safe application. J Agric Food Chem 64:1447–1483. https://doi.org/10.1021/acs.jafc.5b05214

Ogunyemi SO, Zhang M, Abdallah Y, Ahmed T, Qiu W, Ali MA, Li B (2020) The bio-synthesis of three metal oxide nanoparticles (ZnO, MnO2, and MgO) and their antibacterial activity against the bacterial leaf blight pathogen. Front Microbiol 11:588326

Omar AA, Heikal YM, Zayed EM, Shamseldin SA, Salama YE, Amer KE et al (2023) Conferring of drought and heat stress tolerance in wheat (*Triticum aestivum* L.) genotypes and their response to selenium nanoparticles application. Nanomaterials 13(6):998

Onaga G, Kerstin W (2016) Advances in plant tolerance to biotic stresses. In: Abdurakhmonov IY (ed) Plant genomics. InTech, Rijeka, pp 167–228

Ouyang K, Mortimer M, Holden PA, Cai P, Wu Y, Gao C, Huang Q (2020) Towards a better understanding of Pseudomonas putida biofilm formation in the presence of ZnO nanoparticles (NPs): role of NP concentration. Environ Int 137:105485

Pan Q, Zhan J, Liu H, Zhang J, Chen J, Wen P, Huang W (2006) Salicylic acid synthesized by benzoic acid 2-hydroxylase participates in the development of thermotolerance in pea plants. Plant Sci 171(2):226–233

Panpatte DG, Jhala YK, Shelat HN, Vyas RV (2016) Nanoparticles: the next generation technology for sustainable agriculture. In: Microbial inoculants in sustainable agricultural productivity. Springer, New Delhi, pp 289–300

Pathakoti K, Manubolu M, Hwang HM (2017) Nanostructures: current uses and future applications in food science. J Food Drug Anal 25:245–253

Patil CD, Borase HP, Suryawanshi RK, Patil SV (2016) Trypsin inactivation by latex fabricated gold nanoparticles: a new strategy towards insect control. Enzym Microb Technol 92:18–25. https://doi.org/10.1016/j.enzmictec.2016.06.005

Petosa AR, Rajput F, Selvam O, Ohl C, Tufenkji N (2017) Assessing the transport potential of polymeric nanocapsules developed for crop protection. Water Res 111:10–17. https://doi.org/10.1016/j.watres.2016.12.030

Pinto RJ, Daina S, Sadocco P, Neto CP, Trindade T (2013) Antibacterial activity of nanocomposites of copper and cellulose. BioMed Res Int 2013(1–2):280512

Potocnik J (2005) Commission of the European Communities; Brussels: 2005. Nanosciences and nanotechnologies: an action Plan for Europe 2005-2009; pp 1–16

Potocnik J (2011) The European commission: commission recommendation of 18 October 2011 on the definition of nanomaterial. Off J Eur Union 275:38–40

Prakash J, Sun S, Swart HC, Gupta RK (2018) Noble metals-TiO_2 nanocomposites: from fundamental mechanisms to photocatalysis, surface enhanced Raman scattering and antibacterial applications. Appl Mater Today 11:82–135

Prasad TNVKV, Sudhakar P, Sreenivasulu Y, Latha P, Munaswamy V et al (2012) Effect of nanoscale zinc oxide particles on the germination, growth and yield of peanut. J Plant Nutr 35(6):905–927

Prasad R, Bhattacharyya A, Nguyen QD (2017) Nanotechnology in sustainable agriculture: recent developments, challenges, and perspectives. Front Microbiol 8:1014. https://doi.org/10.3389/fmicb.2017.01014

Pretty J (2008) Agricultural sustainability: concepts, principles and evidence. Philos Trans R Soc Lond B Biol Sci 363:447–465. https://doi.org/10.1098/rstb.2007.2163

Pulido-Reyes G, Palomares IMR, Das S, Sakthivel TS, Leganes F, Rosal R, Seal S, Fernández-Piñas F (2015) Untangling the biological effects of cerium oxide nanoparticles: the role of surface valence states. Sci Rep 5:15613

Qiu J, Chen Y, Liu Z, Wen H, Jiang N, Shi H, Kou Y (2023) The application of zinc oxide nanoparticles: an effective strategy to protect rice from rice blast and abiotic stresses. Environ Pollut 331:121925

Rahman M, Chakraborty A, Mazumdar S, Nandi N, Bhuiyan M, Alauddin S, Khan I, Hossain MJ (2020) Effects of poly(vinylpyrrolidone) protected platinum nanoparticles on seed germination and growth performance of Pisumsativum. Nano-Struct Nano-Objects 21:100408

Raja K, Sowmya R, Sudhagar R, Moorthy PS, Govindaraju K, Subramanian K (2019) Biogenic ZnO and Cu nanoparticles to improve seed germination quality in blackgram (*Vignamungo*). Mater Lett 235:164–167

Rajput V, Minkina T, Fedorenko A, Sushkova S, Mandzhieva S, Lysenko V, Duplii N, Fedorenko G, Dvadnenko K, Ghazaryan K (2018) Toxicity of copper oxide nanoparticles on spring barley (Hordeumsativumdistichum). Sci Total Environ 645:1103–1113

Rajput VD, Minkina T, Feizi M, Kumari A, Khan M, Mandzhieva S, Sushkova S, El-Ramady H, Verma KK, Singh A et al (2021a) Effects of silicon and silicon-based nanoparticles on rhizosphere microbiome. Plant Stress Growth Biol 10:791

Rajput VD, Singh A, Minkina T, Rawat S, Mandzhieva S, Sushkova S, Shuvaeva V, Nazarenko O, Rajput P, Ade K et al (2021b) Nano-enabled products: challenges and opportunities for sustainable agriculture. Plan Theory 10:2727

Ramachandraiah K, Han SG, Chin KB (2015) Nanotechnology in meat processing and packaging: potential applications — a review. Asian Australas J Anim Sci 28:290–302

Ramesh M, Palanisamy K, Babu K, Sharma NK (2014) Effects of bulk &nano-titanium dioxide and zinc oxide on physio-morphological changes in Triticumaestivum Linn. J Glob Biosci 3(2):415–422

Rashidi L, Khosravi-Darani K (2011) The applications of nanotechnology in food industry. Crit Rev Food Sci Nutr 51:723–730

Rieger KA, Eagan NM, Schiffman JD (2015) Encapsulation of cinnamaldehyde into nanostructured chitosan films. J Appl Polym Sci 132:41739

Rizwan M, Ali S, Ibrahim M, Farid M, Adrees M, Bharwana SA, Zia-Ur-Rehman M, Qayyum MF, Abbas F (2015) Mechanisms of silicon-mediated alleviation of drought and salt stress in plants: a review. Environ Sci Pollut Res Int 22:15416–15431. https://doi.org/10.1007/s11356-015-5305-x

Roohinejad S, Greiner R (2017) Nanotechnology in agriculture and food science. In: Axelos MA, Van De Voorde M (eds) Nanoscience: relevance for agriculture and the food sector. Wiley, pp 347–362

Sadeghi R, Rodriguez RJ, Yao Y, Kokini JL (2017) Advances in nanotechnology as they pertain to food and agriculture: benefits and risks. Annu Rev Food Sci Technol 8:467–492

Safdar M, Kim W, Park S, Gwon Y, Kim YO, Kim J (2022) Engineering plants with carbon nanotubes: a sustainable agriculture approach. J Nanobiotechnol 20(1):1–30

Sagadevan S, Periasamy M (2014) Recent trends in nanobiosensors and their applications: a review. Rev Adv Mater Sci 36:62–69

Sangwan S, Shameem N, Yashveer S, Tanwar H, Parray JA, Jatav HS, Sharma S, Punia H, Sayyed RZ, Almalki WH, Poczai P (2022) Role of salicylic acid in combating heat stress in plants: insights into modulation of vital processes. Front Biosci-Landmark 27(11):310. https://doi.org/10.31083/j.fbl2711310

Saranya S, Selvi A, Babujanarthanam R, Rajasekar A, Madhavan J (2020) Insecticidal activity of nanoparticles and mechanism of action. In: Siddhardha B, Dyavaiah M, Kasinathan K (eds) Model organisms to study biological activities and toxicity of nanoparticles. Springer, Singapore. https://doi.org/10.1007/978-981-15-1702-0_12

Sedghi M, Hadi M, Toluie SG (2013) Effect of nano zinc oxide on the germination parameters of soybean seeds under drought stress. Ann West Univ Timisoara Ser Biol 16(2):73–78

Sekhon BS (2010) Food nanotechnology–an overview. Nanotechnol Sci Appl 3:1–15

Servin A, Elmer W, Mukherjee A, De la Torre-Roche R, Hamdi H, White JC, Bindraban P, Dimkpa C (2015) A review of the use of engineered nanomaterials to suppress plant disease and enhance crop yield. J Nanopart Res 17:92. https://doi.org/10.1007/s11051-015-2907-7

Shang Y, Hasan MK, Ahammed GJ, Li M, Yin H, Zhou J (2019) Applications of nanotechnology in plant growth and crop protection: a review. Molecules 24(14):2558. https://doi.org/10.3390/molecules24142558. PMID: 31337070; PMCID: PMC6680665

Sharifi-Rad J, Sharifi-Rad M, da Silva JAT (2016) Morphological, physiological and biochemical responses of crops (Zea mays L., Phaseolus vulgaris L.), medicinal plants (Hyssopusofficinalis L., Nigella sativa L.), and weeds (Amaranthusretroflexus L., Taraxacumofficinale F. H. Wigg) exposed to SiO_2 nanoparticles. J Agric Sci Technol 18:1027–1040

Sharma C, Dhiman R, Rokana N, Panwar H (2017) Nanotechnology: an untapped resource for food packaging. Front Microbiol 8:1735

Shaw DS, Honeychurch KC (2022) Nanosensor applications in plant science. Biosensors 12:675. https://doi.org/10.3390/bios12090675

Shenashen M, Derbalah A, Hamza A, Mohamed A, El Safty S (2017) Antifungal activity of fabricated mesoporous alumina nanoparticles against root rot disease of tomato caused by Fusariumoxysporium. Pest Manag Sci 73:1121–1126

Sheng H, Ma J, Pu J, Wang L (2018) Cell wall-bound silicon optimizes ammonium uptake and metabolism in rice cells. Ann Bot 122(2):303–313

Shi S, Wang W, Liu L, Wu S, Wei Y, Li W (2013) Effect of chitosan/nano-silica coating on the physicochemical characteristics of longan fruit under ambient temperature. J Food Eng 118:125–131

Shukla G, Singh A, Chaudhary N, Singh S, Basnal N, Gaurav SS (2024) Metal nanoparticles to improve the heat resilience in wheat (Triticumaestivum L.). Nanotechnol 35:205101. https://doi.org/10.1088/1361-6528/ad27af

Siddiqui MH, Al-Whaibi MH (2014) Role of nano-SiO_2 in germination of tomato (*Lycopersicum esculentum* seeds Mill.). Saudi J Biol Sci 21:13–17. https://doi.org/10.1016/j.sjbs.2013.04.005

Siddiqui MH, Al-Whaibi MH, Faisal M, Al Sahli AA (2014) Nano-silicon dioxide mitigates the adverse effects of salt stress on Cucurbita pepo L. Environ Toxicol Chem 33:2429–2437

Singh R, Singh R, Singh D, Mani JK, Karwasra SS, Beniwal MS (2010) Effect of weather parameters on karnal bunt disease in wheat in Karnal region of Haryana. J Agrometeorol 12:99–101

Singh A, Singh NB, Hussain I, Singh H (2017a) Effect of biologically synthesized copper oxide nanoparticles on metabolism and antioxidant activity to the crop plants Solanumlycopersicum and Brassica oleracea var. botrytis. J Biotechnol 262:11–27

Singh T, Shukla S, Kumar P, Wahla V, Bajpai VK, Rather IA (2017b) Application of nanotechnology in food science: perception and overview. Front Microbiol 8:1501

Singh H, Sharma A, Bhardwaj SK, Arya SK, Bhardwaj N, Khatri M (2021) Recent advances in the applications of nano-agrochemicals for sustainable agricultural development. Environ Sci Process Impacts 23:213–239. https://doi.org/10.1039/D0EM00404A

Sogvar OB, Saba MK, Emamifar A, Hallaj R (2016) Influence of nano-ZnO on microbial growth, bioactive content and postharvest quality of strawberries during storage. Innov Food Sci Emerg Technol 35:168–176

Song Y, Wang BQ, Qiu DY, Xie ZM, Dai S, Li C, Xu SL, Zheng YC, Li S, Jiang M (2021) Melatonin enhances metallic oxide nanoparticle stress tolerance in rice via inducing tetrapyrrole biosynthesis and amino acid metabolism. Environ Sci Nano 8:2310–2323

Tiwari M, Krishnamurthy S, Shukla D, Kiiskila J, Jain A, Datta R, Sharma N, Sahi SV (2016) Comparative transcriptome and proteome analysis to reveal the biosynthesis of gold nanoparticles in arabidopsis. Sci Rep 6:21733

Torabian S, Zahedi M, Khoshgoftar AH (2017) Effects of foliar spray of nano-particles of $FeSO_4$ on the growth and ion content of sunflower under saline condition. J Plant Nutr 40:615–623. https://doi.org/10.1080/01904167.2016.1240187

Tripathi DK, Singh S, Singh VP, Prasad SM, Dubey NK, Chauhan DK (2017) Silicon nanoparticles more effectively alleviated UV-B stress than silicon in wheat (*Triticumaestivum)* seedlings. Plant Physiol Biochem 110:70–81. https://doi.org/10.1016/j.plaphy.2016.06.026

Tsuchiya Y, Yoshimura M, Sato Y, Kuwata K, Toh S, Holbrook-Smith D, Zhang H, McCourt P, Itami K, Kinoshita T et al (2015) Probing strigolactone receptors in strigahermonthica with fluorescence. Science 349:864–868

Tyagi J, Ahmad S, Malik M (2022) Nitrogenous fertilizers: impact on environment sustainability, mitigation strategies, and challenges. Int J Environ Sci Technol 19(11). https://doi.org/10.1007/s13762-022-04027-9

Ubbink J, Krüger J (2006) Physical approaches for the delivery of active ingredients in foods. Trends Food Sci Technol 17:244–254

Van Nhan L, Ma C, Rui Y, Liu S, Li X, Xing B, Liu L (2015) Phytotoxic mechanism of nanoparticles: destruction of chloroplasts and vascular bundles and alteration of nutrient absorption. Sci Rep 5:11618

Vanathi P, Rajiv P, Sivaraj R (2016) Synthesis and characterization of Eichhornia-mediated copper oxide nanoparticles and assessing their antifungal activity against plant pathogens. Bull Mater Sci 39:1165–1170. https://doi.org/10.1007/s12034-016-1276-x

Verma KK, Singh P, Song X-P, Malviya MK, Singh RK, Chen G-L, Solomon S, Li YR (2020) Mitigating climate change for sugarcane improvement: role of silicon in alleviating abiotic stresses. Sugar Tech 22(5):741–749. https://doi.org/10.1007/s12355-020-00831-0

Verma KK, Song XP, Joshi A, Rajput VD, Singh M, Sharma A, Singh RK, Li DM, Arora J, Minkina T, Li YR (2022) Nanofertilizer possibilities for healthy soil, water and food in future: an overview. Front Plant Sci 13:865048. https://doi.org/10.3389/fpls.2022.865048

Verma KK, Song X, Verma CL, Huang H, Singh M, Xu L, Wu J, Khan ZH, Zhang B, Li Y (2023a) Mathematical modeling of climate and fluoride effects on sugarcane photosynthesis with silicon nanoparticles. Plant Physiol Biochem 204:108089. https://doi.org/10.1016/j.plaphy.2023.108089

Verma KK, Singh M, Rajput VD, Verma CL, Burachevskaya M (2023b) Nanofertilizers: a sustainable alternative to conventional means. In: Rajput VD, Singh A, Minkina TM, Verma KK, Singh AK (eds) Nanotechnology for sustainable agriculture: an innovative and eco-friendly approach. Apple Academic Press, Inc., Taylor & Francis. https://doi.org/10.1201/9781003333128

Verma KK, Song XP, Kumari A, Jagadesh M, Singh SK, Bhatt R, Singh M, Seth CS, Li YR (2024a) Climate change adaptation: challenges for agricultural sustainability. Plant Cell Environ:1–12. https://doi.org/10.1111/pce.150

Verma KK, Joshi A, Song X-P, Singh S, Kumari A, Arora J, Singh SK, Solanki MK, Seth CS, Li Y-R (2024b) Synergistic interactions of nanoparticles and plant growth promoting rhizobacteria enhancing soil-plant systems: a multigenerational perspective. Front Plant Sci 15:1376214. https://doi.org/10.3389/fpls.2024.1376214

Vermeulen SJ, Aggarwal PK, Ainslie A, Angelone C, Campbell BM, Challinor AJ, Hansen JW, Ingram JSI, Jarvis A, Kristjanson P et al (2012) Options for support to agriculture and food security under climate change. Environ Sci Pol 15:136–144. https://doi.org/10.1016/j.envsci.2011.09.003

Voke E, Pinals RL, Goh NS, Landry MP (2021) Planta nanosensors: understanding biocorona formation for functional design. ACS Sens 6:2802–2814

Wakeel A, Xu M, Gan Y (2020) Chromium-induced reactive oxygen species accumulation by altering the enzymatic antioxidant system and associated cytotoxic, genotoxic, ultrastructural, and photosynthetic changes in plants. Int J Mol Sci 21:728

Wang Z, Wei F, Liu SY, Xu Q, Huang JY, Dong XY, Yu JH, Yang Q, Zhao YD, Chen H (2010) Electrocatalytic oxidation of phytohormone salicylic acid at copper nanoparticles-modified gold electrode and its detection in oilseed rape infected with fungal pathogen *Sclerotinia sclerotiorum*. Talanta 80:1277–1281. https://doi.org/10.1016/j.talanta.2009.09.023

Wang Q, Ma X, Zhang W, Pei H, Chen Y (2012a) The impact of cerium oxide nanoparticles on tomato (*Solanum lycopersicum* L.) and its implications for food safety. Metallomics 4:1105–1112

Wang Z, Xie X, Zhao J, Liu X, Feng W, White JC, Xing B (2012b) Xylem and Phloem-Based transport of CuO nanoparticles in maize (Zea mays L.). Environ Sci Technol 46:4434–4441

Wang SH, Wang FY, Gao SC (2015) Foliar application with nano-silicon alleviates Cd toxicity in rice seedlings. Environ Sci Pollut Res 22:2837–2845. https://doi.org/10.1007/s11356-014-3525-0

Wang SH, Wang FY, Gao SC, Wang XG (2016) Heavy metal accumulation in different rice cultivars as influenced by foliar application of nano-silicon. Water Air Soil Pollut 227:228. https://doi.org/10.1007/s11270-016-2928-6

Weiss J, Takhistov P, Mcclements DJ (2006) Functional materials in food nanotechnology. J Food Sci 71:R107–R116

Wong MH, Giraldo JP, Kwak SY, Koman VB, Sinclair R, Lew TTS, Bisker G, Liu P, Strano MS (2017) Nitroaromatic detection and infrared communication from wild-type plants using plant nanobionics. Nat Mater 16:264. https://doi.org/10.1038/nmat4771

Wu J, Wang T (2020) Synergistic effect of zinc oxide nanoparticles and heat stress on the alleviation of transcriptional gene silencing in Arabidopsis thaliana. Bull Environ Contam Toxicol 104(1):49–56

Xu C, Peng C, Sun L, Zhang S, Huang H, Chen Y, Shi J (2015) Distinctive effects of TiO2 and CuO nanoparticles on soil microbes and their community structures in flooded paddy soil. Soil Biol Biochem 86:24–33

Yadollahi A, Arzani K, Khoshghalb H (2010) The role of nanotechnology in horticultural crops postharvest management; (Southeast Asia Symposium on Quality and Safety of Fresh and Fresh-Cut Produce). Acta Horticult 875:49–56

Youssef AM, Abdel-Aziz MS, El-Sayed SM (2014) Chitosan nanocomposite films based on Ag-NP and Au-NP biosynthesis by Bacillus Subtilis as packaging materials. Int J Biol Macromol 69:185–191

Yu L, Banerjee IA, Gao X, Nuraje N, Matsui H (2005) Fabrication and application of enzyme-incorporated peptide nanotubes. Bioconjug Chem 16:1484–1487

Yu Y, Zhang S, Ren Y, Li H, Zhang X, Di J (2012) Jujube preservation using chitosan film with nano-silicon dioxide. J Food Eng 113:408–414

Yuan J, Chen Y, Li H, Lu J, Zhao H, Liu M, Nechitaylo GS, Glushchenko NN (2018) New insights into the cellular responses to iron nanoparticles in Capsicum annuum. Sci Rep 8:3228

Zambrano-Zaragoza ML, González-Reza R, Mendoza-Muñoz N, Miranda-Linares V, Bernal-Couoh TF, Mendoza-Elvira S, Quintanar-Guerrero D (2018) Nanosystems in edible coatings: a novel strategy for food preservation. Int J Mol Sci 19(3):705

Zhang Q, Ying Y, Ping J (2022) Recent advances in plant nanoscience. Adv Sci 9:2103414

Zhao L, Li F, Chen G, Fang Y, An X, Zheng Y, Xin Z, Zhang M, Yang Y, Hu Q (2012a) Effect of nanocomposite based packaging on preservation quality of green tea. Int J Food Sci Technol 47:572–578

Zhao L, Peng B, Jose A, Hernandez-Viezcas CR, Youping S, Jose R, Peralta-Videa XT, Genhua N, Lixin J, Varela-Ramirez A, Zhang JY, Gardea-Torresdey JL (2012b) Stress response and tolerance of Zea mays to CeO_2 nanoparticles: cross talk among H_2O_2, heat shock protein, and lipid peroxidation. ACS Nano 6(11):9615–9622. https://doi.org/10.1021/nn302975u

Chapter 3
Formulated Nanomaterials Stimulate Photosynthetic and Antioxidative Enzymatic Activities to Upregulate Productivity in Environmental Stress

Aparna Pandey, Shikha Singh, Varunendra Kumar Singh, Sakshi Pandey, and Sheo Mohan Prasad

3.1 Introduction

Humankind indispensably depends on the crops' productivity for their food needs. With the increasing population size and to meet the rising food demand, it is essential to improve agricultural crop production. Increasing environmental stresses, such as heavy metals, pesticides, salt, drought, and UV-B impose a declining impact on growth and the development of plants (Soni et al. 2022; Verma et al. 2023a). The nano-technology is being explored due to the promising potential of applying nanoparticles or nano-materials to alleviate the prevalent adverse stresses in crop fields. Nanomaterials (NMs) are composed of nanoparticles (NPs) with at least one of their facets. They can be inorganic, i.e., made of metal or metal oxides NPs, organic (chitosan, alginates NPs, etc.), or composite (a mixture of both categories) (Lv et al. 2023). The nano-materials comprise NPs, nano-tubes, etc., and can range in size from 1 to 100 nm and above (Sircar et al. 2022; Kumar et al. 2024a). The nano-size (1 nm to 100 nm) of NPs exploits them to be effectively used in crop stress regulation due to their peculiar characteristics of large surface area to volume size, controlled release into the environment, easy entry into plants, biodegradability and better reactivity inside a plant cell. The NPs are formulated by breaking down the extensive raw materials into nanoscale sizes or combining atoms or molecules into nano-size particles or substances. The size of NPs of 3–5 nm makes them more soluble, which easily cross cell boundaries; hence they are easily absorbed by the plants (Thabet and Alqudah 2024).

A. Pandey · S. Singh · V. K. Singh · S. Pandey · S. M. Prasad (✉)
Ranjan Plant Physiology and Biochemistry Laboratory, Department of Botany, University of Allahabad, Prayagraj, India

C. S. Seth et al. (eds.), *Nano-formulations for Sustainable Agriculture*, Nanotechnology in the Life Sciences, https://doi.org/10.1007/978-3-031-90617-6_3

The charges on the NPs help them to interact with the epidermal cell, which also help in their easy entry into plants; however, the bigger NPs make pores in the cell wall for the invasion, after which they follow both apoplastic and symplastic pathways inside the plant system. The NPs are delivered into plants either via roots or through leaves. The abiotic stress negatively hampers the plant's physiological and biochemical processes by increasing oxidative stress, declining photosynthetic pigments, and photosynthetic efficiency. The NPs, due to the large surface-to-volume ratio, can increase the biochemical reactions (Juárez-Maldonado et al. 2018) and act at the physiological, biochemical, anatomical, and molecular levels, thus have immense capacity in regulating plant stress (Ahmad et al. 2020). Due to this property, NPs could bind to toxic heavy metal ions, thereby inhibiting their entry into plants, thus improve photosynthetic efficiency and anti-oxidant system under different abiotic stresses (Mittal et al. 2020). The NMs or NPs are found to scavenge ROS or induce oxidative stress, which in turn activates the anti-oxidant defense system, thereby, regulate the ROS contents. Some NPs, such as silica and alumina NPs, possess insecticidal properties. The pesticides and fertilizers are combined with NPs to provide stability and controlled release into the soil environment (Nandini et al. 2023).

However, the impact of NPs on plant systems, positive or negative, chiefly depends on their chemical makeup, reactivity, size, and in particular, concentration (Murali et al. 2022). The standardisation of NPs quantity becomes important during delivery into crop plants. Nanoparticles have been found to modulate the activity of antioxidants enzymes, i.e., SOD, CAT, and POD (Siddiqi and Husen 2016). The effects of various NPs, such as ZnO, TiO_2 and SiO_2 have been seen to improve growth (Khodakovskaya et al. 2012; Rajput et al. 2023), water-uptake (Changmei et al. 2002), seed germination (Prasad et al. 2012) and photosynthetic efficiency (Suriyaprabha et al. 2012) in different plants. The Ag NPs and Pt NPs have shown a high DPPH (2,2-Diphenyl-1-picrylhydrazyl) free radical scavenging ability and improved glutathione contents in lettuce seedlings (Jurkow et al. 2020). In addition, they induced chlorophyll synthesis, boosted the non-enzymatic antioxidant contents and increased growth in the lettuce plant (Jurkow et al. 2020).

Thus, the present chapter discusses the roles of different nanomaterials such as metal or metal oxide NPs or organic NPs in the mitigation of prevalent abiotic stresses through the positive regulation of the vital photosynthetic process, enhanced anti-oxidant system and stress-responsive genes in the plants.

3.2 Impact of Different NPs on Physiological and Biochemical Processes for Enhancing Plant Productivity Under Various Stresses

3.2.1 UV-B Stress

Photosynthesis is the basic process that occurs in plants by which plants convert light energy into chemical energy; it profoundly affects the growth, productivity and stress tolerance of plants. The photosynthetic O_2 evolution process is mainly governed by PS II photochemistry, which involves the combined action mediated by light-harvesting pigment complexes, PS II reaction centers (comprising of chlorophyll *a* (Chl *a*) molecules) and the electron transport chain carriers involved in it. The UV-B stress enhances the ROS generation and impacts the production of reducing powers which ultimately reduce the electron transport rate; it also down-regulates expression of the photosynthesis genes and Rubisco activity in plants, which negatively impacts CO_2 fixation (Sebastian et al. 2018; Romanatti et al. 2019). Nevertheless, several studies have reported the mitigating effect of NPs on reduced plant growth under UV-B stress (Soni et al. 2022). As per the findings of Azadi et al. (2021), UV-B stress reduced Chl *a* and Chl *b* contents in *Thymus vulgaris*, however treatment with Ag NPs improved the photosynthetic pigments, which could have assisted improved growth in the test plant.

Shahzad et al. (2024) studied the impact of TiO_2 NPs in rice plants under extended exposure to UV-B stress. TiO_2 NPs application improved the declined growth, photosynthetic pigments content (víz-a-víz Chl *a*, Chl *b*, and Car), PS II photochemistry parameters and the activity of enzymes associated with photosynthesis. In another finding, the treatment of TiO_2 NPs enhanced the photosynthetic O_2 evolution rate in spinach by increasing the rate of ferricyanide reduction (Hill's reaction), which might be due to the binding of TiO_2 to PS II. PS II photochemistry is the oxidation of the PS II reaction center through the conversion of photonic energy into excitation energy. In consonance, TiO_2 NPs treatment could be helping in the photons absorption by the antennae complexes and thus, the channelization of the fluxes to reaction centers, which assisted the electron transfer in the photosynthetic electron transport chain under UV-B stress. Furthermore, this eventually induced the splitting of water, thus O_2 evolution (Mingyu et al. 2007) (Fig. 3.1).

Kataria et al. (2019) have mentioned that TiO_2 NPs can positively regulate the photosynthesis process by stimulating the expression of genes which are associated with dark reactions such as Rubisco, PEP carboxylase and carbonic anhydrase. Flavonoids play different roles in a plant system including protecting from UV radiations and scavenging of singlet O_2. For protection against the UV-B stress NPs could enhanced the expression of flavonoid synthesis genes flavanone-3-hydroxylase (*F3H*) and chalcone synthase (*CHS*) under UV-B stress in *Arabidopsis thaliana* (Wang et al. 2021). The UV-B radiation also leads to damage in DNA, photosynthetic pigments, thylakoids and chloroplasts in the plants (Soni et al. 2022). Furthermore, the application of TiO_2 NPs prevented DNA damage by enhancing the transcription of mitogen-activated protein kinase 6 (*MPK6*) and mitogen-activated protein kinase 3 (*MPK3*) genes involved in stress response and DNA repair genes such as suppressor of gamma-response 1 (*SOG1*), recombinase (*RAD51*) and breast cancer susceptibility gene 1 (*BRCA1*) genes in *Arabidopsis thaliana* under UV-B stress (Wang et al. 2021) (Fig. 3.1).

Plants respond to abiotic stressors commonly by producing toxic free radical ions. To regulate the toxic ROS molecules, plants possess enzymatic [SOD, peroxidase (POD), CAT, glutathione-*S*-transferase (GST), APX, and glutathione reductase (GR)] and non-enzymatic [proline, cysteine (Cys), non-protein thiols (NP-SHs), glutathione (GSH) and ascorbic acid (AsA)] antioxidant systems that show an upsurge in their activities and contents as a defensive response. Tripathi et al. (2017)

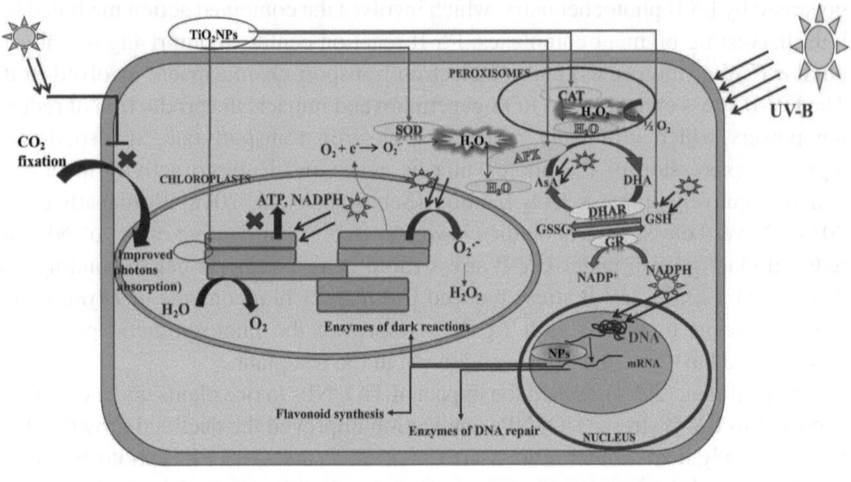

Fig. 3.1 Illustration of the regulatory role of TiO_2 nanoparticles (in pink arrows) on photosynthetic processes and antioxidant systems under damaging impact of UV-B stress (by blue arrows) in plant cell

reported that SiNPs are more effective than Si metal in overcoming UV-B stress. Application of TiO_2 NPs elevated the enzymatic anti-oxidant activities of SOD, CAT, ascorbate peroxidase (APX), glutathione peroxidase (GPX) in spinach under UV-B stress, reducing the oxidative stress biomarkers content and enhanced O_2 evolution rate (Fig. 3.1). The enhanced and ambient UV-B exposure declined wheat seedlings' fresh and dry mass. However, Si NPs treatment reduced the decline in the mass of plants. In addition to this, on the treatment with Si NPs, a boost in enzymatic anti-oxidant system (SOD, CAT, APX and GPX) activity was observed parallel to NO production in wheat plants which suggested involvement of NO signaling in Si NPs mediated UV-B stress regulation in wheat plants (Tripathi et al. 2017).

Dinc et al. (2022) found an ameliorative role of Se NPs under UV radiation stress in micro-alga *Chlorella vulgaris.* The NPs supplementation reduced the oxidative stress, which might be due to an empowered anti-oxidant system that was seen as a reduction in the damaged photosynthetic pigments (Chl *a* and *b*). Also, the cerium oxide (CeO_2) NPs-based silica gels absorbed the UV-B radiations and also, by getting oxidized (from Ce3+ to Ce4+), could minimise the damage in *Chlorella vulgaris* (Sicard et al. 2011). In another study, UV-B stress induced heightened superoxide radicals and hydrogen peroxide contents and reduction in the non-enzymatic anti-oxidant contents [ascorbate (AsA) and glutathione (GSH] was subsided through TiO_2 nano-materials application with improvement in the non-enzymatic antioxidant contents that have shown a demarcation in reduced oxidative stress and thereby improved plant biomass against the UV-B stress (Fig. 3.1). The UV-B treatment declined *Crocus sativus* L. plant's growth, height and leaf numbers, however the pre-treatment of TiO_2 NPs was given to plants before one-month exposure to UV-B stress, recovered the adverse impacts that could be through enhanced anti-oxidant system (Rikabad et al. 2019).

3.2.2 Heavy Metal Stress

Heavy metal (HM) contamination is becoming a more significant problem due to unreasonable mining and waste management, use of fertilizers and agrochemicals, rapid urbanization and industrialization (Zhai et al. 2018; Zhou et al. 2020). There are over ten million contaminated sites worldwide and over half of them are contaminated with HM (Khalid et al. 2017). Applications of NPs to the soil or foliage can alleviate the damaging effects of HM toxicity, enhance photosynthesis and development of plants, and reduce the toxicity induced due to oxidative stress (El-Saadony et al. 2022; El-Ramady et al. 2022) (Table 3.1). Consequently, NPs applications aid in the remediation of contaminated area with HMs. Plants exposed to HMs, after the addition of NPs to them, lowered the level of HM in the soil, controlled HM transfer genes expression in plants, enhanced plant antioxidant systems activity, and hence improved physiological processes (Rui 2021). It was discovered that plants were more resistant to Cd stress when ZnO NPs foliar and soil biochar were used together (Rizwan et al. 2019a). Similarly, rice was less affected by Cd stress when Fe NPs and biochar were applied together (Hussain et al. 2019).

According to studies by Shen et al. (2014) and Keller et al. (2015), nano-Si is useful in cleansing (HMs or reducing their harmful effects while enhancing plant development exposed to HM stress conditions. Furthermore, the application of SiO_2 NPs has been shown to mitigate oxidative damage in contaminated rice cell lines (Cui et al. 2020). Similar to this, rice's resistance to Cd stress can be significantly increased by foliar application of 2.5 mM nano-Si, which controls precipitation of Cd (Wang et al. 2015). Si NPs have also been demonstrated to mitigate Cu, Cd, Zn and Pb HMs' toxicity, and their application was highly successful in lowering HM accumulation than conventional methods (Wang et al. 2016). Si NPs are much more efficient than bulk Si in mitigating the harmful impacts of Pb on rice growth (Liu and Lal 2015). Si NPs stop lead translocation from roots to the shoots of rice and decrease lead precipitation in grains, particularly in cultivars and soils with lead toxicity. Si NPs can also cause significant alterations in plants by regulating the metal transport gene expression, decreasing and chelating active heavy metal ions, boosting antioxidant defense systems, and improving the complexation and co-precipitation of hazardous metals with Si (Adrees et al. 2015).

Ag NPs might also influence how stress genes are expressed. For example, microarray analysis revealed that Ag NPs both up and down-regulated the expression of specific genes. Up-regulated genes were primarily linked to responses to oxidative damage and metal toxicity, while down-regulated genes were linked to hormonal and microbial stimuli (Banerjee and Kole 2016; Choudhary et al. 2024). By preventing plants from absorbing HMs, ZnO NPs also decrease the stress caused by HMs and protect plants from the harmful effects of HMs (Baybordi 2005; Venkatachalam et al. 2017). The use of ZnO NPs can alleviate the oxidative stress brought on by Pb and Cd toxicity. ZnO NPs can also lower lipid peroxidation in growing *Leucaena leucocephala* seedlings while increasing levels of photosynthetic pigment content and total soluble protein (Venkatachalam et al. 2017).

Table 3.1 Different nanoparticles and their impacts on plant photosynthesis and antioxidant system under heavy metals

Nanoparticles	Plant species	Heavy metal	Impact of heavy metal	References
Si NPs	*Zea mays*	As	Maximum quantum efficiency, non-photochemical and photochemical quenching of PS II	Tripathi et al. (2016)
TiO_2 NPs	*Glycine max*	Cd	Enhanced physiological parameters i.e. photosynthetic rate, reduced oxidative stress	Singh and Lee (2016)
SiO_2 NPs	*Pisum sativum*	Cr	Enhanced ascorbate peroxidase (APX) and superoxide dismutase (SOD) and reduced oxidative stress	Tripathi et al. (2015)
Si NPs	*Zea mays*	Al	Activation of antioxidant defense mechanism	de Sousa et al. (2019)
Fe_3O_4 NPs	*Triticum aestivum*	Cd	Oxidative stress homeostasis	Konate et al. (2017)
Se NPs	*Brassica rapa*	Cd	Elevated levels of SOD, plasma glutathione peroxidase (GPX) concentration, plant height, biomass, and chlorophyll in the leaves	Zhang (2019)
Fe_3O_4 NPs	*Solanum lycopersicum*	Cd	Reduced Cd accumulation, Cd toxicity improvement by enhancing nutrient	Rahmatizadeh et al. (2019)
SiO_2 NPs	*Oryza sativa*	As	Reduced oxidative stress	Cui et al. (2020)
ZnO NPs	*Zea mays*	Cd	Reduced absorption of Cd and oxidative stress generated by Cd	Rizwan et al. (2019a)
Si NPs	*Brassica napus* L.	Cr	Regulation of antioxidant defense system, oxidative stress and photosynthesis	Huang et al. (2024)
Si NPs	*Saccharum Officinarum* L.	F	Upregulate diurnal photosynthetic efficiency and tolerate stress	Verma et al. (2023a)
Si NPs	*Triticum aestivum*	Cd	Plant growth and development promoted	Ali et al. (2019)
TiO_2 NPs	*Helianthus annuus*	Cr	By enhancing biochemical defense mechanisms translocation of HM to higher parts was prevented and reduced absorption of toxic element	Kumar et al. (2013)
Cu NPs	*Triticum aestivum*	Cr	Antioxidant defense system enhanced, growth was promoted by preventing the absorption of Cr	Noman et al. (2020)
Haematite NPs	*Zea mays*	Cd	Stimulated the key gene expression of defense mechanisms	Youssef et al. (2020, 2021)
Si NPs	*Triticum aestivum*	Cr	Enhanced leaf cell ultrastructure as evidenced by intact thylakoids and well-shaped chloroplasts in comparison to plants exposed to Cr	Manzoor et al. (2022)

Absorption of Cd and oxidative stress caused by Cd decreased when ZnO NPs were sprayed foliar on maize leaves (Rizwan et al. 2019b). According to Torabian et al. (2016), sunflower leaves treated with ZnO NPs showed a decrease in Na content and enhanced plant growth, chlorophyll content, and photosynthetic index. CuO/ZnO NPs treatment causes improved growth in wheat plants, which might have been brought on by CuO NPs' decreased solubility (Fathi et al. 2017).

3.2.3 Salt Stress

Certain NPs, at concentrations below specific thresholds, have been demonstrated to support the growth and development of plants affected by saline (Etesami et al. 2020; Ali et al. 2021). The primary process that is impacted by salt stress is photosynthesis, which varies based on the kind of plant, the amount of salt present and other elements (Seleiman et al. 2022; Mariyam et al. 2024). According to several researches under salt stress, the majority of NPs were able to boost photosynthesis by raising the concentration of pigments involved in photosynthesis (Abdoli et al. 2020; Singh et al. 2021). Manganese contributes to improving the oxygen evolution and photosynthetic electron transport rate as well as several other areas of the cell, including the mitochondria, chloroplasts, and the structures of certain enzymes (Alejandro et al. 2020). Under abiotic stress, it is discovered that Mn NPs can also aid in maintaining the ideal rate of photosynthesis (Ye et al. 2020). Recently, Ullah et al. (2020) assessed how TiO_2 NPs affected wheat development under salt stress. Their findings demonstrated that the supplementation of NPs enhanced wheat growth under saline conditions, which was explained by the NPs' ability to absorb light, aiding in photosynthesis and facilitating better water absorption by the wheat plant. Baz et al. (2020) worked on varieties of lettuces that were affected by salinity stress; they found that carbon NPs enhanced seed germination in salt stress and promoted photosynthetic pigment accumulation and root growth in the seedlings.

To counter the excess ROS in salinity-stressed plant cells, plants produce antioxidant enzymes (Etesami et al. 2020; Kumar et al. 2024c). Several findings have demonstrated how NPs can enhance the quantity of antioxidant enzymes (Ahmad et al. 2020; Faizan et al. 2021; Fatemi et al. 2021). According to the investigators, some NPs include characteristics of specific antioxidant enzymes: Cesium (Ce), manganese (Mn), copper (Cu), and iron (Fe) NPs, for instance, function similarly to the enzyme peroxidase (POD), and the NPs of cobalt (Co), Fe, and Ce behave like enzyme catalase (CAT) (Rico et al. 2015). The impact of seed-priming pearl millet (*Pennisetum glaucum* L.), which is affected by salinity stress, was examined by Khan et al. (2020) using Ag NPs. According to their findings, the plant's development characteristics significantly increased when NPs were present. This was because the proportion of sodium to potassium decreased and antioxidant enzymes including catalase (CAT), superoxide dismutase (SOD), and glutathione peroxidase (GPX) were increased. In Sami et al. (2020) study, it was also shown that the primary mechanism of silver NPs at low concentrations for plant growth was the enhanced activity of antioxidant enzymes.

Another study revealed that the effects of the nanoparticles on *Dracocephalum moldavica* at different doses under salinity stress under varying doses of TiO_2 NPs enhanced antioxidant content and decreased H_2O_2 (Gohari et al. 2020a). Using carboxylic acid to functionalize multi-walled carbon nanotubes, Gohari et al. (2020b) investigated the effects of these nanotubes at different concentrations on basil seedlings (*Ocimum basilicum* L.) grown in salt stress conditions in varying concentrations. The study's findings demonstrated that 50 mg L^{-1} might increase the amount of both non-enzymatic and enzymatic antioxidants as well as photosynthetic pigments during salinity stress. Furthermore, Ce NPs were able to enhance the antioxidant enzymes' activity to eliminate ROS from cotton plants' cells. Under salinity stress, the NPs also boosted plant growth (Liu et al. 2021). According to recent findings by Gohari et al. (2020a, b), low concentrations of multi-walled carbon nanotubes increased photosynthetic pigments and induced both non-enzymatic (phenolic content) and enzymatic (APX, CAT, and GP) components of the antioxidant defense system in *Ocimum basilicum* plants. Iron has been shown to upregulate important anti-oxidative enzymes, which mediates salt tolerance (Singh and Bhatla 2016). The impact of iron NPs on *Dracocephalum moldavica* L. in salinity stress conditions was examined by Moradbeygi et al. (2020). The experiment's findings indicate that by lowering the amount of antioxidant enzyme activity and raising the concentration of phenolic and flavonoid chemicals, particularly in the roots, NPs were able to enhance the growth of plants in salt. Salinity stress alters gene expression, thereby impacting plant growth by influencing the activity of various genes involved in cellular components and their byproducts. Similarly, nanoparticles (NPs) also exert effects at the molecular level, thereby regulating cellular processes. MicroRNA expression in cells treated with NPs has been analyzed for research in this field (Kumar et al. 2024a). These researchers' findings indicate that the NPs had an impact on the expression of miR398 and miR408, which control seed germination, the development of seedlings and roots, and the production of antioxidants and ROS scavengers. Furthermore, the aforementioned factors are inhibited by overexpression of microRNA. The NP-mediated increase in root development is caused by a decline in miR164 expression, which is connected to the auxin hormone signaling system. The synchronized effects of plant growth-promoting bacteria (*Lactobacillus casei, Bacillus pumilus,* and *Bacillus subtilis,*) and ZnO NPs (0, 20, and 40 mg L^{-1}) on cytosine methylation in *Solanum lycopersicum* L. were inspected in a study with 250 mM NaCl (Hosseinpour et al. 2020). Combination of bacteria and NPs together decreased the DNA hypermethylation brought on by salinity stress, even though salinity stress was still able to raise the polymorphism ratio. DNA methylation changes cell's response to salt stress via methylation and demethylation of genes (Peng and Zhang 2009).

A different study found that when Zn NPs were applied topically to rapeseed (*Brassica napus* L.) plants under salinity stress, the expression of genes involved in stress changed. Some genes, like SKRD2, MYC, and MPK4, were less expressed. However, other genes expressed more, like MYC and SKRD2, which are related to transcription factors and enhance abiotic stress tolerance, and ARP and MPK, which are linked to several physiological, hormonal, and developmental responses

(Hezaveh et al. 2019). Hezaveh et al. (2019) investigated the effects of ZnO NPs on salinity-stressed canola. The stress response genes were affected by the ZnO NPs' reduction of ion leakage and enhancement of the Hill reaction (e.g., ARP expression raised in contrast to SKRD2, MYC, and MPK4 expression declined). These studies demonstrate that Zn NPs have a clear role in reducing salinity stress because Zn is essential for the proper functioning of intra- and intercellular signaling and DNA transcription in salinity stress (Caldelas and Weiss 2017; Sofy et al. 2020).

To understand the NPs' mechanism of action under salinity stress and determine their optimal concentration, future researches should focus on molecular effects. The exposure to silicon Si NPs changed the expression of 29 genes, including stress-related, photosynthetic, kinase/phosphatase, and transcription factor genes under salinity stress (Kim et al. 2014). Siddiqui et al. (2020) have mentioned that Si also contributed to the up-regulation of OsNAC protein production, a stress-response protein that is efficacious. The expression of genes involved in formation of nitric oxide and auxin can be impacted by Si NPs (Tripathi et al. 2021a). Furthermore, a change in the expression of genes including *OsZEP, OsNCED,* and *OsZEP* that are involved in the formation of the abscisic acid (ABA) helps to lessen the effects of Si NPs induced salt stress (Tripathi et al. 2021b). The Si NPs promote greater salt transfer into the vacuole and enhance plant development under saline conditions by up-regulating the expression of the protein OsHMA3. It also boosts the activity of antioxidant enzymes, which shields the plant from the damaging effects of stress (Siddiqui et al. 2020).

3.2.4 Drought Stress

Among the various agricultural problems, drought stress is the most serious, affecting crop growth and development and resulting in reduced yield and productivity (Misra et al. 2006, 2011; Verma et al. 2022a). Drought not only affects the water physically available to plants, but it also disrupts plant antioxidant function by producing ROS and varying the function of stress-induced genes and proteins that regulate physiological, biochemical and morphological modifications in plants, and ultimately, the death of the plant (Choudhary et al. 2025). The NPs reduce the negative effects of drought stress and significantly improve plant growth and development (Ahmed et al. 2021; Verma et al. 2022b; Seleiman et al. 2023). The potential impact of NPs on enhancing photosynthesis and antioxidant defense systems in plants under drought conditions is being explored (Table 3.2).

Plants often suffer decreased photosynthetic rates and chlorophyll decline, leading to reduced growth and productivity under drought stress. Drought stress negatively affects the synthesis of chlorophyll molecules, efficacy of photosystem II (PS II), and electron flow rate, causing a significant reduction in overall plant photosynthetic performance. Nanotechnology has the potential to improve the function of photosynthetic machinery. NPs improve the light immersion in chloroplast and cause enhancement in electron transport rate, efficacy of photosystem II, oxygen

Table 3.2 Frequently used nanoparticles in agriculture and their effects on plants' photosynthesis and antioxidant defense system under drought stress

Applied NPs	Plants	Concentration/ application method	Impacts	Source
Cu NPs	Maize	3.33, 4.44, and 5.55 ppm priming	Increased biomass, photosynthetic pigments, anthocyanin, and seeds per grain	Van Nguyen et al. (2022)
Cerium oxide NPs (Nanoceria)	Sorghum	10 ppm foliar treatment	Declined oxidative stress biomarkers content and upsurge in enzymatic antioxidant activities (SOD, POD and CAT)	Djanaguiraman et al. (2018)
Fe_3O_4 NPs	Wheat	Into soil and foliar application	Increased Fe uptake, photosynthetic activities and growth	Adrees et al. (2020)
Fe_3O_4 NPs	Rice	Into soil and foliar application	Increased nutrient uptake, photosynthetic process and growth	Ahmed et al. (2021)
Se NPs	Pomegranate	20 ppm foliar treatment	Increased photosynthetic parameters, antioxidant levels, osmolytes, ABA and phenolics levels	Zahedi et al. (2021)
Se NPs	Wheat	100 $\mu g\ mL^{-1}$ spray	Enhanced photosynthetic pigments content and antioxidant system activity	El-Saadony et al. (2021)
Se NPs	Wheat	10, 20, 30, and 40 ppm foliar	Increasaed biomass, growth and its related attributes	Ikram et al. (2020)
SiO_2 NPs	Wheat	150 ppm seed treatment	Enhanced photosynthetic pigments content and anti-oxidant system activity (SOD, POD and CAT)	Akhtar and Ilyas (2022)
SiO_2 NPs	Banana	0, 200, 400, and 600 ppm	Improved photosynthetic pigments content, photosynthetic activity, and reduced cell damage.	Mahmoud et al. (2020)
SiO_2 NPs	Coriander	1.5 mM foliar treatment	Alleviated declined plant biomass, increased flavonoid, total phenolic, sugar and relative water content	Afshari et al. (2021)
Si NPs	Hawthorn	Si NPs at 0, 10, 50 and 100 ppm as nutrient solution	Increased photosynthetic pigments content, photosynthetic process and reduced oxidative damage	Ashkavand et al. (2015)
TiO_2 NPs	Linseed	0, 10, 100, 500 ppm irrigation water	Remarkable improvement in physiological process, and reduced H_2O_2 and MDA levels.	Aghdam et al. (2016)

(continued)

Table 3.2 (continued)

Applied NPs	Plants	Concentration/ application method	Impacts	Source
TiO_2 NPs	Wheat	500, 1000, and 2000 mg kg^{-1} soil amendment	TiO_2 NPs significantly increases shoot length, SOD activity, net photosynthetic rate, and reduced H_2O_2 and MDA levels.	Faraji and Sepehri (2020)
TiO_2 and ZnO NPs	Wheat	5 and 10 ppm seed treatment	Increased morphological parameters, photosynthetic pigments. Increase in proline content	El-Bassiouny et al. (2022)
ZnO NPs	Wheat	10 ppm. seed treatment	Enhanced photosynthetic pigments content, proline content and enzymatic antioxidant system activity (SOD, CAT, APX, and DHAR). Decreased oxidative stress biomarkers content	Azmat et al. (2022)
ZnO NPs	Maize	100 ppm exogenously applied	Boost in antioxidant system	Sun et al. (2020)
ZnO NPs	Peanut	10, 20, 40, 50, 100, 300, and 1000 ppm foliar spray	Enhanced antioxidant system activity and plant biomass	Latha et al. (2022)
ZnO NPs	Tomato	25, 50, and 100 ppm green ZnO-NPs foliar application	Improved antioxidant system activity, ascorbic acid and phenols content	El-Zohri et al. (2021)

evolution, and phosphorylation, which strengthens the plant's photosynthetic efficiency in water deficit condition (Shafea et al. 2017). Furthermore, the application of NPs is found to increase calcium, magnesium, sodium, potassium and iron contents and improve gas exchange characteristics, confirming enhanced photosynthesis under drought stress (Faraji and Sepehri 2020). Application of CuO NPs to maize under drought displayed greater tolerance, indicated by higher water content in their leaves and increased photosynthetic pigment contents (Van Nguyen et al. 2021). Carotenoids act as an antioxidant as it is known to protect the chlorophyll molecules from oxidative damage under stress conditions (Emiliani et al. 2018).

The application of CuO NPs significantly improves the carotenoids content, which protects against chlorophyll degradation and maintains high chlorophyll content, thereby increasing photosynthetic efficiency (Van Nguyen et al. 2022). Mahmoud et al. (2020) in banana plants reported that silicon dioxide nanoparticles (SiO_2 NPs) are also helpful in ameliorating drought stress by regulation of ion balance and ROS metabolism (Othmani et al. 2021). According to Ashkavand et al.

(2015), an improvement in photosynthetic rate, relative water content, and chlorophyll and carotenoids contents was evident by silica nanoparticles (SiO_2 NPs) pretreatments in *Crataegus* sp. of arid region's forests. In addition to this, Namjoyan et al. (2020) stated that nano-silicon application in beetroot plants leads to an increase in chlorophyll contents and net photosynthesis under water stress conditions. Further, the application of silicon nanoparticles was found to improve the impacts of drought stress in banana (Mahmoud et al. 2020; Verma et al. 2023b). Apart from this, it has also been reported that 10 nm Se NPs have improved the growth and yield of pomegranate plants and photosynthetic pigment contents under drought stress (Zahedi et al. 2021). Zinc oxide nanoparticles (ZnO NPs) have also been reported to have favorable effects in water deficit conditions. Photosynthetic activity, green pigment content of plants, rate of transpiration, gas exchange through stomata and water use efficiency were found to be increased in maize seedlings under drought stress after application of these nanoparticles (Sun et al. 2020).

Moreover, ZnO NPs aided compensate in the biological characteristics of chloroplasts and mitochondria in water-stressed *Zea mays*, thereby increasing photosynthetic efficiency (Sun et al. 2020). The impact of nano titanium dioxide (TiO_2) on crops' morphological, biochemical, and physiological properties varies for plant species and different application doses under drought stress (Djanaguiraman et al. 2018). Exogenous application of NPs during the growth stage of cucumber plants promoted Rubisco activator activity, green photosynthetic pigments and rate of oxygen production, thereby increasing plant dry mass and photosynthesis (Ghani et al. 2022). Similarly, Gao et al. (2008) reported that exogenously applied nano-TiO_2 stimulates chlorophyll formation and activity of Rubisco and increases the photosynthetic rate in spinach. Foliar feeding of TiO_2 NPs increased photosynthesis, which could lead to increased seed yield of soybean in drought stress (Linh et al. 2020).

In another study, an increase in pigment content was observed in linseed seedlings grown in soil treated with nano TiO_2 under both water surplus and water deficit conditions. In another study, the fabrication of magnetite and iron oxide nanoparticles (Fe_3O_4 NPs) via the biosynthesis process has been found to reduce drought stress in the *Chaetomorpha antenna* (Sreelakshmi et al. 2021). A concomitant increase in pigment contents was observed in desiccated *Setaria italica* plants treated with bare iron oxide NPs and citrate-coated iron oxide NPs, indicating that plants effectively absorbed iron oxide NPs, which can be used for photo assimilation (Sreelakshmi et al. 2021). Addition to this, treatment with Fe_3O_4 NPs having sizes ranging from 40 to 53 nm noticeably amended the growth, and chlorophyll contents of *Fragaria ananassa* under water shortage stage as reported by Mozafari et al. (2017). Cerium oxide nanoparticles (CeO NPs; Nanoceria), which belong to the metal oxide class of nanomaterials having negative charge, are used to improve drought-induced photosynthetic inhibition by sequestering it into chloroplasts (Wu et al. 2017).

Water deficit conditions promote oxidative stress, which leads to the production of ROS, H_2O_2 and malondialdehyde (MDA, a metric of lipid peroxidation) (Kumar et al. 2024b). Some elements and oxides are used as NPs in agriculture to increase fight back against drought. Incorporation of NPs significantly increased the activity of APX, SOD, and CAT enzymes, thereby reducing oxidative damage caused by water loss (Sun et al. 2020). Nano-silica have been widely used to mitigate the adverse effects of various abiotic stresses, including drought (Siddiqui et al. 2020). In response to drought stress, silicon dioxide (SiO_2) NPs at various concentrations (0, 12.5, 25, and 50 ppm) effectively boost the non-enzymatic activities in plants under drought stress by increasing TPC, DPPH, and FRAPS activity (Sutuliene et al. 2021). In a recent study, Cevik (2022) reported decreased MDA content in his research with treatments of TiO_2 NPs to drought-stressed plants. This decrease in MDA contents could indicate increased plant tolerance by TiO_2 NPs treatment.

In contrast, TiO_2 NPs treatment at higher concentrations was found to be toxic on *Zea mays*, and the authors stated that TiO_2 NPs could act as an oxidant, which could cause oxidative stress in plants (Karvar et al. 2022). In their research Van Nguyen et al. (2021) specified that lesser ROS accretion in CuO NPs treated plants resulted from increased SOD and APX enzyme activity, which is helpful to plants to detoxify the excess ROS under water deficit condition. These data were supported by previous studies (Regier et al. 2015; Singh et al. 2017; Tamez et al. 2019) which stated that CuO NPs are helpful to enhance the total antioxidant level of plants together with SOD and POD activities. The application of zinc oxide NPs has been explored to improve the non-enzymatic functions of antioxidant defense system of plants like polyphenols and ascorbate that act analogously to antioxidant enzymes, comprising SOD, CAT, POD, APX and others (El-Zohri et al. 2021). Further, application of nano zinc oxide increased SOD activity and tocopherol content under drought stress conditions (Sedaghi et al. 2021).

Furthermore, treatment with ZnO NPs increases POD and antioxidant polyphenol oxidase (PPO) activities in *Moringa peregrina* species under water deficit conditions, which can significantly enhance drought tolerance in test plant (Foroutan et al. 2018). Selenium (Se) NPs enhance antioxidant activity and total phenolic contents under drought stress. In support of this, exogenously applied Se NPs were found to alleviate the harmful outcome of drought in leaves and fruits of pomegranates by reducing MDA and H_2O_2 contents induced by stress, hence accelerating the potential of anti-oxidative enzymes (Zahedi et al. 2021; Rajput et al. 2022). Further, exogenously applied Se NPs improved the growth metrics of wheat plants by dripping the level of ionic leakage and lipid peroxidation with a reduction in cellular damage caused by drought stress (Ikram et al. 2020). The use of Cu NPs (50 mg/L) caused a significant rise (sixfold) in the transcriptomics of SOD in *Cucumis sativus* plants (Mosa et al. 2018). Contrast to this, ZnO NPs reduced the SOD activity in chickpea plantlets (Burman et al. 2013). Proline is a multi-functional molecule, best known for its osmoprotectants role. Ascorbic acid, a powerful antioxidant is known to down regulate free radicals' level and reducing drought induced damages (Rosales et al. 2006). In the same context, with the use of ZnO NPs, it is possible to prevent the loss of antioxidant enzymes and tocopherols due to the increased amount of ascorbate, which may help in reducing the adverse effects of water stress in soybean during germination (Sedghi et al. 2021).

3.3 Conclusion

With the increasing anthropogenic activities, pollutants, and different abiotic stresses in crop fields and drastic climatic changes, agricultural crop productivity is hampered majorly. The application of various kinds of NMs, such as metal or metal oxide NPs, chitosan-based nanoparticles NPs, etc., have been found to positively regulate different kinds of stresses in crop fields. The treatment of NPs or NMs has been found to ameliorate the photosynthesis processes by improving PS II photochemistry and decline oxidative stress through an empowered anti-oxidant system. The faster uptake of metal-NPs than metals enable them to enter quickly and regulate biochemical and physiochemical reactions and genetic level regulations and combat different plant stresses. They hold promising potential in alleviating different kinds of stresses, such as heavy metal, salt, UV-B, drought, etc., prevalent in the environment. However, certain nanoparticles (NPs) can also adversely affect physiological and biochemical parameters, depending on their concentration and the metal components they contain Moreover, studies need to be conducted to explore the effective methods of NPs delivery, the dosages, and the types of NPs for application in the agricultural fields, which do not impose any negative side-impacts on the environment and thus assist in sustainable development.

Acknowledgement The authors acknowledge the University of Allahabad, Prayagraj for providing the necessary facilities. Aparna Pandey is grateful to the University Grant Commission, New Delhi as SRF (CSIR-UGC-NET Dec. 2018, Ref. No. 627). Shikha Singh, Sakshi Pandey, and Varunendra Kumar Singh are thankful as AU research scholars to the University Grant Commission, New Delhi, India. Sheo Mohan Prasad is thankful to CSIR, New Delhi, India for providing CSIR-Emeritus Scientist project (File No. 21/1166/24/EMR-II)

Declaration of Competing Interest The authors declare no competing interests.

References

Abdoli S, Ghassemi-Golezani K, Alizadeh-Salteh S (2020) Responses of ajowan (*Trachyspermum ammi* L.) to exogenous salicylic acid and iron oxide nanoparticles under salt stress. Environ Sci Pollut Res 27:36939–36953

Adrees M, Ali S, Rizwan M, Zia-ur-Rehman M, Ibrahim M, Abbas F, Farid M, Qayyum MF, Irshad MK (2015) Mechanisms of silicon-mediated alleviation of heavy metal toxicity in plants: a review. Ecotoxicol Environ Saf 119:186–197

Adrees M, Khan ZS, Ali S, Hafeez M, Khalid S, Ur Rehman MZ, Hussain A, Hussain K, Chatha SAS, Rizwan M (2020) Simultaneous mitigation of cadmium and drought stress in wheat by soil application of iron nanoparticles. Chemosphere 238:124681

Afshari M, Pazoki A, Sadeghipour O (2021) Foliar-applied silicon and its nanoparticles stimulate physio-chemical changes to improve growth, yield and active constituents of coriander (*Coriandrum Sativum* L.) Essential oil under different irrigation regimes. Silicon 13:4177–4188

Aghdam MTB, Mohammadi H, Ghorbanpour M (2016) Effects of nanoparticulate anatase titanium dioxide on physiological and biochemical performance of *Linum usitatissimum* (Linaceae) under well-watered and drought stress conditions. Braz J Bot 39(1):139–146

Ahmad P, Alyemeni MN, Al-Huqail AA, Alqahtani MA, Wijaya L, Ashraf M, Kaya C, Bajguz A (2020) Zinc oxide nanoparticles application alleviates arsenic (As) toxicity in soybean plants by restricting the uptake of as and modulating key biochemical attributes, antioxidant enzymes, ascorbate-glutathione cycle and glyoxalase system. Plan Theory 9:825. https://doi.org/10.3390/plants9070825

Ahmed T, Noman M, Manzoor N, Shahid M, Abdullah M, Ali L, Wang G, Hashem A, Al-Arjani AF, Alqarawi AA, Abd Allah EF, Bin L (2021) Nanoparticle based amelioration of drought stress and cadmium toxicity in rice via triggering the stress responsive genetic mechanisms and nutrient acquisition. Ecotoxicol Environ Saf 209:111829

Akhtar N, Ilyas N, Meraj TA, Pour-Aboughadareh A, Sayyed R, Mashwani ZUR, Poczai P (2022) Improvement of plant responses by nano biofertilizer: a step towards sustainable agriculture. Nanomaterials 12:965

Alejandro S, Höller S, Meier B, Peiter E (2020) Manganese in plants: from acquisition to subcellular allocation. Front Plant Sci 11:517877

Ali S, Rizwan M, Hussain A, ur Rehman MZ, Ali B, Yousaf B, Wijaya L, Alyemeni MN, Ahmad P (2019) Silicon nanoparticles enhanced the growth and reduced the cadmium accumulation in grains of wheat (*Triticum aestivum* L.). Plant Physiol Biochem 140:1–8

Ali S, Mehmood A, Khan N (2021) Uptake, translocation, and consequences of nanomaterials on plant growth and stress adaptation. J Nanomater 2021:1–17

Ashkavand P, Tabari M, Zarafshar M, Tomášková I, Struve D (2015) Effect of SiO_2 nanoparticles on drought resistance in hawthorn seedlings. For Res Pap 76(4):350–359

Azadi M, Siavash Moghaddam S, Rahimi A, Pourakbar L, Popović-Djordjević J (2021) Biosynthesized silver nanoparticles ameliorate yield, leaf photosynthetic pigments, and essential oil composition of garden thyme (*Thymus vulgaris* L.) exposed to UV-B stress. J Environ Chem Eng 9:105919. https://doi.org/10.1016/j.jece.2021.105919

Azmat A, Tanveer Y, Yasmin H, Hassan MN, Shahzad A, Reddy M, Ahmad A (2022) Coactive role of zinc oxide nanoparticles and plant growth promoting rhizobacteria for mitigation of synchronized effects of heat and drought stress in wheat plants. Chemosphere 297:133982

Banerjee J, Kole C (2016) Plant nanotechnology: an overview on concepts, strategies, and tools. In: Kole C, Kumar DS, Khodakovskaya MV (eds) Plant nanotechnology. Springer, pp 1–14

Baybordi A (2005) Effect of zinc, iron, manganese and copper on wheat quality under salt stress conditions. J Water Soil 140:150–170

Baz H, Creech M, Chen J, Gong H, Bradford K, Huo H (2020) Water-soluble carbon nanoparticles improve seed germination and post-germination growth of lettuce under salinity stress. Agronomy 10(8):1192

Burman U, Saini M, Kumar P (2013) Effect of zinc oxide nanoparticles on growth and antioxidant system of chickpea seedlings. Toxicol Environ Chem 95:605–612. https://doi.org/10.1080/02772248.2013.803796

Caldelas C, Weiss DJ (2017) Zinc homeostasis and isotopic fractionation in plants: a review. Plant Soil 411:17–46

Cevik S (2022) TiO_2 nanoparticles alleviates the effects of drought stress in tomato seedlings. Bragantia 82:20220203. https://doi.org/10.1590/1678-4499.20220203

Changmei L, Chaoying Z, Junqiang W, Guorong W, Mingxuan T (2002) Research of the effect of nanometer materials on germination and growth enhancement of *Glycine max* and its mechanism. Soybean Sci 21(3):168–171

Choudhary R, Rajput VD, Ghodake G, Ahmad F, Meena M, Rehman RU, Prasad R, Sharma RK, Singh R, Seth CS (2024) Comprehensive journey from past to present to future about seed priming with hydrogen peroxide and hydrogen sulfide concerning drought, temperature, UV and ozone stresses- a review. Plant Soil 500(1):351–373

Choudhary R, Ahmad F, Kaya C, Upadhyay SK, Muneer S, Kumar V, Meena M, Liu H, Upadhyay H, Seth CS (2025) Decrypting proteomics, transcriptomics, genomics, and integrated omics for augmenting the abiotic, biotic, and climate change stress resilience in plants. J Plant Physiol. https://doi.org/10.1016/j.jplph.2025.154430

Cui J, Li Y, Jin Q, Li F (2020) Silica nanoparticles inhibit arsenic uptake into rice suspension cells via improving pectin synthesis and the mechanical force of the cell wall. Environ Sci Nano 7(1):162–171

De Sousa A, Saleh AM, Habeeb TH, Hassan YM, Zrieq R, Wadaan MA, Hozzein WN, Selim S, Matos M, AbdElgawad H (2019) Silicon dioxide nanoparticles ameliorate the phytotoxic hazards of aluminum in maize grown on acidic soil. Sci Total Environ 693:133636

Dinc SK, Vural OA, Kayhan FE, San Keskin NO (2022) Facile biogenic selenium nanoparticle synthesis, characterization and effects on oxidative stress generated by UV in microalgae. Particuology 70:30–42. https://doi.org/10.1016/j.partic.2021.12.005

Djanaguiraman M, Nair R, Giraldo JP, Prasad PVV (2018) Cerium oxide nanoparticles decrease drought-induced oxidative damage in sorghum leading to higher photosynthesis and grain yield. ACS Omega 3:14406–14416

El-Bassiouny HM, Mahfouze HA, Abdallah MM, Bakry BA, El-Enany MA (2022) Physiological and molecular response of wheat cultivars to titanium dioxide or zinc oxide nanoparticles under water stress conditions. Int J Agron. https://doi.org/10.1155/2022/3806574

El-Ramady HR, Verma KK, Rajput VD, Minkina T, Elbehiry F, Elbasiouny HY, Elsakhawy T, Omara AED, Amer MM (2022) Sources of silicon and nano-silicon in soil and plants. In: Silicon and nano-silicon in environmental stress management and crop quality improvement: progress and prospects. Academic Press, Elsevier. https://doi.org/10.1016/B978-0-323-91225-9.00003-0

El-Saadony MT, Saad AM, Najjar AA, Alzahrani SO, Alkhatib FM, Shafi ME, Selem E, Desoky ES, Fouda SE, El-Tahan AM, Hassan MA (2021) The use of biological selenium nanoparticles to suppress *Triticum aestivum* L. crown and root rot diseases induced by *Fusarium* species and improve yield under drought and heat stress. Saudi J Biol Sci 28(8):4461–4471

El-Saadony MT, Saad AM, Soliman SM, Salem HM, Desoky ESM, Babalghith AO, El-Tahan AM, Ibrahim OM, Ebrahim AA, El-Mageed A, Taia A (2022) Role of nanoparticles in enhancing crop tolerance to abiotic stress: a comprehensive review. Front Plant Sci 13:946717

El-Zohri M, Al-Wadaani NA, Bafeel SO (2021) Foliar sprayed green zinc oxide nanoparticles mitigate drought-induced oxidative stress in tomato. Plan Theory 10:2400

Emiliani J, D'andrea L, Lorena Falcone Ferreyra M, Maulión E, Rodriguez E, Rodriguez-Concepción M, Rodriguez E (2018) A role for β,β-xanthophylls in *Arabidopsis* UV-B photoprotection. J Exp Bot 69:4921–4933

Etesami H, Jeong BR, Rizwan M (2020) The use of silicon in stressed agriculture management: action mechanisms and future prospects. In: Deshmukh R, Tripathi DK, Guerriero G (eds) Metalloids in plants: advances and future prospects. Wiley, pp 381–431

Faizan M, Bhat JA, Chen C, Alyemeni MN, Wijaya L, Ahmad P, Yu F (2021) Zinc oxide nanoparticles (ZnO-NPs) induce salt tolerance by improving the antioxidant system and photosynthetic machinery in tomato. Plant Physiol Biochem 161:122–130

Faraji J, Sepehri A (2020) Exogenous nitric oxide improves the protective effects of TiO2 nanoparticles on growth, antioxidant system, and photosynthetic performance of wheat seedlings under drought stress. J Soil Sci Plant Nutr 20:703–714

Fatemi H, Esmaiel Pour B, Rizwan M (2021) Foliar application of silicon nanoparticles affected the growth, vitamin C, flavonoid, and antioxidant enzyme activities of coriander (*Coriandrum sativum* L.) plants grown in lead (Pb)-spiked soil. Environ Sci Pollut Res 28:1417–1425

Fathi A, Zahedi M, Torabian S, Khoshgoftar A (2017) Response of wheat genotypes to foliar spray of ZnO and Fe_2O_3 nanoparticles under salt stress. J Plant Nutr 40(10):1376–1385

Foroutan L, Solouki M, Abdossi A, Fakheri BA (2018) The effects of zinc oxide nanoparticles on enzymatic and osmoprotectants alternations in different Moringa peregrina populations under drought stress. Int J Basic Sci Med 5:178–187

Gao F, Liu C, Qu C, Zheng L, Yang F, Su M, Hong F (2008) Was improvement of spinach growth by nano-TiO_2 treatment related to the changes of Rubisco activase? Biometals 21:211–217

Ghani MI, Saleem S, Rather SA, Rehmani MS, Alamri S, Rajput VD, Kalaji HM, Saleem N, Sial TA, Liu M (2022) Foliar application of zinc oxide nanoparticles: an effective strategy to mitigate drought stress in cucumber seedling by modulating antioxidant defense system and osmolytes accumulation. Chemosphere 289:133202

Gohari G, Mohammadi A, Akbari A, Panahirad S, Dadpour MR, Fotopoulos V, Kimura S (2020a) Titanium dioxide nanoparticles (TiO_2 NPs) promote growth and ameliorate salinity stress effects on essential oil profile and biochemical attributes of *Dracocephalum moldavica*. Sci Rep 10(1):912

Gohari G, Safai F, Panahirad S, Akbari A, Rasouli F, Dadpour MR, Fotopoulos V (2020b) Modified multiwall carbon nanotubes display either phytotoxic or growth promoting and stress protecting activity in *Ocimum basilicum* L. in a concentration-dependent manner. Chemosphere 249:126171

Hezaveh TA, Pourakbar L, Rahmani F, Alipour H (2019) Interactive effects of salinity and ZnO nanoparticles on physiological and molecular parameters of rapeseed (*Brassica napus* L.). Commun Soil Sci Plant Anal 50(6):698–715

Hosseinpour A, Haliloglu K, Tolga Cinisli K, Ozkan G, Ozturk HI, Pour-Aboughadareh A, Poczai P (2020) Application of zinc oxide nanoparticles and plant growth promoting bacteria reduces genetic impairment under salt stress in tomato (*Solanum lycopersicum* L.'Linda'). Agriculture 10(11):521

Huang Q, Ayyaz A, Farooq MA, Zhang K, Chen W, Hannan F, Sun Y, Shahzad K, Ali B, Zhou W (2024) Silicon dioxide nanoparticles enhance plant growth, photosynthetic performance, and antioxidants defence machinery through suppressing chromium uptake in *Brassica napus* L. Environ Pollut 342:123013

Hussain A, Ali S, Rizwan M, ur Rehman MZ, Qayyum MF, Wang H, Rinklebe J (2019) Responses of wheat (*Triticum aestivum*) plants grown in a Cd contaminated soil to the application of iron oxide nanoparticles. Ecotoxicol Environ Saf 173:156–164

Ikram M, Raja NI, Javed B, Hussain M, Hussain M, Ehsan M (2020) Foliar applications of biofabricated selenium nanoparticles to improve the growth of wheat plants under drought stress. Green Process Synth 9(1):706–714

Juárez-Maldonado A, Ortega-Ortíz H, Cadenas-Pliego G, Valdés-Reyna J, Pinedo-Espinoza J, López-Palestina C, Hernández-Fuentes A (2018) Foliar application of Cu nanoparticles modified the content of bioactive compounds in *Moringa oleifera* Lam. Agronomy 8:167. https://doi.org/10.3390/agronomy8090167

Jurkow R, Pokluda R, Sękara A, Kalisz A (2020) Impact of foliar application of some metal nanoparticles on antioxidant system in oakleaf lettuce seedlings. BMC Plant Biol 20(1):290. https://doi.org/10.1186/s12870-020-02490-5

Karvar M, Azari A, Rahimi A, Maddah-Hosseini S, Ahmadi-Lahijani MJ (2022) Titanium dioxide nanoparticles (TiO_2-NPs) enhance drought tolerance and grain yield of sweet corn (*Zea mays* L.) under deficit ırrigation regimes. Acta Physiol Plant 44:1–14

Kataria S, Jain M, Rastogi A, Živčák M, Brestic M, Liu S, Tripathi DK (2019) Role of nanoparticles on photosynthesis. In: Tripathi DK, Ahmad P, Sharma S, Chauhan DK, Dubey NK (eds) Nanomaterials in plants, algae and microorganisms. Elsevier, pp 102–127. https://doi.org/10.1016/B978-0-12-811488-9.00006-8

Keller C, Rizwan M, Davidian JC, Pokrovsky OS, Bovet N, Chaurand P, Meunier JD (2015) Effect of silicon on wheat seedlings (*Triticum turgidum* L.) grown in hydroponics and exposed to 0 to 30 μM Cu. Planta 241:847–860

Khalid S, Shahid M, Niazi NK, Murtaza B, Bibi I, Dumat C (2017) A comparison of technologies for remediation of heavy metal contaminated soils. J Geochem Explor 182:247–268

Khan I, Raza MA, Awan SA, Shah GA, Rizwan M, Ali B, Tariq R, Hassan MJ, Alyemeni MN, Brestic M, Zhang X (2020) Amelioration of salt induced toxicity in pearl millet by seed priming with silver nanoparticles (AgNPs): the oxidative damage, antioxidant enzymes and ions uptake are major determinants of salt tolerant capacity. Plant Physiol Biochem 156:221–232

Khodakovskaya MV, De Silva K, Biris AS, Dervishi E, Villagarcia H (2012) Carbon nanotubes induce growth enhancement of tobacco cells. ACS Nano 6:2128–2135. https://doi.org/10.1021/nn204643g

Kim YH, Khan AL, Waqas M, Shim JK, Kim DH, Lee KY, Lee IJ (2014) Silicon application to rice root zone influenced the phytohormonal and antioxidant responses under salinity stress. J Plant Growth Regul 33:137–149

Konate A, He X, Zhang Z, Ma Y, Zhang P, Alugongo GM, Rui Y (2017) Magnetic (Fe_3O_4) nanoparticles reduce heavy metals uptake and mitigate their toxicity in wheat seedling. Sustainability 9(5):790

Kumar V, Guleria P, Kumar V, Yadav SK (2013) Gold nanoparticle exposure induces growth and yield enhancement in *Arabidopsis thaliana*. Sci Total Environ 461:462–468

Kumar D, Mariyam S, Gupta KJ, Thiruvengadam M, Ghodake GS, Xing B, Seth CS (2024a) Comparative investigation on chemical and green synthesized titanium dioxide nanoparticles against chromium (VI) stress eliciting differential physiological, biochemical, and cellular attributes in Helianthus annuus L. Sci Total Environ 930:172413

Kumar D, Singh R, Upadhyay SK, Verma KK, Tripathi RM, Liu H, Dhankher OP, Tripathi RD, Sahi SV, Seth CS (2024b) Review on interactions between nanomaterials and phytohormones: Novel perspectives and opportunities for mitigating environmental challenges. Plant Sci 340:111964

Kumar D, Biswas JK, Mulla SI, Singh R, Shukla R, Ahanger MA, Shekhawat GS, Verma KK, Siddiqui MW, Seth CS (2024c) Micro and nanoplastics pollution: sources, distribution, uptake in plants, toxicological effects, and innovative remediation strategies for environmental sustainability. Plant Physiol Biochem 213:108795

Latha P, Sudhakar P, Prasad TN (2022) Enhanced activity of antioxidant enzymes by foliar spray of nanoscale zinc oxide under drought stress conditions in peanut (*Arachis hypogaea* L.). Legum Res 1:6. https://doi.org/10.18805/LR-4790

Linh TM, Mai NC, Hoe PT, Lien LQ, Ban NK, Hien LTT, Chau NH, Van NT (2020) Metal-based nanoparticles enhance drought tolerance in soybean. J Nanomater 2020:4056563

Liu R, Lal R (2015) Potentials of engineered nanoparticles as fertilizers for increasing agronomic productions. Sci Total Environ 514:131–139

Liu J, Li G, Chen L, Gu J, Wu H, Li Z (2021) Cerium oxide nanoparticles improve cotton salt tolerance by enabling better ability to maintain cytosolic K^+/Na^+ ratio. J Nanobiotechnol 19(1):153

Lv X, Sha H, Ye Z, Wang Y, Mao B (2023) Nanomaterials in plant management: functions, mechanisms and prospects. Environ Sci Nano 10:3232–3252. https://doi.org/10.1039/D3EN00014A

Mahmoud LM, Dutt M, Shalan AM, El-Kady ME, El-Boray MS, Shabana YM, Grosser JW (2020) Silicon nanoparticles mitigate oxidative stress of in vitro derived banana (*Musa acuminata* 'Grand Nain') under simulated water deficit or salinity stress. S Afr J Bot 132:155–163

Manzoor N, Ali L, Ahmed T, Rizwan M, Ali S, Shahid MS, Schulin R, Liu Y, Wang G (2022) Silicon oxide nanoparticles alleviate chromium toxicity in wheat (*Triticum aestivum* L.). Environ Pollut 315:120391

Mariyam S, Upadhyay SK, Chakraborty K, Verma KK, Duhan JS, Muneer S, Meena M, Sharma RK, Ghodake G, Seth CS (2024) Nanotechnology, a frontier in agricultural science, a novel approach in abiotic stress management and convergence with new age medicine-A review. Sci Total Environ 912:169097

Mingyu S, Fashui H, Chao L, Xiao W, Xiaoqing L, Liang C, Fengqing G, Fan Y, Zhongrui L (2007) Effects of nano-anatase TiO_2 on absorption, distribution of light, and photoreduction activities of chloroplast membrane of spinach. Biol Trace Elem Res 118:120–130. https://doi.org/10.1007/s12011-007-0006-z

Misra AN, Latowski D, Strzalka K (2006) The xanthophyll cycle activity in kidney bean and cabbage leaves under salinity stress. Russ J Plant Physiol 53:113–121

Misra AN, Latowski DK, Strzalka K (2011) Violaxanthin de-epoxidation in aging cabbage (*Brassica oleracea* L.) leaves play as a sensor for photosynthetic excitation pressure. J Life-Sci 5:182–191

Mittal D, Kaur G, Singh P, Yadav K, Ali SA (2020) Nanoparticle-based sustainable agriculture and food science: recent advances and future outlook. Front Nanotechnol 2:579954. https://doi.org/10.3389/fnano.2020.579954

Moradbeygi H, Jamei R, Heidari R, Darvishzadeh R (2020) Investigating the enzymatic and non-enzymatic antioxidant defense by applying iron oxide nanoparticles in *Dracocephalum moldavica* L. plant under salinity stress. Sci Hortic 272:109537

Mosa KA, El-Naggar M, Ramamoorthy K, Alawadhi H, Elnaggar A, Wartanian S, Emy Ibrahim E, Hani H (2018) Copper nanoparticles induced genotoxicty, oxidative stress, and changes in superoxide dismutase (SOD) gene expression in cucumber (*Cucumis sativus*) plants. Front Plant Sci 9:872. https://doi.org/10.3389/fpls.2018.00872

Mozafari A, Havas F, Ghaderi N (2017) Application of iron nanoparticles and salicylic acid in in vitro culture of strawberries (*Fragaria ananassa* D.) to cope with drought stress. Plant Cell Tissue Organ Cult 33:511–523

Murali M, Gowtham HG, Singh SB, Shilpa N, Aiyaz M, Alomary MN, Alshamrani M, Salawi A, Almoshari Y, Ansari MA, Amruthesh KN (2022) Fate, bioaccumulation and toxicity of engineered nanomaterials in plants: current challenges and future prospects. Sci Total Environ 811:152249

Namjoyan SH, Sorooshzadeh A, Alikhani M (2020) Impact of nanosilicon and tebuconazole foliar application on some physiological traits, growth and white sugar yield of sugar beet under drought stress. J Sugar Beet 35(2):157–173

Nandini B, Mawale KS, Giridhar P (2023) Nanomaterials in agriculture for plant health and food safety: a comprehensive review on the current state of agro-nanoscience. 3 Biotech 13:73. https://doi.org/10.1007/s13205-023-03470-w

Noman M, Shahid M, Ahmed T, Tahir M, Naqqash T, Muhammad S, Song F, Abid HMA, Aslam Z (2020) Green copper nanoparticles from a native *Klebsiella pneumoniae* strain alleviated oxidative stress impairment of wheat plants by reducing the chromium bioavailability and increasing the growth. Ecotoxicol Environ Saf 192:110303

Othmani A, Ayed S, Bezzin O, Farookh M, Ayed-Salma O, Slim-Amara H, Younes MB (2021) Effect of silicon supply methods on durum wheat (*Triticum durum* Desf.) response to drought stress. SILICON 13:3047–3057

Peng H, Zhang J (2009) Plant genomic DNA methylation in response to stresses: potential applications and challenges in plant breeding. Prog Nat Sci 19(9):1037–1045

Prasad TNVKV, Sudhakar P, Sreenivasulu Y, Latha P, Munaswamy V, Reddy KR, Sreeprasad TS, Sajanlal PR, Pradeep T (2012) Effect of nanoscale zinc oxide particles on the germination, growth and yield of peanut. J Plant Nutr 35:905–927. https://doi.org/10.1080/01904167.2012.663443

Rahmatizadeh R, Arvin SMJ, Jamei R, Mozaffari H, Reza Nejhad F (2019) Response of tomato plants to interaction effects of magnetic (Fe_3O_4) nanoparticles and cadmium stress. J Plant Interact 14(1):474–481

Rajput VD, Minkina T, Upadhyay SK, Kumari A, Ranjan A, Mandzhieva S, Sushkova S, Singh RK, Verma KK (2022) Nanotechnology in the restoration of polluted soil. Nanomaterials 12:769. https://doi.org/10.3390/nano12050769

Rajput VD, Kumari A, Upadhyay SK, Minkina T, Mandzhieva S, Ranjan A, Sushkova S, Burachevskaya M, Rajput P, Konstantinova E, Singh J, Verma KK (2023) Can nanomaterials improve the soil microbiome and crop productivity? Agriculture 13:231. https://doi.org/10.3390/agriculture13020231

Regier N, Cosio C, von Moos N, Slaveykova VI (2015) Effects of copper-oxide nanoparticles, dissolved copper and ultraviolet radiation on copper bioaccumulation, photosynthesis and oxidative stress in the aquatic macrophyte *Elodea nuttallii*. Chemosphere 128:56–61

Rico CM, Peralta-Videa JR, Gardea-Torresdey JL (2015) Chemistry, biochemistry of nanoparticles, and their role in antioxidant defense system in plants. In: Siddqui M, Al-Whaibi, M, Mohammad F (eds) Nanotechnology and plant sciences: nanoparticles and their impact on plants. Springer, pp 1–17

Rikabad MM, Pourakbar L, Siavash Moghaddam S, Popović-Djordjević J (2019) Agrobiological, chemical and antioxidant properties of saffron (*Crocus sativus* L.) exposed to TiO_2 nanoparticles and ultraviolet-B stress. Ind Crop Prod 137:137–143. https://doi.org/10.1016/j.indcrop.2019.05.017

Rizwan M, Ali S, ur Rehman MZ, Adrees M, Arshad M, Qayyum MF, Ali L, Hussain A, Chatha SAS, Imran M (2019a) Alleviation of cadmium accumulation in maize (*Zea mays* L.) by foliar spray of zinc oxide nanoparticles and biochar to contaminated soil. Environ Pollut 248:358–367

Rizwan M, Ali S, Ali B, Adrees M, Arshad M, Hussain A, ur Rehman MZ, Waris AA (2019b) Zinc and iron oxide nanoparticles improved the plant growth and reduced the oxidative stress and cadmium concentration in wheat. Chemosphere 214:269–277

Romanatti PV, Rocha GA, Veroneze Júnior V, Santos Filho PR, De Souza TC, Pereira FJ, Polo M (2019) Limitation to photosynthesis in leaves of eggplant under UVB according to anatomical changes and alterations on the antioxidant system. Sci Hortic 249:449–454. https://doi.org/10.1016/j.scienta.2019.01.060

Rosales MA, Ruiz JM, Hernández J, Soriano T, Castilla N, Romero L (2006) Antioxidant content and ascorbate metabolism in cherry tomato exocarp in relation to temperature and solar radiation. J Sci Food Agric 86(10):1545–1551

Rui Y (2021) Nanoparticles alleviate heavy metals stress. Available online at: https://encyclopedia.pub/7093. Accessed 2 June 2021

Sami F, Siddiqui H, Hayat S (2020) Impact of silver nanoparticles on plant physiology: a critical review. In: Hayat S, Pichtel J, Faizan M, Fariduddin Q (eds) Sustainable agriculture reviews. Springer, Cham, pp 111–127. https://doi.org/10.1007/978-3-030-33996-8_

Sebastian A, Kumari R, Kiran BR, Prasad MNV (2018) Ultraviolet B induced bioactive changes of enzymatic and non-enzymatic antioxidants and lipids in *Trigonella foenum-graecum* L. (Fenugreek). EuroBiotech J 2:64–71. https://doi.org/10.2478/ebtj-2018-0010

Sedghi M, Jahed PS, Tolouie SG (2021) Zinc oxide nanoparticles alleviate drought stress effects on soybean antioxidant system during germination. Iran J Plant Physiol 11(4):3769–3778

Seleiman MF, Aslam MT, Alhammad BA, Hassan MU, Maqbool R, Chattha MU, Khan I, Gitari HI, Uslu OS, Roy R, Battaglia ML (2022) Salinity stress in wheat: effects, mechanisms and management strategies. Phyton 91(4):9431–9457

Seleiman MF, Al-Selwey WA, Ibrahim AA, Shady M, Alsadon AA (2023) Foliar applications of ZnO and SiO_2 nanoparticles mitigate water deficit and enhance potato yield and quality traits. Agronomy 13:466

Shafea AA, Dawood MF, Zidan MA (2017) Wheat seedlings traits as affected by soaking at titanium dioxide nanoparticles. Environ Earth Ecol 1:102–111

Shahzad R, Harlina PW, Khan SU, Ihtisham M, Khan AH, Husain FM, Karuniawan A (2024) Physio-biochemical and transcriptomics analyses reveal molecular mechanisms of enhanced UV-B stress tolerance in rice induced by titanium dioxide nanoparticles. J Plant Interact 19:2328713. https://doi.org/10.1080/17429145.2024.2328713

Shen X, Xiao X, Dong Z, Chen Y (2014) Silicon effects on antioxidative enzymes and lipid peroxidation in leaves and roots of peanut under aluminum stress. Acta Physiol Plant 36:3063–3069

Sicard C, Perullini M, Spedalieri C, Coradin T, Brayner R, Livage J, Jobbágy M, Bilmes SA (2011) CeO_2 nanoparticles for the protection of photosynthetic organisms immobilized in silica gels. Chem Mater 23:1374–1378. https://doi.org/10.1021/cm103253w

Siddiqi KS, Husen A (2016) Engineered gold nanoparticles and plant adaptation potential. Nanoscale Res Lett 11:400

Siddiqui H, Ahmed KBM, Sami F, Hayat S (2020) Silicon nanoparticles and plants: current knowledge and future perspectives. In: Hayat S, Pichtel J, Faizan M, Fariduddin

Q (eds) Sustainable agriculture reviews. Springer, Cham, pp 129–142. https://doi.org/10.1007/978-3-030-33996-8_7

Singh N, Bhatla SC (2016) Nitric oxide and iron modulate heme oxygenase activity as a long-distance signaling response to salt stress in sunflower seedling cotyledons. Nitric Oxide 53:54–64

Singh J, Lee BK (2016) Influence of nano-TiO_2 particles on the bioaccumulation of Cd in soybean plants (*Glycine max*): a possible mechanism for the removal of Cd from the contaminated soil. J Environ Manag 170:88–96

Singh A, Singh NB, Hussain I, Singh H (2017) Effect of biologically synthesized copper oxide nanoparticles on metabolism and antioxidant activity to the crop plants *Solanum lycopersicum* and *Brassica oleracea* var. botrytis. J Biotechnol 262:11–27

Singh P, Arif Y, Siddiqui H, Sami F, Zaidi R, Azam A, Alam P, Hayat S (2021) Nanoparticles enhance the salinity toxicity tolerance in *Linum usitatissimum* L. by modulating the antioxidative enzymes, photosynthetic efficiency, redox status and cellular damage. Ecotoxicol Environ Saf 213:112020

Sircar A, Rayavarapu K, Bist N, Yadav K, Singh S (2022) Applications of nanoparticles in enhanced oil recovery. Pet Res 7:77–90. https://doi.org/10.1016/j.ptlrs.2021.08.004

Sofy MR, Elhindi KM, Farouk S, Alotaibi MA (2020) Zinc and paclobutrazol mediated regulation of growth, upregulating antioxidant aptitude and plant productivity of pea plants under salinity. Plan Theory 9(9):1197

Soni S, Jha AB, Dubey RS, Sharma P (2022) Application of nanoparticles for enhanced UV-B stress tolerance in plants. Plant Nano Biol 2:100014. https://doi.org/10.1016/j.plana.2022.100014

Sreelakshmi B, Induja S, Adarsh P, Rahul H, Arya S, Aswana S, Haripriya R, Aswathy B, Manoj P, Vishnudasan D (2021) Drought stress amelioration in plants using green synthesized iron oxide nanoparticles. Mater Today: Proc 41:723–727

Sun L, Song F, Guo J, Zhu X, Liu S, Liu F, Li X (2020) Nano-ZnO induced drought tolerance is associated with melatonin synthesis and metabolism in maize. Int J Mol Sci 21:782. https://doi.org/10.3390/ijms21030782

Suriyaprabha R, Karunakaran G, Yuvakkumar R, Rajendran V, Kannan N (2012) Silica nanoparticles for increased silica availability in maize (*Zea mays.* L) seeds under hydroponic conditions. CNANO 8:902–908

Sutuliene R, Ragelienė L, Samuolienė G, Brazaitytė A, Urbutis M, Miliauskienė J (2021) The response of antioxidant system of drought-stressed green pea (*Pisum sativum* L.) affected by watering and foliar spray with silica nanoparticles. Horticulturae 8(1):35

Tamez C, Morelius EW, Hernandez-Viezcas JA, Peralta-Videa JR, Gardea-Torresdey J (2019) Biochemical and physiological effects of copper compounds/nanoparticles on sugarcane (*Saccharum officinarum*). Sci Total Environ 649:554–562

Thabet SG, Alqudah AM (2024) Unraveling the role of nanoparticles in improving plant resilience under environmental stress condition. Plant Soil:1–18. https://doi.org/10.1007/s11104-024-06581-2

Torabian S, Zahedi M, Khoshgoftar AH (2016) Effects of foliar spray of two kinds of zinc oxide on the growth and ion concentration of sunflower cultivars under salt stress. J Plant Nutr 39(2):172–180

Tripathi DK, Singh VP, Prasad SM, Chauhan DK, Dubey NK (2015) Silicon nanoparticles (SiNp) alleviate chromium (VI) phytotoxicity in *Pisum sativum* (L.) seedlings. Plant Physiol Biochem 96:189–198

Tripathi DK, Singh S, Singh VP, Prasad SM, Chauhan DK, Dubey NK (2016) Silicon nanoparticles more efficiently alleviate arsenate toxicity than silicon in maize cultiver and hybrid differing in arsenate tolerance. Front Environ Sci 4:46

Tripathi DK, Singh S, Singh VP, Prasad SM, Dubey NK, Chauhan DK (2017) Silicon nanoparticles more effectively alleviated UV-B stress than silicon in wheat (*Triticum aestivum*) seedlings. Plant Physiol Biochem 110:70–81. https://doi.org/10.1016/j.plaphy.2016.06.026

Tripathi DK, Rai P, Guerriero G, Sharma S, Corpas FJ, Singh VP (2021a) Silicon induces adventitious root formation in rice under arsenate stress with involvement of nitric oxide and indole-3-acetic acid. J Exp Bot 72(12):4457–4471

Tripathi DK, Vishwakarma K, Singh VP, Prakash V, Sharma S, Muneer S, Nikolic M, Deshmukh R, Vaculík M, Corpas FJ (2021b) Silicon crosstalk with reactive oxygen species, phytohormones and other signaling molecules. J Hazard Mater 408:124820

Ullah S, Adeel M, Zain M, Rizwan M, Irshad MK, Jilani G, Hameed A, Khan A, Arshad M, Raza A, Baluch MA (2020) Physiological and biochemical response of wheat (*Triticum aestivum*) to TiO_2 nanoparticles in phosphorous amended soil: a full life cycle study. J Environ Manag 263:110365

Van Nguyen D, Nguyen HM, Le NT, Nguyen KH, Nguyen HT, Le HM, Nguyen AT, Dinh NTT, Hoang SA, Ha CV (2021) Copper nanoparticle application enhances plant growth and grain yield in maize under drought stress conditions. J Plant Growth Regul. https://doi.org/10.1007/s00344-021-10301-w

Van Nguyen D, Nguyen HM, Le NT, Nguyen KH, Nguyen HT, Le HM, Nguyen AT, Dinh NTT, Hoang SA, Ha CV (2022) Copper nanoparticle application enhances plant growth and grain yield in maize under drought stress conditions. J Plant Growth Regul 41:364–375

Venkatachalam P, Jayaraj M, Manikandan R, Geetha N, Rene ER, Sharma NC, Sahi SV (2017) Zinc oxide nanoparticles (ZnONPs) alleviate heavy metal-induced toxicity in *Leucaena leucocephala* seedlings: a physiochemical analysis. Plant Physiol Biochem 110:59–69

Verma KK, Song XP, Yadav G, Degu HD, Parvaiz A, Singh M, Huang HR, Mustafa G, Xu L, Li YR (2022a) Impact of agroclimatic variables on proteo-genomics in sugarcane (*Saccharum* spp.) plant productivity. ACS Omega 7:22997–23008. https://doi.org/10.1021/acsomega.2c01395

Verma KK, Song XP, Joshi A, Tian DD, Rajput VD, Singh M, Arora J, Minkina TM, Li YR (2022b) Recent trends in nano-fertilizer for advancing sustainable agriculture under the era of climate change to ensure food security for future generations. Nanomaterials 12:173. https://doi.org/10.3390/nano12010173

Verma KK, Song X, Verma CL, Huang H, Singh M, Xu L, Wu J, Khan ZH, Zhang B, Li Y (2023a) Mathematical modeling of climate and fluoride effects on sugarcane photosynthesis with silicon nanoparticles. Plant Physiol Biochem 204:108089. https://doi.org/10.1016/j.plaphy.2023.108089

Verma KK, Song X, Degu HD, Guo D, Joshi A, Huang H, Xu L, Singh M, Huang D, Rajput VD, Li Y (2023b) Recent advances in nitrogen and nano-nitrogen fertilizers for sustainable crop production: a mini-review. Chem Biol Technol Agric 10:1–14. https://doi.org/10.1186/s40538-023-00488-3

Wang S, Wang F, Gao S (2015) Foliar application with nano-silicon alleviates Cd toxicity in rice seedlings. Environ Sci Pollut Res 22:2837–2845

Wang S, Wang F, Gao S, Wang X (2016) Heavy metal accumulation in different rice cultivars as influenced by foliar application of nano-silicon. Water Air Soil Pollut 227:1–13

Wang J, Li M, Feng J, Yan X, Chen H, Han R (2021) Effects of TiO_2-NPs pretreatment on UV-B stress tolerance in *Arabidopsis thaliana*. Chemosphere 281:130809. https://doi.org/10.1016/j.chemosphere.2021.130809

Wu H, Tito N, Giraldo JP (2017) Anionic cerium oxide nanoparticles protect plant photosynthesis from abiotic stress by scavenging reactive oxygen species. ACS Nano 11(11):11283–11297

Ye Y, Cota-Ruiz K, Hernández-Viezcas JA, Valdes C, Medina-Velo IA, Turley RS, Peralta-Videa JR, Gardea-Torresdey JL (2020) Manganese nanoparticles control salinity-modulated molecular responses in *Capsicum annuum* L. through priming: a sustainable approach for agriculture. ACS Sustain Chem Eng 8(3):1427–1436

Youssef OA, Tammam AA, El-Bakatoushi RF, Alframawy AM, Emara MM, El-Sadek LM (2020) Hematite nanoparticles influence ultrastructure, antioxidant defenses, gene expression, and alleviate cadmium toxicity in *Zea mays*. J Plant Interact 15(1):54–74

Youssef OA, Tammam AA, El-Bakatoushi RF, Alframawy AM, Emara MM, El-Sadek LM (2021) Uptake of hematite nanoparticles in maize and their role in cell cycle dynamics, PCNA expression and mitigation of cadmium stress. Plant Biol 23(6):1177–1189

Zahedi SM, Hosseini MS, Naghmeh DHM, Peijnenburg W (2021) Mitigation of the effect of drought on growth and yield of pomegranates by foliar spraying of different sizes of selenium nanoparticles. J Sci Food Agric 101(12):5202–5213

Zhai X, Li Z, Huang B, Luo N, Huang M, Zhang Q, Zeng G (2018) Remediation of multiple heavy metal-contaminated soil through the combination of soil washing and in situ immobilization. Sci Total Environ 635:92–99

Zhang S (2019) Mechanism of migration and transformation of nano selenium and mitigates cadmium stress in plants. [master's thesis]. Shandong University

Zhou P, Adeel M, Shakoor N, Guo M, Hao Y, Azeem I, Li M, Liu M, Rui Y (2020) Application of nanoparticles alleviates heavy metals stress and promotes plant growth: an overview. Nanomaterials 11(1):26

Chapter 4
Nanoformulations and Plant Saline Stress Resistance

Christiana Oluwatoyin Ajanaku, Oluwatoyin Adenike Fabiyi, Ayotunde Olayinka Owolabi, Kolawole Oluseyi Ajanaku, and Anuoluwapo Abimbola Akinsiku

4.1 Introduction

The word nano-formulations are a composite form from nanotechnology. The latter is a term which connotes synthesis or design of structural materials into a minutest scale ranging from 1 to 100 nanometers (Bhardwaj et al. 2019; Fabiyi et al. 2020; Kumar et al. 2024a; Mariyam et al. 2024). Nanomaterials (NMs) metamorphosed from the use of these structural materials and are practically smaller in size even than the given scale above. There are a number of specific attributes that comes along with the use of these NMs; these ranges from relatively limited surface area, increasing reactivity as a result of its unnatural surface structure, chemical composition, shape and assemblage. There is need to situate the importance of food in any country as appropriate in this context before exploring nano-formulations further. Food is widely accepted as a basic necessity of life because it is absolutely essential for the maintenance and reproduction of human life (Verma et al. 2024). It serves as

C. O. Ajanaku
Department of Physical Sciences, College of Science and Engineering, Landmark University, Omuaran, Kwara State, Nigeria

Department of Chemistry, College of Science and Engineering, Covenant University, Ota, Ogun State, Nigeria

O. A. Fabiyi (✉)
Department of Crop Protection, Faculty of Agriculture, University of Ilorin, Ilorin, Nigeria

A. O. Owolabi
Department of Agricultural Economics and Extension Landmark University, Omuaran, Kwara State, Nigeria

K. O. Ajanaku · A. A. Akinsiku
Department of Chemistry, College of Science and Engineering, Covenant University, Ota, Ogun State, Nigeria

C. S. Seth et al. (eds.), *Nano-formulations for Sustainable Agriculture*, Nanotechnology in the Life Sciences, https://doi.org/10.1007/978-3-031-90617-6_4

a vital sustaining element to the health of an individual and its seen as the basic requirement of life before considering the other two basic needs of life (shelter and clothing) especially at the household level. It is in this light that food security becomes an important factor in considering the health of a nation. A projection was made on global population to increase to about 9.8 billion by the year 2050, while it is expected to increase to 11.2 billion by the year 2100 as indicated in Fig. 4.1 (United Nations 2015).

In order to meet the demands of the growing population, it is important to implement farming techniques that will facilitate increase in crop production, as well as protect the environment (Fabiyi et al. 2024a). Sustainable development goals 2 and 12 refer to zero hunger and responsible consumption and production, respectively. Good health, food security, nutrition and well-being are possible with good economic growth. This can be made feasible by achieving the goal of sustainable agriculture and food security. Food security can be sustained by good agricultural practices that will control biotic and abiotic factors like drought, salinity, floods, or deforestation in addition to application of fertilizers, pesticides and advanced technology that will promote food production (Fabiyi et al 2025). Food security encompasses economic and physical access to adequate, secure and nutritive food to meet the dietary needs and food preferences of citizens of a nation in order to secure an active and healthy life (FAO 2018; Jagadesh et al. 2024).

The Global Food Security Index (GFSI) is the relative measurement tool used in determining the level of food security status from one country to another based on the level of availability, affordability, qualitative and safety of food to the inhabitants of a country. It affords each country an opportunity to measure its status quo based on these indices that spells out the food security level and serves as a veritable international standard for measuring food security status of a nation. Data from various regions worldwide indicate that the number of food-insecure people in countries such as those in South America and many parts of Africa, including Liberia, Nigeria, Uganda, Somalia, Sudan, Malawi, Tanzania, and Zambia, increased

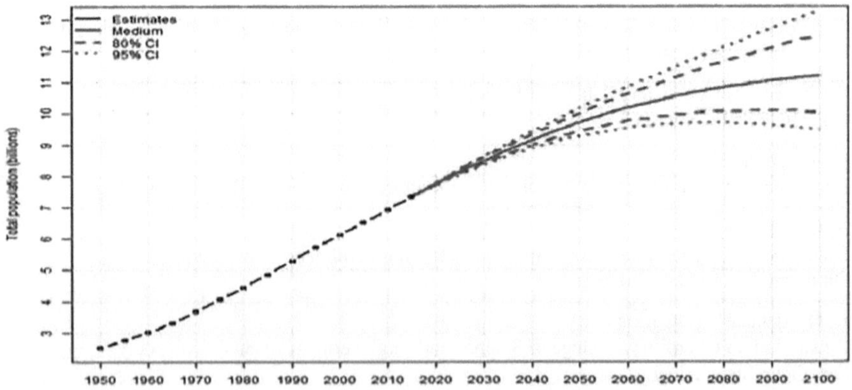

Fig. 4.1 Median variant projections of world population 2015–2100. (Source: https://en.wikipedia.org/wiki/Projections_of_population_growth#cite_note-UN-WPP-2015-4)

from an estimated 804 million in 2016 to nearly 821 million in 2017 (FAO 2018). When a country is not food secured, it manifests in diverse ways in forms like chronic food deprivation, inadequate access to healthy food, poor physiological adaptations due to food restrictions will and can eventually lead to all manner of health impairment like being underweight, overweight and obese. Available data showed that Africa constitute the major continent where a higher proportion of people are known to not being food secured. Data from the United Nations shows that the number of people suffering from hunger rose by about 38 million to 815 million (11%) of the world's population in 2016.

Report given by Squires and Glenn (2004) on the global impart of salinity, revealed about 30% of agricultural land becoming saline. According to the estimate of Bot et al. (2000), million hectares of land has been affected by soil salinity and has affected the economy to about twelve billion dollars per annum. According to FAO (2000), salinization occurs mostly in extremely dry soils and semi-tropical areas where irrigation is practiced with poor quality water. Developing and developed regions both experience soil salinity as shown in Fig. 4.2. Diverse strategies have been employed to mitigate its effect on crop production. Improper agricultural practices in dry regions lead to land deprivation through increasing concentration of salt.

Some countries that are affected by soil salinity includes: Bangladesh, where crop production is hindered as a result of climate change, poor irrigation water source and increasing excess soluble salts (Miah et al. 2020); Pakistan, affected by highly concentrated salt rates whereby the biological, chemical and physical properties of soils are affected (Syed et al. 2021); India, where a large portion of the land is under soil deprivation, salinity, water-logging, wind erosion and other environmental forces (Godfray et al. 2010; Kumar and Sharma 2020). Omar and Moussa (2016) reported that Egypt depends majorly on the water from Nile river; accounting for 93% of the country's conventional water resources. Application of shallow

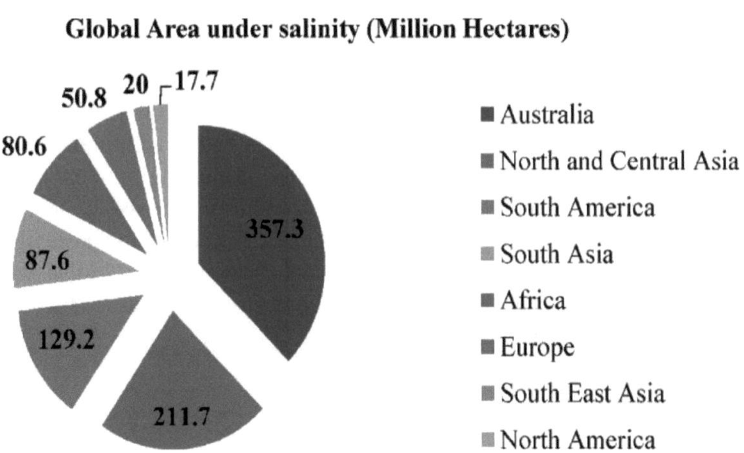

Fig. 4.2 Global distribution of salt-affected areas. (Source: Szabolcs 1993; Khan et al. 2011)

water, waste water and re-use of drainage water has been employed to fill the gap between water demand and water availability in Egypt.

Increasing deposits of salts in soil and water results in salinity, which limits plant growth and crop production (Zaki et al. 2011). This could either have occurred naturally or as a result of environmental activities, such as: improper irrigation management, temperature, sea level intrusion, excess irrigation, poor drainage, climate change and irregular rainfall. This is prevalent in desert areas or semi-arid areas. Some environmental factors that affect crop production include: light, temperature, availability of water and alkalinity.

4.2 Typical Abiotic Factors Affecting Plants

Typical abiotic stress types as reported by Safdar et al. (2019) include:

4.2.1 *Drought*

Insufficient moisture as a result of absence of rainfall, low water potential or low soil moisture content can result in drought. Consequently, there is a high rate of water loss from the aerial parts of the plant than water uptake from the roots. In order to save some water, plants close their stomata or the tiny pores on leaves. Symptoms that accompany drought stress includes: leaf scorching, yellowing of leaves, leaf rolling or wilting of plant leaves (Seleiman et al. 2021; Kumari et al. 2022).

4.2.2 *Temperature*

Plant functions can be affected by both high and low temperatures. While low temperatures can result in chilling injury, which affects cell structure and metabolic processes, high temperatures can cause heat stress, damaging proteins and membranes (Choudhary et al. 2025). Plants are constantly exposed to temperature changes throughout their lifecycle. As a result, there is needing to constantly adapt to various environmental changes. One of such changes is the global warming effect; whereby ocean and land temperatures are on the increase. This has been observed to affect the crop yields, over time.

4.2.3 Salinity

Presence of soluble ions in the soil, such as potassium, sulphate, sodium and chloride is known as salinity. Potassium and sulphate salts are considered to be more beneficiary to plants, while sodium and chloride ions are not measured as plant nutrients. Among the cations in the soil, an excess of sodium ions is referred to as sodicity (Qadir et al. 2007). Salinity can be brought about as a result of natural phenomenon, such as mineral weathering processes, water table quality or topography (Sverdrup and Warfvinge 1988). This is often referred to as primary salinization. Irrigation without proper drainage and water logging are referred to as secondary salinity. Consequently, salts are gradually accumulated in the soil, without being washed out, thereby mitigating growth of plant (Ayers and Westcot 1985). The effect of these salts in the soil by environmental and agricultural practices can upset the growth and development of plants negatively by affecting the availability of essential nutrients for plants as well as its water uptake. This results in ion toxicity, imbalance of nutrients as well as dehydration. Other abiotic factors include: Deposit of minerals, poor irrigation practices, effect of light, excess application of fertilizers, industrial wastes, deforestation, overgrazing.

4.2.4 Mineral Deposits

Natural activities such as weathering of rocks, mineral deposits, evaporation of water gradually accumulates salt deposits over a period of time, which adversely affects crop production.

4.2.5 Irrigation Practices

Too much irrigation and using low quality water can lead to increase in salt content in the soil. Poor drainage system can lead to accumulation of salts.

4.2.6 Light

Too little or too much light can be harmful. Low light levels can hinder photosynthesis, and high light levels can cause photo-inhibition, which is the damage of photosynthetic pigments by light energy.

4.2.7 Excess Fertilizers

Essential nutrients required for plant growth, such as potassium, nitrogen and phosphorus, are encapsulated in form of fertilizers. They enhance efficient nutrient uptake, strengthens plants against diseases and pests, thereby improving crop yield, when applied in proper quantity. Application of fertilizers in excess can result in too much deposits of salt, limits uptake of nutrients, cause toxicity of excess ions and environmental pollution.

4.2.8 Industrial Wastes

An accumulation of heavy metals, as a result of industrial effluents in the soil can harm cell membranes, interfere with plant growth, and disturb a number of metabolic processes.

4.2.9 Deforestation

This practice exposes the soil to adverse weather conditions particularly in arid areas; causing salt migration in upper and lower layers.

4.2.10 Overgrazing

Excessive overgrazing caused by animal husbandry in search of fodder has resulted in soil exposure and increasing salinization. As pasture diminishes, it often results in desertification.

4.3 Influence of Abiotic Factors on Plants Growth and Development

Plant growth is hindered in situation where there is deficiency of water supply due to minimization of extension of cell wall and turgor. The cell enlargement is more affected than cell division under this situation. In mild cases, an increase in respiration is observed, while in severe conditions, respiration decreases. Plants exposed to drought stress often undergo osmotic adjustment to adapt to these challenging conditions. The plant maintains rise in sugar content in roots compared to the shoots and experiences more growth in the former. Some of other effects of increasing temperature, such as inhibition of protein synthesis, changes in respiration and photosynthesis, cell mitigation. Low temperatures can affect crop growth by reducing nutrient uptake and minimizing metabolic processes. Heat stress is induced as a result of high temperature in plants, reduction of photosynthetic and metabolic processes, generation of ATP, electron transport rate and consequently, crop production (Wahid et al. 2007; Choudhary et al. 2024; Fabiyi and Ogundele 2025). In addition, the reproductive parts of a plant are affected resulting in minimized fruiting and yields (Prasad et al. 2008; Verma et al. 2020).

One of the effects of salinization is degradation of land, which eventually affects crop production and security of food. Consequently, global agricultural revenue is affected (Wichelns and Qadir 2014). The soil structure deteriorates as a result of sodification, thereby resulting in penetration of air in the soil as well as permeation of water. This leads to minimizing oxygen and water availability to plants (Stavi et al. 2021; Kumar et al. 2024b). There are two main effects of salinity on plants. First, the osmotic effect whereby, concentration of salt in the soil or water is higher than the concentration of salt inside the plant's cells. Consequently, a difference in osmotic pressure is created making it difficult for the plant to take up water from the soil. As a result, the plant experiences dehydration and wilting, leading to reduced growth and productivity. Secondly, there is ionic stress which occurs when high levels of salt in the soil or water lead to an excess accumulation of certain ions, such as sodium (Na^+) and chloride (Cl^-), inside the plant's cells. These ions can interfere with normal cellular processes, disrupt ion balance, and cause toxicity symptoms such as leaf burn, chlorosis, and necrosis. Ionic stress can also disrupt nutrient uptake and transport, leading to nutrient imbalances in the plant. Various crops react to salinity in diverse ways; while halophytes grow under saline conditions, glycophytes mostly exhibit reduction in growth and total yield. Secondary stress is induced as a result of ion toxicity and osmotic stress, resulting in diminished development in plants. Other effects include reduced photosynthetic activities arising from stomatal closure and continuous growth of oxidative species in plants (Hasanuzzaman and Fujita 2022).

4.4 Plant Response Mechanism

Plants make use of three main strategies in handling drought stress. These are: avoidance of stress, stress escape and stress tolerance. In avoiding stress, some plants make use of mechanisms that involves seasonal growth before commencement of the dry part of the year, reducing plant life cycle, self-reproduction, early flowering, etc. This has an implication of minimized growth period of plant and consequently plant productivity. In addition, reducing transpiration loss by stomatal closure, as well as high water intake from root system is also employed. Other plant mechanisms include: solute accumulation, antioxidant defence as well as osmotic adjustment among others. Plants respond to moderately high temperatures, by increasing growth of parts of plant, as well as make some changes like lengthening of the petiole, leaf hyponasty for monocots and dicot plants (Bawa et al. 2020; Rajput et al. 2021a). At the same time, plant organs development is influenced by increasing temperature. This is to protect the organs from the negative effects of the temperature rise (Zhu et al. 2021). For temperatures resulting from climate change, the flowering stage of the plant is enhanced (Zhao et al. 2020), while for male plants, sterility is inhibited. Opening of the stomata is observed for high temperatures. These morphological and physiological changes come along with cooling of the leaf parts (Kostaki et al. 2020). For high temperature changes, the process of light and photosynthesis is also affected, by minimizing the content of the chlorophyll (Mishra et al. 2021). To minimize the negative effects of excess temperature, Good irrigation practices, heat and cold-tolerant crops can be employed (Lobell et al. 2013).

Plants respond to saline stress in two ways. Firstly, cell expansion is inhibited and there is closure of stomata resulting in growth reduction, independent of ion concentration. Secondly, as the concentration of ions build up, the metabolic processes slow down leading to cell death. This is accompanied by series of molecular and physiological mechanisms such as: tissue tolerance, ionic tolerance and osmotic tolerance (Isayenkov and Maathuis 2019). Stomata closure reduces carbon accumulation. Transpiration is also limited so as to reduce loss of water. Chlorophyll content is reduced as a result of salinity stress; generally, limiting photosynthetic activities. To protect the over minimization of the photosynthetic electron chain, photorespiration has been adopted by plant. In addition, water cycle is another protective mechanism (Acosta-Motos et al. 2017).

Salt stress can affect the uptake of essential nutrients required by plants for growth and development, where sodium and chloride ions are in competition with phosphorus, nitrogen and potassium ions (Balasubramaniam et al. 2023). However, certain plants are known to be salt-tolerant; in such cases, growth in plants are not affected. Accumulation of osmolytes and compartmentalization of salts in vacuoles assist in mechanism for salt tolerance in plants (Acosta-Motos et al. 2017). Recent developments have introduced the application of NMs for better agricultural practices. Nanomaterials have relieved osmotic and ionic stress, enabled ROS detoxification, improved plant tolerance to salt stress by protecting photosynthesis in plants. Nanomaterials that have been employed include: zinc oxide, silicon oxide, titanium oxide, carbon quantum dot NPs (Li et al. 2022).

4.5 Brief Overview of the Current Challenges Faced in Crop Protection and Saline Stress

Low yield or reductions in cultivation of crops in agriculture have been triggered by a combination of biotic and abiotic factors. These factors range from environmental pollution, climatic variations, water and soil nutrients demands and agricultural land availability. Among these limiting factors is a hazardous indicator to crop growth and protection called saline stress which falls under environmental factors. This is also known as soil salinization. It literally connotes extreme accumulation of salts dissolved in the soil which inhibits growth and development of the plants (Safdar et al. 2019). High salt concentrations in the soil cause soil stress, leading to several negative outcomes, including reduced seed germination, inhibited seedling growth, and decreased water uptake by plant roots, this subsequently leads to retention of salts in plant tissues such that it becomes poisonous to the plant growth. Moreover, its impaired effect on seed germination arises from the osmotic built up at the external part of the seed which hinders absorption of water by the seed. Studies reviewed further shows that saline stress in plants also inhibits the process of photosynthesis in plant growth. The reduction in vegetative part of plants linked to high concentration of salts in plants was related to a significant delayed process of photosynthesis in these plants (Mandal 2019). However other studies revealed that this finding of salt concentration in plants linked with delayed photosynthesis process does not suffice for majority of other plant species (Safdar et al. 2019). The implication of saline stress on photosynthesis process in plants has been attributed to the specie of plant in view as well as salt concentration. This connotes that when there is low concentration of salts in the plants there is probability that the photosynthesis rate may be activated. Other limitations soil salinity brings encompasses weakening of cells metabolism, slowing down plant nutrition rate, osmotic stress, reduction in microbial activities thereby slowing down the rate of water and nutrient uptake in plants from the soil (Van den Burg et al. 2024).

Seeds need adequate water and nutrition in order to germinate. Seed germination requires enzymatic hydrolysis by α-amylase in order to break down the carbohydrate stored in the endosperm so as to release the energy required for the growth and development of the growing embryo (Weiss and Ori 2007). Presence of salinity reduces the uptake of quality water and nutrients (Munns et al. 2020). This is as a result of the osmotic pressure in the soil being reduced compared to the osmotic pressure in the seed. Consequently, the period of seed germination is prolonged. After germination, viability of the embryo is hindered due to the accumulation of excess chloride and sodium ions (El Sabagh et al. 2021).

Some ions are beneficial to plants, these ae: potassium, sulphate, nitrate ions. However, some are not referred to as nutritious. These are: chloride and sodium ions. Increasing concentration of such ions in soil solution lowers its osmotic pressure, thereby reducing the absorption of water from the soil. This results in osmotic stress, which hinders growth and crop yield. Also, absorption of sodium or chloride ions by the plant can lead to cytotoxicity, resulting in stunted growth of plant. High levels of sodium ions leads to competition of cations, whereby essential nutrients like magnesium, calcium ions may be unavailable; resulting in nutritional deficiencies (Atta et al. 2021).

Reactive oxygen species are primarily produced in cytosol, endoplasma, mitochondria, and chloroplasts. Water deficiency in addition to saline stress causes the oxygen species to have a damaging effect on plant organs. Earlier reports have shown damages done to crops like *Zea mays, Oryza sativa* and *Solanum lycopasicum* (Kaveh et al. 2011; Khodarahmpour et al. 2012; Rahman et al. 2016). Effect on

Crop Yield: Studies have shown how abiotic stress, especially, salinity can affect crop yield. Inadequate water uptake, reduced transpiration, stomatal closure, lower photosynthetic activities and oxidative damage due to salinity stress have contributed to a drop in crop production. The table below shows the extent to which salinity has affected several crop yield (Table 4.1).

4.5.1 An Obstacle to Plant Development

The amount of soluble salts in the soil, or soil salinity, is a significant environmental issue that affects agricultural productivity across a wide range of global regions. To ensure sustainable agricultural practices, it is essential to comprehend its causes, effects, and possible management solutions (Zhao et al. 2020; Kumar et al. 2024c). Salinity stress in plant is a condition whereby salts in the soil, such as carbonates, nitrates, chlorides, phosphates, sulphates, reach a concentration such that it becomes harmful to the growth and development of the plant. This is often accompanied by diverse alterations in the photosynthetic, biochemical, enzymatic and physiological activities of the plant resulting in delayed growths and reduced yield of the crop. Plants under salinity stress usually undergo limited water uptake, imbalance of nutrients, osmotic stress, ion toxicity, reactive oxygen species and hormonal imbalance (Wang et al. 2011; Kumar et al. 2020).

Irrigation salinity and dry land salinity are the main types of salinity. Irrigation salinity takes place as a result of accumulation of salts on the soil surface from rigorous irrigation practices with groundwater and leaching of irrigation water. Accumulation of concentrated salts on non-irrigated soil, is referred to as dry land salinity. Processes linked with dry land salinity includes: ground water movement, deep drainage and ground water discharge. Dry land salinity can be primary or secondary in classification: primary salinity occurs as a result of high concentration of salts in the soil or water, in arid or semi-arid regions, attributable to low rainfall and high temperature. Secondary salinity is as a result of activities such as: mining, industrial processes and irrigation (Atta et al. 2023). Methods through which salinity can be measured in soil or water include: Total dissolved solid measurement,

Table 4.1 Yield reduction due to salinity stress on different crops (Atta et al. 2023)

Crops	Yield loss (%)	Source
Maize (*Zea mays*)	50–60	Zorb et al. (2018)
Wheat (*Triticum aestivum)*	10–50	Yadav et al. (2020)
Sorghum (*Sorghum bicolor*)	15–50	Kenneth and Neeltje (2002)
Rice (*oryza sativa*)	30–50	Hoang et al. (2015)
Groundnut (*Arachis hypogaea*)	25–50	Zorb et al. (2018)
Barley (*Hordeum vulgare*)	10–60	Abd El-Monem et al. (2013)
Cotton (*Gossypium* spp.)	10–20	Zorb et al. (2018)

electrical conductivity measurement, chloride ion concentration measurement, sodium ion adsorption measurement, soil sampling and laboratory analysis.

Some of the causes of salinity in soil include: weathering of rocks that contain soluble salts is a natural process; tidal inundation near the coast; evaporitic processes that leave concentrated salts behind in arid areas; human endeavours; Using salted water for irrigation: Using salted water on a regular basis can cause the soil to become salted; Poor drainage: Salts can accumulate in the soil due to inadequate drainage, which prevents them from leaching (Singh et al. 2022). Some of the effects of salinity on soil include: Decreased water availability: Plant roots have harder time absorbing water from the soil because salt produces an osmotic pressure gradient. Plant growth is stunted and water stress results from this. Unbalanced nutrient intake: excessive salinity can interfere with plants' ability to absorb vital nutrients, resulting in deficits that hinder growth and development (Tanveer et al. 2020). Changes in soil composition: Salts can disrupt soil particles, leading to a loose soil structure, reduced water infiltration, and increased erosion. Decreased microbial activity: Excessive salinity can harm soil health and nutrient cycling by inhibiting the growth and activity of beneficial soil microorganisms (Ding et al. 2020). Some of the effects of saline stress on plant include**:** osmotic, oxidative, metabolic stresses and growth inhibition.

Plants that have trouble getting water and nutrients grow in an unnatural way. Stress induced by abiotic agents can cause hormonal alterations that favour survival over growth, which reduces the rate of cell division and elongation of shoots. Salt toxicity and water stress make it difficult for seeds to sprout. Salinity stress can also induce leaf injury. The browning and dying of leaf tips and margins can be caused by salt-induced tissue death. Overall crop yields are greatly decreased as a result of hampered plant growth and development (Zheng et al. 2023). Exposure of plants to high levels of salt, results in a number of disrupted physiological and biochemical processes resulting in reduced photosynthesis, nutrient imbalance, impaired growth, and oxidative damage. High concentration of ions in the soil affects the osmotic balance in the root of crops. Increasing absorption of sodium ions by plant cells causes a disruption in cell activities such as enzymatic activities, maintenance of the membrane potential as well as cell tugor (Yadav et al. 2020).

Abiotic factors like drought, temperature, salinity have adverse effect on plant crop productivity. However, increase in the oxygen reactive species in some plants, is as a result of salinity. Oxidative stress is observed when there is an imbalance between production of reactive oxygen species and its detoxification by mechanisms of antioxidants. This results in mutations of genetic parts of the plant like the proteins, RNA, DNA and finally, cell mutation (Mishra et al. 2023). Metabolic activities in plants are affected by abiotic factors. Some of these activities include: reduced photosynthesis, nutrient assimilation, respiration. Saline stress induces synthesis of abscisic acid in plants which results in stomata closure. Consequently, photosynthetic activity is hindered. Unbalanced nutrient uptake and translocation: Increase in uptake of ions can cause deficiencies or imbalances in nutrients that can hinder a plant's ability to grow and develop.

4.6 Nanomaterials and Their Potential in Plant Saline Stress Management

Application of nanotechnology to agriculture is a method employed for treatment and detection of plant diseases which fuel availability of plant nutrients thereby improving crop production, health and soil fertility. It has been established earlier that soil salinization could be as a result of a natural course or biological, physical and organic agricultural activities such as climatic variations. While on the other hand it could be due to unnatural courses such as poor irrigation practices, inadequate land clearing practices, poor drainage, erosion and upward movement of salts from the sub-soil to the surface soil. In time past, farmers have used several measures in reducing the fertility of agricultural land for optimum production by using crop rotation method, shifting cultivation, mixed cropping, these contemporary times, a lot of researches have been evolving towards the deployment of nanotechnology in developing improved materials, tools and devices that can control salty soils.

Application of nanotechnology in agriculture can be viewed as a new approach through which researchers can use in understanding and creating a new tool that can help in the identification and management of plant diseases. Nanomaterials could be classified into different categories: metals, metal oxides, metal sulfides, carbon, graphene oxides, lipids or polymers. Some examples of metals like Zn, C, Fe, Ag, Au, Ni, for metal oxides, i.e., Al_2O_3, CeO_2, MnO_2, TiO_2, Fe_2O_3, for metal sulphides, such as ZnS, CuS, AgS, FeS_2 (Sharma et al. 2019). Myriads of opportunities and possibilities are embedded in the use of NMs especially with regard to precision agriculture. It has scientific proofs of conserving resources and has more energy efficiency (Bhardwaj et al. 2019).

The roles of NMs in reducing plant saline stress cannot be overemphasized. These include bringing a geometric progression to agricultural production by improving soil fertility, ensuring adequate supply of nutrients to the plants and eliminating the osmotic imbalance causing soil salinization. When saline stress has been eliminated in a plant, there are other features within the plants that experiences radical transformation. These features include excellent germination of seeds, large growth of plant vegetative parts, healthy formation of the shoot and root regions of the plants, enhanced chlorophyll content, coloration and photosynthetic rate (Babaei et al. 2017). Farmers capacity is enhanced to improving the quality of agricultural products maximally in terms of reduction of farm input materials and resources (seed, water, fuel, etc). Significant attainment of the genetic potential of desired crop is speedily realized as this new phase of NMs usage is embraced. Soil health balance is much more pronounced as a result of reduced number of pesticides as it also aids reduction in over-dependence of agriculture in climatic variations. The importance of these unique materials in curbing saline stress reviewed in past studies include building defence systems of plants, helps in scaling up hormonal concentrations, ion homeostasis, gene expressions and antioxidants enzyme activities have been facilitated efficiently.

Plants treated with NMs exhibit fortification against various abiotic stresses and enhance improved photosynthetic activities, transpiration rate, stomatal conductance, water efficiency and proline content. Nanomaterials stimulate activities of antioxidant enzymes, accumulation of free amino acids, nutrients and osmolytes (Khan et al. 2017). Application of NPs to the leaves of plants has been shown to increase the amount of chlorophyll and photosynthetic responses. Previous studies show addition of zinc NPs to plant under salinity stress to exhibit improved photosynthetic activities as well as improved fluorescence activity (Cao et al. 2018).

4.6.1 Accumulation of Phenolic Compounds

Earlier studies reveal that addition of NPs to plants have contributed to an increase of phenolic compounds in plants. Experimental results show increase in phenolic and flavonoid metabolites particularly in the root of plants. Copper and zinc NPs exhibit similar properties by increasing glutathione, polyphenols, flavonoids and phenolic contents (Khan et al. 2020; Rajput et al. 2021a).

4.6.2 Improved Nutrient Uptake

Nutritional deficiencies in plant are one of the effects of salt stress, as a result of imbalance in nutrient uptake. Addition of NPs, such as silicon dioxide and copper has been observed to promote growth and balance of sodium/potassium ratio by increasing the amount of potassium in leaves (Rehman et al. 2024).

4.6.3 Strengthens Antioxidant Defence

Some NPs show certain characteristics similar to some antioxidant enzymes by removing oxidative species. Earlier studies reveal CAT–like attributes of cobalt and cerium NPs, while copper, iron and manganese show POD-like behaviour (Rico et al. 2015). The growth features of plants were enhanced as a result of the antioxidant activities.

4.6.4 Nanomaterials Mitigate Plants Saline Stress

Nanomaterials have emerged as promising agents in mitigating saline stress in plants, offering new avenues for enhancing plant growth and productivity in saline environments. There are several ways in which the saline stress in plants could be alleviated using NMs. Some of the few ways include:

4.6.5 Osmotic Adjustment

There are certain NMs that helps plants to maintain osmotic balance in saline conditions. They do it by modulating the levels of osmolytes, such as proline and glycine, betaine, which protect plants by stabilizing proteins and membranes. The growth of plant at earlier stages are usually affected to quite a significant level when water sustainability NPs are employed. It has also been established that NPs promote plant growth; and when higher doses of NPs are employed, this can impair a negative impact also on the plant (Verma et al. 2022).

4.6.6 Nutrient and Water Uptake

The efficiency of nutrient uptake in plants in saline conditions can be improved using NMs. The creation of NFs has the role of releasing nutrients in a controlled manner, making them more available to plants and this helps in mitigating the adverse effects of salt stress. Ahmad and Akhtar (2019) gave insight on the applications of different types of NMs in controlling salt stress in plants. It was shown that the NMs were able to enhance water uptake in plants, protected crops from invasion by diseases and pests and improved precision farming.

4.6.7 Antioxidant Defence System

Reactive oxidative species can accumulate in plants due to salinity stress. Nanomaterials with antioxidant properties like carbon based NPs or metal NPs can enhance the antioxidants defence mechanism in plants, by scavenging the oxidative species. Plant physiological function is maintained by reducing the oxidative damage by employing copper, iron, manganese and celenium NPs which act in similarity to antioxidant enzyme glutathione peroxidase, while iron celenium and cobalt NPs act similar to enzyme catalase (Yang et al. 2017; Khalid et al. 2022).

4.6.8 Ion Homeostasis

NMs can assist in maintaining ion homeostasis by enhancing the selectivity of ions, reducing the uptake of toxic ions, like Na^+, and improving the uptake of beneficial ions (like K^+), thus alleviating the ionic stress component of salinity. This results in increased osmotic potential within the plant. Tomato growth performance was enhanced as well as its sodium/potassium ratio, on foliar application of copper NPs. Likewise, the soya bean leaf seedling growth was enhanced on application of silicon dioxide NPs, by increasing concentration of potassium ion in the leaf (Khalid et al. 2022). Mushtaq et al. (2018) illustrated the application of SiO_2 NPs for both fresh and dried parts of tomato plant. They reported an increase in chlorophyll content, enhanced seed germination and improved proline accumulation. Laware and Raskar (2014) illustrated how the application of titanium oxide NPs in onions improved seedling growth and seed germination. It was also employed to induce the release of antioxidant enzymes. Another example of effect of NPs, cerium oxide on *Brassica napus* L was studied by Rossi et al. (2016). It was reported to have improved nutrient uptake, proline, the leaf pigment chlorophyll increased, as well as photosynthetic responses. In addition, Soliman et al. (2015) identified the effect of iron magnetic and zinc oxide NPs on Moringa plants. There was improvement in the macro and micro nutrients, enzymes, non-enzymes and antioxidant within the Moringa plant.

4.7 Transformations of Nanoparticles

Nanoparticles have various modifications in their interactions with environment. These modifications are of immense importance to their effectiveness on plants. Nanomaterials are synthesized at the nano scale using diverse techniques which could be biological, physical, or chemical in nature. These materials are characterized based on their sizes, shapes and surface areas. This will enable the effective determination of the efficiency of the NMs. Aggregation or dispersion of the NMs will depend on the amount of ions present and the pH of the water or soil. Presence of ions can bring about electrostatic repulsion or attraction of the NMs, while the organic substances can stabilize or destabilize the NMs. The coatings of NMs are carefully selected in order to enhance the effectiveness of NMs as well as their uptake. The coatings are modified based on factors like temperature and ions present in the soil (Alallam et al. 2023).

Chemical changes that take place in the environment can affect the impact of nanomaterials in the biological systems of plants. This is due to oxidation and reduction reactions that are likely to take place with metal NPs. Therefore, it is important that the NMs are designed to enhance their strength, reactivity and conductivity in its application (Staron et al. 2020). Interactions of NMs with the components in the soil such as organic matter, microbes and minerals can lead to alterations in the microflora of the soil and its fertility (Yang et al. 2017). The minute nature of NMs allows them to dissolve in soil water and then absorbed into the plant cells, through the root system. They get to the xylem by means of the cortex and the pericycle. This allows for interactions and modifications in the morphological and physiological activities of the plant (Burman and Kumar 2018). Chemical processes which initiate oxidative reactive species, reactions in plants, transportation, lipid peroxidation, guide the reactions of NMs in plants (Rehman et al. 2024). Nanomaterials move from one part of the plant to another.

4.8 Applications of Nanoparticles

Development of NMs in agriculture has given a promising outlook on the growth and development of crops (Fabiyi et al. 2021, 2023a, b). Recent studies reveal the effects of various NPs on plants. Silver NPs applied to wheat, tomato, carrot and rice enabled their growth and yield, zinc oxide, alumina and zinc enhanced germination and growth of cucumber, corn cabbage and lettuce. One of the essential micronutrients required for growth in plant is zinc. However, studies have shown that it protects the plant cells against oxidative species. Zinc has a biochemical role in the formation of chlorophyll and in the manufacture of indoleacetic acid from tryptophan (Fabiyi et al. 2018, 2024b; Ghidan and Al Antary 2020; Fabiyi 2021; Fabiyi and Olatunji 2022). Nano particles are efficient transport systems of agricultural inputs to the target sites. Silicon NPs transport fertilizers and pesticides to the desired site, as well as assist in water retention by acting as a nano-zeolite. It also acts as a nanosensor by detecting metals present in plants. Multi-walled carbon nanotubes, are employed in studying accumulation of pesticides in roots and shoots of plants.

Plants have always been seen to adjust to abiotic factors such as: drought, heat and salinity. Certain NPs have been employed in managing saline stress. For example, drought stress on wheat was minimized with the application of titanium NPs. Other NPs utilized in drought stress include: sodium and silicon silicate (Pei et al. 2010). According to Mohammadi et al., plant membrane disruption, ion leakage has been minimized by application of titanium oxide NPs. Other types of NPs that aid absorption of nutrients, growth of amino acids, proline, improve activity of antioxidant enzymes and enhance tolerance of plants against the severe environment, include nano silicon oxide and nano zinc oxide (Elsakhawy et al. 2018). Mesoporous silicon NPs can be applied to crops in various ways. It could be applied as a supplement to the soil, or in a powdered form to the nutrient solution where it would be absorbed by the roots or in a liquid form to the leafy parts of the crop which were taken up by epidermal cells or stomata. Suriyaprabha et al. (2014) earlier study showed that the soil application of silicon NPs was more effective than foliage application. Silicon NPs improved the germination of tomato with higher fresh and dry weight of seedlings (Roychoudhury 2020). Calcium borate NPs improved the growth of roots and shoots of lettuce. Fruit settings of strawberry was enhanced as a result of application of zinc oxide NPs. Titanium dioxide, silver NPs, multi-walled carbon nanotubes have enhanced growth of crops such as: soybean, cotton, barley, maize and spinach (Liu et al. 2021). Several types of NPs have been formulated to alleviate salinity stress in plants. Some of these NPs include: silica, chitosan, zinc, iron oxide, silver, copper oxide, titanium oxide NPs (Junedi et al. 2023).

4.9 Synthesis and Characterization of Nano Formulations for Plant Saline Stress

The introduction of nano-formulations has brought advancement in lowering the effects of saline stress on plants, as well as transferring nutrients in a targeted manner. Basically, nano materials can be synthesized physically, chemically or biologically (Dulta et al. 2020). The reduction of bulky materials into a nano scale is referred to as the physical method; while the chemical synthesis involves the application of chemical reactions in synthesizing NPs (Table 4.2).

4.9.1 Nano Agricultural Inputs

These are a novel class of agricultural materials that harness the power of nanotechnology with new functionalities to address modern agricultural challenges. These advanced materials combine the unique properties of NPs with targeted delivery systems to enhance plant growth, productivity, and resilience in a sustainable manner. Smart nano-agri inputs (SNAIs), provide more control and customization, enabling customized solutions based on unique crop needs and environmental conditions (Wang et al. 2020). Other benefits include: optimized photosynthesis, certain NPs, like metallic or carbon nanotubes, can act as light antennas, gathering solar radiation and focusing it on the pigments in plant cells that are engaged in photosynthetic reactions. Examples of Smart nano-agri inputs are as NFs, nanosensors and nanopesticides, and nano encapsulated biopesticides encasing natural biocontrol agents or targeted pesticides for controlled release and increased efficacy against particular pests and diseases (Fabiyi 2025). NFs consist of nano-sized materials having micro or macro nutrients like: nano-urea, nano-iron, nano-zeolite, nano-potash, nano-apartite (Chhipa 2017; Xin et al. 2020).

Macronutrients such as carbon, calcium, magnesium, phosphorus, potassium and nitrogen are encapsulated by various NMs. Studies have shown enhancement in

Table 4.2 Nanoparticles synthesized from the various method (Ealia and Saravanakumar 2017).

Category	Method	Nanoparticles
Bottom-up	Sol-gel	Carbon, metal and metal oxide based
	Spinning	Organic polymers
	Chemical Vapour Deposition (CVD)	Carbon and metal based
	Pyrolysis	Carbon and metal oxide based
	Biosynthesis	Organic polymers and metal based
Top-down	Mechanical milling	Metal, oxide and polymer based
	Nanolithography	Metal based
	Laser ablation	Carbon based and metal oxide based
	Sputtering	Metal based
	Thermal decomposition	Carbon and metal oxide based

absorption of fertilizers as well as reduced outflow of fertilizers. Chitosan NPs loaded with nitrogen, potassium and phosphorus have shown increase in absorption of the essential elements in coffee plant (Liu et al. 2021). Micronutrients such as: iron, zinc, copper, boron, manganese when enclosed in NPs, such as chitosan, copper, calcium, and zinc oxide NPs have enabled effective accessibility of micronutrients to plants. Micronutrients are involved in physiological and chemical processes in plants such as photosynthesis, enzyme activities. Nano encapsulated fertilizers are enhanced to resist degradation by evaporation, photolysis, microbial decomposition or weathering (Liu et al. 2021).

Nano-encapsulated nutrients provide a slow-release mechanism that improves nutrient use efficiency and minimizes leaching losses, particularly beneficial in saline soils. The nano fertilizers are employed by modifying the normal traditional fertilizer, or by extracting different vegetative or reproductive parts of the plant. Scientists have successfully introduced carbon nanotubes into plant photosynthesis's site, the chloroplast, and have seen an increase in electron flow and total photosynthetic activity as a result. Research is being conducted to determine whether it is possible to use NPs to directly supply chloroplasts with protective molecules or vital nutrients, thus improving their resilience and functionality (Elhefnawy and Elsheery 2023).

4.10 Challenges of Nanoformulations

Nanomaterials have been applied as a form of nano-remediation in agriculture, by enhancing ability of plants against abiotic stress, managing plant diseases, improving nutrient absorption, inhibiting loss of nutrients, waste water management. However, elevated concentrations of nanomaterials (NMs) can have adverse effects due to the generation of free radicals. The molecular mechanisms involved in the interactions between NPs and plant cells needs to be adequately studied. For instance, the influence of NPs involved in anti-oxidative activities in plants which enables plants to withstand any type of stress. This will enable the designing of specific NPs to meet specific targets (Rajput et al. 2021a, b). The current cost of producing and implementing NPs-based agricultural solutions on a large scale prevents their widespread adoption. For instance, treatment with silver or gold NPs is not economically reasonable (Rajput et al. 2021a, b).

Intake of vegetables and fruits encapsulated with NPs can result in ingestion of NPs into certain organs in the body such as blood, spleen, heart, liver or brain. Accessibility of NPs into the nucleus may result in damage to the DNA (Rajput et al. 2021b). Neural inflammation has been reported on exposure to ferric oxide NPs and titanium oxide NPs. In addition, further studies that enhances understanding of the interactions between NMs and plant cells and how it relates to biochemical pathways is necessary in order to identify which NPs will be most suitable under various conditions. Continuous application of NMs in crop sustainability can have effect on the public and environmental health, thereby creating the need for policies and regulations for use and sustainability. Careful consideration and risk assessment are required for the long-term consequences of introducing NPs into plant systems as well as any possible environmental effects of using them in agriculture.

4.11 Future Prospects of Nanoformulations

Encapsulating essential nutrients in NPs can promote uptake of nutrients in plants thereby enhancing crop growth, crop yield and improving plant ability to withstand salinity stress. NPs can protect nutrients from leaching and volatilization, increasing their availability to plants and reducing nutrient losses in the soil (Farooq et al. 2022). Sustainable soil management: Nanoformulations assist by mitigating the negative impacts of salinity stress on crop production and support long-term agricultural resilience in salt-affected regions. Nanoparticles can interact with soil components to enhance soil aggregation, water retention, and nutrient availability, thereby creating a conducive environment for plant growth in saline soils (Hussain et al. 2018).

Innovation in crop breeding: Development of salt-tolerant varieties through nanotechnology-based approaches can be driven through nano-formulations. Salt-tolerant crop varieties with improved yield potential and resilience to salinity stress, can be developed by integrating nanoformulations with modern breeding techniques, such as genome editing and marker-assisted selection. Nanoparticles can be used to deliver stress-responsive genes, proteins, and signalling molecules to plants, enabling the targeted manipulation of plant stress responses and enhancing salt tolerance in crops (Rajput et al. 2021a, b). There is need for practicable communication between researchers, industry and policy makers, to promote the production, development and deployment of nanoformulations for plant salinity on a large scale. The future of nanoformulations for plant salinity hinges on their successful commercialization and scalability in agricultural systems. By addressing regulatory, economic, and technical challenges, stakeholders can unlock the full potential of nanoformulations for sustainable agriculture and food production in saline environments.

4.12 Conclusions

Nanomaterials offer diverse mechanisms for managing plant saline stress, covering areas such as nutrient uptake enhancement; ROS scavenging, to gene expression modulation and water balance regulation. The unique properties available in NMs has enabled researchers and agronomists develop innovative strategies to enhance plant resilience to salinity stress and improve agricultural sustainability. Modern technologies such as geographic information systems (GIS), global positioning systems (GPS), satellite-based navigation systems, and remote sensing can be utilized to improve nutrient uptake and enhance productivity Further research into the long-term effects and environmental implications of NMs use in agriculture in cereals and other specific crops is essential to ensure the safe and effective application of nanotechnology in managing plant saline stress.

References

Abd El-Monem AA, El-Habbasha SF, Hozayn M (2013) Mitigation salinity stress effects on barley (Hordeum vulgare L.) growth, yield and some physiological aspects by hemin. J Appl Sci Res 9(3):2411–2421

Acosta-Motos JR, Ortuño MF, Bernal-Vicente A, Diaz-Vivancos P, Sanchez-Blanco MJ, Hernandez JA (2017) Plant responses to salt stress: adaptive mechanisms. Agronomy 7(1):18

Ahmad I, Akhtar MS (2019) Use of nanoparticles in alleviating salt stress. Salt Stress, Microbes, and Plant Interactions: Causes and Solution 1:199–215

Alallam B, Doolaanea AA, Alfatama M, Lim V (2023) Phytofabrication and characterisation of zinc oxide nanoparticles using pure curcumin. Pharmaceuticals 16(2):269

Atta K, Pal AK, Jana K (2021) Effects of salinity, drought and heavy metal stress during seed germination stage in ricebean [Vigna umbellata (Thunb.) Ohwi and Ohashi]. Plant Physiol Rep 26:109–115. https://doi.org/10.1007/s40502-020-00542-4

Atta K, Mondal S, Gorai S, Singh AP, Kumari A, Ghosh T et al (2023) Impacts of salinity stress on crop plants: improving salt tolerance through genetic and molecular dissection. Front Plant Sci 14:1241736

Ayers RS, Westcot DW (1985) Water quality for agriculture. FAO, Rome. FAO Irrigation and Drainage Paper no. 29

Babaei K, Sharifia RS, Pirzadb A, Khalilzadeha R (2017) Effects of bio fertilizer and nano Zn-Fe oxide on physiological traits, antioxidant enzymes activity and yield of wheat (Triticum aestivum L.) under salinity stress. J Plant Interact 12(1):381–389

Balasubramaniam T, Shen G, Esmaeili N, Zhang H (2023) Plants' response mechanisms to salinity stress. Plan Theory 12(12):2253

Bawa G, Feng L, Chen G et al (2020) Gibberellins and auxin regulate soybean hypocotyl elongation under low light and high-temperature interaction. Physiol Plant 170:345–356

Bhardwaj AK, Hamed LMM, Sharma N, Rajwar D, Meti S, Nagaraja MS (2019) Engineered polymeric and nano-materials for taming salty soils and waters used for crop production. In: Research developments in saline agriculture. Springer

Bot A, Nachtergaele F, Young A (2000) Land resource potential and constraints at regional and country levels. FAO Rome, World Soil Resources Report 90, p 114

Burman U, Kumar P (2018) Plant response to engineered nanoparticles. In: Nanomaterials in plants, algae, and microorganisms. Elsevier, London, pp 103–118

Cao K, Yu J, Xu D, Ai K, Bao E, Zou Z (2018) Exposure to lower red to far-red light ratios improve tomato tolerance to salt stress. BMC Plant Biol 18:1–12

Chhipa H (2017) Nanofertilizers and nanopesticides for agriculture. Environ Chem Lett 15:15–22

Choudhary R, Rajput VD, Ghodake G, Ahmad F, Meena M, Rehman RU, Prasad R, Sharma RK, Singh R, Seth CS (2024) Comprehensive journey from past to present to future about seed priming with hydrogen peroxide and hydrogen sulfide concerning drought, temperature, UV and ozone stresses- a review. Plant Soil 500:351–373

Choudhary R, Ahmad F, Kaya C, Upadhyay SK, Muneer S, Kumar V, Meena M, Liu H, Upadhyay H, Seth CS (2025) Decrypting proteomics, transcriptomics, genomics, and integrated omics for augmenting the abiotic, biotic, and climate change stress resilience in plants. J Plant Physiol. https://doi.org/10.1016/j.jplph.2025.154430

Ding Z, Kheir AM, Ali MG, Ali OA, Abdelaal AI, Lin XE et al (2020) The integrated effect of salinity, organic amendments, phosphorus fertilizers, and deficit irrigation on soil properties, phosphorus fractionation and wheat productivity. Sci Rep 10(1):2736

Dulta K, Thakur K, Chauhan P, Chauhan PK (2020) Biogenic nanoparticles: a new paradigm for treating infectious diseases in the era of antibiotic resistance. In: Nanotechnological approaches in food microbiology. CRC Press, pp 129–164

Ealia SAM, Saravanakumar MP (2017, November) A review on the classification, characterisation, synthesis of nanoparticles and their application. In: IOP conference series: materials science and engineering, vol 263, No 3. IOP Publishing, p 032019

Elsakhawy T, Omara AED, Alshaal T, El-Ramady H (2018) Nanomaterials and plant abiotic stress in agroecosystems. Environment, Biodiversity and Soil Security 2(2018):73–94

El Sabagh A, Islam MS, Skalicky M, Ali Raza M, Singh K, Hossain M et al (2021) Salinity stress in wheat (Triticum aestivum L.) in the changing climate: adaptation and management strategies. Front Agron 3. https://doi.org/10.3389/fagro.2021.661932

Elhefnawy SM, Elsheery NI (2023) Use of nanoparticles in improving photosynthesis in crop plants under stress. In: Photosynthesis. Academic, pp 105–135

Fabiyi OA (2021) Sustainable management of meloidogyne incognita infecting carrot: green synthesis of silver nanoparticles with cnidoscolus aconitifolius: (Daucus carota). Vegetos 34(2):277–285. Published by Springer

Fabiyi OA (2023a) The perspective of nanotechnology in plant protection. In: Avila-Quezada GD, Rai M (eds) Nanophytopathology, pp 23–38. https://doi.org/10.1201/9781003344513-3

Fabiyi OA (2023b) Nano clay materials and their role in plant disease management. In: Avila-Quezada GD, Rai M (eds) Nanophytopathology, pp 145–156. https://doi.org/10.1201/9781003344513-11

Fabiyi OA (2025) Nanomaterials of cellulose origin: novel strategies for plant disease control. In: Rajan R, Ahmad F, Khoshriya A, Seth CS (eds) Sustainable era of nanomaterials. Sustainable development and biodiversity. Springer, Singapore, pp 379–397

Fabiyi OA, Ogundele AV (2025) Biosensors for monitoring climatic stressors in agricultural crops. In: Uddin I (eds) Smart nanosensors. Springer, Singapore, pp 229–245

Fabiyi OA, Olatunji GA (2022) Application of green synthesis in silver nanoparticles for root knot nematode management on tomato. Indian J Nematol 52(2):164–173. https://doi.org/10.5958/0974-4444.2022.00020.8

Fabiyi OA, Olatunji GA, Saadu AO (2018) Suppression of Heterodera sacchari in rice with agricultural waste-silver nano particles. J Solid Waste Manage 44(2):87–91

Fabiyi OA, Alabi RO, Ansari RA (2020) Nanoparticles synthesis and their application in the management of phytonematodes: an overview. In: Ansari RA, Rizvi R, Mahmood I (eds) Management of phytonematodes: recent advances and future challenges, Springer, Singapore, pp 125–140

Fabiyi OA, Claudius-Cole AO, Olatunji GA, Abubakar DO, Adejumo OA (2021) Evaluation of the in vitro response of meloidogyne incognita to silver nano particle liquid from agricultural wastes. Agrivita J Agric Sci 43(3):524–534

Fabiyi OA, Ansari RA, Rizvi R, Sumbul A, Mahmood, I (2024a) Influence of arbuscular mycorrhizal fungi on soil health amelioration and plant fitness under hostile environment. In: Ansari RA, Rizvi R, Mahmood I (eds) Mycorrhizal symbiosis and agroecosystem restoration. Springer, Singapore, pp 227–248

Fabiyi OA, Lateef A, Gueguim-Kana EB, Beukes LS, Matyumza N, Bello TT, Olatunji GA (2024b) Characterization and nematicidal potential of copper, iron and zinc nanoparticles synthesized from Tridax procumbens L. extract on Meloidogyne incognita infected cabbage plants. Eur J Plant Pathol 168(4):683–695. https://doi.org/10.1007/s10658-023-02792-y

Fabiyi OA, Wawata IG, Akinsipo OB, Ogundele AV (2025) Smart nanosensors: a new approach in crop protection and pest management. In: Uddin I (eds) Smart nanosensors. Springer, Singapore, pp 247–266

FAO (2018) The state of food security and nutrition in the World. Building climate resilience for food security and nutrition. A Global Report on Food Crises

FAO (Food and Agricultural Organization) (2000) Global network on integrated soil management for sustainable use of salt affected soils. Available from: https://www.fao.org/agl. Accessed on 13/03/2016

Farooq MA, Hannan F, Islam F, Ayyaz A, Zhang N, Chen W et al (2022) The potential of nanomaterials for sustainable modern agriculture: present findings and future perspectives. Environ Sci Nano 9(6):1926–1951

Ghidan AY, Al Antary T (2020) Applications of nanotechnology in agriculture. IntechOpen. https://doi.org/10.5772/intechopen.88390

Godfray HCJ, Beddington JR, Crute IR, Haddad L, Lawrence D, Muir JF et al (2010) Food security: the challenge of feeding 9 billion people. Science 327:812–818. https://doi.org/10.1126/science.1185383

Hasanuzzaman M, Fujita M (2022) Plant responses and tolerance to salt stress: physiological and molecular interventions. Int J Mol Sci 23(9):4810

Hoang TML, Moghaddam L, Williams B, Khanna H, Dale J, Mundree SG (2015) Development of salinity tolerance in rice by constitutive-overexpression of genes involved in the regulation of programmed cell death. Front Plant Sci 6. https://doi.org/10.3389/fpls.2015.00175

Hussain A, Rizwan M, Ali Q, Ali S, Seed M, Rehman MZ, Irshad MK, Azam A, Ok YS (2018) Stabilization of chromium-contaminated soils using biochar, zeolite, and rock phosphate: effects on plant growth, chromium uptake, and soil microbial activity. Int J Phytoremediation 20(7):690–697

Isayenkov SV, Maathuis FJ (2019) Plant salinity stress: many unanswered questions remain. Front Plant Sci 10:435515

Jagadesh M, Dash M, Kumari A, Singh SK, Verma KK, Kumar P, Bhatt R, Sharma SK (2024) Revealing the hidden world of soil microbes: metagenomic insights into plant, bacteria, and fungi interactions for sustainable agriculture and ecosystem restoration. Microbiol Res 285:127764. https://doi.org/10.1016/j.micres.2024.127764

Junedi MA, Mukhopadhyay R, Manjari KS (2023) Alleviating salinity stress in crop plants using new engineered nanoparticles (ENPs). Plant Stress 100184

Kaveh H, Nemati H, Farsi M, Jartoodeh SV (2011) How salinity affect germination and emergence of tomato lines. J Biol Environ Sci 5:159–163

Kenneth K, Neeltje C (2002) Agricultural drainage water management in arid and semi-arid areas. FAO Irrigation and Drainage Paper, Rome, p 61

Khalid MF, Iqbal Khan R, Jawaid MZ, Shafqat W, Hussain S, Ahmed T et al (2022) Nanoparticles: the plant saviour under abiotic stresses. Nanomaterials 12(21):3915

Khan AA, McNeilly T, Azhar F (2011) Stress tolerance in crop plants. Review. Int J Agric Biol 3:250–255. [Google Scholar]

Khan MN, Mobin M, Abbas ZK, AlMutairi KA, Siddiqui ZH (2017) Role of nanomaterials in plants under challenging environments. Plant Physiol Biochem 110:194–209

Khan I, Raza MA, Awan SA, Shah GA, Rizwan M, Ali B, Tariq R, Hassan MJ, Alyemeni MN, Brestic M et al (2020) Amelioration of salt induced toxicity in pearl millet by seed priming with silver nanoparticles (AgNPs): the oxidative damage, antioxidant enzymes and ions uptake are major determinants of salt tolerant capacity. Plant Physiol Biochem 156:221–232

Khodarahmpour Z, Ifar M, Motamedi M (2012) Effects of NaCl salinity on maize (Zea mays L.) at germination and early seedling stage. Afr J Biotechnol 11:298–304

Kostaki KI, Coupel-Ledru A, Bonnell VC et al (2020) Guard cells integrate light and temperature signals to control stomatal aperture. Plant Physiol 182:1404–1419

Kumar P, Sharma PK (2020) Soil salinity and food security in India. Front Sustain Food Syst 4:533781L

Kumar A, Singh S, Gaurav AK, Srivastava S, Verma JP (2020) Plant growth-promoting bacteria: biological tools for the mitigation of salinity stress in plants. Front Microbiol 11:1216

Kumar D, Mariyam S, Gupta KJ, Thiruvengadam M, Ghodake GS, Xing B, Seth CS (2024a) Comparative investigation on chemical and green synthesized titanium dioxide nanoparticles against chromium (VI) stress eliciting differential physiological, biochemical, and cellular attributes in Helianthus annuus L. Sci Total Environ 930:172413

Kumar D, Singh R, Upadhyay SK, Verma KK, Tripathi RM, Liu H, Dhankher OP, Tripathi RD, Sahi SV, Seth CS (2024b) Review on interactions between nanomaterials and phytohormones: novel perspectives and opportunities for mitigating environmental challenges. Plant Sci 340:111964

Kumar D, Biswas JK, Mulla SI, Singh R, Shukla R, Ahanger MA, Shekhawat GS, Verma KK, Siddiqui MW, Seth CS (2024c) Micro and nanoplastics pollution: sources, distribution, uptake in plants, toxicological effects, and innovative remediation strategies for environmental sustainability. Plant Physiol Biochem 213:108795

Kumari A, Lakshmi GA, Krishna GK, Patni B, Prakash S, Bhattacharyya M, Singh SK, Verma KK (2022) Climate change and its impact on crops: a comprehensive investigation for sustainable agriculture. Agronomy 12:3008. https://doi.org/10.3390/agronomy12123008

Laware SL, Raskar S (2014) Effect of titanium dioxide nanoparticles on hydrolytic and antioxidant enzymes during seed germination in onion. Int J Curr Microbiol App Sci 3(7):749–760

Li Z, Zhu L, Zhao F, Li J, Zhang X, Kong X et al (2022) Plant salinity stress response and nano-enabled plant salt tolerance. Front Plant Sci 13:843994

Liu C, Zhou H, Zhou J (2021) The applications of nanotechnology in crop production. Molecules 26(23):7070. https://doi.org/10.3390/molecules26237070

Lobell DB, Schlenker W, Costa-Roberts J (2013) Climate trends and global crop production since 1980. Science 341(6145):1155–1158

Mandal AK (2019) Modern technologies for diagnosis and prognosis of salt-affected soils and poor-quality waters. In: Dagar JC et al (eds) Research developments in saline agriculture. ICAR-Central Soil Salinity Research Institute, Karnal. https://doi.org/10.1007/978-981-13-5832-6_4

Mariyam S, Upadhyay SK, Chakraborty K, Verma KK, Duhan JS, Muneer S, Meena M, Sharma RK, Ghodake G, Seth CS (2024) Nanotechnology, a frontier in agricultural science, a novel approach in abiotic stress management and convergence with new age medicine-A review. Sci Total Environ 912:169097

Miah MY, Kamal MZU, Salam MA, Islam MS (2020) Impact of salinity intrusion on agriculture of Southwest Bangladesh. Int J Agric Policy Res

Mishra D, Shekhar S, Chakraborty S, Chakraborty N (2021) Wheat 2-Cys peroxiredoxin plays a dual role in chlorophyll biosynthesis and adaptation to high temperature. Plant J 105:1374–1389

Mishra N, Jiang C, Chen L, Paul A, Chatterjee A, Shen G (2023) Achieving abiotic stress tolerance in plants through antioxidative defense mechanisms. Front Plant Sci 14:1110622

Munns R, Passioura JB, Colmer TD, Byrt CS (2020) Osmotic adjustment and energy limitations to plant growth in saline soil. New Phytol 225:1091–1096. https://doi.org/10.1111/nph.15862

Mushtaq A, Jamil N, Rizwan S, Mandokhel F, Riaz M, Hornyak GL, Malghani MN, Shahwani MN (2018) Engineered silica nanoparticles and silica nanoparticles containing controlled release fertilizer for drought and saline areas. Iop Conf Ser Mater Sci Eng 414:012029. https://doi.org/10.1088/1757-899X/414/1/012029

Omar M, Moussa A (2016) Water management in Egypt for facing the future challenges. J Adv Res 7:403–412

Pei ZF, Ming DF, Liu D, Wan GL, Geng XX, Gong HJ, Zhou WJ (2010) Silicon improves the tolerance to water-deficit stress induced by polyethylene glycol in wheat (Triticum aestivum L.) seedlings. J Plant Growth Regul 29:106–115

Prasad PVV, Staggenborg SA, Ristic Z (2008) Impacts of drought and/or heat stress on physiological, developmental, growth, and yield processes of crop plants. In: Response of crops to limited water. Springer, Dordrecht, pp 301–355

Qadir M, Oster JD, Schubert S, Noble AD, Sahrawat KL (2007) Phytoremediation of sodic and saline-sodic soils. Adv Agron 96:197–247. https://doi.org/10.1016/s0065-2113(07)96006-x

Rahman A, Nahar K, Hasanuzzaman M, Fujita M (2016) Calcium supplementation improves Na+/K+ ratio, antioxidant defense and glyoxalase systems in salt-stressed rice seedlings. Front Plant Sci 7. https://doi.org/10.3389/fpls.2016.00609

Rajput VD, Minkina T, Kumari A, Harish, Singh VK, Verma KK, Mandzhieva S, Sushkova S, Srivastava S, Keswani C (2021a) Coping with the challenges of abiotic stress in plants: new dimensions in the field application of nanoparticles. Plan Theory 10(6):1221. https://doi.org/10.3390/plants10061221

Rajput VD, Singh A, Minkina TM, Shende SS, Kumar P, Verma KK, Bauer T, Gorobtsova O, Deneva S, Sindireva A (2021b) Potential applications of nanobiotechnology in plant nutrition and protection for sustainable agriculture. In: Ingle AP (ed) Nanotechnology in plant growth promotion and protection: recent advances and impacts, Ist edn. Wiley

Rehman A, Khan S, Sun F, Peng Z, Feng K, Wang N et al (2024) Exploring the nano-wonders: unveiling the role of nanoparticles in enhancing salinity and drought tolerance in plants. Front Plant Sci 14:1324176

Rico CM, Peralta-Videa JR, Gardea-Torresdey JL (2015) Chemistry, biochemistry of nanoparticles, and their role in antioxidant defense system in plants. Nanotechnology and plant sciences: nanoparticles and their impact on plants:1–17

Rossi L, Zhang W, Lombardini L, Ma X (2016) The impact of cerium oxide nanoparticles on the salt stress responses of Brassica napus L. Environ Pollut 219:28–36

Roychoudhury A (2020) Silicon-nanoparticles in crop improvement and agriculture. Int J Recent Adv Biotechnol Nanotechnol 3(1):2582–1571

Safdar H, Amin A, Shafiq Y, Ali A, Yasin R, Shoukat A, Hussan MU, Sarwar MI (2019) A review: impact of salinity on plant growth. J Nat Sci 17(1):34–40

Seleiman MF, Al-Suhaibani N, Ali N, Akmal M, Alotaibi M, Refay Y, Dindaroglu T, Abdul-Wajid HH, Battaglia ML (2021) Drought stress impacts on plants and different approaches to alleviate its adverse effects. Plants (Basel) 10(2):259. https://doi.org/10.3390/plants10020259. PMID: 33525688; PMCID: PMC7911879

Sharma VK, Sayes CM, Guo B, Pillai S, Parsons JG, Wang C et al (2019) Interactions between silver nanoparticles and other metal nanoparticles under environmentally relevant conditions: a review. Sci Total Environ 653:1042–1051

Singh P, Kumar V, Sharma J, Saini S, Sharma P, Kumar S et al (2022) Silicon supplementation alleviates the salinity stress in wheat plants by enhancing the plant water status, photosynthetic pigments, proline content and antioxidant enzyme activities. Plan Theory 11(19):2525

Soliman AS, El-Feky SA, Darwish E (2015) Alleviation of salt stress on Moringa peregrina using foliar application of nano fertilizers. Journal of Horticulture and Forestry 7(2):36–47

Squires V, Glenn E (2004) Salinization, desertification, and soil erosion: the role of food, agriculture, forestry and fisheries in human nutrition. Oxford

Staron A, Dlugosz O, Pulit-Prociak J, Banach M (2020) Analysis of the exposure of organisms to the action of nanomaterials. Materials (Basel) 13(2):349. https://doi.org/10.3390/ma13020349

Stavi I, Thevs N, Priori S (2021) Soil salinity and sodicity in drylands: a review of causes, effects, monitoring, and restoration measures. Front Environ Sci 9:712831

Suriyaprabha R, Karunakaran G, Kavitha K, Yuvakkumar R, Rajendran V, Kannan N (2014) Application of silica nanoparticles in maize to enhance fungal resistance. IET Nanobiotechnol 8(3):133–137

Sverdrup H, Warfvinge P (1988) Weathering of primary silicate minerals in the natural soil environment in relation to a chemical weathering model. Water Air Soil Pollut 38:387–408

Syed A, Sarwar G, Shah SH, Muhammad S (2021) Soil salinity research in 21st century in Pakistan: its impact on availability of plant nutrients, growth and yield of crops. Commun Soil Sci Plant Anal 52(3):183–200

Szabolcs I (1993) Soil salinity and sodicity as particular plant/crop stress factors. In: Handbook of plant and crop stress, 2nd edn. CRC Press, Boca Raton. [Google Scholar]

Tanveer K, Gilani S, Hussain Z, Ishaq R, Adeel M, Ilyas N (2020) Effect of salt stress on tomato plant and the role of calcium. J Plant Nutr 43(1):28–35

United Nations, Department of Economic and Social Affairs, Population Division (2015) World population prospects: the 2015 revision. United Nations, New York. Available from: https://en.wikipedia.org/wiki/Projections_of_population_growth#cite_note-UN-WPP-2015-4

Van den Burg S, Deolu-Ajayi AO, Nauta R, Cervi WR, van der Werf A, Poelman M, Wilbers GJ, Snethlage J, Alphen MV, van der Meer IM (2024) Knowledge gaps on how to adapt crop production under changing saline circumstances in The Netherlands. Sci Total Environ 915:170118

Verma KK, Singh P, Song X-P, Malviya MK, Singh RK, Chen G-L, Solomon S, Li YR (2020) Mitigating climate change for sugarcane improvement: role of silicon in alleviating abiotic stresses. Sugar Tech 22(5):741–749. https://doi.org/10.1007/s12355-020-00831-0

Verma KK, Song XP, Joshi A, Tian DD, Rajput VD, Singh M et al (2022) Recent trends in nano-fertilizers for sustainable agriculture under climate change for global food security. Nanomaterials 12(1):173

Verma KK, Song XP, Kumari A, Jagadesh M, Singh SK, Bhatt R, Singh M, Seth CS, Li YR (2024) Climate change adaptation: challenges for agricultural sustainability. Plant Cell Environ 2024:1–12. https://doi.org/10.1111/pce.15078

Wahid A, Gelani S, Ashraf M, Foolad MR (2007) Heat tolerance in plants: an overview. Environ Exp Bot 61(3):199–223

Wang M, Xu Q, Yuan M (2011) The unfolded protein response induced by salt stress in arabidopsis. The Unfolded Protein Response and Cellular Stress, Part A, 319–328

Wang C, He Z, Liu Y, Zhou C, Jiao J, Li P et al (2020) Chitosan-modified halloysite nanotubes as a controlled-release nanocarrier for nitrogen delivery. Appl Clay Sci 198:105802

Weiss D, Ori N (2007) Mechanisms of cross talk between gibberellin and other hormones. Plant Physiol 144(3):1240–1246. https://doi.org/10.1104/pp.107.100370

Wichelns D, Qadir M (2014) Achieving sustainable irrigation requires executive management of salts, soil salinity, and shallow groundwater. Agric Water Manag 157:31–38. https://doi.org/10.1016/j.agwat.2014.08.016

Xin X, Judy JD, Sumerlin BB, He Z (2020) Nano-enabled agriculture: from nanoparticles to smart nanodelivery systems. Environ Chem 17(6):413–425

Yadav S, Modi P, Dave A, Vijapura A, Patel D, Patel M (2020) Effect of abiotic stress on crops. Sustainable Crop Production 3(17):5–16

Yang J, Cao W, Rui Y (2017) Interactions between nanoparticles and plants: phytotoxicity and defense mechanisms. J Plant Interact 12(1):158–169

Zaki R, Wali A, Mosa M (2011) Sedimentological and hydrochemical spectrum of recent continental sabkha and signs of its capabilities to generate hydrocarbons: a case study in northwest El Fashn area, Western Desert. Egypt Carbonates Evaporites 26:273–286. https://doi.org/10.1007/s13146-011-0062-5

Zhao J, Lu Z, Wang L, Jin B (2020) Plant responses to heat stress: physiology, transcription, noncoding RNAs, and epigenetics. Int J Mol Sci 22(1):117

Zheng Y, Cao X, Zhou Y, Li Z, Yang Y, Zhao D et al (2023) Effect of planting salt-tolerant legumes on coastal saline soil nutrient availability and microbial communities. J Environ Manag 345:118574

Zhu T, Fonseca De Lima CF, De Smet I (2021) The heat is on: how crop growth, development, and yield respond to high temperature. J Exp Bot 72(21):7359–7373

Zorb C, Geilfus CM, Dietz KJ (2018) Salinity and crop yield. Plant Biol 21:31–38. https://doi.org/10.1111/plb.12884

Chapter 5
Engineered Nanomaterials and Their Potential Involvement in Relieving Cold and Heat Stress in Plants

Ayorinde Victor Ogundele and Oluwatoyin Adenike Fabiyi

5.1 Introduction

The population of the world is expected to rise continuously, reaching 9.1 billion by 2050, but agricultural output is not keeping up with this growth. Increasing productivity is difficult, especially when it is anticipated that the amount of land under cultivation would either stay the same or decline as demand for non-agricultural uses grows (Verma et al. 2024a). While more funding and technical developments could make food more readily available, better access and use of food resources might not always follow. Furthermore, the impending danger of climate change highlights the importance of heat and cold stressors in maintaining agricultural output. These pressures are thought to have a direct or indirect impact on 70% of the decrease in agricultural output (Ahluwalia et al. 2021; Mariyam et al. 2024). Plants experience a variety of environmental circumstances over their life cycle, and in order to survive and guarantee successful reproduction, they must be able to adapt to these situations. Extreme temperature changes are occurring more frequently as a result of the recent increase in climatic variability (Yanagi 2024; Choudhary et al. 2024). Plant development and production are negatively impacted by a variety of morphological, physiological, biochemical, and molecular alterations brought on by cold and heat stress (Fig. 5.1).

Producing plants that are resistant to these kinds of stresses can be a useful strategy for tackling the problem of the world's food production. It has proven difficult for conventional breeding techniques to increase agricultural plants' stress tolerance

A. V. Ogundele
Laboratory of Natural Products & Drug Discovery, Centre CEBIM, Department of Basic Sciences, Faculty of Medicine, Universidad de La Frontera, Temuco, Chile

O. A. Fabiyi (✉)
Department of Crop Protection, Faculty of Agriculture, University of Ilorin, Ilorin, Nigeria

C. S. Seth et al. (eds.), *Nano-formulations for Sustainable Agriculture*,
Nanotechnology in the Life Sciences, https://doi.org/10.1007/978-3-031-90617-6_5

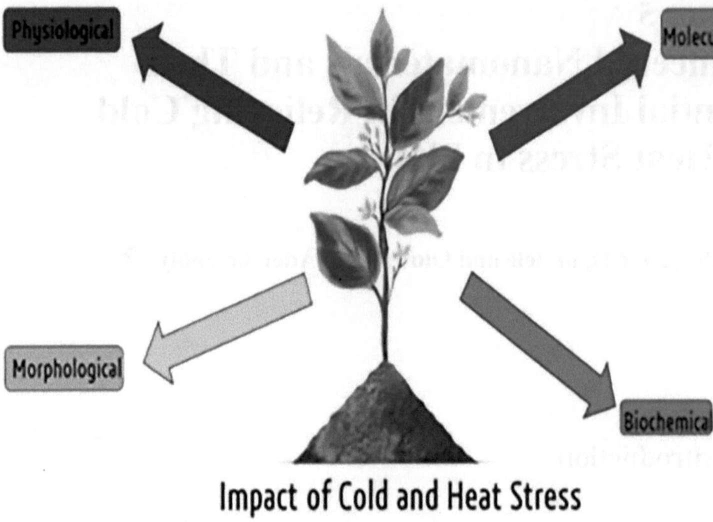

Fig. 5.1 Influence of cold and heat stress on plant morphological, physio-biochemical and molecular responses

through interspecific or intergeneric hybridization. The complex nature of stress tolerance characteristics, the low genetic diversity of yield components under stress, and the lack of suitable selection criteria impede these traditional methods. Thus, it is essential to investigate different approaches for growing crops that can withstand stress. A range of conventional breeding techniques, including polyploidy, hybridization, selection, and mutation, have been utilized to genetically modify agricultural plants. Crop development has had some accomplishments in the past, but current yields have plateaued, which has led to food insecurity and poverty in many developing countries. Therefore, it is imperative to investigate new approaches and incorporate them with current conventional and cutting-edge breeding instruments.

The worldwide requirement in the current setting is to increase food production with limited resources while reducing the use of pesticides and fertilizers to reduce environmental degradation (Fabiyi, 2023; Fabiyi et al., 2025). New agricultural technology aiming at transforming contemporary farming techniques have emerged in response to this desire (Verma et al. 2023a Fabiyi, 2025). According to Sheikh et al. (2024), one of the most promising developments in plant biotechnology and agriculture is the use of engineered nanomaterials (NMs). The use of nanoparticles (NPs) or nanodevices can have a range of effects, both beneficial and detrimental, on plant growth at different developmental stages. The special qualities of NMs are embraced by nanotechnology, which makes agricultural research easier for crop enhancement initiatives and provides ways to lessen the effects of heat and cold stress (Das et al. 2024; Kumar et al. 2024a; Fabiyi and Ogundele, 2025). When compared to the same quantity of material in bigger sizes, these materials have a comparatively greater surface area. Materials' strength or electrical characteristics can be affected by NPs, which can also increase a material's chemical reactivity.

These particles show increased reactivity and possible metabolic activity because of their high surface-to-volume ratio (Wang et al. 2024).

5.2 Applications of Engineered Nanomaterials in Plant Stress Mitigation

One peculiar effect of oxidative stress brought on by abiotic stimuli is the excessive production of reactive oxygen species (ROS) by many cellular constituents (Choudhary et al. 2025). In addition to causing harm, reactive oxygen species (ROS) also act as initiators of defense mechanisms by triggering signalling pathways inside cells and controlling the expression of several genes (Hancock et al. 2001; Verma et al. 2024b). Plants are subject to adverse effects from abiotic stresses, especially heat and cold stress; even though they have enzymatic and non-enzymatic antioxidant systems that constantly neutralize harmful ROS. This is because plants' cellular defense mechanisms cannot keep up with the destructive potential of these stresses (Kumar et al. 2024b). Therefore, in order to maintain their survival under harsh climatic conditions, plants must react to stress signals swiftly and properly before the initiation of stress-induced damage. By boosting the activities of antioxidant enzymes and encouraging the build-up of osmolytes, free amino acids, and nutrients. NMs significantly contribute to the defense of plants against cold and heat shocks (Al-Khayri et al. 2023) (Fig. 5.2).

The understanding of the interaction between NMs and plants in situations of cold and heat stress will be the main emphasis of the next sections, given the evidence that is currently available. The atmospheric build-up of greenhouse gases is responsible for the increase in ambient temperature. The average global temperature has increased steadily as a result of this temperature rise, and estimates suggest that it will climb by a further 2 °C by the year 2100 (Malhi et al. 2021). As a result, there

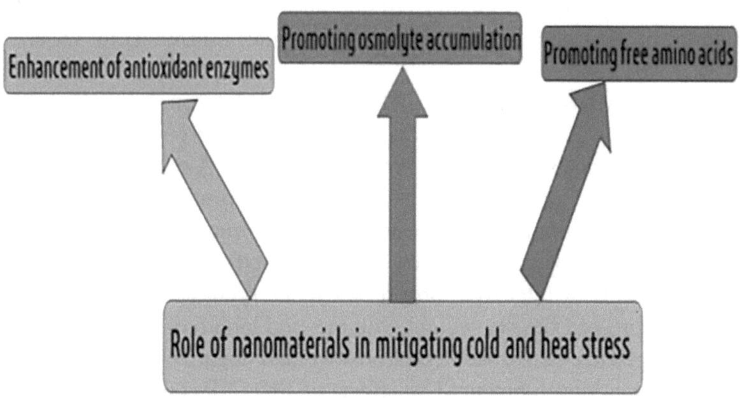

Fig. 5.2 Strategies of nanoformulated NPs in mitigating cold and heat stress

have been notable reductions in world agricultural productivity. Crops are affected differently by the main greenhouse gases (Mukhtar et al. 2020). Although increasing carbon dioxide (CO_2) is linked to improved photosynthesis, which boosts plant growth and production, higher temperatures counteract this effect by making crop respiration and evapotranspiration more intense. Thus, the length of crop development cycles is shortened, insect infestation and weed population expansion are stimulated, and the microbial composition and enzymatic activities in the soil are negatively impacted (Battaglia et al. 2022). In order to survive heat stress, plants change both structurally and biochemically. This causes oxidative stress due to the overproduction of reactive oxygen species (ROS), which has a substantial effect on the growth, development, and yield of the plant (Hasanuzzaman et al. 2013).

Plants respond to temperature stressors by initiating signaling pathways that lead to the synthesis of osmolytes. Osmotic control helps maintain cell turgidity, while other secondary metabolites affect the antioxidant defence system (Arif et al. 2020). Elevated temperatures negatively impact photosynthetic activities by hindering the production of ATP, the oxygen-evolving complex, PSII cofactors, and carbon absorption (Allakhverdiev et al. 2008). On the other hand, freezing damage and decreased fluidity are experienced by cellular membranes under low-temperature stress (Los and Murata 2004). By improving photophosphorylation, oxygen evolution, water splitting via the CP43 protein, nitrogen metabolism, photosynthetic efficiency, activity of antioxidant enzymes, lowering lipid peroxidation, and restoring the structural integrity of chloroplasts and the nucleus, NMs have notably shown beneficial effects on plants under temperature stress (Younis et al. 2020; Verma et al. 2023b).

Depending on how much cold stress they can withstand different plant species can respond and adapt differently. By lessening ion leakage caused by membrane degradation, titanium dioxide nanoparticles (TiO_2 NMs) show promise in mitigating the negative effects of extreme low-temperature stress (Mohammadi et al. 2013). Photosynthesis, an essential plant activity, is seriously threatened by cold stress. Under cold stress, plant photosystem face a variety of difficulties, such as lower amounts of chlorophyll, slowed transpiration rates, difficulty assimilating CO_2, and degradation of the photosynthesis-critical enzyme rubisco (Liu et al. 2012) (Fig. 5.3).

The effects of Mo (<100 nm) and SiO_2 (10–20 nm) NPs on rice seed germination were investigated by Adhikari and Sivakoti (2020). They discovered that when these NPs were present, rice seedlings germinated effectively. A growth of 101.4 mg L^{-1} of Mo and 106 mg L^{-1} of SiO_2 NPs was shown to be optimal. Mo NPs impeded root elongation and development at concentrations reaching 50 mg L^{-1}, sometimes resulting in root necrosis, while SiO_2 NPs exhibited no harmful effects on rice growth. Both kinds of NPs were absorbed by rice seedlings; however, SiO_2 NPs significantly increased the dry weight, volume, and length of the roots and shoots. This study emphasizes the dual effects-positive and negative—that NMs exposure may have on plant development. The effects of SiO_2 NPs on tall wheatgrass seed germination and seedling growth were investigated by Azimi et al. (2014). Three treatments were used to evaluate various NPs concentrations (580 ppm) as control,

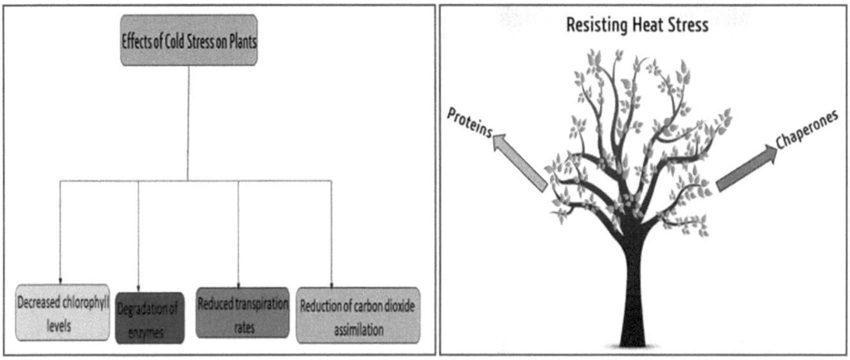

Fig. 5.3 Role of nanoparticles on stressed plants in response to adverse climatic condition

pre-chilling the seed before applying the NPs and NPs application prior to pre-chilling. The outcomes demonstrated that SiO_2 NPs, particularly at concentrations of 40 and 60 ppm, greatly increased seed germination rates. While high quantities of NPs decreased seedling weight, they also improved seedling development. SiO_2 NPs and seed pre-chilling together decreased tall wheatgrass seed dormancy overall (Azimi et al. 2014). Within the plant photosystem, NMs increase the amount of the enzyme Rubisco (Jalil and Ansari 2019), increase the ability of chloroplasts to absorb light (Ze et al. 2011), and prevent the generation of reactive oxygen species (ROS) (Giraldo et al. 2014; Kumar et al. 2024c).

TiO_2 NMs have been shown to boost the activity of antioxidant enzymes and the expression of genes encoding chlorophyll-binding proteins and Rubisco (Mohammadi et al. 2014; Hasanpour et al. 2015; Verma et al. 2022a). Reduced levels of dehydroascorbate reductase, glutathione reductase, and monodehydroascorbate reductase activities are seen in plants under low-temperature stress, and these activities aid in scavenging reactive oxygen species. Furthermore, there is an upregulation of the *MeCu/ZnSOD* and *MeAPX2* genes, which results in a decrease in oxidative stress, which includes lipid peroxidation, H_2O_2 production, and chlorophyll degradation (Xu et al. 2014). This leads to an increase in stress tolerance. It's interesting to note that plants exposed to NMs in the presence of exceptionally low temperatures have shown enhanced biochemical traits and development patterns under particular stress conditions (Kohan-Baghkheirati and Geisler-Lee 2015). Wu et al. (2017) discovered that in *Arabidopsis thaliana* L., CeO_2 NPs (nanoceria) improve photosynthesis and ROS scavenging subjected to extremes of temperature, light, and darkness. Specifically, low-ratio Ce^{3+}/Ce^{4+} nanoceria dramatically lower leaf ROS concentrations. Under cold stress, these NPs boosted PS II quantum yield, carbon absorption, and rubisco carboxylation after entering chloroplasts through non-endocytic routes. Nevertheless, during cold stress, photosynthesis was not protected by nanoceria with a high Ce^{3+}/Ce^{4+} ratio; instead, they elevated ROS levels (Wu et al. 2017).

Plants that are exposed to high temperatures for extended periods of time are at risk of heat stress, which can cause irreparable harm to their growth and development. High temperatures have the potential to cause oxidative stress and reactive oxygen species (ROS), which can lead to protein degradation, membrane ion leakage, lipid membrane degradation, decreased amounts of chlorophyll, and decreased rates of photosynthesis (Prasad et al. 2011; Verma et al. 2022b). Selenium NPs, when applied at low concentrations, improve plant growth overall, chlorophyll content, and hydration capacity, hence mitigating the effects of heat stress (Haghighi et al. 2014). At low concentrations, Se NMs have antioxidative qualities; but, at greater concentrations, they can cause oxidative stress (Hasanuzzaman et al. 2013). Plants produce heat shock proteins and molecular chaperones in response to heat stress. These proteins are essential for stress resistance because they keep other proteins stable (Fig. 5.3).

Research has demonstrated that multiwall carbon nanotubes (CNTs) enhance the production of heat shock proteins such as *HSP90*, hence promoting resistance to heat stress (Khodakovskaya et al. 2011). In a similar vein, maize exposed to cerium oxide nanoparticles (CeO_2 NMs) produces an excessive amount of H_2O_2 particles, which in turn causes *HSP70* to be upregulated (Zhao et al. 2012). Furthermore, via controlling stomatal opening mechanisms, the application of titanium dioxide nanoparticles (TiO_2 NMs) to plants mitigates the impacts of heat stress (Qi et al. 2013). More focus is being paid to the role that bulk and Se-NPs play in lowering heat stress by altering stress-responsive pathways. Deciphering the genetic and molecular pathways behind thermotolerance in crops requires an understanding of the genes activated by Se and their function in heat stress tolerant. It has been discovered that exogenous Se-NPs increase the expression of the transcription factor dehydration-responsive element-binding (*DREB1A*), which starts downstream signaling pathways and ensuing mechanisms for stress tolerance (Iqbal et al. 2022).

Heat stress increased ascorbate peroxidase activity in the Giza-171 wheat genotype as compared to control circumstances. In addition, compared to control treatments, applied pressures produced higher catalase (CAT) activity. It's interesting to note that, in comparison to the control, the Giza-171 wheat genotype showed the greatest CAT activity under heat stress. But when compared to the control, wheat plants exposed to heat stress and given 10 ppm Se-NPs did not exhibit a statistically significant change in CAT activity (Omar et al. 2023). The application of selenium nanoparticles to mitigate heat stress in sorghum plants was studied by Djanaguiraman et al. (2018). They observed increased antioxidant enzyme activities and decreased oxidant levels by spraying these NPs at 10 ppm during booting, which eventually boosted pollen germination and seed output. Higher selenium nanoparticle concentrations, however, may cause oxidative stress, therefore care must be used (Djanaguiraman et al. 2018). Furthermore, Se NPs enhanced pollen germination rates in heat-stressed environments, leading to a notable rise in seed yield. Selenium NPs have been found to act as antioxidants in plant systems at lower concentrations, but to cause oxidative stress at higher concentrations (Hartikainen et al. 2000).

Iqbal et al. (2019) examined how wheat crop development was impacted by heat stress using silver NPs made from extract from *Moringa oleifera* L. During the trifoliate stage, wheat plants were treated with a range of doses (25–100 ppm). Growth parameters were lowered after 3 days of heat stress (35–40 °C for 3 hours each day). In contrast to controls, the application of silver NPs, especially at 50 and 75 ppm, reduced the impacts of heat stress and improved plant growth and development (Iqbal et al. 2019). In Egypt, a country with hot environment, Hassan et al. (2018) studied the cultivation of two wheat varieties: Sids1, which is heat-tolerant, and Gimmeza7, which is heat-sensitive. To evaluate their effect on heat stress, they applied iron oxide (50 nm) and zinc oxide (80 nm) NPs at different quantities. The Gimmeza7 cultivar had the highest grain output when treated with 1.40 mg L^{-1} iron oxide and 52.50 ppm zinc oxide NPs, resulting in the best wheat survival under heat stress. Hassan et al. (2018) linked this impact to elevated antioxidant enzyme activity.

5.3 Challenges and Opportunities

Creative solutions are required to address the issues of falling agricultural output in the face of an increasing global population and climate change. In agricultural biotechnology, engineered NMs have become more promising instruments, providing special chances to alleviate heat and cold stress in plants. Their use is not without difficulties, though. Enhancing crop resilience by conventional breeding approaches is hampered by the complexity of stress tolerance characteristics. It takes substantial genetic expertise and resources to recognize these features and incorporate them into breeding programs. Moreover, standard methods of enhancing stress tolerance are restricted in their efficiency by limited genetic variety within crop species. It is impossible to ignore environmental issues raised by the extensive use of nanotechnology in agriculture. Thorough risk assessment and mitigation strategies are required due to the possible consequences of NMs buildup in soil and water ecosystems, as well as effects on non-target animals. The many regulatory frameworks that oversee nanotechnology provide further challenges to its extensive adoption, since there are ambiguities pertaining to safety evaluations, mandatory labeling, and disclosure guidelines.

The use of nanotechnology in agriculture has social and economic ramifications as well, especially in underdeveloped nations. Cost and infrastructural limitations may prevent certain people from accessing nanotechnology-enabled solutions, hence widening the socioeconomic divide already present. Engineered NPs provide a plethora of potential for plant stress control despite these obstacles. Nanomaterials can give focused solutions for crop stress management while reducing environmental effect, making precision agriculture an attractive area to pursue. Through targeted genetic alterations for stress tolerance features, genetic engineering—aided by nanotechnology—fast-tracks the creation of robust crop cultivars. Utilizing NMs-based formulations for nutrient delivery systems and pest and disease control can also lead to sustainable agricultural practices. Nanotechnology reduces pollution and increases environmental sustainability by lowering dependency on agrochemical inputs. Furthermore, the use of synthetic NMs has the potential to increase agricultural resistance to pressures brought on by climate change. Nanomaterials improve overall stress tolerance and help agricultural systems become more resilient to climate change by modifying the physiological responses of plants to temperature extremes. Nanotechnology is crucial to the effort to provide food security for all people on the planet. Food security for a growing population in the face of climate uncertainty is addressed by nanotechnology by boosting agricultural yields and resilience to environmental shocks.

5.4 Conclusions

Precision agriculture is another appealing option because NMs will deliver targeted solutions for crop stress management while reducing environmental impact to a minimum. Genetic engineering through nanotechnology speeds up the development of resilient crop varieties. The implications of declining agricultural productivity and looming threats of climate change call for innovative solutions, and NMs now show promise as tools in agricultural biotechnology because they offer new ways of mitigating cold and heat stresses in plants. Even though there are many challenges that need to be faced, from complex stress tolerance traits to limited genetic variability, environmental concerns, regulatory uncertainties, and socio-economic implications, the potentials for NMs in plant stress management are tremendous. NMs-based fertilizer delivery systems, pest and disease management formulations make sustainable agriculture techniques more feasible. Through the reduction of agrochemical input reliance, nanotechnology fosters environmental sustainability and reduces pollution. Additionally, NMs can be engineered to increase agricultural resilience to pressures brought on by climate change. By improving overall stress tolerance and regulating a plant's physiological reactions to temperature extremes, NMs help agricultural systems become more resilient to climate change. Nanotechnology is a key component in the effort to achieve global food security. Nanotechnology increases crop yields and tolerance to environmental shocks, addressing the challenge of feeding a growing population in the face of climate unpredictability. Therefore, it's critical to carry out further research and development in the field of nanotechnology-enabled agriculture in order to create food production systems that are robust and sustainable in the face of constantly changing environmental difficulties.

References

Adhikari, T. and Sivakoti R. (2020). Nano fertilizer: its impact on crop growth and soil health. The Journal of Research, PJTSAU, 47(3). https://epubs.icar.org.in/index.php/TJRP/article/view/97462

Ahluwalia O, Singh PC, Bhatia R (2021) A review on drought stress in plants: implications, mitigation and the role of plant growth promoting rhizobacteria. Resour Environ Sustain 5:100032. https://doi.org/10.1016/j.resenv.2021.100032

Al-Khayri JM, Rashmi R, Surya Ulhas R, Sudheer WN, Banadka A, Nagella P et al (2023) The role of nanoparticles in response of plants to abiotic stress at physiological, biochemical, and molecular levels. Plan Theory 12. https://doi.org/10.3390/plants12020292

Allakhverdiev SI, Kreslavski VD, Klimov VV, Los DA, Carpentier R, Mohanty P (2008) Heat stress: an overview of molecular responses in photosynthesis. Photosynth Res 98(1):541–550. https://doi.org/10.1007/s11120-008-9331-0

Arif Y, Singh P, Siddiqui H, Bajguz A, Hayat S (2020) Salinity induced physiological and biochemical changes in plants: an omic approach towards salt stress tolerance. Plant Physiol Biochem 156:64–77. https://doi.org/10.1016/j.plaphy.2020.08.042

Azimi R, Borzelabad MJ, Feizi H, Azimi A (2014) Interaction of SiO nanoparticles with seed prechilling on germination and early seedling growth of tall wheatgrass (Agropyron Elongatum L.). Pol J Chem Technol 16(3):25–29. https://doi.org/10.2478/pjct-2014-0045

Battaglia ML, Thomason WE, Fike JH, Evanylo GK, Stewart RD, Gross CD et al (2022) Corn and wheat residue management effects on greenhouse gas emissions in the mid-atlantic USA. Land 11. https://doi.org/10.3390/land11060846

Choudhary R, Rajput VD, Ghodake G, Ahmad F, Meena M, Rehman RU, Prasad R, Sharma RK, Singh R, Seth CS (2024) Comprehensive journey from past to present to future about seed priming with hydrogen peroxide and hydrogen sulfide concerning drought, temperature, UV and ozone stresses- a review. Plant Soil. https://doi.org/10.1007/s11104-024-06499-9

Choudhary R, Ahmad F, Kaya C, Upadhyay SK, Muneer S, Kumar V, Meena M, Liu H, Upadhyay H, Seth CS (2025) Decrypting proteomics, transcriptomics, genomics, and integrated omics for augmenting the abiotic, biotic, and climate change stress resilience in plants. J Plant Physiol. https://doi.org/10.1016/j.jplph.2025.154430

Das JM, Bala A, Upadhyay J, Borah R (2024) Chapter 4 - Role of nanotechnology in crop management. In: Pudake RN, Tripathi RM, Gill SS (eds) Nanotechnology for abiotic stress tolerance and management in crop plants. Academic Press. https://doi.org/10.1016/B978-0-443-18500-7.00004-1

Djanaguiraman M, Belliraj N, Bossmann SH, Prasad PVV (2018) High-temperature stress alleviation by selenium nanoparticle treatment in grain sorghum. ACS Omega 3(3):2479–2491. https://doi.org/10.1021/acsomega.7b01934

Fabiyi OA (2025). Nanomaterials of cellulose origin: novel strategies for plant disease control In Rajan R., Ahmad F, Khoshariya A., Seth C.S.(eds). Sustainable era of nanomaterials, sustainable development and biodiversity vol 40, Springer Singapore. 379–397 Pp

Fabiyi OA and Ogundele AV (2025). Biosensors for monitoring climatic stressors in agricultural crops. In: Uddin I, (eds). Smart nanosensors. Springer, Singapore. 229–245 Pp

Fabiyi OA (2023) The perspective of nanotechnology in plant protection. In :Nanophytopathology. Avila-Quezada, G.D and Rai, M. (eds). 23–38 Pp

Fabiyi OA, Wawata IG, Akinsipo OB and Ogundele AV (2025) Smart nanosensors: A new approach in crop protection and pest management. In: Uddin, I. (eds). Smart nanosensors. Springer, Singerpore. 247–266. Pp

Giraldo JP, Landry MP, Faltermeier SM, McNicholas TP, Iverson NM, Boghossian AA et al (2014) Plant nanobionics approach to augment photosynthesis and biochemical sensing. Nat Mater 13(4):400–408. https://doi.org/10.1038/nmat3890

Haghighi M, Abolghasemi R, Teixeira da Silva JA (2014) Low and high temperature stress affect the growth characteristics of tomato in hydroponic culture with Se and nano-Se amendment. Sci Hortic 178:231–240. https://doi.org/10.1016/j.scienta.2014.09.006

Hancock JT, Desikan R, Neill SJ (2001) Role of reactive oxygen species in cell signalling pathways. Biochem Soc Trans 29(Pt 2):345–350. https://doi.org/10.1042/0300-5127:0290345

Hasanuzzaman, M., Nahar, K., & Fujit, M. (2013). Extreme Temperature Responses, Oxidative Stress and Antioxidant Defense in Plants. InTech. https://doi.org/10.5772/54833

Hartikainen H, Xue T, Piironen V (2000) Selenium as an anti-oxidant and pro-oxidant in ryegrass. Plant Soil 225(1):193–200. https://doi.org/10.1023/A:1026512921026

Hasanpour H, Maali-Amir R, Zeinali H (2015) Effect of TiO2 nanoparticles on metabolic limitations to photosynthesis under cold in chickpea. Russ J Plant Physiol 62(6):779–787. https://doi.org/10.1134/S1021443715060096

Hassan SN, Salah El Din AT, Hendawey HM, Borai HI, Mahdi AA (2018) Magnetite and zinc oxide nanoparticles alleviated heat stress in wheat plants. Curr Nanomater 3:32–43. https://doi.org/10.2174/2405461503666180619160923

Iqbal M, Raja NI, Mashwani Z-U-R, Hussain M, Ejaz M, Yasmeen F (2019) Effect of silver nanoparticles on growth of wheat under heat stress. Iran J Sci Technol Trans A: Sci 43(2):387–395. https://doi.org/10.1007/s40995-017-0417-4

Iqbal M, Shafiq F, Anwar S, Akram NA, Ashraf MA, Raza SH et al (2022) In: Hossain MA, Ahammed GJ, Kolbert Z, El-Ramady H, Islam T, Schiavon M (eds) Selenium and nano-selenium-mediated heat stress tolerance in plants BT - selenium and nano-selenium in environmental stress management and crop quality improvement. Springer. https://doi.org/10.1007/978-3-031-07063-1_8

Jalil SU, Ansari MI (2019) Chapter 35 - Nanoparticles and abiotic stress tolerance in plants: synthesis, action, and signaling mechanisms. In: Khan MIR, Reddy PS, Ferrante A, Khan NABT-PSM (eds) Plant signaling molecules 1st edition role and regulation under stressful environments. Elsevier. https://doi.org/10.1016/B978-0-12-816451-8.00034-4

Khodakovskaya MV, de Silva K, Nedosekin DA, Dervishi E, Biris AS, Shashkov EV et al (2011) Complex genetic, photothermal, and photoacoustic analysis of nanoparticle-plant interactions. Proc Natl Acad Sci 108(3):1028–1033

Kohan-Baghkheirati E, Geisler-Lee J (2015) Gene expression, protein function and pathways of arabidopsis thaliana responding to silver nanoparticles in comparison to silver ions, cold, salt, drought, and heat. Nanomaterials 5:436–467. https://doi.org/10.3390/nano5020436

Kumar D, Mariyam S, Gupta KJ, Thiruvengadam M, Ghodake GS, Xing B, Seth CS (2024a) Comparative investigation on chemical and green synthesized titanium dioxide nanoparticles against chromium (VI) stress eliciting differential physiological, biochemical, and cellular attributes in Helianthus annuus L. Sci Total Environ 930:172413. https://doi.org/10.1016/j.scitotenv.2024.172413

Kumar D, Singh R, Upadhyay SK, Verma KK, Tripathi RM, Liu H, Dhankher OP, Tripathi RD, Sahi SV, Seth CS (2024b) Review on interactions between nanomaterials and phytohormones: novel perspectives and opportunities for mitigating environmental challenges. Plant Sci 340:111964. https://doi.org/10.1016/j.plantsci.2023.111964

Kumar D, Biswas JK, Mulla SI, Singh R, Shukla R, Ahanger MA, Shekhawat GS, Verma KK, Siddiqui MW, Seth CS (2024c) Micro and nanoplastics pollution: sources, distribution, uptake in plants, toxicological effects, and innovative remediation strategies for environmental sustainability. Plant Physiol Biochem 213:108795. https://doi.org/10.1016/j.plaphy.2024.108795

Liu YF, Qi MF, Li TL (2012) Photosynthesis, photoinhibition, and antioxidant system in tomato leaves stressed by low night temperature and their subsequent recovery. Plant Sci 196:8–17. https://doi.org/10.1016/j.plantsci.2012.07.005

Los DA, Murata N (2004) Membrane fluidity and its roles in the perception of environmental signals. Biochim Biophys Acta (BBA) – Biomembranes 1666(1):142–157. https://doi.org/10.1016/j.bbamem.2004.08.002

Malhi GS, Kaur M, Kaushik P (2021) Impact of climate change on agriculture and its mitigation strategies: a review. Sustainability 13. https://doi.org/10.3390/su13031318

Mariyam S, Upadhyay SK, Chakraborty K, Verma KK, Duhan JS, Muneer S, Meena M, Sharma RK, Ghodake G, Seth CS (2024) Nanotechnology, a frontier in agricultural science, a novel approach in abiotic stress management and convergence with new age medicine-a review. Sci Total Environ 912:169097. https://doi.org/10.1016/j.scitotenv.2023.169097

Mohammadi, Rahmat, Maali-Amiri, R., & Abbasi, A. (2013). Effect of TiO2 Nanoparticles on Chickpea Response to Cold Stress. Biol Trace Elem Res 152(3):403–410. https://doi.org/10.1007/s12011-013-9631-x

Mohammadi R, Maali-Amiri R, Mantri NL (2014) Effect of TiO2 nanoparticles on oxidative damage and antioxidant defense systems in chickpea seedlings during cold stress. Russ J Plant Physiol 61(6):768–775. https://doi.org/10.1134/S1021443714050124

Mukhtar T, Rehman SU, Smith D, Sultan T, Seleiman MF, Alsadon AA et al (2020) Mitigation of heat stress in Solanum lycopersicum L. by ACC-deaminase and exopolysaccharide producing bacillus cereus: effects on biochemical profiling. Sustainability 12. https://doi.org/10.3390/su12062159

Omar AA, Heikal YM, Zayed EM, Shamseldin SAM, Salama YE, Amer KE et al (2023) Conferring of drought and heat stress tolerance in wheat (Triticum aestivum L.) genotypes and their response to selenium nanoparticles application. Nanomaterials 13. https://doi.org/10.3390/nano13060998

Prasad PVV, Pisipati SR, Momčilović I, Ristic Z (2011) Independent and combined effects of high temperature and drought stress during grain filling on plant yield and chloroplast EF-Tu expression in spring wheat. J Agron Crop Sci 197(6):430–441. https://doi.org/10.1111/j.1439-037X.2011.00477.x

Qi M, Liu Y, Li T (2013) Nano-TiO2 improve the photosynthesis of tomato leaves under mild heat stress. Biol Trace Elem Res 156(1):323–328. https://doi.org/10.1007/s12011-013-9833-2

Sheikh ZN, Raina S, Bhat UH, Sharma N, Saffeullah P, Mohammad F et al (2024) Deciphering the potential role of nano-based biosensors in growth, development and stress management of crop plants. S Afr J Bot 167:542–553. https://doi.org/10.1016/j.sajb.2024.02.035

Verma KK, Song XP, Joshi A, Tian DD, Rajput VD, Singh M, Arora J, Minkina TM, Li YR (2022a) Recent trends in nano-fertilizer for advancing sustainable agriculture under the era of climate change to ensure food security for future generations. Nanomaterials 12:173. https://doi.org/10.3390/nano12010173

Verma KK, Song XP, Joshi A, Rajput VD, Singh M, Sharma A, Singh RK, Li DM, Arora J, Minkina T, Li YR (2022b) Nanofertilizer possibilities for healthy soil, water and food in future: an overview. Front Plant Sci 13:865048. https://doi.org/10.3389/fpls.2022.865048

Verma KK, Singh M, Rajput VD, Verma CL, Burachevskaya M (2023a) Nanofertilizers: a sustainable alternative to conventional means. In: Rajput VD, Singh A, Minkina TM, Verma KK, Singh AK (eds) Nanotechnology for sustainable agriculture: an innovative and eco-friendly approach. Apple Academic Press, Inc., Taylor & Francis. https://doi.org/10.1201/9781003333128

Verma KK, Song X, Degu HD, Guo D, Joshi A, Huang H, Xu L, Singh M, Huang D, Rajput VD, Li Y (2023b) Recent advances in nitrogen and nano-nitrogen fertilizers for sustainable crop production: a mini-review. Chem Biol Technol Agric 10:1–14. https://doi.org/10.1186/s40538-023-00488-3

Verma KK, Song XP, Kumari A, Jagadesh M, Singh SK, Bhatt R, Singh M, Seth CS, Li YR (2024a) Climate change adaptation: challenges for agricultural sustainability. Plant Cell Environ:1–12. https://doi.org/10.1111/pce.15078

Verma KK, Joshi A, Song X-P, Singh S, Kumari A, Arora J, Singh SK, Solanki MK, Seth CS, Li Y-R (2024b) Synergistic interactions of nanoparticles and plant growth promoting rhizobacteria enhancing soil-plant systems: a multigenerational perspective. Front Plant Sci 15:1376214. https://doi.org/10.3389/fpls.2024.1376214

Wang J, Wu H, Wang Y, Ye W, Kong X, Yin Z (2024) Small particles, big effects: how nanoparticles can enhance plant growth in favorable and harsh conditions. J Integrat Plant Biol. https://doi.org/10.1111/jipb.13652

Wu H, Tito N, Giraldo JP (2017) Anionic cerium oxide nanoparticles protect plant photosynthesis from abiotic stress by scavenging reactive oxygen species. ACS Nano 11(11):11283–11297. https://doi.org/10.1021/acsnano.7b05723

Xu J, Yang J, Duan X, Jiang Y, Zhang P (2014) Increased expression of native cytosolic Cu/Zn superoxide dismutase and ascorbate peroxidase improves tolerance to oxidative and chilling stresses in cassava (Manihot esculenta Crantz). BMC Plant Biol 14(1):208. https://doi.org/10.1186/s12870-014-0208-4

Yanagi M (2024) Climate change impacts on wheat production: reviewing challenges and adaptation strategies. Adv Resour Res 4(1):89–107. https://doi.org/10.50908/arr.4.1_89
Younis, A. A., Khattab, H., & Emam, M. M. (2020). Impacts of silicon and silicon nanoparticles on leaf ultrastructure and TaPIP1 and TaNIP2 gene expressions in heat-stressed wheat seedlings. Biologia plantarum, 64(1)
Ze Y, Liu C, Wang L, Hong M, Hong F (2011) The regulation of TiO2 nanoparticles on the expression of light-harvesting complex II and photosynthesis of chloroplasts of arabidopsis thaliana. Biol Trace Elem Res 143(2):1131–1141. https://doi.org/10.1007/s12011-010-8901-0
Zhao L, Peng B, Hernandez-Viezcas JA, Rico C, Sun Y, Peralta-Videa JR et al (2012) Stress response and tolerance of Zea mays to CeO2 nanoparticles: cross talk among H2O2, heat shock protein, and lipid peroxidation. ACS Nano 6(11):9615–9622. https://doi.org/10.1021/nn302975u

Chapter 6
Nanomaterials Mediated Mitigation of Toxic Metals in Plants: A Way Forward Towards Precision Agriculture

Ayorinde Victor Ogundele and Oluwatoyin Adenike Fabiyi

6.1 Introduction

The rising accumulation of heavy metals (HMs) in the food and water supply chain is a significant public health concern. Due to their high density and non-biodegradable nature, HMs, such as Hg(II), Cr(VI), Pb(II), and Cd(II) are among the most problematic environmental pollutants (Yu et al. 2020; Wawata and Fabiyi, 2024). Soil, which supports various life forms from microbes to complex organisms like humans, is a major reservoir for these contaminants. The disruption of natural nutrient cycles by different human activities has led to the accumulation of unwanted substances, including HMs, in the soil. Achieving sustainable development is hindered by the current levels of soil pollution. As populations grow, soil conservation becomes increasingly important, especially given the challenges of diminishing land area and food and shelter shortages. Most known HMs are highly toxic to both plants and animals, with cadmium (Cd), lead (Pb), arsenic (As), and nickel (Ni) being particularly notable examples. HMs persist in the environment and resist biodegradation, allowing them to easily move through the food chain. These properties present a significant risk to human health, making them a major concern. Although some HMs serve as trace nutrients in biological systems and are required in small amounts, the majority are harmful to life even at very low concentrations (Rahman et al. 2022).

Over the past decade, the security of crop production has been significantly threatened by the rising contamination of arable land with toxic HMs. Heavy metals are released into the environment through multiple sources, including metallurgy,

A. V. Ogundele
Laboratory of Natural Products & Drug Discovery, Centre CEBIM, Department of Basic Sciences, Faculty of Medicine, Universidad de La Frontera, Temuco, Chile

O. A. Fabiyi (✉)
Department of Crop Protection, Faculty of Agriculture, University of Ilorin, Ilorin, Nigeria

C. S. Seth et al. (eds.), *Nano-formulations for Sustainable Agriculture*,
Nanotechnology in the Life Sciences, https://doi.org/10.1007/978-3-031-90617-6_6

electroplating, chemical plants, industrial wastewater, agricultural runoff, mining and ore extraction, coal combustion, fertilizer application, smelting, and alloy production (Fig. 6.1).

These metal-containing industrial effluents can leach into groundwater, causing contamination. The process of removing heavy metals from environmental matrices such as soil and water is highly complex (Engwa et al. 2019). Studies have indicated that high HM concentrations damage crops by harming bio-macromolecules, disrupting ion balance, generating excessive reactive oxygen species (ROS) and toxic radicals, inducing oxidative stress, and damaging the photosynthetic apparatus. To mitigate these adverse effects, various strategies have been developed, including marker-assisted breeding, transgenic crop engineering, and conventional breeding (Manzoor et al. 2022). Additionally, numerous treatment methods have been devised to address HM contamination. These include physical techniques such as adsorption, coagulation, evaporation, and filtration; chemical methods like precipitation, oxidation, ion exchange, and electrochemical processes; and biological approaches including biodegradation and phytoremediation (Minju et al. 2015) (Fig. 6.2).

However, these methods often face significant drawbacks, such as high costs, operational complexity, and resulting secondary pollution (Cai et al. 2019). For example, while chemical precipitation boasts excellent removal efficiency, its installation costs are significantly high (Zhang et al. 2018). Among the various methods available, adsorption is widely favoured due to its affordability, high removal efficiency, strong practicality, broad applicability, and ease of operation (Ihsanullah et al. 2016). Numerous adsorbents have been explored for heavy metal removal,

Fig. 6.1 Sources of heavy metals

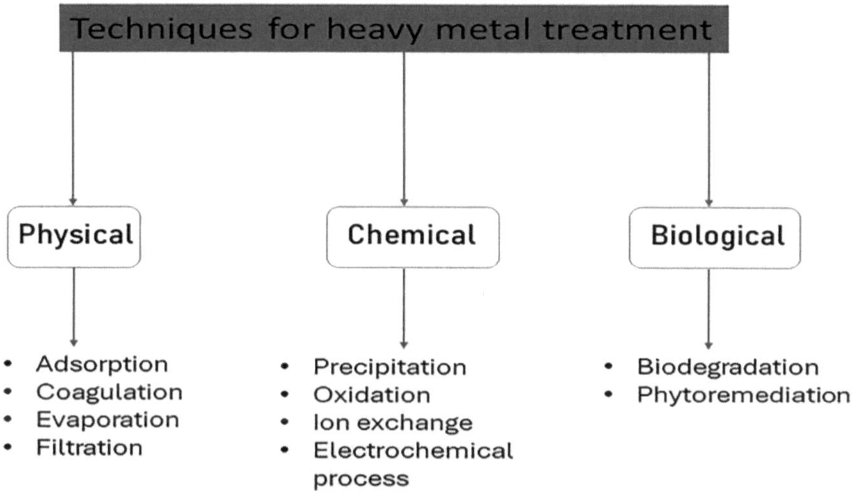

Fig. 6.2 Technical strategies for heavy metal treatments

including activated carbon, activated cellulose, modified cellulose, silica, among others. Some commonly employed adsorbents are activated carbon, activated alumina, biochar, clay minerals, carbon films, silica gel, zeolite, biomass, and more (Znad et al. 2022).

However, many suffer from limited efficiency due to their non-selectivity and poor affinity for the targeted heavy metals. Absorbency plays a crucial role in heavy metal adsorption, emphasizing the importance of selecting suitable and efficient adsorbents with large surface areas, numerous surface interactions, abundant adsorption sites, cost-effectiveness, and rapid adsorption rates. Nanoscale materials have emerged as promising adsorbents for efficient heavy metal remediation due to the unique properties associated with their nano sizes (Kumar et al. 2024a). Various types of nanoparticles have been studied for their suitability as nano-adsorbents, including metal oxide nanoparticles, polymeric nanocomposites, carbon-based nanomaterials, quantum dots, clay nanocomposites, among others. Their small dimensions and extensive surface areas make them effective adsorbents for remediating various environmental contaminants (Choudhury et al. 2022).

Research on the use of nanomaterials (NMs) for addressing heavy metal contamination is gaining traction and scholarly significance, evident in the growing number of publications each year. While nanomaterials have been extensively studied for their capability to remediate heavy metals, there is a lack of systematic and current research specifically dedicated to nanomaterials-mediated mitigation of toxic metals in plants for heavy metal removal.

6.2 Effects of Heavy Metal Toxicity on Soil, Plants, Harvested Fruits and Vegetables

Heavy metals are characterized by their high density and significant atomic weight (Ismanto et al. 2023). They naturally occur in the environment but are predominantly released into the soil, water, and air through human activities like industrial processes, mining, agriculture, and the use of various products. Their toxic properties enable them to accumulate in living organisms, posing severe environmental risks (Ali et al. 2019; Verma et al. 2023a). They adversely affect soil microorganisms essential for maintaining soil fertility and overall ecosystem health (Naz et al. 2022). HMs has diverse and profound effects on soil microorganisms, impacting the overall health and productivity of soil ecosystems. The presence of HMs in the soil primarily hinders the growth and development of microorganisms by damaging their cells, leading to a reduced microbial population. This decline in microbial biomass disrupts crucial activities such as the breakdown of organic matter and alters the nitrogen cycle, including nitrification and denitrification processes, thereby decreasing the soil's overall nutrient availability and contaminant decomposition (Verma et al. 2024a). Metals like lead, mercury, and cadmium are particularly detrimental to soil microorganisms (Kumar et al. 2022). Apart from inhibiting growth, toxic HMs disrupts enzymatic activity, essential for vital metabolic processes. These processes encompass nutrient transformation and the breakdown of organic compounds into forms usable by plants. Excessive HMs can imbalance these cycles and nutrient availability, diminishing soil fertility, productivity, and overall plant growth and development (Bortoloti and Baron 2022; Verma et al. 2023b).

Soil microorganisms are crucial for maintaining ecological balance and resilience, contributing to processes like disease suppression, soil structure formation, and nutrient cycling (Fabiyi et al. 2023; Fabiyi et al. 2024a; Jagadesh et al. 2024). Certain microbial groups are highly sensitive to high concentrations of HMs, leading to reduced population diversity and negative effects on ecosystem stability and functioning (Kumari et al. 2023). Some microorganisms, however, develop tolerance to heavy metal stress through bioaccumulation, accumulating HMs in their tissues (Kumar et al. 2024b). This accumulation affects the microbes and higher trophic levels in the food web. Additionally, heavy metals can disrupt symbiotic relationships between microorganisms, plants, and mycorrhizal fungi, impairing nutrient acquisition and limiting plant growth and development (Meena and Sarita 2019). Soil microbes are also essential in maintaining soil pH balance. Extreme pH levels can disrupt these microorganisms, and HMs present in soil can alter pH levels (Verma et al. 2024b). Certain heavy metals, such as aluminium, can acidify the soil, lowering its pH and affecting the viability and function of soil microorganisms, which have varying tolerances to pH changes. These fluctuations in soil pH also influence the mobility and bioavailability of HMs, exacerbating their effects on microbes. For instance, lead contamination acidifies soil (Collin et al. 2022), inhibiting nitrogen fixation, photosynthesis (Porter and Sheridan 1981), mycorrhizal fungi activity, and soil enzymes. Short-term additions of nitrogen and phosphorus are particularly sensitive to microbial communities and enzymatic activities (Zi et al. 2022; Verma et al. 2023c) (Fig. 6.3).

Research has shown that cadmium contamination reduces soil pH, negatively impacting bacteria, fungi, and earthworms, and disrupting nitrogen cycling (Ukalska-Jaruga et al. 2022). High levels of copper and zinc slightly increase soil pH, which can disrupt bacterial communities involved in decomposition. Arbuscular mycorrhizal fungi can accumulate zinc (Sazykin et al. 2023). While mercury contamination does not directly affect soil pH, certain soil bacteria and fungi can convert mercury into toxic methylmercury, which bio-accumulates in the food chain, posing risks to higher organisms, particularly in marine ecosystems (Harding et al. 2018). Chromium contamination, although not significantly affecting soil pH, can be toxic to soil bacteria, fungi, and other microorganisms, inhibiting their growth and activity (Ali et al. 2023). The impact of HMs on soil pH and microorganisms depends on factors such as metal concentration, soil type, exposure duration, and specific microorganisms present. Additionally, microorganism tolerance and response to HMs vary among species. Soil microbes are also essential for maintaining soil structure and stability. HMs adversely affect soil aggregation, leading to compaction, reduced water infiltration, and poor root growth, which ultimately impacts soil health and plant growth (Wang et al. 2022).

Soil contamination by HMs is a major concern in today's world. The physical properties of soil, such as its composition, porosity, bulk density, consistency, temperature, colour, and resistivity, are all impacted by HMs. Heavy metals like

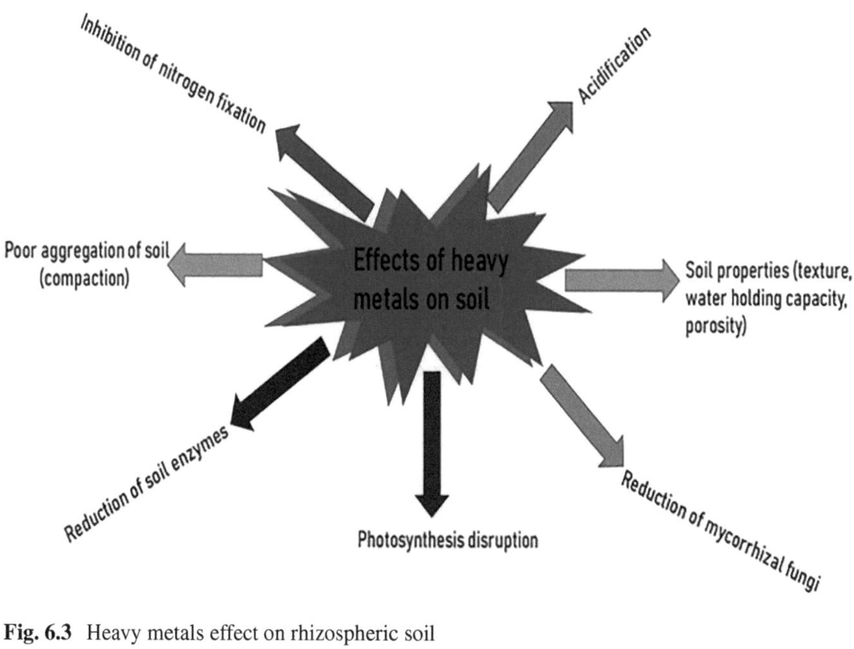

Fig. 6.3 Heavy metals effect on rhizospheric soil

chromium, copper, lead, zinc, cadmium, and nickel are key contributors to soil pollution (Kumar et al. 2024c). These metals indirectly influence soil enzymatic activities by altering the microbial communities responsible for enzyme production (Huang et al. 2009). HMs, characterized by their high atomic weights and densities, can significantly deteriorate the physical properties of soil. Metals, such as As, Pb, Cr, Hg, and Cd are particularly harmful when present in excess. Their presence can adversely affect soil structure, texture, porosity, and water-holding capacity in the following ways:

6.2.1 *Disruption of Soil Structure*

HMs can interfere with the formation and stability of soil aggregates, which are essential for maintaining healthy soil structure, porosity, and water infiltration. Metals like Cd, Pb, and Cu can destabilize aggregates, leading to soil compaction (Alloway 2012).

6.2.2 *Alteration of Soil Texture*

Accumulation of HMs in the soil can alter particle size distribution, affecting soil texture. Metals such as Ni and Zn can change the ratios of sand, silt, and clay particles, modifying the soil texture (Wang et al. 2024).

6.2.3 *Reduction of Soil Porosity*

Contamination by HMs can clog soil pores, reducing soil porosity and impeding the circulation of air, water, and nutrients. Cr, Cd, and Pb are known to block soil pores, limiting water percolation and root growth (Hamdi et al. 2025).

6.2.4 *Impact on Water-Holding Capacity*

High levels of HMs can diminish soil porosity and pore connectivity, decreasing the soil water-holding capacity and increasing runoff. Metals such as Cu and Zn have been found to reduce the soil's ability to retain water, leading to increased runoff and less water available for plants (Imfeld et al. 2020).

6.2.5 Acceleration of Soil Erosion

Heavy metal contamination can destabilize soil aggregates, making the topsoil more susceptible to erosion by wind or water. The accumulation of HMs in the topsoil can enhance soil erosion rates (Xie et al. 2016).

Heavy metals significantly impact the chemical properties of soil in several ways:

6.2.6 pH Alteration

Heavy metals can alter soil pH, either increasing or decreasing it. Cd and Al tend to lower soil pH, leading to increased acidity, while Ni and Cr can raise soil pH, making it more alkaline. These pH changes affect nutrient availability, microbial activity, and overall soil health. The soil pH and organic content are crucial in determining the accumulation and movement of heavy metals (Hu et al. 2018).

6.2.7 Nutrient Imbalance

Heavy metals can disrupt the balance of essential nutrients in the soil. Metals like Ca, Pb, and Zn can compete with and inhibit the uptake of essential nutrients such as Mg, Ca, Fe, and Mn by plants, leading to nutrient deficiencies and impacting plant growth and productivity. Lead, in particular, can reduce microbial diversity, nutrient availability, and soil fertility (Dotaniya et al. 2020).

6.2.8 Soil Organic Matter Degradation

Heavy metal contamination can affect the content and decomposition rates of soil organic matter. High levels of metals like Zn and Cu can inhibit microbial activity responsible for organic matter decomposition, resulting in the accumulation of organic matter and disrupting nutrient cycling and soil fertility (Martin Calvarro et al. 2014).

6.2.9 Redox Reactions

Heavy metals can influence redox reactions in the soil, affecting nutrient availability and mobility. Mn and Fe can undergo redox reactions, impacting the oxidation and reduction states of the soil. These reactions play a significant role in nutrient transformations, such as the conversion of nitrogen and sulfur compounds, which in turn affects soil fertility (Kicińska et al. 2022).

6.2.10 Adsorption and Desorption

Heavy metals undergo adsorption and desorption processes in the soil, influencing their mobility and availability. Soil properties, including clay minerals and organic matter content, are crucial in heavy metal adsorption. Pb and Cd have a high affinity for soil particles and can become strongly adsorbed, reducing their bioavailability to plants (Liu et al. 2023).

6.2.11 Plants

The impact of HMs on plants is a significant issue, raising concerns about agricultural productivity. HMs can accumulate in the soil either naturally or through human activities. Harmful human activities include industrial pollution from oil refineries, leakage from septic systems, oil spills, illegal chemical dumping, mining activities, and the use of contaminated inputs such as fertilizers, pesticides, herbicides, insecticides, and irrigation water. This accumulation poses a serious threat to the environment and agricultural systems. The presence of toxic heavy metals in the soil can profoundly and negatively affect the growth, development, and overall productivity of plants (Asiminicesei et al. 2024). The accumulation of HMs occurs in the edible parts of plants, such as vegetables, grains, and fruits. Crops with elevated HM levels may become unsuitable for consumption, diminishing the quality of harvested produce. To render these foods safe for consumption, extensive processing may be necessary, leading to significant economic burdens and posing potential risks to human health (Mariyam et al. 2024).

Plants exposed to HMs may contain reduced levels of essential nutrients, such as vitamins and minerals, which further compound the issue. Even after undergoing extensive processing, crops may still exhibit off-flavours or metallic tastes, potentially resulting in adverse health effects including organ damage, HM poisoning, and chronic health issues if consumed over an extended period (Najmi et al. 2023). Moreover, it contributes to an increase in toxicity levels. This affects the physiological and biochemical processes of plants, consequently reducing chlorophyll production, hindering enzymatic activity, and impeding nutrient absorption, thereby resulting in inhibited plant growth, development, and decreased crop yields (Rajput et al. 2022) (Fig. 6.4).

Additionally, HMs induce oxidative stress in plants, leading to the generation of reactive oxygen species (ROS) that can harm cellular structures, disrupt normal metabolic functions, and cause harm to aquatic organisms. Oxidative stress can result in damage to cell membranes, degradation of proteins, and alterations in

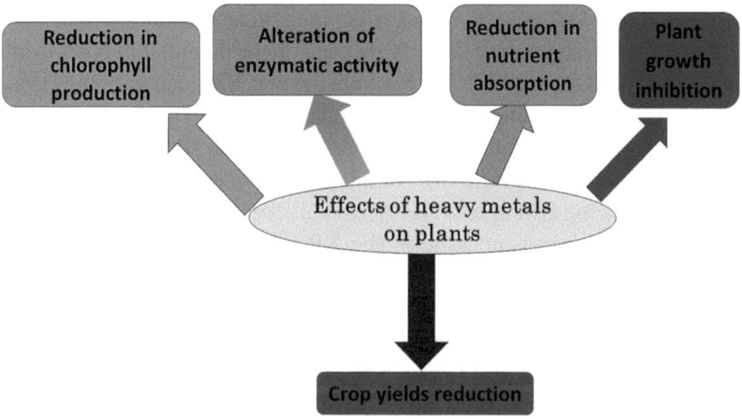

Fig. 6.4 Heavy metals impact on plant growth and development

DNA, ultimately impacting plant growth and agricultural productivity (Ejaz et al. 2023). HMs can induce nutritional imbalances in plants. Certain heavy metals, such as Pb and Cd, possess the ability to mimic essential nutrients, competing with them for uptake by plant roots. This competition may result in nutrient deficiencies as HMs are preferentially absorbed, leading to inadequate nutrient uptake by crops. Nutrient deficiencies can impair plant growth, disrupt reproductive processes, and reduce both the quality and quantity of crop yields. HMs present in the soil can also affect the availability and mobility of other crucial nutrients. They can bind to soil particles or form insoluble compounds, diminishing the availability of nutrients for plants. This reduced availability of nutrients for crop uptake and utilization can further impede plant growth and development (Gladkov et al. 2023). Furthermore, HMs indirectly impact plants by interfering with beneficial interactions between plants and soil microorganisms. Mycorrhizal fungi form symbiotic associations with plant roots, facilitating essential functions such as nutrient and water uptake and retention, which influence plant growth and development (Fabiyi et al. 2024b; Sazykin et al. 2023). This interference diminishes nutrient uptake by plants and undermines their ability to withstand environmental stresses. Water stress, particularly prevalent in arid and semi-arid regions with limited water sources, negatively affects plant physiological processes like photosynthesis (Choudhary et al. 2024).

6.3 Uses of Nanomaterials to Mitigate Metal Toxicity in Soil and Plants

Nanomaterials possess unique properties, such as small size and large surface area, making them valuable in various industries, including textiles, paints, cosmetics, medical products, and agriculture (Fabiyi, 2023; Corsi et al. 2018). When these

NMs enter soil and sediment, they can affect both plant and animal life (Yang et al. 2010). Recent research has increasingly focused on the potential of NMs to alleviate abiotic stress (Awad et al. 2019). Studies indicate that NMs can mitigate HMs toxicity in both soil and plants and influence the uptake of HMs by plants (Hussain et al. 2019). The response varies depending on the NMs form, size, application method, plant species, and soil type. Some studies have found that NMs enhance plant growth while reducing HM uptake (Hussain et al. 2021; El-Ramady et al. 2022). For instance, a study on iron nanoparticles (Fe-NPs) showed their impact on reducing the uptake of chromium (Cr) and cadmium (Cd) by wheat over a 7-day incubation period. NMs can immobilize harmful HMs in soil or sediments by altering metal speciation, converting water-soluble and exchangeable fractions into non-bioavailable forms through surface adsorption (Feizi 2018).

The global population is growing at an alarming rate, making food security a critical issue in terms of both quantity and quality. Recently, the contamination of crops with HMs has raised significant health concerns (Rajput et al. 2021). HMs contamination can adversely affect human health, disrupting metabolic processes and potentially leading to severe illness or even death. When evaluating HMs toxicity in plants, it is crucial to consider factors such as seed germination, plant growth, and biomass production. The use of NMs has diverse effects on plant growth and biomass under HM stress. For instance, radish plants exposed to both TiO_2 and Cd showed concentration-dependent effects on root length and seed germination after 5 days (Manesh et al. 2018). Combining Fe_2O_3 NMs with ascorbic acid significantly improved seed germination and shoot-root length in *Brassica juncea* compared to ascorbic acid alone (Praveen et al. 2018). The application of nanohydroxyapatite (nHAP) enhanced growth and biomass in rice plants and *Brassica chinensis* under Cd toxicity (Huang et al. 2017). Similarly, Fe NMs reduced Cd uptake, promoting better growth and biomass in wheat (Hussain et al. 2019).

TiO_2 NMs increased root length, plant height, and fresh weight in hydroponically grown rice plants under Cd stress compared to controls (Ji et al. 2017). ZnO NMs also improved growth and photosynthesis in wheat facing Cd stress (Adrees et al. 2021). Liu et al. (2014) found that foliar spraying with nanosized silica sols boosted grain yield and overall development in rice across hydroponic, potted, and field conditions (Liu et al. 2014). Seed priming with ZnO and Fe NMs led to increased plant height and dry biomass in plants under Cd stress (Rizwan et al. 2019b). When rice was cultivated hydroponically under Cd stress, the foliar application of SiO_2 NMs resulted in a significant increase in biomass compared to Cd treatment alone. Similarly, SiO_2 NMs enhanced stem-root length and biomass under Cr stress in both pea and rice plants (Wang et al. 2016). In pot studies, the application of iron (Fe) NMs significantly boosted wheat biomass and photosynthesis under Cd stress (Adrees et al. 2020). Various NMs have been applied externally to different plant species under HM stress to enhance chlorophyll content. For example, TiO_2 NMs increased both the gross photosynthetic value and chlorophyll content in rice under Cd stress (Ji et al. 2017). In *Brassica juncea*, supplementation with Fe_2O_3 NMs elevated protein and chlorophyll levels in leaves when compared to As treatment alone (Praveen et al. 2018).

SiO_2 NMs also significantly improved chlorophyll fluorescence indices (Fv/Fm, Fv/F0, and qP) in pea leaves while reducing NPQ under Cr stress (Tripathi et al. 2015). Furthermore, foliar application of SiO_2 NMs to rice resulted in significant increases in chlorophyll a and b content under Cd stress (Wang et al. 2016). Fe NMs greatly enhanced wheat growth and photosynthetic activity in pot studies under Cd stress (Adrees et al. 2020). Numerous studies indicate that exogenous application of NMs can reduce HM uptake in many plant species (Zhou et al. 2021). For instance, TiO_2 NMs decreased Cd concentrations in rice roots and leaves, while SiO_2 NMs reduced Cr concentrations in pea shoots and roots (Ji et al. 2017). ZnO NMs were found to reduce Cd absorption in wheat by immobilizing the metal ions (Rizwan et al. 2019a). Topical application of TiO_2 and Si NMs to rice plants under Cd stress limited Cd uptake in aboveground parts (Lyu et al. 2018). These observations suggest that the effects of NMs on metal uptake in plants can vary depending on the plant type and cultivar (Hussain et al. 2021).

When nano-silica was sprayed on hybrid and conventional rice cultivars grown in multimetal-contaminated soil, it demonstrated varied effects on plant growth and Cd uptake (Wang et al. 2016). Different varieties of the same species also exhibited diverse responses to HM exposure. For instance, (Michálková et al. 2017) reported that the application of nanoscale zero-valent iron (nZVI) to soil reduced the concentrations, soil bioavailability, and levels of Cd, Pb, and Zn in sunflowers. In rice cultivated in metal-infused soil, the application of inorganic and organic nano-Si altered the absorption profiles of Cd, Zn, Pb, and Cu (Liu et al. 2015). Liang et al. (2018) found that applying C_{60} fullerene to rice varieties exposed to Cd, Pb, and Cu showed different toxicity levels, indicating that NM impacts on metal uptake can vary with co-contamination compared to single-metal exposure. Another study revealed that wheat exposed to combined or independent Cr and Cd stress showed reduced Cd–Cr absorption (Liu and Lal 2015). Most research on NMs under HM stress has been conducted in labs and greenhouses. Field studies, such as those by (Hussain et al. 2021), showed that co-applying ZnO, Fe, and Si NMs enhanced wheat growth and reduced Cd toxicity in a Cd-contaminated area. Foliar application of lower concentrations of Zn and Ti NMs to sunflower plants in the field also improved physiological traits, with Zn showing a greater effect (Huang et al. 2018). In a pot experiment with Cd-contaminated silt, (Singh and Lee 2016) observed that increasing doses of nZVI progressively raised Cd concentrations in ramie stems, leaves, and roots. Similarly, TiO_2 NMs increased Cd levels in soybean plants grown in highly Cd-contaminated soil (Moameri and Abbasi Khalaki 2019). Adding nano-silica to soil polluted with Cd and Pb increased metal concentrations in *Secale montanum*, especially in roots treated with 500.0 mg kg^{-1} NMs (Rajput et al. 2020).

The use of NMs is essential for advancing modern agriculture and agrotechnology, as they impact the food cycle by transferring from plants to animals (Staroń et al. 2020). Studies have shown that NMs can be absorbed and stored in the edible parts of plants, with potential hazards to human health if they end up in food (Tripathi et al. 2017; Verma et al. 2022). NMs can migrate to plant aerial parts after root interaction, accumulating at cellular or subcellular levels (Nair et al. 2010). For instance, CuO NMs are well-adsorbed by plant roots and then distributed through plant tissues after modifications like solubilization, biotransformation, and bioaccumulation (Rajput et al. 2018). Metal-origin NMs translocate from roots to shoots, but their effects on plant growth and health vary, with some, like Ag and CuO NMs, acting as phytotoxicants and hindering germination and root growth (Zhu et al. 2012). The transmission of NMs depends on crop variety and NM properties, with some plants unable to accumulate certain NMs. For example, *Cucurbita pepo* and *Raphanus raphanistrum* cannot accumulate Au NMs, while *Oryza sativa* shoots can (Vélez et al. 2021). Agricultural NMs must be safe for consumers, beneficial to soil microbes, and environmentally benign. NMs are being researched for their potential to enhance plant productivity, control viruses, nematodes and phytopathogens, and monitor soil and plant health (Delfani et al. 2014; Fabiyi and Olatunji 2024).

NMs benefit agriculture and remediation by increasing crop yields with lower NMs substance quantities, reducing soil and water pollution, and aiding soil restoration and food production (Cai et al. 2020). Studies on soil microorganisms and NMs are still emerging, but they play crucial roles in biogeochemical cycles, enhancing soil health, plant output, and plant-soil interactions. NMs-regulated microbes perform major tasks like xenobiotic breakdown, contamination prevention, and environmental remediation (Schloter et al. 2018). NMs used to enhance plant nutrients, known as nano-fertilizers, offer benefits such as lower transportation and application costs, soil salt removal, and synchronization with soil nutrient status and plant development stages (Kah et al. 2018). NFs improve nutrient availability, absorption efficiency, crop yields and nutrient values (Usman et al. 2020). Siderophores strongly attract NMs metals like Ag, Cu, and Zn, suggesting NMs will increase in number and dissolve through metal-siderophore interactions (Bayat et al. 2018). Artificial root exudates may enhance the absorption rate of free Cu^{2+} in soil, accelerating Cu NM breakdown (Huang et al. 2017). NMs impact broad bean development and soil salinity stress and may enhance natural soil qualities. Iron oxide NMs can alter soil bulk density, porosity, chemical and physical properties, and they can address groundwater, surface water, and sewage contamination by hazardous metal ions, inorganic and organic solutes, and microorganisms (Cha et al. 2013).

6.4 Application of Nano-sensors in Mitigating the Effect of Toxic Metals (Precision Agriculture)

Nano-sensors are advanced sensors comprising elements connected to a nanoscale transducer, designed to acquire, process, and analyze data at the atomic level (Ghaffar et al. 2020). These sensors can detect analytes at very low concentrations using a compact physicochemical transducer linked to sensitized elements (Scognamiglio et al. 2013). Nano-sensors offer several advantages, including functionalization, immobilization, and miniaturization, which enhance the analytical performance of nanoscale materials by integrating different components into complex structures (Usman et al. 2020). Compared to traditional sensors, nano-sensors provide increased detection specificity and have applications in various fields such as soil nutrient monitoring, bioprocessing, environmental control, and agriculture (Fabiyi et al. 2025). By enabling the rational management of water, land, fertilizers, and pesticides, nano-sensor technology can enhance crop yields through early detection and rapid decision-making (Mandal et al. 2020). Heavy metals can be detected using nanomaterials produced through biochemical and green synthesis methods. Various NMs like nanoclusters, antibodies, and aptamers, and nanodevices, such as FRET or fluorescence resonance energy transfer, quantum dots, and cantilever biosensors are currently employed for detecting heavy metals in plants and soil (Fabiyi and Ogundele 2025). (Fig. 6.5a, b, c).

In a ground-breaking study, Wong and collaborators developed CoPhMoRe-based Bombolitin II nanosensors (BSWNTs) and embedded them within the leaf tissues of wild-type spinach plants, transforming these plants into functional nanodevices. These plants acted as autonomous samplers and preconcentrators of signalling molecules from underground water, detecting various analytes. An infrared (IR) communication system then transmits the collected data to a user's smartphone. Additionally, the researchers utilized two near-infrared (NIR) fluorescent nanosensors implanted in the leaf mesophyll of spinach plants. The first sensor, a conjugation of peptide Bombolitin II with single-walled carbon nanotubes (SWNTs),

Fig. 6.5 Biosensor applications in greenhouse (**a**), field (**b**) and monitoring of plant growth development (**c**)

detected nitroaromatics by emitting IR fluorescence at 1100 nm with a temporal resolution of 5–15 minutes after the introduction of picric acid (PA, 2,4,6-trinitrophenol; 400 μM) to the roots. The second sensor, a PVA-activated SWNT (P-SWNT), served as a fixed control signal. Upon the uptake of nitroaromatic pollutants by the roots, these pollutants travelled to the leaf tissues and accumulated in the mesophyll, where the SWNT nanosensors were located. This study highlights the substantial potential of integrating nanomaterials with living plants to create a robust, plant-based chemical monitoring system for underground water, offering real-time data communication (Wong et al. 2017).

Due to the increasing importance of detecting HMs, electrochemical techniques have demonstrated their capability to enhance detection and validate their superiority over traditional methods. Multi-nanoparticle (MNP) assemblies, carbon nanotube (CNT) networks, and biological receptors have been widely utilized to enhance the versatility of nanomaterial-based electrochemical detection (Numan et al. 2021). Wang et al. (2019) reported the use of biosensors to detect the secretion of vitronectin-like (VN) proteins as biomarkers of plant cell damage through changes in electrochemical impedance, aiding in the identification of cadmium and lead stress in plant species such as Arabidopsis and soybeans (Choudhary et al. 2025). Guo et al. (2020) developed a smartphone-enabled detection platform capable of sensing Hg(II) at concentrations as low as 2.53 nm. Another method for mercury and cadmium detection involved using a lateral flow assay with UCNPs, AuNPs, aptamers, DNA probes, and antibody probes, achieving detection limits of 5 ppb for mercury and 0.18 ng/mL for cadmium. Nano-sensors can detect heavy metal phytotoxicity by measuring parameters like chlorophyll content and plant water status (Thirugnanasambandan 2021). Arsenic is a heavy metal contaminant that poses significant health risks to organisms in ecosystems. Lew and his colleagues investigated the natural arsenic accumulation ability of the plant *Pteris cretica* using nano-sensors embedded in plant tissue. These nano-sensors can monitor real-time changes within the plant during heavy metal uptake. The study demonstrated that the nano-sensors could detect arsenic levels ranging from 0.6 to 0.2 ppb after 2 weeks (Lew et al. 2021).

6.5 Disadvantages of Nanomaterials in Toxic Metal Remediation

Nanomaterials have shown substantial promise for removing heavy metals from various environmental settings. However, their growing use brings potential risks to humans, animals, and the environment due to the inherent toxicity of some nanomaterials. Therefore, the toxicity and impact of nanomaterials on the bioavailability of contaminants like heavy metals must be thoroughly evaluated during their application in metal remediation (Yang et al. 2019). The health implications of using nanomaterials for heavy metal removal require careful consideration. According to statistics, the global production of engineered nanomaterials in 2010 was estimated to be between 260,000 and 309,000 tons, with 9–17% ending up in soils, 2% in water bodies, and 0.1–1.5% in the atmosphere (Jośko et al. 2014). Nanomaterials that persist in the environment for extended periods can biodegrade or bioaccumulate in food chains, spreading through trophic levels and posing ecotoxicological risks (Ameen et al. 2021).

Plants act as both sinks and entry points for nanomaterials, facilitating their accumulation in the food chain. The relevance of nanomaterials to phytoremediation, including their effects on biomass production and plant responses, has led to increased focus on their general effects on plant uptake. Despite the potential of nanotechnology to solve issues that cannot be addressed by larger-scale products, the safety of nanomaterials remains a significant barrier to their widespread application. Consequently, thorough studies are necessary before their use. Recent reports indicate that nanomaterials pose significant challenges for agriculture. Deliberate application of nanomaterials can lead to their accumulation or increased concentration in soil, altering its properties. Soils containing nanomaterials have been reported to exhibit altered pH, a critical parameter affecting soil nutrient availability, microbiology, health, and plant growth (Shi et al. 2018). For instance, nanoscale zerovalent iron (nZVI) modification in soil can substantially raise pH levels. Similar findings were reported in studies on the effects of CuO nanomaterials on soil pH, where CuO nanomaterials consumed H^+ from the soil, producing Cu ions and $Cu(OH)^+$ complexes, particularly in acidic soils. Additionally, nanomaterials of Ag, Au, Ti, and Zn have been reported to alter soil pH adversely, affecting soil microbiota and nematode populations (García-Gómez et al. 2018). The detrimental effects of nanomaterials in soil are influenced by their type and concentration, soil type, and soil enzymatic activity. Elevated concentrations of nanomaterials can lead to decreased dehydrogenase activity, affecting soil nutrient balance and fertility. Absorption and internalization of NMs can significantly impact mycelium and disrupt normal microbial functions (Ameen et al. 2021).

Three main mechanisms are involved in binding and transporting nanomaterials in soil: surface complexation, hydrophobic partitioning, and ion exchange, which also couple pollutants to colloid surfaces (Gao et al. 2017). Continuous emissions of nanomaterials into the environment make them ubiquitous, likely penetrating the food chain at various trophic levels and exerting toxicological effects on various aquatic and terrestrial animals, as well as humans (Cedervall et al. 2012). Nanomaterials act as pollutants in water bodies, where they are ingested by lower aquatic organisms. Studies testing metal nanomaterial toxicity on species like *Danio rerio* (zebrafish), *Daphnia pulex* (Daphnia), and the microalgae *Pseudokirchneriella subcapitata* found Daphnia and microalgae more susceptible to nanomaterial exposure due to their particle filter-feeding habits. Once nanomaterials enter biological systems, they influence biochemical processes at the molecular and tissue levels, causing toxicological effects such as DNA damage, oxidative stress from ROS, protein folding disruption, and cell death (Wen Yang et al. 2021). Therefore, before implementing nanomaterials for toxic metal remediation, these environmental concerns must be prioritized to develop nanomaterials that promote sustainable applications.

6.6 Conclusions

The application of NMs in mitigating HMs toxicity in soil and plants has emerged as a promising and multifaceted approach to address the increasing concerns of food security and environmental health. NMs, due to their unique properties such as high surface area and reactivity, can immobilize HMs in soil by altering metal speciation and converting bioavailable fractions into non-bioavailable forms. This immobilization significantly reduces HM uptake by plants, as demonstrated in numerous studies where NMs like iron nanoparticles (Fe-NPs), titanium dioxide (TiO_2), zinc oxide (ZnO), and silica (SiO_2) have enhanced plant growth and reduced metal accumulation in various crops under HM stress. Research has shown that NMs can effectively improve seed germination, root and shoot growth, biomass production, and photosynthetic activity in plants exposed to HMs such as Cd, Cr, As, and Pb. For instance, the application of Fe-NPs has been reported to enhance wheat growth and reduce Cd uptake, while TiO_2 NMs have improved root length and chlorophyll content in rice under Cd stress. Additionally, nano-hydroxyapatite (nHAP) has been found to boost growth and biomass in rice and Brassica species under Cd toxicity. The potential of NMs extends beyond growth enhancement to include the reduction of HM concentrations in plant tissues, thereby mitigating the risks of HMs transfer to the food chain. For example, the foliar application of SiO_2 NMs has shown significant reductions in Cr concentrations in pea and rice plants, while ZnO NMs have immobilized Cd in wheat, reducing its bioavailability.

Moreover, the integration of NMs with advanced nano-sensor technologies offers a novel approach to real-time monitoring and management of HMs contamination in agricultural systems. Nano-sensors can detect low concentrations of HMs with high specificity, enabling timely and precise interventions to prevent HM uptake and ensure crop safety. Innovative studies have demonstrated the use of nano-sensors embedded in plant tissues to monitor HM levels and plant responses, highlighting the potential for developing plant-based chemical monitoring systems. Despite the promising results, most research on NMs and HMs mitigation has been conducted in controlled laboratory and greenhouse settings. There is a pressing need for extensive field studies to validate these findings under diverse environmental conditions and to understand the long-term implications of NM applications in agriculture. Furthermore, it is crucial to evaluate the safety of NMs for soil microbes, plants, and consumers to ensure that their benefits outweigh any potential risks. The strategic use of NMs in agriculture presents a powerful tool for mitigating HM toxicity, enhancing crop productivity, and safeguarding food quality. By reducing HMs bioavailability and improving plant resilience, NMs can contribute significantly to sustainable agricultural practices and environmental remediation. Continued interdisciplinary research and field validation will be essential to fully harness the potential of NMs and to develop safe, effective, and scalable solutions for modern agriculture.

References

Adrees M, Khan ZS, Ali S, Hafeez M, Khalid S, ur Rehman MZ et al (2020) Simultaneous mitigation of cadmium and drought stress in wheat by soil application of iron nanoparticles. Chemosphere 238:124681. https://doi.org/10.1016/j.chemosphere.2019.124681

Adrees M, Khan ZS, Hafeez M, Rizwan M, Hussain K, Asrar M et al (2021) Foliar exposure of zinc oxide nanoparticles improved the growth of wheat (Triticum aestivum L.) and decreased cadmium concentration in grains under simultaneous Cd and water deficient stress. Ecotoxicol Environ Saf 208:111627. https://doi.org/10.1016/j.ecoenv.2020.111627

Ali, H., Khan, E., & Ilahi, I. (2019). Environmental Chemistry and Ecotoxicology of Hazardous Heavy Metals: Environmental Persistence, Toxicity, and Bioaccumulation. Journal of Chemistry, 2019(1), 6730305. https://doi.org/10.1155/2019/6730305

Ali S, Mir RA, Tyagi A, Manzar N, Kashyap AS, Mushtaq M et al (2023) Chromium toxicity in plants: signaling, mitigation, and future perspectives. Plan Theory 12. https://doi.org/10.3390/plants12071502

Alloway BJ (ed) (2012) Heavy metals in soils: trace metals and metalloids in soils and their bioavailability, vol 22. Springer Science & Business Media

Ameen F, Alsamhary K, Alabdullatif JA, ALNadhari S (2021) A review on metal-based nanoparticles and their toxicity to beneficial soil bacteria and fungi. Ecotoxicol Environ Saf 213:112027. https://doi.org/10.1016/j.ecoenv.2021.112027

Asiminicesei D-M, Fertu DI, Gavrilescu M (2024) Impact of heavy metal pollution in the environment on the metabolic profile of medicinal plants and their therapeutic potential. Plan Theory 13. https://doi.org/10.3390/plants13060913

Awad YM, Vithanage M, Niazi NK, Rizwan M, Rinklebe J, Yang JE et al (2019) Potential toxicity of trace elements and nanomaterials to Chinese cabbage in arsenic- and lead-contaminated soil amended with biochars. Environ Geochem Health 41(4):1777–1791. https://doi.org/10.1007/s10653-017-9989-3

Bayat H, Kolahchi Z, Valaey S, Rastgou M, Mahdavi S (2018) Novel impacts of nanoparticles on soil properties: tensile strength of aggregates and compression characteristics of soil. Arch Agron Soil Sci 64(6):776–789. https://doi.org/10.1080/03650340.2017.1393527

Bortoloti GA, Baron D (2022) Phytoremediation of toxic heavy metals by Brassica plants: a biochemical and physiological approach. Environ Adv 8:100204. https://doi.org/10.1016/j.envadv.2022.100204

Cai C, Zhao M, Yu Z, Rong H, Zhang C (2019) Utilization of nanomaterials for in-situ remediation of heavy metal(loid) contaminated sediments: a review. Sci Total Environ 662:205–217. https://doi.org/10.1016/j.scitotenv.2019.01.180

Cai L, Cai L, Jia H, Liu C, Wang D, Sun X (2020) Foliar exposure of Fe3O4 nanoparticles on Nicotiana benthamiana: evidence for nanoparticles uptake, plant growth promoter and defense response elicitor against plant virus. J Hazard Mater 393:122415. https://doi.org/10.1016/j.jhazmat.2020.122415

Cedervall T, Hansson L-A, Lard M, Frohm B, Linse S (2012) Food chain transport of nanoparticles affects behaviour and fat metabolism in fish. PLoS One 7(2):e32254. Retrieved from https://doi.org/10.1371/journal.pone.0032254

Cha C, Shin SR, Annabi N, Dokmeci MR, Khademhosseini A (2013) Carbon-based nanomaterials: multifunctional materials for biomedical engineering. ACS Nano 7(4):2891–2897. https://doi.org/10.1021/nn401196a

Choudhary R, Rajput VD, Ghodake G, Ahmad F, Meena M, Rehman RU et al (2024) Comprehensive journey from past to present to future about seed priming with hydrogen peroxide and hydrogen sulfide concerning drought, temperature, UV and ozone stresses-a review. Plant Soil:1–23. https://doi.org/10.1007/s11104-024-06499-9

Choudhary R, Ahmad F, Kaya C, Upadhyay SK, Muneer S, Kumar V, Meena M, Liu H, Upadhyay H, Seth CS (2025) Decrypting proteomics, transcriptomics, genomics, and integrated omics for augmenting the abiotic, biotic, and climate change stress resilience in plants. J Plant Physiol. https://doi.org/10.1016/j.jplph.2025.154430

Choudhury TR, Rahman MS, Liba SI, Islam A, Quraishi SB, Begum BA et al (2022) Adsorptive removal of chromium from aqueous solutions using flax (Linum usitatissimum): kinetics and equilibrium studies. Environ Chem Ecotoxicol 4:132–139. https://doi.org/10.1016/j.enceco.2022.02.004

Collin S, Baskar A, Geevarghese DM, Ali MNVS, Bahubali P, Choudhary R et al (2022) Bioaccumulation of lead (Pb) and its effects in plants: a review. J Hazard Mater Lett 3:100064. https://doi.org/10.1016/j.hazl.2022.100064

Corsi I, Winther-Nielsen M, Sethi R, Punta C, Della Torre C, Libralato G et al (2018) Ecofriendly nanotechnologies and nanomaterials for environmental applications: key issue and consensus recommendations for sustainable and ecosafe nanoremediation. Ecotoxicol Environ Saf 154:237–244. https://doi.org/10.1016/j.ecoenv.2018.02.037

Delfani M, Baradarn Firouzabadi M, Farrokhi N, Makarian H (2014) Some physiological responses of black-eyed pea to iron and magnesium nanofertilizers. Commun Soil Sci Plant Anal 45(4):530–540. https://doi.org/10.1080/00103624.2013.863911

Dotaniya ML, Dotaniya CK, Solanki P, Meena VD, Doutaniya RK (2020) Lead contamination and its dynamics in soil–plant system. In: Gupta DK, Chatterjee S, Walther C (eds) BT - lead in plants and the environment. https://doi.org/10.1007/978-3-030-21638-2_5

Ejaz U, Khan SM, Khalid N, Ahmad Z, Jehangir S, Fatima Rizvi Z et al (2023) Detoxifying the heavy metals: a multipronged study of tolerance strategies against heavy metals toxicity in plants. Front Plant Sci 14. Retrieved from https://www.frontiersin.org/journals/plant-science/articles/10.3389/fpls.2023.1154571

El-Ramady HR, Verma KK, Rajput VD, Minkina T, Elbehiry F, Elbasiouny HY, Elsakhawy T, Omara AED, Amer MM (2022) Sources of silicon and nano-silicon in soil and plants. In: Silicon and nano-silicon in environmental stress management and crop quality improvement: progress and prospects. Academic Press, Elsevier. https://doi.org/10.1016/B978-0-323-91225-9.00003-0

Engwa GA, Ferdinand PU, Nwalo FN, Unachukwu MN (2019) In: Karcioglu O, Arslan B (eds) Mechanism and health effects of heavy metal toxicity in humans. IntechOpen. https://doi.org/10.5772/intechopen.82511

Fabiyi OA, Adebisi O, Falore S and Bello TT (2023) Assessment of Actinomycetes and Pseudomonas species on Meloidogyne incognita population and growth of carrot plants in disparate soils. Indian Phytopathology. 76: 593–604

Fabiyi OA, Adebisi OO, Falore SO and Claudius-Cole AO (2024a) Potential inhibition of entomopathogenic nematodes and plant growth promoting bacteria with exposure to selected herbicides and insecticides. Vegetos. 37(4): 1503–1512

Fabiyi OA, Ansari RA, Rizvi R and Mahmood I (2024b) Influence of arbuscular mycorrhizal fungi on soil health amelioration and plant fitness under hostile environment. In: Ansari, R.A., Rizvi, R and Mahmood, I. (eds). Mycorrhizal symbiosis and agroecosystem restoration. Springer, Singapore. 227–248 Pp.

Fabiyi OA (2023) The perspective of nanotechnology in plant protection. In: Nanophytopathology.. Avila-Quezada, G.D. and Rai, M. (eds). 145–156 Pp. Taylor and Francis, USA.

Fabiyi OA and Olatunji GA (2024) Nematicidal potency of silver nanoparticles and other extracts from Borreria verticillata (L) Mey on Meloidogyne incognita infected tomato plants. Indian Phytopathology. 77(2): 311–321

Fabiyi OA and Ogundele AV (2025) Biosensors for monitoring climatic stressors in agricultural crops.. In: Uddin, I. (eds). Smart nanosensors. Springer, Singapore. 229–245. Pp

Fabiyi OA, Wawata IG, Akinsipo OB and Ogundele AV (2025) Smart nanosensors: A new approach in crop protection and pest management. In: Uddin, I. (eds). Smart nanosensors. Springer, Singapore. 247–266 Pp

Feizi M, Jalali M, Renella G (2018) Nanoparticles and modified clays influenced distribution of heavy metals fractions in a light-textured soil amended with sewage sludges. J Hazard Mater 343:208–219. https://doi.org/10.1016/j.jhazmat.2017.09.027

Gao J, Wang Y, Du Y, Zhou L, He Y, Ma L et al (2017) Construction of biocatalytic colloidosome using lipase-containing dendritic mesoporous silica nanospheres for enhanced enzyme catalysis. Chem Eng J 317:175–186. https://doi.org/10.1016/j.cej.2017.02.012

García-Gómez C, Fernández MD, García S, Obrador AF, Letón M, Babín M (2018) Soil pH effects on the toxicity of zinc oxide nanoparticles to soil microbial community. Environ Sci Pollut Res 25(28):28140–28152. https://doi.org/10.1007/s11356-018-2833-1

Ghaffar N, Farrukh MA, Naz S (2020) Applications of nanobiosensors in agriculture. In: Javad S (ed) BT - Nanoagronomy. https://doi.org/10.1007/978-3-030-41275-3_10

Gladkov EA, Tereshonok DV, Stepanova AY, Gladkova OV (2023) Plant–microbe interactions under the action of heavy metals and under the conditions of flooding. Diversity 15. https://doi.org/10.3390/d15020175

Guo Z, Kang Y, Liang S, Zhang J (2020) Detection of Hg(II) in adsorption experiment by a lateral flow biosensor based on streptavidin-biotinylated DNA probes modified gold nanoparticles and smartphone reader. Environ Pollut 266:115389. https://doi.org/10.1016/j.envpol.2020.115389

Hamdi FM, Ganbat N, Altaee A, Samal AK, Ibrar I, Zhou JL, Sharif AO (2025) Hybrid and enhanced electrokinetic system for soil remediation from heavy metals and organic matter. J Environ Sci 147:424–450. https://doi.org/10.1016/j.jes.2023.11.005

Harding G, Dalziel J, Vass P (2018) Bioaccumulation of methylmercury within the marine food web of the outer Bay of Fundy, Gulf of Maine. PLoS One 13(7):e0197220. Retrieved from https://doi.org/10.1371/journal.pone.0197220

Hu W, Wang H, Dong L, Huang B, Borggaard OK, Bruun Hansen HC et al (2018) Source identification of heavy metals in peri-urban agricultural soils of southeast China: an integrated approach. Environ Pollut 237:650–661. https://doi.org/10.1016/j.envpol.2018.02.070

Huang S, Peng B, Yang Z, Chai L, Zhou L (2009) Chromium accumulation, microorganism population and enzyme activities in soils around chromium-containing slag heap of steel alloy factory. Trans Nonferrous Metals Soc China 19(1):241–248. https://doi.org/10.1016/S1003-6326(08)60259-9

Huang Y, Zhao L, Keller AA (2017) Interactions, transformations, and bioavailability of nano-copper exposed to root exudates. Environ Sci Technol 51(17):9774–9783. https://doi.org/10.1021/acs.est.7b02523

Huang D, Qin X, Peng Z, Liu Y, Gong X, Zeng G et al (2018) Nanoscale zero-valent iron assisted phytoremediation of Pb in sediment: impacts on metal accumulation and antioxidative system of Lolium perenne. Ecotoxicol Environ Saf 153:229–237. https://doi.org/10.1016/j.ecoenv.2018.01.060

Hussain A, Ali S, Rizwan M, ur Rehman MZ, Qayyum MF, Wang H, Rinklebe J (2019) Responses of wheat (Triticum aestivum) plants grown in a Cd contaminated soil to the application of iron oxide nanoparticles. Ecotoxicol Environ Saf 173:156–164. https://doi.org/10.1016/j.ecoenv.2019.01.118

Hussain A, Rizwan M, Ali S, ur Rehman MZ, Qayyum MF, Nawaz R et al (2021) Combined use of different nanoparticles effectively decreased cadmium (Cd) concentration in grains of wheat grown in a field contaminated with Cd. Ecotoxicol Environ Saf 215:112139. https://doi.org/10.1016/j.ecoenv.2021.112139

Ihsanullah, Abbas A, Al-Amer AM, Laoui T, Al-Marri MJ, Nasser MS et al (2016) Heavy metal removal from aqueous solution by advanced carbon nanotubes: critical review of adsorption applications. Sep Purif Technol 157:141–161. https://doi.org/10.1016/j.seppur.2015.11.039

Imfeld G, Meite F, Wiegert C, Guyot B, Masbou J, Payraudeau S (2020) Do rainfall characteristics affect the export of copper, zinc and synthetic pesticides in surface runoff from headwater catchments? Sci Total Environ 741:140437. https://doi.org/10.1016/j.scitotenv.2020.140437

Ismanto A, Hadibarata T, Widada S, Indrayanti E, Ismunarti DH, Safinatunnajah N et al (2023) Groundwater contamination status in Malaysia: level of heavy metal, source, health impact, and remediation technologies. Bioprocess Biosyst Eng 46(3):467–482. https://doi.org/10.1007/s00449-022-02826-5

Jagadesh M, Dash M, Kumari A, Singh SK, Verma KK, Kumar P, Bhatt R, Sharma SK (2024) Revealing the hidden world of soil microbes: metagenomic insights into plant, bacteria, and fungi interactions for sustainable agriculture and ecosystem restoration. Microbiol Res 285:127764. https://doi.org/10.1016/j.micres.2024.127764

Ji Y, Zhou Y, Ma C, Feng Y, Hao Y, Rui Y et al (2017) Jointed toxicity of TiO2 NPs and Cd to rice seedlings: NPs alleviated Cd toxicity and Cd promoted NPs uptake. Plant Physiol Biochem 110:82–93. https://doi.org/10.1016/j.plaphy.2016.05.010

Jośko I, Oleszczuk P, Futa B (2014) The effect of inorganic nanoparticles (ZnO, Cr2O3, CuO and Ni) and their bulk counterparts on enzyme activities in different soils. Geoderma 232–234:528–537. https://doi.org/10.1016/j.geoderma.2014.06.012

Kah M, Kookana RS, Gogos A, Bucheli TD (2018) A critical evaluation of nanopesticides and nanofertilizers against their conventional analogues. Nat Nanotechnol 13(8):677–684. https://doi.org/10.1038/s41565-018-0131-1

Kicińska A, Pomykała R, Izquierdo-Diaz M (2022) Changes in soil pH and mobility of heavy metals in contaminated soils. Eur J Soil Sci 73(1):e13203. https://doi.org/10.1111/ejss.13203

Kumar H, Ishtiyaq S, Varun M, Favas PJ, Ogunkunle CO, Paul MS (2022) Bioremediation: plants and microbes for restoration of heavy metal contaminated soils. In: Bioenergy crops. CRC Press, pp 37–70

Kumar D, Mariyam S, Gupta KJ, Thiruvengadam M, Ghodake GS, Xing B, Seth CS (2024a) Comparative investigation on chemical and green synthesized titanium dioxide nanoparticles against chromium (VI) stress eliciting differential physiological, biochemical, and cellular attributes in Helianthus annuus L. Sci Total Environ 930:172413. https://doi.org/10.1016/j.scitotenv.2024.172413

Kumar D, Singh R, Upadhyay SK, Verma KK, Tripathi RM, Liu H et al (2024b) Review on interactions between nanomaterials and phytohormones: novel perspectives and opportunities for mitigating environmental challenges. Plant Sci 340:111964. https://doi.org/10.1016/j.plantsci.2023.111964

Kumar D, Biswas JK, Mulla SI, Singh R, Shukla R, Ahanger MA et al (2024c) Micro and nanoplastics pollution: sources, distribution, uptake in plants, toxicological effects, and innovative remediation strategies for environmental sustainability. Plant Physiol Biochem 108795. https://doi.org/10.1016/j.plaphy.2024.108795

Kumari A, Dash M, Singh SK, Jagadesh M, Mathpal B, Mishra PK, Pandey SK, Verma KK (2023) Soil microbes: a natural solution for mitigating the impact of climate change. Environ Monit Assess 195:1436. https://doi.org/10.1007/s10661-023-11988-y

Lew TTS, Park M, Cui J, Strano MS (2021) Plant nanobionic sensors for arsenic detection. Adv Mater 33(1):2005683. https://doi.org/10.1002/adma.202005683

Liang C, Xiao H, Hu Z, Zhang X, Hu J (2018) Uptake, transportation, and accumulation of C60 fullerene and heavy metal ions (Cd, Cu, and Pb) in rice plants grown in an agricultural soil. Environ Pollut 235:330–338. https://doi.org/10.1016/j.envpol.2017.12.062

Liu R, Lal R (2015) Potentials of engineered nanoparticles as fertilizers for increasing agronomic productions. Sci Total Environ 514:131–139. https://doi.org/10.1016/j.scitotenv.2015.01.104

Liu C, Wei L, Zhang S, Xu X, Li F (2014) Effects of nanoscale silica sol foliar application on arsenic uptake, distribution and oxidative damage defense in rice (Oryza sativa L.) under arsenic stress. RSC Adv 4(100):57227–57234. https://doi.org/10.1039/C4RA08496A

Liu J, Cai H, Mei C, Wang M (2015) Effects of nano-silicon and common silicon on lead uptake and translocation in two rice cultivars. Front Environ Sci Eng 9(5):905–911. https://doi.org/10.1007/s11783-015-0786-x

Liu X, Ju Y, Mandzhieva S, Pinskii D, Minkina T, Rajput VD et al (2023) Sporadic Pb accumulation by plants: influence of soil biogeochemistry, microbial community and physiological mechanisms. J Hazard Mater 444:130391. https://doi.org/10.1016/j.jhazmat.2022.130391

Lyu Y, Yu Y, Li T, Cheng J (2018) Rhizosphere effects of Loliumperenne L. and Beta vulgaris var. cicla L. on the immobilization of Cd by modified nanoscale black carbon in contaminated soil. J Soils Sediments 18(1):1–11. https://doi.org/10.1007/s11368-017-1724-2

Mandal N, Adhikary S, Rakshit R (2020) Nanobiosensors: recent developments in soil health assessment. In: Rakshit A, Ghosh S, Chakraborty S, Philip V, Datta A (eds) BT - Soil analysis: recent trends and applications. https://doi.org/10.1007/978-981-15-2039-6_15

Manesh RR, Grassi G, Bergami E, Marques-Santos LF, Faleri C, Liberatori G, Corsi I (2018) Co-exposure to titanium dioxide nanoparticles does not affect cadmium toxicity in radish seeds (Raphanus sativus). Ecotoxicol Environ Saf 148:359–366. https://doi.org/10.1016/j.ecoenv.2017.10.051

Manzoor N, Ali L, Ahmed T, Noman M, Adrees M, Shahid MS et al (2022) Recent advancements and development in nano-enabled agriculture for improving abiotic stress tolerance in plants. Front Plant Sci 13. https://doi.org/10.3389/fpls.2022.951752

Mariyam S, Upadhyay SK, Chakraborty K, Verma KK, Duhan JS, Muneer S et al (2024) Nanotechnology, a frontier in agricultural science, a novel approach in abiotic stress management and convergence with new age medicine-A review. Sci Total Environ 912:169097. https://doi.org/10.1016/j.scitotenv.2023.169097

Martin Calvarro L, de Santiago-Martín A, Quirós Gómez J, González-Huecas C, Quintana JR, Vázquez A et al (2014) Biological activity in metal-contaminated calcareous agricultural soils: the role of the organic matter composition and the particle size distribution. Environ Sci Pollut Res 21(9):6176–6187. https://doi.org/10.1007/s11356-014-2561-0

Meena K, Sarita S (2019) Bioremediation options for heavy metal pollution. J Health Pollut 9(24):191203. https://doi.org/10.5696/2156-9614-9.24.191203

Michálková Z, Martínez-Fernández D, Komárek M (2017) Interactions of two novel stabilizing amendments with sunflower plants grown in a contaminated soil. Chemosphere 186:374–380. https://doi.org/10.1016/j.chemosphere.2017.08.009

Minju N, Venkat Swaroop K, Haribabu K, Sivasubramanian V, Senthil Kumar P (2015) Removal of fluoride from aqueous media by magnesium oxide-coated nanoparticles. Desalin Water Treat 53(11):2905–2914. https://doi.org/10.1080/19443994.2013.868837

Moameri M, Abbasi Khalaki M (2019) Capability of Secale montanum trusted for phytoremediation of lead and cadmium in soils amended with nano-silica and municipal solid waste compost. Environ Sci Pollut Res 26(24):24315–24322. https://doi.org/10.1007/s11356-017-0544-7

Nair R, Varghese SH, Nair BG, Maekawa T, Yoshida Y, Kumar DS (2010) Nanoparticulate material delivery to plants. Plant Sci 179(3):154–163. https://doi.org/10.1016/j.plantsci.2010.04.012

Najmi A, Albratty M, Al-Rajab AJ, Alhazmi HA, Javed SA, Ahsan W et al (2023) Heavy metal contamination in leafy vegetables grown in Jazan Region of Saudi Arabia: assessment of possible human health hazards. Int J Environ Res Public Health 20. https://doi.org/10.3390/ijerph20042984

Naz M, Dai Z, Hussain S, Tariq M, Danish S, Khan IU et al (2022) The soil pH and heavy metals revealed their impact on soil microbial community. J Environ Manag 321:115770. https://doi.org/10.1016/j.jenvman.2022.115770

Numan A, Gill AAS, Rafique S, Guduri M, Zhan Y, Maddiboyina B et al (2021) Rationally engineered nanosensors: a novel strategy for the detection of heavy metal ions in the environment. J Hazard Mater 409:124493. https://doi.org/10.1016/j.jhazmat.2020.124493

Perea Vélez YS, Carrillo-González R, del González-Chávez MCA (2021) Interaction of metal nanoparticles–plants–microorganisms in agriculture and soil remediation. J Nanopart Res 23(9):206. https://doi.org/10.1007/s11051-021-05269-3

Porter JR, Sheridan RP (1981) Inhibition of nitrogen fixation in alfalfa by arsenate, heavy metals, fluoride, and simulated acid rain. Plant Physiol 68(1):143–148. https://doi.org/10.1104/pp.68.1.143

Praveen A, Khan E, Ngiimei DS, Perwez M, Sardar M, Gupta M (2018) Iron oxide nanoparticles as nano-adsorbents: a possible way to reduce arsenic phytotoxicity in indian mustard plant (Brassica juncea L.). J Plant Growth Regul 37(2):612–624. https://doi.org/10.1007/s00344-017-9760-0

Rahman SU, Nawaz MF, Gul S, Yasin G, Hussain B, Li Y, Cheng H (2022) State-of-the-art OMICS strategies against toxic effects of heavy metals in plants: a review. Ecotoxicol Environ Saf 242:113952. https://doi.org/10.1016/j.ecoenv.2022.113952

Rajput VD, Minkina T, Sushkova S, Tsitsuashvili V, Mandzhieva S, Gorovtsov A et al (2018) Effect of nanoparticles on crops and soil microbial communities. J Soils Sediments 18(6):2179–2187. https://doi.org/10.1007/s11368-017-1793-2

Rajput V, Minkina T, Mazarji M, Shende S, Sushkova S, Mandzhieva S et al (2020) Accumulation of nanoparticles in the soil-plant systems and their effects on human health. Ann Agric Sci 65(2):137–143. https://doi.org/10.1016/j.aoas.2020.08.001

Rajput VD, Singh A, Minkina T, Rawat S, Mandzhieva S, Sushkova S, Shuvaeva V, Nazarenko O, Rajput P, Komariah, Verma KK, Singh AK, Rao M, Upadhyay SK (2021) Nano-enabled products: challenges and opportunities for sustainable agriculture. Plan Theory 10:2727. https://doi.org/10.3390/plants10122727

Rajput VD, Minkina T, Upadhyay SK, Kumari A, Ranjan A, Mandzhieva S, Sushkova S, Singh RK, Verma KK (2022) Nanotechnology in the restoration of polluted soil. Nanomaterials 12:769. https://doi.org/10.3390/nano12050769

Rizwan M, Ali S, ur Rehman MZ, Malik S, Adrees M, Qayyum MF et al (2019a) Effect of foliar applications of silicon and titanium dioxide nanoparticles on growth, oxidative stress, and cadmium accumulation by rice (Oryza sativa). Acta Physiol Plant 41(3):35. https://doi.org/10.1007/s11738-019-2828-7

Rizwan M, Noureen S, Ali S, Anwar S, ur Rehman MZ, Qayyum MF, Hussain A (2019b) Influence of biochar amendment and foliar application of iron oxide nanoparticles on growth, photosynthesis, and cadmium accumulation in rice biomass. J Soils Sediments 19(11):3749–3759. https://doi.org/10.1007/s11368-019-02327-1

Sazykin I, Khmelevtsova L, Azhogina T, Sazykina M (2023) Heavy metals influence on the bacterial community of soils: a review. Agriculture 13. https://doi.org/10.3390/agriculture13030653

Schloter M, Nannipieri P, Sørensen SJ, van Elsas JD (2018) Microbial indicators for soil quality. Biol Fertil Soils 54(1):1–10. https://doi.org/10.1007/s00374-017-1248-3

Scognamiglio V, Pezzotti I, Pezzotti G, Cano J, Manfredonia I, Buonasera K et al (2013) A new embedded biosensor platform based on micro-electrodes array (MEA) technology. Sensors Actuators B Chem 176:275–283. https://doi.org/10.1016/j.snb.2012.09.101

Shi J, Ye J, Fang H, Zhang S, Xu C (2018) Effects of copper oxide nanoparticles on paddy soil properties and components. Nanomaterials 8. https://doi.org/10.3390/nano8100839

Singh J, Lee B-K (2016) Influence of nano-TiO2 particles on the bioaccumulation of Cd in soybean plants (Glycine max): a possible mechanism for the removal of Cd from the contaminated soil. J Environ Manag 170:88–96. https://doi.org/10.1016/j.jenvman.2016.01.015

Staroń A, Długosz O, Pulit-Prociak J, Banach M (2020) Analysis of the exposure of organisms to the action of nanomaterials. Materials 13. https://doi.org/10.3390/ma13020349

Thirugnanasambandan T (2021) Advances in nanotechnology for bio-sensing in agriculture and food. In: Pudake RN, Jain U, Kole C (eds) BT - Biosensors in agriculture: recent trends and future perspectives. https://doi.org/10.1007/978-3-030-66165-6_3

Tripathi DK, Singh VP, Prasad SM, Chauhan DK, Dubey NK (2015) Silicon nanoparticles (SiNp) alleviate chromium (VI) phytotoxicity in Pisum sativum (L.) seedlings. Plant Physiol Biochem 96:189–198. https://doi.org/10.1016/j.plaphy.2015.07.026

Tripathi DK, Shweta, Singh S, Singh S, Pandey R, Singh VP et al (2017) An overview on manufactured nanoparticles in plants: uptake, translocation, accumulation and phytotoxicity. Plant Physiol Biochem 110:2–12. https://doi.org/10.1016/j.plaphy.2016.07.030

Ukalska-Jaruga A, Siebielec G, Siebielec S, Pecio M (2022) The effect of soil amendments on trace elements' bioavailability and toxicity to earthworms in contaminated soils. Appl Sci 12. https://doi.org/10.3390/app12126280

Usman M, Farooq M, Wakeel A, Nawaz A, Cheema SA, ur Rehman H et al (2020) Nanotechnology in agriculture: current status, challenges and future opportunities. Sci Total Environ 721:137778. https://doi.org/10.1016/j.scitotenv.2020.137778

Verma KK, Song XP, Joshi A, Rajput VD, Singh M, Sharma A, Singh RK, Li DM, Arora J, Minkina T, Li YR (2022) Nanofertilizer possibilities for healthy soil, water and food in future: an overview. Front Plant Sci 13:865048. https://doi.org/10.3389/fpls.2022.865048

Verma KK, Song X, Verma CL, Huang H, Singh M, Xu L, Wu J, Khan ZH, Zhang B, Li Y (2023a) Mathematical modeling of climate and fluoride effects on sugarcane photosynthesis with silicon nanoparticles. Plant Physiol Biochem 204:108089. https://doi.org/10.1016/j.plaphy.2023.108089

Verma KK, Singh M, Rajput VD, Verma CL, Burachevskaya M (2023b) Nanofertilizers: a sustainable alternative to conventional means. In: Rajput VD, Singh A, Minkina TM, Verma KK, Singh AK (eds) Nanotechnology for sustainable agriculture: an innovative and eco-friendly approach. Apple Academic Press, Inc., Taylor & Francis. https://doi.org/10.1201/9781003333128

Verma KK, Song X, Degu HD, Guo D, Joshi A, Huang H, Xu L, Singh M, Huang D, Rajput VD, Li Y (2023c) Recent advances in nitrogen and nano-nitrogen fertilizers for sustainable crop production: a mini-review. Chem Biol Technol Agric 10:1–14. https://doi.org/10.1186/s40538-023-00488-3

Verma KK, Song XP, Kumari A, Jagadesh M, Singh SK, Bhatt R, Singh M, Seth CS, Li YR (2024a) Climate change adaptation: challenges for agricultural sustainability. Plant Cell Environ:1–12. https://doi.org/10.1111/pce.15078

Verma KK, Joshi A, Song X-P, Singh S, Kumari A, Arora J, Singh SK, Solanki MK, Seth CS, Li Y-R (2024b) Synergistic interactions of nanoparticles and plant growth promoting rhizobacteria enhancing soil-plant systems: a multigenerational perspective. Front Plant Sci 15:1376214. https://doi.org/10.3389/fpls.2024.1376214

Wang P, Lombi E, Zhao F-J, Kopittke PM (2016) Nanotechnology: a new opportunity in plant sciences. Trends Plant Sci 21(8):699–712. https://doi.org/10.1016/j.tplants.2016.04.005

Wang X, Cheng M, Yang Q, Wei H, Xia A, Wang L et al (2019) A living plant cell-based biosensor for real-time monitoring invisible damage of plant cells under heavy metal stress. Sci Total Environ 697:134097. https://doi.org/10.1016/j.scitotenv.2019.134097

Wang G, Ren Y, Bai X, Su Y, Han J (2022) Contributions of beneficial microorganisms in soil remediation and quality improvement of medicinal plants. Plants 11. https://doi.org/10.3390/plants11233200

Wang Y, Zhang Z, Li Y, Liang C, Huang H, Wang S (2024) Available heavy metals concentrations in agricultural soils: relationship with soil properties and total heavy metals concentrations in different industries. J Hazard Mater 471:134410. https://doi.org/10.1016/j.jhazmat.2024.134410

Wawata, I.G and Fabiyi, O.A. (2024). Sustainable application of nanomaterials in the removal of heavy metals from water.. In: Uddin, I. (eds). Sustainable nanomaterials. Sustainable materials and technology. Springer, Singapore. 21–44Pp.

Wong MH, Giraldo JP, Kwak S-Y, Koman VB, Sinclair R, Lew TTS et al (2017) Nitroaromatic detection and infrared communication from wild-type plants using plant nanobionics. Nat Mater 16(2):264–272. https://doi.org/10.1038/nmat4771

Xie Z, Jiang Y, Zhang H, Wang D, Qi S, Du Z, Zhang H (2016) Assessing heavy metal contamination and ecological risk in Poyang Lake area, China. Environ Earth Sci 75(7):549. https://doi.org/10.1007/s12665-015-5240-7

Yang, Wenrong, Ratinac KR, Ringer SP, Thordarson P, Gooding, JJ & Braet F (2010) Carbon Nanomaterials in Biosensors: Should You Use Nanotubes or Graphene? Angewandte Chemie International Edition, 49(12):2114–2138. https://doi.org/10.1002/anie.200903463

Yang J, Hou B, Wang J, Tian B, Bi J, Wang N et al (2019) Nanomaterials for the removal of heavy metals from wastewater. Nanomaterials 9. https://doi.org/10.3390/nano9030424

Yang W, Wang L, Mettenbrink EM, DeAngelis PL, Wilhelm S (2021) Nanoparticle toxicology. Ann Rev Pharmacol Toxicol 61:269–289. https://doi.org/10.1146/annurev-pharmtox-032320-110338

Yu G, Liu J, Long Y, Chen Z, Sunahara GI, Jiang P et al (2020) Phytoextraction of cadmium-contaminated soils: comparison of plant species and low molecular weight organic acids. Int J Phytoremediation 22(4):383–391. https://doi.org/10.1080/15226514.2019.1663488

Zhang Y, Wang L, Zhang N, Zhou Z (2018) Adsorptive environmental applications of MXene nanomaterials: a review. RSC Adv 8(36):19895–19905. https://doi.org/10.1039/C8RA03077D

Zhou P, Adeel M, Shakoor N, Guo M, Hao Y, Azeem I et al (2021) Application of nanoparticles alleviates heavy metals stress and promotes plant growth: an overview. Nanomaterials 11. https://doi.org/10.3390/nano11010026

Zhu Z-J, Wang H, Yan B, Zheng H, Jiang Y, Miranda OR et al (2012) Effect of surface charge on the uptake and distribution of gold nanoparticles in four plant species. Environ Sci Technol 46(22):12391–12398. https://doi.org/10.1021/es301977w

Zi H, Hu L, Wang C (2022) Differentiate responses of soil microbial community and enzyme activities to nitrogen and phosphorus addition rates in an alpine meadow. Front Plant Sci 13. Retrieved from https://www.frontiersin.org/journals/plant-science/articles/10.3389/fpls.2022.829381

Znad H, Awual MR, Martini S (2022) The utilization of algae and seaweed biomass for bioremediation of heavy metal-contaminated wastewater. Molecules 27. https://doi.org/10.3390/molecules27041275

Chapter 7
Nanomaterials Strategies for Enhancing Plant Adaptation in the Adverse Agroclimatic Variables

Yogesh Dashrath Naik, Anil Namdev Kale, Viabhav Kumar Upadhayay, Adita Sharma, and Ajay Veer Singh

7.1 Introduction

Climate change impacts food security, livelihoods, agriculture, supply chains, and prices, potentially reducing production in tropical and temperate regions (Raj et al. 2022). The rise in temperature, atmospheric CO_2 levels, and changes in rainfall, drought occurrences, water availability and extreme weather events profoundly affect agriculture (Seppelt et al. 2022; Verma et al. 2024). These environmental shifts create challenges for plant growth and development due to abiotic stress, which is a prominent ecological consequence of climate change. Abiotic stresses have frequently been reported as leading to a substantial 50–70% reduction in the productivity of major crops (Ghosh and Xu 2014; Kumari et al. 2022). This

The original version of the chapter has been revised. A correction to this chapter can be found at https://doi.org/10.1007/978-3-031-90617-6_21

Y. D. Naik · A. N. Kale
Department of Agricultural Biotechnology and Molecular Biology, College of Basic Sciences & Humanities, Dr. Rajendra Prasad Central Agricultural University, Pusa, Samastipur, Bihar, India

V. K. Upadhayay (✉)
Department of Microbiology, College of Basic Sciences & Humanities, Dr. Rajendra Prasad Central Agricultural University, Pusa, Samastipur, Bihar, India

A. Sharma
College of Fisheries, (Dr. Rajendra Prasad Central Agricultural University, Pusa, Bihar), Dholi, Muzaffarpur, Bihar, India

A. V. Singh
Department of Microbiology, College of Basic Sciences & Humanities, G.B. Pant University of Agriculture & Technology, Pantnagar, Uttarakhand, India

C. S. Seth et al. (eds.), *Nano-formulations for Sustainable Agriculture*, Nanotechnology in the Life Sciences, https://doi.org/10.1007/978-3-031-90617-6_7

disproportionately affects people who are already struggling with poverty and food insecurity, which will face even more challenges. It could lead to millions more people not having enough food to eat (Raj et al. 2022). In real-world conditions, these climatic changes often occur simultaneously, leading to unpredictable physiological responses and interactions among plants (Kumari et al. 2023).

Efforts to enhance stress tolerance in crops have involved various methods, notably breeding stress-tolerant varieties and genetically modifying plants. Traditional breeding methods like wide hybridization and mutation breeding are important but time-consuming. Genetic modification offers a more targeted approach but faces global limitations due to biosafety concerns about environmental and biodiversity impacts (Kumar et al. 2024a). Understanding the molecular mechanisms of stress response in plants is crucial for developing stress-tolerant crops. However, the widespread adoption of genetically modified plants is hindered by regulatory restrictions and public apprehensions. A promising approach for enhancing plant resistance to environmental stresses is through priming. Chemical priming has shown significant potential in boosting plant tolerance to abiotic stresses by activating existing defence pathways without genetic modifications. Pre-treating plants with natural or synthetic compounds improves their response under adverse conditions like drought, heat, salinity, and heavy metals, compared to untreated plants (Savvides et al. 2016). Priming enhances plant defence mechanisms by regulating redox signalling, reactive oxygen species (ROS) accumulation and the expression of stress-responsive genes (Balmer et al. 2015). Widely used priming agents include phytohormones, amino acids, polyamines, melatonin, reactive oxygen-nitrogen-sulfur species fungicides, and synthetic compounds like NOSH-aspirin (Ioannou et al. 2020). Alternatively, nanomaterials (NMs) can be used as coatings for seeds, which can improve stress tolerance and promote growth even under normal conditions.

In recent times, there has been a rise in research focusing on the application of nanotechnology in agriculture (Tripathi et al. 2023). NMs applied in agriculture can be categorized into three main groups, such as inorganic, which includes metal and metal oxide nanoparticles (NPs), such as silver, gold and titanium dioxide (TiO_2), organic, primarily consisting of naturally produced NPs, and combined NPs such as clay (He et al. 2019). Generally, NMs enhance various aspects of crop protection and improve nutrition due to their small size, large surface area, precise dosage, slow release and other unique properties (Duhan et al. 2017; Kumar et al. 2024b). This nanotechnology holds promise for sustainable agriculture and has applications in various fields such as engineering, food technology and biotechnology. Additionally, nanoscale materials enhance plant tolerance to environmental stressors such as high temperature, drought, while also promoting soil structure (Kah et al. 2019; Verma et al. 2022a). Furthermore, this chapter illustrates the imperative role of NMs in mitigating the detrimental impacts of abiotic stressors on crop growth, thereby providing viable solutions to ensure effective crop cultivation in the face of climate change challenges.

7.2 Nanomaterials: Types and Agricultural Applications

NMs are composed of extremely small individual units, and possess unique properties due to their size, making them highly valuable in various manufacturing applications. NPs have gained significant attention in technological advancements due to their adjustable physicochemical properties, such as anti-microbial properties, catalytic activity, thermal conductivity, and light absorption and scattering, which provide superior performance compared to bulk materials (Jeevanandam et al. 2018; Verma et al. 2023a). NMs can be classified into five categories based on their size, origin, structural configuration, pore size, and potential toxicity (Mekuye and Abera 2023). Size refers to their dimensions, typically between 1 and 100 nm. Origin differentiates whether they are naturally occurring or artificially synthesized. Structural configuration involves the atomic arrangement within the materials. Pore size pertains to the dimensions of the pores in the materials, affecting their surface area and applications. Potential toxicity considers the possible health and environmental risks associated with the NMs (Verma et al. 2022b; Mekuye and Abera 2023).

However, advanced NMs can also be categorized into four main types: super, smart, active, and swarm NMs, each with distinct properties and applications (Chausali et al. 2023). These categories highlight the diverse and significant potential of NMs in various fields. In agriculture, the development of nanoscale agrochemicals including nano fertilizers (NFs), nanopesticides, nanosensors, and nanoformulations has transformed conventional farming methods, enhancing their sustainability and efficiency (El-Saadony et al. 2022; Upadhayay et al. 2023a, b). Although their potential is significant, the exact mechanisms by which NPs interact with plants remain largely unclear. Additionally, nanotechnology offers multiple applications in agriculture, including wastewater treatment, improving the quality of polluted soil, and enhancing crop productivity through sensors that detect pathogens (Neme et al. 2021; Rajput et al. 2021a). For example, nanobiosensors are versatile tools that support the development of advanced agricultural farms and demonstrate the practical and potential applications of nanotechnology in precision control and management of agricultural inputs (Upadhayay et al. 2023b). Additionally, nanotechnology-based applications have highlighted the increasing need for NPs in food biotechnology, food processing, food safety, pathogen detection in food, and extending the shelf life of food (Rajput et al. 2021b; Neme et al. 2021).

7.3 Impact of Climate Change on Crop Productivity

The relationship between climate change and agricultural production is highly associated. Numerous studies have elucidated the multifarious ways in which escalating global temperatures adversely impact the agricultural sector. The most salient mechanisms include alterations in average temperature and precipitation patterns, as well as an increase in the frequency and severity of extreme weather events such as heat waves, floods, and droughts (Francini and Sebastiani 2019; Verma et al. 2023b; Janni et al. 2024). These deleterious environmental conditions have been shown to significantly impair crop performance, leading to yield losses as high as 70% in some instances (Francini and Sebastiani 2019; Rajput et al. 2023). Temperature emerges as a critical environmental determinant governing plant growth and development. Prolonged exposure to temperatures exceeding the optimal thermal range can induce physiological stress, hindering crucial processes such as flowering and fruit development, ultimately resulting in substantial yield reductions (Kan et al. 2023). Beyond a certain critical threshold, excessively high temperatures may even prove lethal to plants. Extreme heat events are typically characterized by their intensity, frequency, and duration, as measured by the maximum temperatures recorded. Elevated temperatures can trigger heat stress in crop plants, perturbing their physiology, biochemical processes, and genetic regulation pathways. Each crop species exhibits a unique temperature tolerance threshold, surpassing which can lead to adverse effects on crop health and productivity. Notably, the reproductive stages of plants are particularly vulnerable to heat stress, with significant implications for yield potential (Kumar et al. 2024c).

Empirical studies have quantified the detrimental impact of terminal heat stress on wheat (*Triticum aestivum*) yields, reporting an 18.1% decrease under current climatic conditions (Dubey et al. 2020). Similarly, recent projections estimate paddy yield reductions of up to 10% in major rice-producing regions (Lyman et al. 2013). Soybean cultivation faces even more severe yield declines, with long-term projections indicating an 18.11% decrease and short-term estimates showing a 5.51% reduction for every 1 °C rise in maximum temperature (Sharma et al. 2023). Furthermore, high temperatures exceeding 35 °C during chickpea reproductive development have been linked to substantial grain yield reductions, with potential losses reaching up to 39% (Devasirvatham and Tan 2018). In addition to heat stress, drought conditions pose a significant challenge to agricultural productivity. These stressors often coincide with other detrimental factors like salinity, heat, and pathogen pressures. In response, plants undergo a series of physiological and morphological adaptations, including osmotic adjustments, reduced photosynthesis rates and transpiration, inhibited growth, heightened production of ROS, and altered stress signalling pathways (Ahluwalia et al. 2021). While these adaptive responses are crucial for plant survival, they can also lead to lasting damage and compromise long-term crop health and productivity.

7.4 Mitigating Climate Change Impacts on Crops with Nanomaterials

Over the past few decades, the Earth's average surface temperature has been rising rapidly due to the accumulation of greenhouse gases like CO_2, which trap heat in the atmosphere. This has led to a range of severe consequences, including floods, droughts, pollution, the spread of diseases and decreased crop growth and productivity (Chausali et al. 2023). Nanotechnology offers promising solutions across many sectors, including agriculture, environment, food, medicine, catalysis, and material science. In agriculture, specifically, nanotechnology has provided advanced tools such as NFs, nanopesticides and other NMs. These innovations not only enhance productivity but also help restore ecological balance (Verma et al. 2023c). Additionally, nanosensors play a crucial role in modern agriculture by detecting agro-climatic conditions essential for improving crop productivity (Tripathi et al. 2018). This integration of nanotechnology in agriculture presents a viable strategy for mitigating the adverse impacts of climate change on crop production. Different NMs used to enhance productivity under adverse conditions have been summarized in Table 7.1.

7.4.1 Heat Stress Tolerance

Heat stress is a major challenge for plant growth and survival. NMs have shown potential in improving plant tolerance to heat stress. For instance, the application of silver NPs synthesized using moringa extract has significantly enhanced plant biomass and leaf area in wheat under heat-stress conditions (Iqbal et al. 2019). These green-synthesized AgNPs helped wheat crops to maintain tissue water content and increased chlorophyll levels compared to the control (Mariyam et al. 2024). Furthermore, the application synthesized AgNPs activated the antioxidant defence system, thereby improving heat stress tolerance. Various genes involved in regulating responses to high temperatures are influenced by different NPs. For example, pre-treating alfalfa seedlings with ZnO NPs before subjecting them to heat stress, changes the ultrastructure of mitochondria, chloroplasts and cell walls, thereby preventing heat-induced damage and promoting better plant growth. ZnO NPs significantly mitigate the heat stress in *Arabidopsis* exposed to 37 °C (Wu and Wang 2020). Foliar application of nano-ZnO significantly enhanced mungbean performance in both non-heat and heat-stressed conditions. Therefore, using nano-ZnO as a foliar spray is recommended as an effective method to protect the crop from heat stress-related damage (Kareem et al. 2022). Conversely, research found that *TaPIP1* and *TaNIP2* genes were overexpressed in wheat exposed to heat stress when silicon was applied, as opposed to Si NPs, enhancing the plants relative water content (Younis et al. 2020).

The use of selenium NPs with the bacterial strain *Lactobacillus casei*, enhanced heat stress tolerance in chrysanthemums by activating the antioxidant defence system (Seliem et al. 2020). Additionally, the application of titanium dioxide (TiO_2) improved heat tolerance in tomatoes by regulating stomatal oscillation (Qi et al. 2013). A recent study, evaluated the effects of 200 ppm TiO_2 NP pre-treatment on safflower seeds under salt and heat stress. TiO_2 NP pre-treatment improved plumule weights and reduced MDA content in cultivars under heat stress (Dinler et al. 2024). In summary, NMs, including ZnO NPs, Si NPs, Ag NPs, and TiO_2 NPs, show

Table 7.1 Summary of different nanomaterials used in crop plants for reducing the impact of biotic and abiotic stresses

Crop	Nanomaterials	Source	Biotic/Abiotic stresses	Reference
Maize	Copper nanoparticle (Cu NPs)	–	Drought	Van Nguyen et al. (2022)
	Copper oxide nanoparticles (CuO NPs)	–	Salinity	Shafiq et al. (2024)
	Green-synthesized lignin nanoparticles (LNPs)	–	Salinity	Javaid et al. (2024)
Rice	Silver nanoparticles (Ag NPs)	*Moringa oleifera*	Fungal disease	Ejaz et al. (2023)
	Zinc oxide nanoparticles (ZnO NPs)	–	Blast disease	Mazhar et al. (2023)
	Silica nanoparticles (Si NPs)	–	Drought	Ahmad and Hassim (2024)
	Zinc oxide nanoparticles (ZnO NPs)	–	Drought	Mazhar et al. (2022)
	Silicon and iron (Si–NPs & Fe)	–	Lead	Ghouri et al. (2024a)
	Genome doubling and titanium dioxide (TiO_2)	–	Lead	Ghouri et al. (2024b)
	Zinc oxide (ZnO)	–	Cold stress	Song et al. (2021)
Wheat	Iron nanoparticles (Fe NPs)	–	Cadmium and drought	Adrees et al. (2020)
	Zinc oxide nanoparticles (ZnO NPs)	–	Drought	Kausar et al. (2023)
	Silver nanoparticles (Ag NPs)	*M. oleifera*	Heat	Iqbal et al. (2019)
	Selenium nanoparticles (Se NPs)	*M. azedarach*	Spot blotch disease	Shahbaz et al. (2023)
	Titanium dioxide nanoparticles (TiO_2 NPs)	*Moringa oleifera*	Yellow stripe rust	Satti et al. (2022)
	Zinc oxide and titanium dioxide (ZnO and TiO_2)	–	Heat	Thakur et al. (2021)
	Silver and zinc nanoparticles (Ag and ZnO)	Plant leaf extract	Heat	Shukla et al. (2024)
	Selenium (SeNPs)/silica nanoparticles (SiO_2 NPs)		Salinity	Wang et al. (2024a)
Pepper	Zinc oxide nanoparticles (ZnO NP)	*Trichoderma longibrachiatum*	Root-knot nematodes	Ghareeb et al. (2024)
Soybean	Silicon nanoparticles (SiO_2 NPs)	–	Salinity	Wang et al. (2024b)
Sugarcane	Nano-selenium (Se)	–	Leaf scald disease	Shi et al. (2023)
Mungbean	Zinc oxide nanoparticles (ZnO NPs)	–	Heat	Kareem et al. (2022)
Safflower	Titanium dioxide nanoparticle (TiO_2 NPs)	–	Salinity and heat	Dinler et al. (2024)

promise in enhancing plant resilience to heat stress by modulating cellular structures, activating antioxidant defense, and regulating physiological processes. Further exploration in this field offers innovative solutions to bolster crop resilience amidst rising temperatures and climate change challenges.

7.4.2 Drought Stress Tolerance

Drought is the most critical environmental challenge that restricts plant growth and leads to more significant yield losses than other abiotic stresses. Therefore, nanotechnology offers promising applications in agriculture, with NPs synthesized from metals or metal oxides showing potential to enhance crop resilience. NMs positively impact plant growth and development by increasing antioxidant activity and reducing oxidative damage, though their effectiveness varies based on origin, size, concentration, and timing of application. They enhance drought tolerance by decreasing hydrogen peroxide and malondialdehyde levels, maintaining photosynthetic efficiency, and improving electron transmission and light absorption in chloroplasts (Rasheed et al. 2022). For instance, studies reveal the efficacy of zinc oxide nanoparticles (ZnO NPs) in mitigating drought stress in sorghum (*Sorghum bicolor*), leading to enhanced grain yield and nutrient acquisition (Dimkpa et al. 2019). Similarly, kinetin-capped zinc oxide nanoparticles (Kn-ZnO NPs) demonstrated significant mitigation of drought stress in mung bean (*Vigna radiata*) by bolstering the antioxidant defense system and increasing osmolyte levels (Ajmal et al. 2023). Silicon nanoparticles (Si NPs) applied via foliar spraying exhibited marked enhancement of drought tolerance in plants, notably at a concentration of 60 ppm, elevating physiological responses, antioxidant enzyme activities (APOX, CAT and SOD), proline content, and upregulating stress-related genes (*DREB2, MYB33, MYB3R, WRKY 19* and *SnRK 2.4*) (Boora et al. 2023).

The application of 100 ppm calcium oxide nanoparticles (CaO NPs) enhanced the photosynthetic capacity of rapeseed plants under drought conditions by modulating key physiological parameters such as stomatal conductance, photosystem II activity, and nutrient accumulation. Furthermore, CaO NPs were found to impact primary and secondary metabolic pathways, including N-metabolism, hormonal regulation, and flavonoid biosynthesis (Ayyaz et al. 2024). Foliar spraying with Se NPs significantly improved growth attributes in tomato plants under drought stress, likely due to enhanced antioxidant activity, i.e., ascorbate, glutathione, catalase, peroxidase, increased nutrient concentrations, and higher levels of phenylalanine ammonia-lyase, proline, non-protein thiols, and flavonoids (Neysanian et al. 2024). Graphene oxide NMs mitigated drought stress in soybeans by increasing the activities of antioxidant enzymes (SOD, CAT, POD and APX) and drought-related hormones (JA, SA and ABA), while reducing relative conductivity, MDA, and H_2O_2 accumulation. Additionally, GO treatment enhanced the expression of drought-related genes *GmP5CS, GmGOLS, GmDREB1* and *GmNCED1* (Zhao et al. 2022).

The application of chitosan nanoparticles (C-NPs) significantly improved various physiological and biochemical responses in *Salvia abrotanoides* when subjected to drought stress (Attaran Dowom et al. 2022). This included decreased electrolyte conductivity by 35% and increased relative water content by 12.65%, along with elevated levels of carotenoids (68%), soluble sugar (58%), chlorophyll (63%), proline (49%), protein (45.2%), flavonoid (36.4%), phenol (23.1%). Additionally, CNPs treatment led to heightened antioxidant enzyme activities, such as an 86% increase in superoxide dismutase, a 72.8% increase in polyphenol oxidase, and a 75.7% increase in guaiacol peroxidase, indicating its potential to enhance drought tolerance in regions prone to drought. The study by Mishra et al. (2023) examined the impact of calcium phosphate-urea nanoparticles (CaP-U NPs) on finger millet growth under irrigated and drought conditions, revealing notable increases in key enzyme activities such as superoxide dismutase, guaiacol peroxidase and ascorbate peroxidase. In addition, significant improvements in plant growth parameters were also observed, including shoot length, root length, shoot and root fresh/dry weight, and chlorophyll content. ZnO NPs application, particularly at 100 ppm, significantly improved growth parameters, photosynthetic activity, and antioxidant defense mechanisms, mitigating oxidative damage in cucumber induced by drought and also maintained phenol and mineral nutrient levels and increased osmolyte accumulation (Ghani et al. 2022).

Noor et al. (2022) showed that iron oxide nanoparticles (Fe_3O_2 NPs) synthesized from cumin extracts (1.2 mM) showed remarkable effectiveness in alleviating the effects of drought stress on wheat, with a notable increase in chlorophyll a, b and carotenoid levels (72%, 265% and 96%), along with improvements in proline (127%), superoxide dismutase (115%), peroxidase (43.8%) and ascorbate peroxidase (44.6%) activities. In addition, the application of this NM also resulted in a greater reduction in lipid peroxidation and electrolyte leakage, as well as an increase in soluble sugar and total Fe content in roots and shoots. Application of nano-ZnO (100 mg L^{-1}) alleviated photosynthetic pigment degradation, preserved stomatal function and increased net photosynthetic rate, thereby improving water use efficiency and increasing drought tolerance in maize (*Zea mays*).

Application of nano-vermicompost (100 mg kg^{-1} soil), significantly enhanced growth, mitigated chlorophyll decline, reduced ROS accumulation and improved membrane stability in tomato plants, while upregulating the antioxidant system and increasing osmolyte accumulation in drought-stressed conditions (Ahanger et al. 2021). Administration of exogenous ZnO and SiO_2 NPs elicited improvements in leaf gas exchange, leaf relative water content, photosynthetic pigments, leaf green index, total yield, and harvest index in potatoes subjected to diverse water deficit treatments. Notably, the highest water use efficiency (WUE) was achieved when potato plants were irrigated with 50% ETc and treated with 100 ppm of ZnO NPs, demonstrating superior performance compared to fully irrigated plants (Al-Selwey et al. 2023).

Karvar et al. (2022) found that applying titanium dioxide nanoparticles (TiO_2-NPs) at 50 ppm increased leaf superoxide dismutase (SOD), ascorbate peroxidase (APX) and catalase (CAT) antioxidant activities by 17%, 10% and 24%, respectively, compared to the control, contributing to improved drought tolerance in sweet corn under deficit irrigation. Fatima et al. (2023) explored the bioengineering of ZnO NPs using dog-rose (*Rosa canina*) aqueous extract and their conjugation with abscisic acid (ABA) to enhance drought tolerance in maize. The study found that among the various treatment methods evaluated, ABA-ZnO NPs, applied through seed imbibition, significantly increased root length, relative water content, chlorophyll and carotenoid content, proline levels, and antioxidant activities. Additionally, it improved cob length and weight while reducing stomatal opening and leaf damage index compared to the drought control. The application of chitosan NPs improved drought tolerance in periwinkle (*Catharanthus roseus*), evidenced by enhancements in stomatal conductance, total chlorophyll content, and proline accumulation, alongside increased activity of catalase and ascorbate peroxidase. Additionally, it led to elevated alkaloid accumulation, linked with up to a 5.6-fold increase in gene expression of geissoschizine synthase, strictosidine synthase, peroxidase 1 and deacetylvindoline-4-O-acetyltransferase (Ali et al. 2021).

7.4.3 Salinity Stress Tolerance

Salinity is a major agricultural stressor threatening food security for the projected 9.3 billion people by 2050. Climate change, seawater intrusion, and agricultural practices like irrigation and fertilization are accelerating soil salinization, hindering plant growth, yield, and product quality. This harsh environment disrupts plant health in three main ways: osmotic stress limits water uptake, ionic stress disrupts nutrient balance, and excess ROS damage cellular components (Li et al. 2022). Fortunately, research suggests NMs offer a powerful tool to combat salinity stress. They work through several mechanisms, including regulating ion transport and uptake, enhancing water retention, and boosting antioxidant activity in plants (Junedi et al. 2023). Studies have demonstrated the effectiveness of specific NPs against salinity. Foliar application of ZnO-NPs significantly improved photosynthetic pigments, antioxidant enzyme activity, and water balance in okra (*Abelmoschus esculentus*) across various salinity levels (Alabdallah and Alzahrani 2020). Similarly, treatment with silicon nanoparticle-based biochar (Si-BC2) enhanced growth, chlorophyll content, and water retention in wheat under both normal and saline conditions (Gill et al. 2024).

Chitosan nanoparticles (CsNPs) applied to common bean (*Phaseolus vulgaris*) plants mitigated the negative effects of salinity by improving yield, pigment content, and the antioxidant system, while reducing oxidative stress markers (Alenazi et al. 2024). Copper nanoparticles (Cu NPs) also proved effective in Fenugreek (*Trigonella foenum*), enhancing growth parameters, photosynthetic pigments, antioxidant activity, and reducing oxidative stress markers (Fouda et al. 2024). Studies with cerium oxide nanomaterials (CeO_2 NMs) on maize revealed improved salt tolerance, maintained ion balance, enhanced photosynthesis, reduced ROS levels, and increased expression of genes related to antioxidant defense (Liu et al. 2022).

Finally, bio-fabricated selenium nanoparticles (bioSeNPs) applied to rapeseed (*Brassica napus*) under salinity stress showcased multifaceted benefits. These NPs enhanced seed germination, seedling growth, water uptake efficiency, photosynthetic efficiency, and antioxidant activity, highlighting their potential to improve crop resilience and productivity (El-Badri et al. 2022). The study investigates the potential of green-synthesized lignin nanoparticles (LNPs) to mitigate salt stress in maize by characterizing LNPs and assessing their impact on seedlings under NaCl stress. LNPs effectively reduced Na^+ accumulation and lipid peroxidation, enhanced lignin absorption, photosynthetic performance, and plant growth, while upregulating salt-tolerant genes (*ZmNHX3, CBL, ZmHKT1* and *MAPK1*) and preserving cellular structures (Javaid et al. 2024).

7.4.4 Cold Stress Tolerance

Cold stress exerts a significant impact on crop viability and yield stability, with plants exhibiting varying degrees of tolerance to chilling, particularly those originating from temperate regions. The process of cold acclimation, marked by biochemical and physiological adaptations, is instrumental in enhancing freezing tolerance in plants. Although agronomic practices and management techniques can mitigate crop losses in cooler climates, they often fall short in providing comprehensive support to plants facing severe cold conditions (Adhikari et al. 2022; Choudhary et al. 2024). The utilization of nanomaterials presents a promising avenue for alleviating abiotic stresses. For instance, foliar application of ZnO-NPs has shown efficacy in reducing chilling stress in rice (*Oryza sativa*) seedlings, resulting in notable enhancements in growth parameters and chlorophyll content, alongside mitigation of oxidative stress markers. Moreover, this treatment has been found to upregulate the expression of genes associated with antioxidative defense systems and chilling response mechanisms, underscoring its potential in alleviating cold stress (Song et al. 2021).

Similarly, treatments involving TiO_2 NPs and spermine have been observed to bolster the resilience of licorice plants against cold stress, leading to reductions in oxidative damage levels typified by decreases in MDA and H_2O_2 contents (Ghabel and Karamian 2020). In chickpea, TiO_2 treatments not only averted oxidative damage in both sensitive and tolerant genotypes but also curtailed membrane damage indices under cold stress conditions (Mohammadi et al. 2013). Moreover, foliar applications of chitosan nanoparticles (C-NPs) at a concentration of 400 ppm have been shown to significantly enhance the growth and cold stress resistance of banana plants. This approach resulted in substantial increases in fresh and dry weights compared to untreated controls, underscoring its viability and economic feasibility in bolstering banana plant resilience to cold stress (Wang et al. 2021).

In a study conducted by Song et al. (2021), the synergistic application of arbuscular mycorrhiza fungus (AMF) inoculation with a foliar spray containing ZnO-NPs + Se-NPs demonstrated a significant enhancement in chili pepper growth and productivity under cold stress conditions across two consecutive winter seasons. This combined approach elicited increased fruit weight, yield, and improvements in various biochemical attributes, while concurrently activating antioxidative enzyme activity and attenuating oxidative stress markers. Furthermore, foliar application of ZnO nanoparticles (ZnO-NPs) at optimal concentrations has been shown to enhance low-temperature stress tolerance and stimulate early growth in fragrant rice cultivars. This treatment led to improvements in root dry weight, root-shoot ratio, total dry weight, chlorophyll a/b ratio, antioxidant enzyme activity, MDA content, and N-metabolizing enzyme activity, emphasizing its potential in mitigating the adverse effects of cold stress (Mai et al. 2024).

7.4.5 Heavy Metals Stress Tolerance

Heavy metal contamination poses a significant environmental concern, impeding plant growth and posing risks to human health. These metals persist in the environment for long durations as they are tough to degrade. Therefore, addressing heavy metal pollution in soil is essential for achieving a safe and sustainable environment (Rajput et al. 2022). Plants dealing with heavy metal contamination struggle with lacking vital nutrients and minerals, which disrupt their normal functions and slow down their growth (Gulzar and Mazumder 2022). Various strategies, including breeding, genetic engineering and cultural practices, have been developed to manage this stress. Recently, nanotechnology has emerged as a promising approach to enhance crop productivity under heavy metal stress conditions. Engineered NMs can greatly reduce the harmful effects of heavy metals on plants due to their large surface area and high reactivity, allowing them to quickly penetrate contaminated areas (Prakash and Chandran 2023).

Numerous studies have explored the mechanisms by which nanoparticles alleviate heavy metal stress. Green synthesis of zero-valent iron nanoparticles (ZVI NPs) using plant extracts is environmentally friendly. With a reducing potential and effectively detoxifies heavy metal compounds such as arsenic, cadmium, lead, copper and zinc in wastewater (Thilakan et al. 2022). ZnO-NPs in tomato plants effectively reduced the adverse effects of cadmium (Cd) exposure (Faizan et al. 2021). ZnO-NPs treatment also increased protein content and the activities of carbonic anhydrase and nitrate reductase in both stressed and non-stressed plants. Similarly, Cd stress negatively impacts carrot plants. However, Ag NPs and IAA (indole-3-acetic acid) can counteract these effects by detoxifying ROS (Choudhary et al. 2025). The application of Ag-NPs and IAA also promotes plant growth by lowering malondialdehyde levels (Faiz et al. 2022). Recent research showed that green synthesized zinc oxide nanoparticles (ZnO-NPs) combined with *Bacillus* spp. improved the physiological and biochemical activities of rice seedlings in heavy metal-contaminated water, enhancing seed germination, growth, and various biochemical parameters (Akhtar et al. 2022). Additionally, in peppermint, exposure to TiO_2-NP significantly improved shoot dry weight, increased lateral branches and reduced Cd uptake in plant tissues. TiO_2 NPs also boosted plastid pigment content, chlorophyll fluorescence, and various biochemical parameters while decreasing oxidative stress markers (Mohammadi et al. 2023).

Foliar application of TiO_2 NPs enhanced growth and antioxidant defense in Cd-stressed peppermint. Likewise, the use of iron oxide (FeO) NPs along with *Bacillus subtilis* enhanced the production of stress-reducing polyamines in pumpkin seedlings exposed to stress such as arsenic toxicity (Mushtaq et al. 2020). Furthermore, garden pea seedlings experience the harmful effects of chromium (Cr) phytotoxicity. However, silicon NPs protect pea seedlings from Cr toxicity by decreasing Cr accumulation and oxidative stress, strengthening the antioxidant defence system and enhancing nutrient uptake (Tripathi et al. 2015). In conclusion, NPs offer promising solutions for mitigating metal stress in crop plants. Different kinds of NPs, such as Ag-NPs, ZnO-NPs, MgO-NPs and Si-NPs have been proven to lower oxidative stress and boost the activity of antioxidant enzymes. This ultimately improves plant growth and increases their ability to tolerate heavy metal stress across various plant species.

7.5 Nanomaterials in Addressing Biotic Stresses

Climate fluctuations exert a profound influence on biotic stressors, including pathogen and pest infestations, necessitating innovative strategies for mitigation. NPs endowed with antimicrobial or antibacterial properties have emerged as pivotal tools in this endeavor, serving as potent biostimulants or NFs to counteract biotic stresses (Guzmán et al. 2022; Priya et al. 2023). A suite of non-toxic metal or metal oxide NMs, encompassing silver, copper oxides, selenium, zinc oxide, titanium dioxide, and lipid-based NPs, have been judiciously harnessed in agricultural domains to combat biotic stressors, as corroborated by Pascoli et al. (2018) and Satti et al. (2022). The distinctive physicochemical characteristics and nano-scale dimensions of these materials confer enhanced solubility and expedited translocation within plants, thereby accentuating their efficacy in alleviating biotic stresses (Salajegheh et al. 2020), while concurrently reducing reliance on conventional pesticide applications. Introducing AgNPs into rice cultivation has elicited promising results in curtailing the synthesis of antioxidative enzymes and nonenzymatic chemicals.

Notably, Aflatoxin B1, a potent hepatotoxin, accrues in plants under stressful conditions, posing significant health hazards. However, NPs integration has shown promise in mitigating this accumulation, as evidenced by Ejaz et al. (2023). Moreover, NPs-based interventions have demonstrated efficacy in combating fungal maladies such as rice blast disease, a scourge on global rice yields. Mazhar et al. (2023) reported a substantial reduction in fungal lesions following foliar application of ZnO NPs, underscoring their potency in mitigating rice blast infestations. Similarly, TiO_2 has been enlisted to combat rust disease in wheat, with Satti et al. (2022) documenting a notable decline in disease severity. This phenomenon was accompanied by the upregulation of stress-responsive genes, as delineated by proteome analyses. Furthermore, the deployment of Se NPs in wheat has shown promise in suppressing spot blotch disease, augmenting plant physiological and biochemical attributes while inhibiting fungal proliferation (Shahbaz et al. 2023).

However, the efficacy of nano-selenium-based nanomaterials in bolstering leaf scald disease resistance and enhancing sugarcane juice quality (Shi et al. 2023). Moreover, using *Trichoderma longibrachiatum* filtrate for zinc NPs synthesis has emerged as a potent biocontrol modality against soil-borne pathogens, presenting significant potential in safeguarding sweet pepper and agricultural crop yields (Ghareeb et al. 2024). Likewise, moringa chitosan nanoparticles (M-CsNPs) have proven effective in combating rice blast disease, elucidating enhanced plant vigor and modulation of the rice microbiome (Hafeez et al. 2024). In field settings, seed treatments with mesoporous silica nanoparticles (MSNPs) and chitosan-coated MSNs (CTS-MSNs) have yielded promising outcomes, mitigating *Fusarium virguliforme* disease progression while concomitantly augmenting chlorophyll content (O'Keefe et al. 2024). Lastly, Noman et al. (2024) underscored the potential of bio Fe NPs and SINCs as biostimulants and bioprotectants for fostering growth and suppressing *Fusarium* wilt in watermelon.

7.6 Role of Nanomaterials in CO_2 Sequestration

Greenhouse gas emissions, car emissions, and industrial emissions are the principal causes of increased CO_2 concentrations in the atmosphere, leading to global warming (Anderson et al. 2016). The rise in global temperature is also due to an increase in atmospheric CO_2 concentrations. Ultimately, this directly affects agricultural production worldwide. This indicates a need to reduce CO_2 concentrations in the atmosphere. Various ways have been utilized to lower the rising atmospheric CO_2 concentration in the atmosphere; the primary options are CO_2 capture, utilization, and storage (Youns et al. 2023). The generation of methane and methanol from CO_2 as a source of fuel is one commercial example of reducing the increased CO_2 concentration in the atmosphere. Instead of these strategies, nanotechnology has been employed to capture atmospheric CO_2 (Kumar et al. 2020). NMs have high CO_2 capture efficiency and have a significant role in CO_2 sequestration (Youns et al. 2023). Several types of NMs have been utilized to capture atmospheric CO_2. NMs play a key role in CO_2 capture because of their large surface area and remarkable properties. NMs are employed as adsorbents to lower ambient CO_2 levels. Calcium hydroxide, copper NPs, green NMs, carbon NMs, and graphene-based NMs assist in CO_2 adsorption (Khdary et al. 2018; Yurdacan and Sari 2021). The combination of Zr and CaO was also used as a nano-absorbent for carbon capture (Lu et al. 2009). Certain NMs oxides, such as aluminum oxides, contribute to reducing CO_2 levels (Lu et al. 2009).

Certain NPs organic hybrid components additionally help in CO_2 collection. Some NMs, such as boron nitride, are better CO_2 adsorbents and have been employed as sorbents (Sun et al. 2013). Immobilized Ag NPs were also used for effective electrochemical CO_2 reduction (Kim et al. 2015). Nickel NPs have no substantial effect on CO_2 capture (Ramsden et al. 2018). Liu et al. (2024) demonstrated the efficacy of sulfur-doped porous nanocarbons synthesized from coconut shell and potassium persulfate as potent sorbents for CO_2 capture, with significant uptake of 3.37 mmol/g at 0 °C and 1 bar, and 2.56 mmol/g at 25 °C and 1 bar. The study by Zare et al. (2019) evaluated the viability of utilizing stable water-based ZnO, TiO_2, and MWCNT nanofluids as CO_2 absorbents in a polypropylene hollow fiber membrane contactor, revealing ZnO nanofluid's effectiveness in enhancing CO_2 absorption efficiency. The AFSNP-HPAM nano-fluid, integrating amine-functionalized silica NPs derived from Moonj weed grass and hydrolyzed polyacrylamide, effectively showcased its capability in enhancing CO_2 absorption and bolstering thermal stability, thereby indicating promising prospects for subsurface carbon utilization and storage (Pandey et al. 2022). This study by Elhambakhsh et al. (2020) assessed the potential of Fe_3O_4 nanoparticles for CO_2 capture, revealing significant performance improvements with Fe_3O_4@SiO_2-lysine nanoparticles, achieving an 88% increase in capture efficiency compared to the base solvent. A bryophyte-inspired AAO, carbon, and MgO composite, created through electroanodization and hydrothermal methods, showed optimal CO_2 capture efficiency at a 4 mM $MgCl_2$ concentration due to the ideal density of MgO structures (Cortés-Valadez et al. 2024).

7.7 Conclusion and Future Perspectives

The environmental challenges posed by urbanization, severe weather conditions and global pollution are significant factors contributing to abiotic stress, raising concerns about the survival of plants worldwide. In the context of an increasing population, abiotic stressors such as salinity, extreme temperatures, heavy metals and moisture stress further impact global food security. Nanotechnology emerges as a contemporary solution for enhancing abiotic stress tolerance in plants. NPs, like TiO_2, SiO_2, and Ag have shown promise in mitigating abiotic stress damage and enhancing plant growth and crop yield. They activate plant defense mechanisms, penetrate plant tissues easily, and improve morphological, physiological, and biochemical processes, leading to increased productivity. Despite their potential, research in this area is still in its early stages and somewhat inconsistent. The widespread use of NPs raises concerns regarding their potential environmental impacts and the accumulation of NPs in edible plant parts. Therefore, there is a pressing need for focused research to develop robust evaluation methodologies for assessing the effects of NPs on both biotic and abiotic components of ecosystems. Moreover, understanding the impact of NPs on human health and establishing acceptable limits is imperative. Future research endeavors should prioritize the development of NPs that are cost-effective, non-toxic, environmentally friendly, and biodegradable to facilitate the transition of nanotechnology from laboratory settings to agricultural fields.

References

Adhikari L, Baral R, Paudel D, Min D, Makaju SO, Poudel HP, Acharya JP, Missaoui AM (2022) Cold stress in plants: strategies to improve cold tolerance in forage species. Plant Stress 4:100081

Adrees M, Khan ZS, Ali S, Hafeez M, Khalid S, ur Rehman MZ, Hussain A, Hussain K, Chatha SAS, Rizwan M (2020) Simultaneous mitigation of cadmium and drought stress in wheat by soil application of iron nanoparticles. Chemosphere 238:124681

Ahanger MA, Qi M, Huang Z, Xu X, Begum N, Qin C, Zhang C, Ahmad N, Mustafa NS, Ashraf M, Zhang L (2021) Improving growth and photosynthetic performance of drought stressed tomato by application of nano-organic fertilizer involves up-regulation of nitrogen, antioxidant and osmolyte metabolism. Ecotoxicol Environ Saf 216:112195

Ahluwalia O, Singh PC, Bhatia R (2021) A review on drought stress in plants: implications, mitigation and the role of plant growth promoting rhizobacteria. Resour Environ Sustain 5:100032

Ahmad A, Hassim MFN (2024) Effects of silica nanoparticles on morpho-histological and antioxidant activities of rice seedlings under drought stress. S Afr J Bot 168:497–508

Ajmal M, Ullah R, Muhammad Z, Khan MN, Kakar HA, Kaplan A, Okla MK, Saleh IA, Kamal A, Abdullah A, Abdul Razak S (2023) Kinetin capped zinc oxide nanoparticles improve plant growth and ameliorate resistivity to Polyethylene Glycol (PEG)-induced drought stress in *Vigna radiata* (L.) R. Wilczek (Mung Bean). Molecules 28(13):5059

Akhtar N, Khan S, Rehman SU, Rha ES, Jamil M (2022) Combined effect of zinc oxide nanoparticles and bacteria on osmolytes and antioxidative parameters of rice (*Oryza sativa* L.) plant grown in heavy metal-contaminated water. Adsorpt Sci Technol 2022:4148765

Alabdallah NM, Alzahrani HS (2020) The potential mitigation effect of ZnO nanoparticles on [Abelmoschus esculentus L. Moench] metabolism under salt stress conditions. Saudi J Biol Sci 27(11):3132–3137

Alenazi MM, El-Ebidy AM, El-shehaby OA, Seleiman MF, Aldhuwaib KJ, Abdel-Aziz HMM (2024) Chitosan and Chitosan nanoparticles differentially alleviate salinity stress in *Phaseolus vulgaris* L. plants. Plants 13(3):398

Ali EF, El-Shehawi AM, Ibrahim OHM, Abdul-Hafeez EY, Moussa MM, Hassan FAS (2021) A vital role of chitosan nanoparticles in improvisation the drought stress tolerance in *Catharanthus roseus* (L.) through biochemical and gene expression modulation. Plant Physiol Biochem 161:166–175

Al-Selwey WA, Alsadon AA, Ibrahim AA, Labis JP, Seleiman MF (2023) Effects of zinc oxide and silicon dioxide nanoparticles on physiological, yield, and water use efficiency traits of potato grown under water deficit. Plan Theory 12(1):218

Anderson TR, Hawkins E, Jones PD (2016) CO_2, the greenhouse effect and global warming: from the pioneering work of Arrhenius and Callendar to today's earth system models. Endeavour 40(3):178–187

Attaran Dowom S, Karimian Z, Mostafaei Dehnavi M, Samiei L (2022) Chitosan nanoparticles improve physiological and biochemical responses of *Salvia abrotanoides* (Kar.) under drought stress. BMC Plant Biol 22(1):364

Ayyaz A, Batool I, Zhang K, Hannan F, Sun Y, Qin T, Athar H-R, Zafar ZU, Farooq MA, Zhou W (2024) Unravelling mechanisms of CaO nanoparticle-induced drought tolerance in *Brassica napus*: an analysis of metabolite and nutrient profiling. Environ Sci: Nano 11(6):2550–2567

Balmer A, Pastor V, Gamir J, Flors V, Mauch-Mani B (2015) The 'prime-ome': towards a holistic approach to priming. Trends Plant Sci 20(7):443–452

Boora R, Sheoran P, Rani N, Kumari S, Thakur R, Grewal S (2023) Biosynthesized Silica Nanoparticles (Si NPs) helps in mitigating drought stress in wheat through physiological changes and upregulation of stress genes. SILICON 15(13):5565–5577

Chausali N, Saxena J, Prasad R (2023) Nanotechnology as a sustainable approach for combating the environmental effects of climate change. J Agric Food Res 12:100541

Choudhary R, Rajput VD, Ghodake G, Ahmad F, Meena M, Rehman RU et al (2024) Comprehensive journey from past to present to future about seed priming with hydrogen peroxide and hydrogen sulfide concerning drought, temperature, UV and ozone stresses-a review. Plant Soil 500:1–23. https://doi.org/10.1007/s11104-024-06499-9

Choudhary R, Ahmad F, Kaya C, Upadhyay SK, Muneer S, Kumar V, Meena M, Liu H, Upadhyay H, Seth CS (2025) Decrypting proteomics, transcriptomics, genomics, and integrated omics for augmenting the abiotic, biotic, and climate change stress resilience in plants. J Plant Physiol. https://doi.org/10.1016/j.jplph.2025.154430

Cortés-Valadez PJ, Baños-López E, Hernández-Rodríguez YM, Cigarroa-Mayorga OE (2024) Bryophyte-bioinspired nanoporous AAO/C/MgO composite for enhanced CO_2 capture: the role of MgO. Nanomaterials 14(8):658

Devasirvatham V, Tan DK (2018) Impact of high temperature and drought stresses on chickpea production. Agronomy 8(8):145

Dimkpa CO, Singh U, Bindraban PS, Elmer WH, Gardea-Torresdey JL, White JC (2019) Zinc oxide nanoparticles alleviate drought-induced alterations in sorghum performance, nutrient acquisition, and grain fortification. Sci Total Environ 688:926–934

Dinler BS, Cetinkaya H, Koc FN, Gül V, Sefaoğlu F (2024) Effects of titanium dioxide nanoparticles against salt and heat stress in safflower cultivars. Acta Botanica Brasilica 38:e20230136

Dubey R, Pathak H, Chakrabarti B, Singh S, Gupta DK, Harit RC (2020) Impact of terminal heat stress on wheat yield in India and options for adaptation. Agric Syst 181:102826

Duhan JS, Kumar R, Kumar N, Kaur P, Nehra K, Duhan S (2017) Nanotechnology: the new perspective in precision agriculture. Biotechnol Rep 15:11–23

Ejaz M, Raja NI, Khan SA, Mashwani ZUR, Hanif A, Iqbal M, Hussain M, Syed A, Iqbal RK, Qureshi H, Anwar T (2023) Biosynthesized silver nanoparticles ameliorate biotic stress in rice (*Oryza sativa*) by intricating biochemical and mineral profile. Pak J Bot 55(6):2019–2028

El-Badri AM, Batool M, Mohamed IAA, Wang Z, Wang C, Tabl KM, Khatab A, Kuai J, Wang J, Wang B, Zhou G (2022) Mitigation of the salinity stress in rapeseed (Brassica napus L.) productivity by exogenous applications of bio-selenium nanoparticles during the early seedling stage. Environ Pollut 310:119815

Elhambakhsh A, Zaeri MR, Mehdipour M, Keshavarz P (2020) Synthesis of different modified magnetic nanoparticles for selective physical/chemical absorption of CO2 in a bubble column reactor. J Environ Chem Eng 8(5):104195

El-Saadony MT, Saad AM, Soliman SM, Salem HM, Desoky ESM, Babalghith AO, El-Tahan AM, Ibrahim OM, Ebrahim AA, Abd El-Mageed TA, Elrys AS (2022) Role of nanoparticles in enhancing crop tolerance to abiotic stress: a comprehensive review. Front Plant Sci 13:946717

Faiz S, Yasin NA, Khan WU, Shah AA, Akram W, Ahmad A, Ali A, Naveed NH, Riaz L (2022) Role of magnesium oxide nanoparticles in the mitigation of lead-induced stress in Daucus carota: modulation in polyamines and antioxidant enzymes. Int J Phytoremediation 24(4):364–372

Faizan M, Faraz A, Mir AR, Hayat S (2021) Role of zinc oxide nanoparticles in countering negative effects generated by cadmium in *Lycopersicon esculentum*. J Plant Growth Regul 40:101–115

Fatima A, Safdar N, Ain N, Yasmin A, Chaudhry G-S (2023) Abscisic acid-loaded ZnO nanoparticles as drought tolerance inducers in *Zea mays* L. with physiological and biochemical attributes. J Plant Growth Regul 42(11):7280–7293

Fouda HM, Saied E, Abdelmouty ES, Osman MS (2024) Ameliorative role of copper nanoparticle in alleviating salt-induced oxidative stress in fenugreek (*Trigonella foenum-graecum* L.) plants. Biocatal Agric Biotechnol 57:103095

Francini A, Sebastiani L (2019) Abiotic stress effects on performance of horticultural crops. Horticulturae 5(4):67

Ghabel VK, Karamian R (2020) Effects of TiO2 nanoparticles and spermine on antioxidant responses of Glycyrrhiza glabra L. to cold stress. Acta Bot Croat 79(2):137–147

Ghani MI, Saleem S, Rather SA, Rehmani MS, Alamri S, Rajput VD, Kalaji HM, Saleem N, Sial TA, Liu M (2022) Foliar application of zinc oxide nanoparticles: an effective strategy to mitigate drought stress in cucumber seedling by modulating antioxidant defense system and osmolytes accumulation. Chemosphere 289:133202

Ghareeb RY, Belal EB, El-Khateeb NMM, Shreef BA (2024) Utilizing bio-synthesis of nanomaterials as biological agents for controlling soil-borne diseases in pepper plants: root-knot nematodes and root rot fungus. BMC Plant Biol 24(1):110

Ghosh D, Xu J (2014) Abiotic stress responses in plant roots: a proteomics perspective. Front Plant Sci 5:66977

Ghouri F, Sarwar S, Sun L, Riaz M, Haider FU, Ashraf H, Lai M, Imran M, Liu J, Ali S, Liu X (2024a) Silicon and iron nanoparticles protect rice against lead (Pb) stress by improving oxidative tolerance and minimizing Pb uptake. Sci Rep 14(1):5986

Ghouri F, Shahid MJ, Zhong M, Zia MA, Alomrani SO, Liu J, Sun L, Ali S, Liu X, Shahid MQ (2024b) Alleviated lead toxicity in rice plant by co-augmented action of genome doubling and TiO_2 nanoparticles on gene expression, cytological and physiological changes. Sci Total Environ 911:168709

Gill S, Ramzan M, Naz G, Ali L, Danish S, Ansari MJ, Salmen SH (2024) Effect of silicon nanoparticle-based biochar on wheat growth, antioxidants and nutrients concentration under salinity stress. Sci Rep 14(1):6380

Gonzalez Guzman M, Cellini F, Fotopoulos V, Balestrini R, Arbona V (2022) New approaches to improve crop tolerance to biotic and abiotic stresses. Physiol Plant 174(1):e13547

Gulzar ABM, Mazumder PB (2022) Helping plants to deal with heavy metal stress: the role of nanotechnology and plant growth promoting rhizobacteria in the process of phytoremediation. Environ Sci Pollut Res 29(27):40319–40341

Hafeez R, Guo J, Ahmed T, Jiang H, Raza M, Shahid M, Ibrahim E, Wang Y, Wang J, Yan C, An Q, White JC, Li B (2024) Bio-formulated chitosan nanoparticles enhance disease resistance against rice blast by physiomorphic, transcriptional, and microbiome modulation of rice (Oryza sativa L.). Carbohydr Polym 334:122023

He X, Deng H, Hwang HM (2019) The current application of nanotechnology in food and agriculture. J Food Drug Anal 27(1):1–21

Ioannou A, Gohari G, Papaphilippou P, Panahirad S, Akbari A, Dadpour MR, Krasia-Christoforou T, Fotopoulos V (2020) Advanced nanomaterials in agriculture under a changing climate: the way to the future? Environ Exp Bot 176:104048

Iqbal M, Raja NI, Mashwani ZUR, Hussain M, Ejaz M, Yasmeen F (2019) Effect of silver nanoparticles on growth of wheat under heat stress. Iran J Sci Technol Trans A: Sci 43:387–395

Janni M, Maestri E, Gullì M, Marmiroli M, Marmiroli N (2024) Plant responses to climate change, how global warming may impact on food security: a critical review. Front Plant Sci 14:1297569

Javaid MH, Chen N, Yasin MU, Fan X, Neelam A, Rehman M, Haider Z, Bukhari SAH, Munir R, Ahmad I, Gan Y (2024) Green-synthesized lignin nanoparticles enhance Zea mays resilience to salt stress by improving antioxidant metabolism and mitigating ultrastructural damage. Chemosphere 359:142337

Jeevanandam J, Barhoum A, Chan YS, Dufresne A, Danquah MK (2018) Review on nanoparticles and nanostructured materials: history, sources, toxicity and regulations. Beilstein J Nanotechnol 9(1):1050–1074

Junedi MA, Mukhopadhyay R, Manjari KS (2023) Alleviating salinity stress in crop plants using new engineered nanoparticles (ENPs). Plant Stress 9:100184

Kah M, Tufenkji N, White JC (2019) Nano-enabled strategies to enhance crop nutrition and protection. Nat Nanotechnol 14(6):532–540

Kan Y, Mu XR, Gao J, Lin HX, Lin Y (2023) The molecular basis of heat stress responses in plants. Mol Plant 16(10):1612–1634

Kareem HA, Saleem MF, Saleem S, Rather SA, Wani SH, Siddiqui MH, Alamri S, Kumar R, Gaikwad NB, Guo Z, Niu J (2022) Zinc oxide nanoparticles interplay with physiological and biochemical attributes in terminal heat stress alleviation in mungbean (*Vigna radiata* L.). Front Plant Sci 13:842349

Karvar M, Azari A, Rahimi A, Maddah-Hosseini S, Ahmadi-Lahijani MJ (2022) Titanium dioxide nanoparticles (TiO2-NPs) enhance drought tolerance and grain yield of sweet corn (*Zea mays* L.) under deficit irrigation regimes. Acta Physiol Plant 44(2):14

Kausar A, Hussain S, Javed T, Zafar S, Anwar S, Hussain S et al (2023) Zinc oxide nanoparticles as potential hallmarks for enhancing drought stress tolerance in wheat seedlings. Plant Physiol Biochem 195:341–350

Khdary NH, Ghanem MA, Abdesalam ME, Al-Garadah MM (2018) Sequestration of CO_2 using Cu nanoparticles supported on spherical and rod-shape mesoporous silica. J Saudi Chem Soc 22(3):343–351

Kim C, Jeon HS, Eom T, Jee MS, Kim H, Friend CM, Min BK, Hwang YJ (2015) Achieving selective and efficient electrocatalytic activity for CO2 reduction using immobilized silver nanoparticles. J Am Chem Soc 137(43):13844–13850

Kumar R, Mangalapuri R, Ahmadi MH, Vo DVN, Solanki R, Kumar P (2020) The role of nanotechnology on post-combustion CO_2 absorption in process industries. Int J Low-Carbon Technol 15(3):361–367

Kumar D, Mariyam S, Gupta KJ, Thiruvengadam M, Ghodake GS, Xing B, Seth CS (2024a) Comparative investigation on chemical and green synthesized titanium dioxide nanoparticles against chromium (VI) stress eliciting differential physiological, biochemical, and cellular attributes in Helianthus annuus L. Sci Total Environ 930:172413. https://doi.org/10.1016/j.scitotenv.2024.172413

Kumar D, Singh R, Upadhyay SK, Verma KK, Tripathi RM, Liu H et al (2024b) Review on interactions between nanomaterials and phytohormones: novel perspectives and opportunities for mitigating environmental challenges. Plant Sci 340:111964. https://doi.org/10.1016/j.plantsci.2023.111964

Kumar D, Biswas JK, Mulla SI, Singh R, Shukla R, Ahanger MA et al (2024c) Micro and nanoplastics pollution: sources, distribution, uptake in plants, toxicological effects, and innovative remediation strategies for environmental sustainability. Plant Physiol Biochem 108795. https://doi.org/10.1016/j.plaphy.2024.108795

Kumari A, Lakshmi GA, Krishna GK, Patni B, Prakash S, Bhattacharyya M, Singh SK, Verma KK (2022) Climate change and its impact on crops: a comprehensive investigation for sustainable agriculture. Agronomy 12:3008. https://doi.org/10.3390/agronomy12123008

Kumari A, Dash M, Singh SK, Jagadesh M, Mathpal B, Mishra PK, Pandey SK, Verma KK (2023) Soil microbes: a natural solution for mitigating the impact of climate change. Environ Monit Assess 195:1436. https://doi.org/10.1007/s10661-023-11988-y

Li Z, Zhu L, Zhao F, Li J, Zhang X, Kong X, Wu H, Zhang Z (2022) Plant salinity stress response and nano-enabled plant salt tolerance. Front Plant Sci 13:843994

Liu Y, Cao X, Yue L, Wang C, Tao M, Wang Z, Xing B (2022) Foliar-applied cerium oxide nanomaterials improve maize yield under salinity stress: reactive oxygen species homeostasis and rhizobacteria regulation. Environ Pollut 299:118900

Liu C, Zhi Y, Yu Q, Tian L, Demir M, Colak SG, Farghaly AA, Wang L, Hu X (2024) Sulfur-enriched nanoporous carbon: a novel approach to CO2 adsorption. ACS Applied Nano Materials 7(5):5434–5441

Lu H, Khan A, Pratsinis SE, Smirniotis PG (2009) Flame-made durable doped-CaO nanosorbents for CO2 capture. Energy Fuel 23(2):1093–1100

Lyman NB, Jagadish KS, Nalley LL, Dixon BL, Siebenmorgen T (2013) Neglecting rice milling yield and quality underestimates economic losses from high-temperature stress. PLoS One 8(8):e72157

Mai Y, Ren Y, Deng S, Ashraf U, Tang X, Duan M, Mo Z (2024) Influence of ZnO nanoparticles on early growth stage of fragrant rice at low temperature (LT) stress. J Soil Sci Plant Nutr 24(1):1301–1317

Mariyam S, Upadhyay SK, Chakraborty K, Verma KK, Duhan JS, Muneer S et al (2024) Nanotechnology, a frontier in agricultural science, a novel approach in abiotic stress management and convergence with new age medicine-A review. Sci Total Environ 912:169097. https://doi.org/10.1016/j.scitotenv.2023.169097

Mazhar WM, Ishtiaq M, Hussain I, Parveen A, Hayat Bhatti K, Azeem M, Thind S, Ajaib M, Maqbool M, Sardar T, Muzammil K (2022) Seed nano-priming with Zinc Oxide nanoparticles in rice mitigates drought and enhances agronomic profile. PLoS One 17(3):e0264967

Mazhar MW, Ishtiaq M, Maqbool M, Hussain SA (2023) Foliar application of zinc oxide nanoparticles improves rice yield under biotic stress posed by *Magnaporthe oryzae*. Arch Phytopathol Plant Protect 56(14):1093–1111

Mekuye B, Abera B (2023) Nanomaterials: an overview of synthesis, classification, characterization, and applications. Nano Select 4(8):486–501

Mishra D, Chitara MK, Upadhayay VK, Singh JP, Chaturvedi P (2023) Plant growth promoting potential of urea doped calcium phosphate nanoparticles in finger millet (*Eleusine coracana* (L.) Gaertn.) under drought stress. Front Plant Sci 14:1137002

Mohammadi R, Maali-Amiri R, Abbasi A (2013) Effect of TiO2 nanoparticles on chickpea response to cold stress. Biol Trace Elem Res 152:403–410

Mohammadi H, Kazemi Z, Aghaee A, Hazrati S, Golzari Dehno R, Ghorbanpour M (2023) Unraveling the influence of TiO_2 nanoparticles on growth, physiological and phytochemical characteristics of Mentha piperita L. in cadmium-contaminated soil. Sci Rep 13(1):22280

Mushtaq T, Shah AA, Akram W, Yasin NA (2020) Synergistic ameliorative effect of iron oxide nanoparticles and Bacillus subtilis S4 against arsenic toxicity in Cucurbita moschata: polyamines, antioxidants, and physiochemical studies. Int J Phytoremediation 22(13):1408–1419

Neme K, Nafady A, Uddin S, Tola YB (2021) Application of nanotechnology in agriculture, postharvest loss reduction and food processing: food security implication and challenges. Heliyon 7(12)

Neysanian M, Iranbakhsh A, Ahmadvand R, Ardebili ZO, Ebadi M (2024) Selenium nanoparticles conferred drought tolerance in tomato plants by altering the transcription pattern of microRNA-172 (miR-172), bZIP, and CRTISO genes, upregulating the antioxidant system, and stimulating secondary metabolism. Protoplasma 261(4):735–747

Noman M, Ahmed T, Shahid M, Nazir MM, Azizullah, Li D, Song F (2024) Salicylic acid-doped iron nano-biostimulants potentiate defense responses and suppress Fusarium wilt in watermelon. J Adv Res 59:19–33

Noor R, Yasmin H, Ilyas N, Nosheen A, Hassan MN, Mumtaz S, Khan N, Ahmad A, Ahmad P (2022) Comparative analysis of iron oxide nanoparticles synthesized from ginger (*Zingiber officinale*) and cumin seeds (*Cuminum cyminum*) to induce resistance in wheat against drought stress. Chemosphere 292:133201

O'Keefe TL, Deng C, Wang Y, Mohamud S, Torres-Gómez A, Tuga B, Huang C-H, Alvarez Reyes WR, White JC, Haynes CL (2024) Chitosan-coated mesoporous silica nanoparticles for suppression of *Fusarium virguliforme* in soybeans (*Glycine max*). ACS Agric Sci Technol 4(5):580–592

Pandey N, Chaturvedi KR, Sharma T, Ojha U (2022) A polymeric suspension of amine functionalized silica nanoparticles derived from Moonj grass for the carbon capture and storage applications. J Macromol Sci A 59(2):133–147

Pascoli M, Lopes-Oliveira PJ, Fraceto LF, Seabra AB, Oliveira HC (2018) State of the art of polymeric nanoparticles as carrier systems with agricultural applications: a minireview. Energy Ecol Environ 3:137–148

Prakash P, Chandran S (2023) Nano-phytoremediation of heavy metals from soil: a critical review. Pollutants 3(3):360–380

Priya S, Kumar A, Upadhayay VK, Chaudhary A, Parveen H, Kumar G (2023) Impact of nanoparticles on biotic stress tolerance. In: Advances in nanotechnology for smart agriculture. CRC Press, pp 197–220

Qi M, Liu Y, Li T (2013) Nano-TiO2 improve the photosynthesis of tomato leaves under mild heat stress. Biol Trace Elem Res 156:323–328

Raj S, Roodbar S, Brinkley C, Wolfe DW (2022) Food security and climate change: differences in impacts and adaptation strategies for rural communities in the Global South and North. Front Sustain Food Syst 5:691191

Rajput VD, Minkina T, Kumari A, Harish, Singh VK, Verma KK, Mandzhieva S, Sushkova S, Srivastava S, Keswani C (2021a) Coping with the challenges of abiotic stress in plants: new dimensions in the field application of nanoparticles. Plan Theory 10:1221. https://doi.org/10.3390/plants10061221

Rajput VD, Singh A, Minkina TM, Shende SS, Kumar P, Verma KK, Bauer T, Gorobtsova O, Deneva S, Sindireva A (2021b) Potential applications of nanobiotechnology in plant nutrition and protection for sustainable agriculture. In: Ingle AP (ed) Nanotechnology in plant growth promotion and protection: recent advances and impacts, 1st edn. Wiley

Rajput VD, Minkina T, Upadhyay SK, Kumari A, Ranjan A, Mandzhieva S, Sushkova S, Singh RK, Verma KK (2022) Nanotechnology in the restoration of polluted soil. Nanomaterials 12:769. https://doi.org/10.3390/nano12050769

Rajput VD, Kumari A, Upadhyay SK, Minkina T, Mandzhieva S, Ranjan A, Sushkova S, Burachevskaya M, Rajput P, Konstantinova E, Singh J, Verma KK (2023) Can nanomaterials improve the soil microbiome and crop productivity? Agriculture 13:231. https://doi.org/10.3390/agriculture13020231

Ramsden JJ, Sokolov IJ, Malik DJ (2018) Questioning the catalytic effect of Ni nanoparticles on CO2 hydration and the very need of such catalysis for CO2 capture by mineralization from aqueous solution. Chem Eng Sci 175:162–167

Rasheed A, Li H, Tahir MM, Mahmood A, Nawaz M, Shah AN, Aslam MT, Negm S, Moustafa M, Hassan MU, Wu Z (2022) The role of nanoparticles in plant biochemical, physiological, and molecular responses under drought stress: a review. Front Plant Sci 13

Salajegheh M, Yavarzadeh M, Payandeh A, Akbarian MM (2020) Effects of titanium and silicon nanoparticles on antioxidant enzymes activity and some biochemical properties of *Cuminum cyminum* L. under drought stress. Banat's J Biotechnol 11(21):19–25

Satti SH, Raja NI, Ikram M, Oraby HF, Mashwani ZUR, Mohamed AH, Singh A, Omar AA (2022) Plant-based titanium dioxide nanoparticles trigger biochemical and proteome modifications in *Triticum aestivum* L. under biotic stress of *Puccinia striiformis*. Molecules 27(13):4274

Savvides A, Ali S, Tester M, Fotopoulos V (2016) Chemical priming of plants against multiple abiotic stresses: mission possible? Trends Plant Sci 21(4):329–340

Seliem MK, Hafez Y, El-Ramady H (2020) Using nano-selenium in reducing the negative effects of high temperature stress on *Chrysanthemum morifolium* Ramat. J Sustain Agric Sci 46(3):47–60

Seppelt R, Klotz S, Peiter E, Volk M (2022) Agriculture and food security under a changing climate: an underestimated challenge. Iscience 25(12)

Shafiq H, Shani MY, Ashraf MY, De Mastro F, Cocozza C, Abbas S, Ali N, Tahir A, Iqbal M, Khan Z, Gul N (2024) Copper oxide nanoparticles induced growth and physio-biochemical changes in maize (*Zea mays* L.) in saline soil. Plan Theory 13(8):1080

Shahbaz M, Akram A, Mehak A, Haq EU, Fatima N, Wareen G, Fitriatin BN, Sayyed RZ, Ilyas N, Sabullah MK (2023) Evaluation of selenium nanoparticles in inducing disease resistance against spot blotch disease and promoting growth in wheat under biotic stress. Plan Theory 12(4):761

Sharma RK, Dhillon J, Kumar P, Mulvaney MJ, Reed V, Bheemanahalli R, Cox MS, Kukal MS, Reddy KN (2023) Climate trends and soybean production since 1970 in Mississippi: empirical evidence from ARDL model. Sci Total Environ 905:167046

Shi MT, Zhang TJ, Fang Y, Pan CP, Fu HY, Gao SJ (2023) Nano-selenium enhances sugarcane resistance to *Xanthomonas albilineans* infection and improvement of juice quality. Ecotoxicol Environ Saf 254:114759

Shukla G, Singh A, Chaudhary N, Singh S, Basnal N, Gaurav SS (2024) Metal nanoparticles to improve the heat resilience in wheat (*Triticum aestivum* L.). Nanotechnology 35(20):205101

Song Y, Jiang M, Zhang H, Li R (2021) Zinc oxide nanoparticles alleviate chilling stress in rice (*Oryza Sativa* L.) by regulating antioxidative system and chilling response transcription factors. Molecules 26(8):2196

Sun Q, Li Z, Searles DJ, Chen Y, Lu G, Du A (2013) Charge-controlled switchable CO_2 capture on boron nitride nanomaterials. J Am Chem Soc 135(22):8246–8253

Thakur S, Asthir B, Kaur G, Kalia A, Sharma A (2021) Zinc oxide and titanium dioxide nanoparticles influence heat stress tolerance mediated by antioxidant defense system in wheat. Cereal Res Commun:1–12

Thilakan D, Patankar J, Khadtare S, Wagh NS, Lakkakula J, El-Hady KM, Islam S, Islam MR, Khan MS, Alafaleq NO, Tarique M (2022) Plant-derived iron nanoparticles for removal of heavy metals. Int J Chem Eng 2022:1517849

Tripathi DK, Singh VP, Prasad SM, Chauhan DK, Dubey NK (2015) Silicon nanoparticles (SiNp) alleviate chromium (VI) phytotoxicity in *Pisum sativum* (L.) seedlings. Plant Physiol Biochem 96:189–198

Tripathi M, Kumar S, Kumar A, Tripathi P, Kumar S (2018) Agro-nanotechnology: a future technology for sustainable agriculture. Int J Curr Microbiol Appl Sci 7:196–200

Tripathi S, Mahra S, Tiwari K, Rana S, Tripathi DK, Sharma S, Sahi S (2023) Recent advances and perspectives of nanomaterials in agricultural management and associated environmental risk: a review. Nanomaterials 13(10):1604

Upadhayay VK, Chitara MK, Mishra D, Jha MN, Jaiswal A, Kumari G, Ghosh S, Patel VK, Naitam MG, Singh AK, Pareek N, Taj G, Maithani D, Kumar A, Dasila H, Sharma A (2023a) Synergistic impact of nanomaterials and plant probiotics in agriculture: a tale of two-way strategy for long-term sustainability. Front Microbiol 14:1133968

Upadhayay VK, Naitam MG, Ghosh S, Patel VK, Sharma A, Chandra AK, Taj G (2023b) Techniques for integrating nanotechnology in agriculture. In: Advances in nanotechnology for smart agriculture. CRC Press, pp 19–40

Van Nguyen D, Nguyen HM, Le NT, Nguyen KH, Nguyen HT, Le HM, Nguyen AT, Dinh NTT, Hoang SA, Van Ha C (2022) Copper nanoparticle application enhances plant growth and grain yield in maize under drought stress conditions. J Plant Growth Regul 41(1):364–375

Verma KK, Song XP, Joshi A, Rajput VD, Singh M, Sharma A, Singh RK, Li DM, Arora J, Minkina T, Li YR (2022a) Nanofertilizer possibilities for healthy soil, water and food in future: an overview. Front Plant Sci 13:865048. https://doi.org/10.3389/fpls.2022.865048

Verma KK, Song XP, Joshi A, Tian DD, Rajput VD, Singh M, Arora J, Minkina TM, Li YR (2022b) Recent trends in nano-fertilizer for advancing sustainable agriculture under the era of climate change to ensure food security for future generations. Nanomaterials 12:173. https://doi.org/10.3390/nano12010173

Verma KK, Song X, Degu HD, Guo D, Joshi A, Huang H, Xu L, Singh M, Huang D, Rajput VD, Li Y (2023a) Recent advances in nitrogen and nano-nitrogen fertilizers for sustainable crop production: a mini-review. Chem Biol Technol Agric 10:1–14. https://doi.org/10.1186/s40538-023-00488-3

Verma KK, Song X, Verma CL, Huang H, Singh M, Xu L, Wu J, Khan ZH, Zhang B, Li Y (2023b) Mathematical modeling of climate and fluoride effects on sugarcane photosynthesis with silicon nanoparticles. Plant Physiol Biochem 204:108089. https://doi.org/10.1016/j.plaphy.2023.108089

Verma KK, Singh M, Rajput VD, Verma CL, Burachevskaya M (2023c) Nanofertilizers: a sustainable alternative to conventional means. In: Rajput VD, Singh A, Minkina TM, Verma KK, Singh AK (eds) Nanotechnology for sustainable agriculture: an innovative and eco-friendly approach. Apple Academic Press, Inc., Taylor & Francis. https://doi.org/10.1201/9781003333128

Verma KK, Song XP, Kumari A, Jagadesh M, Singh SK, Bhatt R, Singh M, Seth CS, Li YR (2024) Climate change adaptation: challenges for agricultural sustainability. Plant Cell Environ:1–12. https://doi.org/10.1111/pce.15078

Wang A, Li J, Al-Huqail AA, Al-Harbi MS, Ali EF, Wang J, Ding Z, Rekaby SA, Ghoneim AM, Eissa MA (2021) Mechanisms of chitosan nanoparticles in the regulation of cold stress resistance in banana plants. Nanomaterials 11(10):2670

Wang M, Mu C, Lin X, Ma W, Wu H, Si D, Ge C, Cheng C, Zhao L, Li H, Zhou D (2024a) Foliar application of nanoparticles reduced cadmium content in wheat (*Triticum aestivum* L.) grains via long-distance "leaf–root–microorganism" regulation. Environ Sci Technol 58(16):6900–6912

Wang P, Zhang H, Hu X, Xu L, An X, Jin T, Ma R, Li Z, Chen S, Du S, Wei G (2024b) Comparing the potential of silicon nanoparticles and conventional silicon for salinity stress alleviation in soybean (*Glycine max* L.): growth and physiological traits and rhizosphere/endophytic bacterial communities. J Agric Food Chem 72:10781–10793

Wu J, Wang T (2020) Synergistic effect of zinc oxide nanoparticles and heat stress on the alleviation of transcriptional gene silencing in *Arabidopsis thaliana*. Bull Environ Contam Toxicol 104(1):49–56

Younis AA, Khattab H, Emam MM (2020) Impacts of silicon and silicon nanoparticles on leaf ultrastructure and TaPIP1 and TaNIP2 gene expressions in heat stressed wheat seedlings. Biol Plant 64(1)

Youns YT, Manshad AK, Ali JA (2023) Sustainable aspects behind the application of nanotechnology in CO_2 sequestration. Fuel 349:128680

Yurdacan HM, Sari MM (2021) Functional green-based nanomaterials towards sustainable carbon capture and sequestration. In: Sustainable materials for transitional and alternative energy. Gulf Professional Publishing, pp 125–177

Zare P, Keshavarz P, Mowla D (2019) Membrane absorption coupling process for CO_2 capture: application of water-based ZnO, TiO2, and multi-walled carbon nanotube nanofluids. Energy Fuel 33(2):1392–1403

Zhao L, Wang W, Fu X, Liu A, Cao J, Liu J (2022) Graphene oxide, a novel nanomaterial as soil water retention agent, dramatically enhances drought stress tolerance in soybean plants. Front Plant Sci 13:810905

Chapter 8
Engineered Nanomaterials Enhancing Sustainability and Tolerance in Advanced Agri-farming Systems

Ravinder Kumar, Surekha, Vishnu D. Rajput, Baljeet Singh Saharan, and Joginder Singh Duhan

8.1 Introduction

The ever-increasing global population presents a significant challenge for agriculture, such as producing enough food sustainably while facing declining arable land, climate change, and resource scarcity. Precisely diagnosing plant diseases or applying pest management strategies may help minimize crop losses (Verma et al. 2024). Rapid, accurate, highly sensitive plant disease diagnostic techniques are essential to prevent the spread of diseases in large areas, minimize crop loss, and improve agricultural yield. The total loss, globally, due to pests ranges from 50% in wheat to over and above 80% in cotton. Other crops that suffer significant production losses due to pests are soybean (26–29%) as well as maize, rice, and potatoes (31%, 37%, and 40%) (Oerke 2006; Kashyap et al. 2016).

Climate stressors are wreaking havoc on global agriculture, with profound implications for food production and demand. Food security is put at risk by continuously changing global climate conditions that result in prolonged climate extremes like drought and flooding, intense precipitation, abrupt temperature swings, soil salinization, land degradation, and decreased fertility, soil compaction, reduced

R. Kumar (✉) · J. S. Duhan
Department of Biotechnology, Chaudhary Devi Lal University, Sirsa, India

Surekha
Department of Botany, Government National College, Sirsa, India

V. D. Rajput
Academy of Biology and Biotechnology, Southern Federal University, Rostov-on-Don, Russia

B. S. Saharan
Department of Microbiology, Chaudhary Charan Singh Haryana Agricultural University, Hisar, India

C. S. Seth et al. (eds.), *Nano-formulations for Sustainable Agriculture*, Nanotechnology in the Life Sciences, https://doi.org/10.1007/978-3-031-90617-6_8

microbial diversity, and war (Kumari et al. 2023). These factors have a significant negative yield impact on global agriculture. The long-term, intense use of agrochemicals, or chemical pesticides and fertilizers, exposes agriculturally appropriate land to heavy metals like Cr, Cd, Hg, Pb, Cu, and Zn (Alengebawy et al. 2021). With an increasing population comes overuse of natural resources and unsustainable farming methods that further tax the land and soil. Because of one or more abiotic pressures, around 90% of the arable land is in high danger. Abiotic stressors have the potential to significantly reduce agricultural yields in main food crops grown worldwide by around 51–82% (Kumar et al. 2024a). Therefore, to generate better and more climate-resilient crops and to implement policies to deal with changing climatic circumstances, mechanistic knowledge of abiotic stress in global agriculture is essential (Kumari et al. 2022; Nehra et al. 2024).

Traditional farming practices struggle to meet this demand, necessitating innovative solutions. Engineered nanomaterials (ENMs) offer promising tools for revolutionizing agriculture by enhancing sustainability and crop tolerance to various stressors (Kumar et al. 2024b). Combining antimicrobial activity with nanotechnology can be a useful tool for reducing the indiscriminate use of agrochemicals while preventing the development of antibiotic resistance and halting the spread of infectious diseases (Kumar et al. 2017; Kumar et al. 2018; Khezerlou et al. 2018). According to Kumar et al. (2022a, b, c, d), nanostructured formulations can improve antimicrobial agent sensitivity, pace, and precision, facilitating crop protection and detection. The effectiveness of fungicides and insecticides can be increased using nanoformulations, decreasing the dosage needed and the negative environmental impact (Kah et al. 2019).

Additionally, biopesticides, such as chemicals produced from plants or beneficial microbes, can be delivered by ENMs with improved stability and targeted administration, reducing off-target effects and advancing the sustainability of pest control techniques (Ge et al. 2019). By altering plant physiological systems and boosting stress resistance, ENMs provide novel ways to increase crop tolerance to abiotic stresses. For instance, nano-scale formulations of certain substances, including calcium and silicon, have been demonstrated to improve plant resistance to salt and drought stress by encouraging root development and controlling stomatal closure (Gupta et al. 2020). According to Gogos et al. (2018), incorporating nano-scale amendments, such as nano-clays and nano-humus, might enhance soil structure and water retention capacity, mitigating erosion and augmenting plant development. Furthermore, ENMs functionalized with elements that promote plant development, including mycorrhizal fungi and rhizobacteria, can increase nutrient cycling and soil microbial activity, which will enhance soil fertility and plant production (Dimkpa and Bindraban 2016; Kalia and Kaur 2019).

This chapter explores the potential of ENMs in advanced agri-farming systems, focusing on their applications in sustainable agriculture activities like plant protection, agrochemical delivery, biosensors, and targeted delivery (with reduced leaching loss of pesticides and fertilizers) and to combat various stresses (salinity, temperature, drought, heavy metals).

8.2 ENMs from Biopolymers for Plant Protection and Targeted Delivery

Traditional pesticides and herbicides often harm human health and the environment. ENMs offer alternatives for targeted and controlled delivery of these chemicals, reducing their environmental impact and promoting biological control. Nanoparticles encapsulate pesticides, releasing them slowly and specifically at the target site, minimizing exposure to non-target organisms and reducing environmental impact (Kahrilas et al. 2013). Biopolymeric nanomaterials from tamarind seeds can load pesticides for targeted delivery, such as gums, carrageenan, galactomannans, and polysaccharides. Biopolymeric systems not only offer advantages like enhanced stability/shelf life, decreased toxicity, and better pesticide solubility, but they are also easily obtainable, affordable, biocompatible, biodegradable, and safe (meaning they release associated active compounds gradually without posing a risk to the environment), and low carbon footprint. Nanoparticles can deliver beneficial nutrients from agrochemicals like pesticides and fungicides, enhancing natural pest control mechanisms and reducing reliance on chemical pesticides, delivering pesticides and herbicides more effectively and minimizing environmental impact (Chandrasekaran et al. 2022). Furthermore, biopolymeric nanoformulations enhance soil aeration and microbial activity to support plant development, benefiting the environment (Kumar et al. 2022a, b, c, d).

Overusing pesticides has also resulted in pest resistance, a serious agricultural issue (Duhan et al. 2017). Guar gum nanoemulsions (NEs) of 186.5–394.1 nm containing the chemical fungicide mancozeb were created utilizing a green chemistry approach and evaluated using various physio-chemical techniques. Almost 98% of the commercial mancozeb was released in under 2 hours, compared to just around 43% (Kumar et al. 2023). Chitosan-carrageenan NPs loaded with mancozeb shown high antifungal effectiveness against *A. solani, A. alternata, S. lycopersici,* and *Sclerotinia Sclerotiorum* in tomato and potato plants, both in vitro and in vivo (Kumar et al. 2021).

Guar gum nanoemulsions NEs (GG-1.5) loaded with 1.5 mg/mL mancozeb showed an 84.5% inhibition against *A. alternata*, comparable to commercial mancozeb (86.5 ± 0.7%). *Sclerotiorum* and *S. lycopersici* exhibited the highest mycelial inhibition of all the species. In addition to the traits associated with plant growth (germination percentage, root/shoot ratio, and dry biomass), NEs showed increased antifungal activity in tomatoes and potatoes grown in pots. Using the mycelium inhibition method, Nano CSGA-M-1.0 (chitosan-gum acacia with 1.0 mg/mL mancozeb) at 1.5 ppm showed maximum inhibition (83.8 ± 0.7%) against *Alternaria solani*. In contrast, *Sclerotinia sclerotiorum* showed a 100% inhibition at 1.0 and 1.5 ppm. For both fungi, commercial mancozeb inhibited 84.6 ± 0% and 100%, respectively. NCs were discovered to have good antibacterial action in pot house settings. The disease control effectiveness (DCE, expressed as a percentage) of pathogen-treated plants with CSGA-M-1.0 was 64.6 ± 5.0 and 60.2 ± 1.4%, respectively, against early blight and stem rot diseases (Kumar et al. 2022a, b, c, d).

fourth-instar larvae, the hydrothermally synthesized MgO nanostructures performed better than the other synthesis techniques.

8.3.3 Iron-Based ENMs

All living things require iron (Fe) as a necessary nutrient (Zuo and Zhang 2011). Numerous crops suffer from iron shortage, but peanuts (*Arachis hypogaea*) are especially vulnerable (Sánchez-Alcalá et al. 2014). While there is often a high concentration of iron in the soil, a significant amount is bound to soil particles (Mimmo et al. 2014; Bindraban et al. 2015). Because iron is mostly found as insoluble Fe^{3+}, particularly in high pH and aerobic soils, these soils typically lack accessible iron in Fe^{2+} (Ye et al. 2015). Fe-deficient soils result in Fe-deficient plants since plants typically take up Fe^{2+} from the soil (Kobayashi and Nishizawa 2012).

Rui et al. (2016) found that peanut plants have more chlorophyll after being exposed to Fe_2O_3 NPs. When peanut plants were treated with 2, 10, and 1000 $mg \cdot kg^{-1}$ Fe_2O_3 NPs, their leaves had more chlorophyll than the control group. The Fe_2O_3 NPs promoted the growth of peanuts by regulating phytohormone contents and antioxidant enzyme activity. The Fe contents in peanut plants with Fe_2O_3 NPs and EDTA-Fe treatments were higher than in the control group, as large amounts of Fe_2O_3 NPs adhered to soil particles, reducing nutrient loss. In a similar study, Dola et al. (2022) found that under comparison to the untreated plants, the 200 ppm nanoparticles enhanced the soybean seed output by 40.12% and 32.60% under drought and well-watered situations, respectively. In addition, compared to the untreated control, nano-iron raised the oil content of soybean seeds by 10.14% and 7.87% under drought and well-watered situations, respectively.

8.3.4 Zinc-Based ENMs

These materials exhibit unique physicochemical properties, such as high surface area-to-volume ratio, tunable electronic properties, and biocompatibility, making them highly suitable for agricultural applications. Zinc nanomaterials can be synthesized through various methods, including sol-gel synthesis, chemical vapor deposition, and green synthesis using plant extracts or microorganisms (Sharma et al. 2020). Zinc nano-fertilizers offer a promising solution to this challenge by enabling targeted and controlled delivery of zinc to plants, enhancing nutrient uptake and utilization efficiency (Dimkpa and Bindraban 2016). Furthermore, zinc nano-fertilizers can be functionalized with other nutrients or biostimulants to create multifunctional formulations that address multiple nutrient deficiencies simultaneously, thereby promoting balanced plant nutrition and sustainable intensification of agricultural production (Sharma et al. 2020).

8.3.5 Silver-Based ENMs

Strong antimicrobial action of silver nanoparticles is demonstrated against various plant diseases, such as bacteria, fungi, and viruses. They cause microbial cell membranes to rupture, obstruct cellular functions, and stop the spread and multiplication of pathogens. AgNPs may also scavenge reactive oxygen species (ROS) produced under abiotic stress conditions and have antioxidative qualities (Mawale et al. 2024; Usman et al. 2024; Mariyam et al. 2024). Plants with this antioxidative potential can better withstand stress and lessen the effects of oxidative damage. To suppress *Aspergillus flavus*, rice plants were exposed to silver nanoparticles, which caused physiological, biochemical, and oxidative changes (Sultana et al. 2021).

Siver ENMs can be biosynthesized using a variety of biomaterials like plants, bacteria, fungi, and biopolymers (Bansal et al. 2014). *In vitro* studies using silver ENMs synthesized by microwave-aided synthesis from Kinnow plants, *Alternaria solani, Thanatephorus cucumeris*, and *Botryodiplodia theobromae* demonstrated effective potential against all fungal diseases (Bansal et al. 2017). Another researcher conducted a study on the biogenesis of silver nanoparticles (NPs) utilizing *Microbacterium mitrae*, their potential antibacterial activity against the same three phytopathogens, and their potential against phytopathogens. Using sister chromatid exchange analysis, the genotoxicity of NPs was investigated in human cells; no adverse effects were noted up to a dose of 100 μg/ml (Duhan et al. 2022).

Because of their extraordinary mobility, nutrients may be distributed throughout a plant quickly (Nongbet et al. 2024; Adetuyi and Adetunji 2024; Jalil and Ansari 2019). By reducing biotic stress, the application of biosynthesized AgNPs dramatically lowered the synthesis of antioxidative enzymes and non-enzymatic chemicals. Similarly, rice grains from rice plants solely exposed to biotic stress had the highest level of aflatoxin B1 (25.1 μg/kg FW). However, when biosynthesized AgNP was applied during biotic stress, the level of aflatoxin B1 (10.6 μg/kg) was dramatically reduced at 50 mg/kg AgNPs. For the rice biochemical reaction, aflatoxin B1, the administration of 50 mg/kg biosynthesized AgNPs under biotic stress was the most appropriate dose (Ejaz et al. 2023).

AgNPs showed high antifungal activity (95%) against *Fusarium oxysporum* f. sp. *ciceri in vitro* at 100 μg/ml concentration. Pot experiments showed maximum and minimum reduction in wilt incidence (Kaur et al. 2018b). Since pre-Columbian times, farmers have used biochar, a valuable carbon black material, to improve soil fertility and promote crop development. The pyrolysis of different feedstock materials produces bulk-BCs, which are mechanically transformed into the final nanobiochar product with added value via the ball milling technique in the nano-synthesis process. Unlike bulk biochar, nano-biochar has superior physicochemical properties, such as highly precise surface area, strong catalytic activity, and exceptionally stable and amazing nanostructures (Brar et al. 2024). Table 8.1 elaborates on some engineered nanomaterials for plant protection and targeted delivery of pesticides.

Table 8.1 Engineered nanomaterials for plant protection and targeted delivery of pesticides

ENM/pesticide	Target Pest/plant	Disease/Toxicity	Release behaviour	Reference
Chitosan and tripolyphosphate/ Hexaconazole	*R. solani*	Vero cell lines	–	Chauhan et al. (2017)
Chitosan and Pectin/ Carbendazim	*F. oxysporum* and *A. parasiticus*/Cucumber, Maize, Tomato	*E. coli* and *S. aureus*	–	Kumar et al. (2017)
Chitosan-carrageenan/ Mancozeb	*A. solani, A. alternata, S. lycopersici, Sclerotinia Sclerotiorum*/Potato and tomato	Showed good antifungal efficacy *in vitro* and *in vivo*	Slow in steady release	Kumar et al. (2021)
Chitosan-gum acacia/ Mancozeb	Early blight and stem rot (*Alternaria solani, Sclerotinia Sclerotiorum*)/Potato	Inhibition of 83.8 and 100%, respectively	–	Kumar et al. (2022a)
Chitosan/Mancozeb	Tomato *(Solanum lycopersicum)*	Inhibited the disease *in vitro* Fusarium wilt	–	Kumar et al. (2022b)
Chitosan-gum acacia/ Mancozeb	*Alternaria alternata*	Inhibition of disease was 85.2%.	10 hours for ENMs and 2 hours commercial mancozeb	Kumar et al. (2022c)
Cu-chitosan and Zn chitosan NPs	*Rhizoctonia solani* and *Trichoderma logibrachiatum*	Cotton seedlings dumping off disease/Cu-chitosan nanocomposite were more potent against *R. solani*	–	Abd-Elsalam et al. (2018)
Ch-CuO and Ch-ZnO nanocomposite/Copperoxy-chloride (CuOCl)	*Fusarium oxysporum* f. sp. *ciceri* (FOC)	Good antifungal efficacy for Fusarium wilt disease and promoted the growth of chickpea plant.	–	Kaur et al. (2018)
Starch and alginate/Thiram (Fungicide)		–	The entrapment efficacy of thiram was more than 90%, and the case II diffusion mechanism observed	Singh et al. (2009)

ENM/pesticide	Target Pest/plant	Disease/Toxicity	Release behaviour	Reference
Chitosan-silver Microbeads/pesticide		–	Pesticide removal for extended periods	Saifuddin et al. (2011)
Cu/Zn loaded PVA-starch nanocomposite/ micronutrient.		–	The releases of micronutrient and (Cu/Zn) from polymeric nanocomposite was controlled.	Kumar et al. (2018)
Cu-chitosan	*Pyricularia grisea*	75% control of the disease		Sathiyabama and Manikandan (2018)
Cu-chitosan nanocomposite	*R. solani* and *S. rolfsii*	The growth of both *S. rolfsii* and *R. solani* was inhibited, and destruction in the hyphae was confirmed.		Rubina et al. (2017)
Chitosan-guar gum	*A. solani, A. alternata, S. lycopersici, Sclerotinia Sclerotiorum*	Mycelial inhibition in ENMs was greater compared to commercial fungicides	The commercial mancozeb was released within 2 hours while ENMs enhanced the time to 8–10 hours.	Kumar et al. (2023)

8.4 Engineered Nanomaterials for Abiotic Stress Tolerance

Climate change and resource scarcity expose crops to abiotic stresses like drought, salinity, and extreme temperatures. ENMs can enhance plant tolerance to these stresses and combat various stresses like salinity, temperature, drought, and heavy metals. Enhancing plant resistance to drought, salinity, and other abiotic stresses, ensuring better yields under harsh conditions (Verma et al. 2023a, b). Production is impacted, and crop species are exposed to more abiotic stress. For instance, drought affects 45% of arable land (Bot et al. 2000), and between 1950 and 2000, the percentage of the globe under aridity stress rose from 17% to 27% (Thornton et al. 2014).

Additionally, salinity is responsible for significant losses in crop production (Ahmad 2010) since the majority of crop species are susceptible to salt stress (1.0–1.8 dS m^{-1}), which can reduce yield by up to 50%, depending on the amount of salt present (Panta et al. 2014). Zinc-engineered nanomaterials offer innovative strategies for mitigating the adverse effects of abiotic stress on crops, thereby enhancing their resilience and productivity in challenging environments (Kumar et al. 2024c). For instance, zinc oxide nanoparticles have been shown to alleviate drought stress by enhancing antioxidant enzyme activity, regulating stomatal closure, and modulating osmotic balance in plants (Hussain et al. 2018). Similarly, zinc nanoparticles can mitigate salinity stress by improving ion homeostasis, osmotic adjustment, and water uptake in salt-stressed plants (Sharma et al. 2020). Furthermore, zinc nanomaterials can ameliorate heavy metal toxicity in contaminated soils by immobilizing toxic metals through adsorption or precipitation reactions, thereby reducing their bioavailability to plants and minimizing phytotoxicity (Singh et al. 2018a).

8.4.1 NMs for Heavy Metals Remediation

Reduced yields, stunted development, and decreased plant production might result from high metal levels in the soil. Furthermore, crops exposed to heavy metals may reduce nutritional quality, exhibiting lower concentrations of vital minerals such as zinc and iron. As a result, microbial activity and soil fertility may be negatively impacted by heavy metals (Qin et al. 2020; Ansari et al. 2022).

8.4.2 ENMs for Drought Stress/Water Management

The yield loss that occurred when water was lowered by around 40% varied from about 21% in wheat to 39% in maize, according to a meta-study based on data from field tests (Daryanto et al. 2016). The drought in European nations has been anticipated to result in 10% decline in wheat and maize yields (Webber et al. 2018). When comparing species for their resilience to drought, it is important to remember that, similar to other abiotic stressors, the amount of production lost due to drought varies depending on the cultivar (Mickky et al. 2020; Verma et al. 2022a). Furthermore, drought degrades crop quality. For example, reduced seed oil content in rapeseed seeds has been shown in water-shortage situations (Hussain et al. 2018).

Superabsorbent nanoparticles can store water and release it slowly to plants during drought. Water translocation, xylem humidity, and rising turgor pressure are all impacted by ENM, which raises relative water content and water usage efficiency (Verma et al. 2022b). SiO_2 NPs have been shown to improve plant germination and developmental features. This effect may be linked to an increased K/Na ratio, which decreases Na^+ absorption and promotes the production of antioxidant enzymes (Alsaeedi et al. 2018; Farhangi-Abriz and Torabian 2018; El-Ramady et al. 2022). According to Almutairi (2016), and Verma et al. (2022c) strong correlations have been observed between the application of Si NPs and improved seed germination and growth in tomato-stressed, high-salinity plants, as well as higher expression of genes related to salt tolerance. When saline water was applied after applying Si treatments, *Capsicum annuum* plants grew more than when Si NPs were not applied (Tantawy et al. 2015).

There is a current water shortage, and in the upcoming years, plants may experience regular droughts, drastically lowering the output of important food crops. In the last several years, nanotechnology has been increasingly popular in practically every industry, and it is presently being thoroughly investigated in many agricultural domains, including stress reduction. Numerous research suggest that the use of engineered nanomaterials (NMs) for the development of plant tolerance and the alleviation of drought stress is extremely promising. Natural killers enhance the drought resistance of plants by acting on many morphological, physiological, and even molecular levels. By interacting with certain targeted molecules in plant systems, nanoparticles (NPs) can alter gene expression levels and increase resilience to drought stress (Mariyam et al. 2024). Though the impact depends on their type, size, method, and administration time, their ability to achieve this at extremely low concentrations is enticing. The beneficial impacts of ENPs on agronomical parameters, yield, productivity, relative water content, and the photosynthetic and antioxidant machinery of plants growing in water-deficient environments have been demonstrated in several crops thus far (Grewal et al. 2024). By interacting with certain targeted molecules in plant systems, NPs can alter gene expression levels and increase resilience to drought stress. Though the impact depends on their type, size, method, and administration time, their ability to achieve this at extremely low concentrations is enticing. The beneficial impacts of ENPs on agronomical parameters, yield, productivity, relative water content, and the photosynthetic and antioxidant machinery of plants growing in water-deficient environments have been demonstrated in many crops thus far (Grewal et al. 2024; Choudhary et al. 2024).

8.4.3 ENMs in Salinity Stress

One form of abiotic stress that is common and significantly slows down plant development is salinity. Due to the low osmotic potential surrounding the seeds, which limits water intake, soil salinity decreases seed germination (Alkharabsheh et al. 2021; Taha et al. 2021). In legumes, oxidative stress caused by sodium chloride (NaCl) significantly reduces nodulation, reduces seed nutritional quality, and hinders development (Ahmad et al. 2008). Enzymatic and non-enzymatic antioxidant defense systems are among the strategies plants can use to lessen the impact of salinity-related oxidative damage. Proline is a recognized osmoregulatory stress-related molecule, whereas ascorbate and carotenoids are essential non-enzymatic defense mechanisms against salt (Anoop and Gupta 2003).

Applying NPs is crucial to reducing the negative effects of salt on plants as an abiotic stressor. When AgNPs are applied to *Lathyrus sativus* under salt stress, germination percentage, shoot and root length, and seedling fresh weight and dry weight all improve during the germination stage. This improves osmotic control and lessens the harmful impacts of salinity (Hojjat and Kamyab 2017; Khan et al. 2023). According to Noman et al. (2020), adding Cu NPs to the soil improved plant growth and yield while reducing oxidative stress in wheat. When applied to *S. hortensis*, Ag NPs decrease the effects of salt stress on plant shoot length and germination percentage while increasing plant tolerance to salt stress (Nejatzadeh 2021).

The metabolic response of *S. lycopersicum* to ZnO NPs under salinity stress was studied by Alharby et al. (2016). They discovered that the NPs can counteract the negative effects of salinity stress by controlling tolerance-related proteins and enzymes, primarily by upregulating the expression of the glutathione peroxidase and superoxide dismutase genes (Choudhary et al. 2025). These findings agree with those of Haripriya et al. (2018), who discovered that finger millet is less affected by salt stress when ZnO NPs are sprayed topically. Through transferring grain nitrogen and restoring total nitrogen content, ZnO NPs treatment in soil-grown Sorghum can also enhance drought-stress resistance (Dimkpa et al. 2019). Abiotic stress tolerance by some ENMs is depicted in Table 8.2, along with its effects on plants.

8.5 ENMs for Precision Agriculture/Biosensors

ENMs can be used as a biosensor in pest and pesticide detection, monitor nutrient levels in plants and soil, and precisely calculate the amount of fertilizer required by the plants.

Table 8.2 Engineered nanomaterials for abiotic (drought, salinity, heavy metals) stress tolerance

Nanomaterial	Abiotic Stress	Application/Effect	Reference
Zinc Oxide NPs	Drought	Enhance antioxidant activity, regulate stomatal closure	Hussain et al. (2018)
Copper NPs	Salinity	Improve ion homeostasis, reduce oxidative stress	Pérez-Labrada et al. (2019)
Iron Oxide NPs	Heavy Metals	Immobilize toxic metals, reduce phytotoxicity	Mathur et al. (2022)
Carbon Nanotubes	Drought	Enhance water uptake, improve osmotic adjustment	Gupta et al. (2016)
Silica NPs	Salinity	Reduce sodium uptake, enhance water retention	Ijaz et al. (2023), Ahmad et al. (2019)
Cerium Oxide NPs	Heavy Metals	Scavenge reactive oxygen species, alleviate oxidative stress	Shahbaz et al. (2022), Singh et al. (2018)
Copper NPs	Drought	Regulate stomatal conductance, enhance antioxidant activity	Zhang et al. (2016)
Titanium Dioxide NPs	Salinity	Enhance photosynthetic efficiency, reduce oxidative damage	Rizwan et al. (2019), Chen et al. (2017)
Selenium NPs	Heavy Metals	Bind to toxic metals, reduce bioavailability	Saad et al. (2022), Li et al. (2017)
Graphene Oxide	Drought	Improve water retention, regulate stomatal conductance	Zhao et al. (2020)
Silver NPs	Heavy Metals	Enhance metal sequestration, reduce toxicity	Gonzalez-Melendi et al. (2019)
Iron Oxide NPs	Drought	Improve soil water retention, enhance root growth	Wang et al. (2017)
Carbon Nanotubes	Salinity	Stimulate antioxidant defense, reduce lipid peroxidation	Lu et al. (2020), Liu et al. (2018)
Silica NPs	Heavy Metals	Enhance metal immobilization, promote phytostabilization	Song et al. (2016)
Cerium Oxide NPs	Drought	Regulate water loss, enhance drought tolerance	Kumar et al. (2019)
Copper NPs	Salinity	Improve membrane stability, reduce ion toxicity	Fouda et al. (2024)
Titanium Dioxide NPs	Heavy Metals	Enhance metal sorption, mitigate metal toxicity	Zhang et al. (2020), Liu et al. (2020)
Selenium NPs	Drought	Modulate osmolyte accumulation, enhance stress tolerance	Basit et al. (2023)
Graphene Oxide	Salinity	Regulate ion uptake, reduce osmotic stress	Zahedi et al. (2023)

8.5.1 Sensors, Monitoring, and Pest Detection

Traditional pathogen detection techniques take much time and effort and are not as accurate. However, molecular approaches improved the sensitivity and specificity and shortened the detection time for pathogens, but they require sophisticated laboratory equipment and highly qualified personnel. In this case, plant pathogens may be quickly and accurately diagnosed on-site thanks to biosensors based on nanotechnology. Nano-enabled sensors can monitor nutrient levels in soil and plants in real-time, enabling precision fertilization and reducing overapplication (Wong et al. 2021). Nanobiosensors can detect pests and diseases quickly and accurately, enabling timely intervention and minimizing pesticide use (Krishnan et al. 2022).

8.5.2 Nanomaterials for Nutrient Delivery and Management

One critical challenge in agriculture is inefficient nutrient use, leading to significant waste and environmental contamination. ENMs can improve nutrient delivery and management in several ways, such as improving nutrient use efficiency and targeted delivery to crops and reducing waste and environmental pollution. NPs encapsulate fertilizers, releasing them slowly and directly to plant roots, maximizing uptake and minimizing soil runoff and leaching (Liu et al. 2020; Rajput et al. 2023). Functionalized ENMs can attach to specific nutrients or plant parts, ensuring delivery to the desired location and reducing unintended side effects (Andres et al. 2020).

8.6 Conclusions and Future Prospectives

The integration of ENMs holds immense potential for enhancing sustainability and tolerance in advanced agri-farming systems. Through their unique physicochemical properties and versatile applications, ENMs offer innovative solutions to address modern agriculture's key challenges, including nutrient management, pest and disease control, and abiotic stress tolerance. One of the key findings from the chapters is the effectiveness of ENMs in enhancing nutrient uptake and utilization efficiency in plants. Moreover, the targeted delivery and controlled release of nutrients enabled by ENMs offer a sustainable approach to optimize fertilizer use and minimize environmental impacts. Additionally, ENMs have shown great potential in improving pest and disease management in agri-farming systems. Nanoformulations of pesticides and biopesticides have exhibited enhanced efficacy, prolonged activity, and reduced off-target effects compared to conventional formulations. Furthermore, developing smart nanomaterials capable of targeted delivery and controlled release of active ingredients holds promise for sustainable pest management strategies that minimize pesticide use and environmental contamination.

Moreover, ENMs have emerged as effective tools for enhancing abiotic stress tolerance in crops, particularly under drought, salinity, and heavy metal toxicity conditions. Nanomaterials such as metal oxides, zinc, and copper, and polymeric nanomaterials like chitosan, gum acacia, etc., have demonstrated the ability to mitigate stress-induced damage, improve water and nutrient uptake, and enhance plant resilience to environmental challenges. By harnessing the unique properties of ENMs, researchers, and practitioners can develop tailored solutions to address specific stressors and optimize crop performance in diverse agroecosystems. Looking ahead, future research in the field of engineered nanomaterials for agri-farming systems should focus on addressing key knowledge gaps and translating laboratory findings into practical applications. Long-term studies evaluating the environmental fate, toxicity, and ecosystem impacts of ENMs are essential to ensure their safe and sustainable use in agriculture.

References

Abd-Elsalam KA, Vasil'kov AY, Said-Galiev EE, Rubina MS, Khokhlov AR, Naumkin AV et al (2018) Bimetallic blends and chitosan nanocomposites: novel antifungal agents against cotton seedling damping-off. Eur J Plant Pathol 151(1):57–72. https://doi.org/10.1007/s10658-017-1349-8

Abinaya S, Kavitha HP (2023) Magnesium oxide nanoparticles: effective antilarvicidal and antibacterial agents. ACS Omega 8(6):5225–5233. https://doi.org/10.1021/acsomega.2c01450

Adetuyi BO, Adetunji CO (2024) Application of plant-based nanobiopesticides for mitigation of several biotic stress. In: Handbook of agricultural biotechnology, Volume 1: Nanopesticides, p 239

Ahmad B (2010) Effects of salinity on yield and component characters in canola (*Brassica napus* L.) cultivars. Not Sci Biol 2:81–83. https://doi.org/10.15835/nsb.2.1.3560

Ahmad P, John R, Sarwat M, Umar S (2008) Responses of proline, lipid peroxidation and antioxidative enzymes in two varieties of *Pisum sativum* L. under salt stress. Int J Plant Prod 2:353–366. https://doi.org/10.22069/ijpp.2012.626

Ahmad I et al (2019) Silica nanoparticles enhance drought tolerance in rice seedlings: a comparative study of rice varieties. Plant Physiol Biochem 139:558–567

Alengebawy A, Abdelkhalek ST, Qureshi SR, Wang MQ (2021) Heavy metals and pesticides toxicity in agricultural soil and plants: ecological risks and human health implications. Toxics 9(3):42. https://doi.org/10.3390/toxics9030042

Alharby HF, Metwali EMR, Fuller MP, Aldhebiani AY (2016) The alteration of mRNA expression of SOD and GPX genes and proteins in tomato (*Lycopersicon esculentum* Mill) under stress of NaCl and/or ZnO nanoparticles. Saudi J Biol Sci 23(6):773–781. https://doi.org/10.1016/j.sjbs.2016.04.012

Alkharabsheh HM, Seleiman MF, Hewedy OA, Battaglia ML, Jalal RS, Alhammad BA et al (2021) Field crop responses and management strategies to mitigate soil salinity in modern agriculture: a review. Agronomy 11(11):2299. https://doi.org/10.3390/agronomy11112299

Almutairi ZM (2016) Effect of nano-silicon application on the expression of salt tolerance genes in germinating tomato (*Solanum lycopersicum* L.) seedlings under salt stress. Plant OMICS 9:106–114

Alsaeedi A, El-Ramady H, Alshaal T, El-Garawani M, Elhawat N, Al-Otaibi A (2018) Exogenous nanosilica improves germination and growth of cucumber by maintaining K^+/Na^+ ratio under elevated Na+ stress. Plant Physiol Biochem 125:164–171. https://doi.org/10.1016/j.plaphy.2018.02.006

Andres LC, Souza DC, Ferreira DS, Pereira DS, Oliveira HC (2020) Chitosan-coated nanoemulsions for targeted delivery of nutrients to plants. Carbohydr Polym 249:117005

Anoop N, Gupta AK (2003) Transgenic indica rice cv IR-50 overexpressing *Vigna aconitifolia* Δ^1-pyrroline −5- carboxylate synthetase cDNA shows tolerance to high salt. J Plant Biochem Biotechnol 12(2):109–116. https://doi.org/10.1007/BF03263170

Ansari MS, Tauseef A, Haris M, Hussain T, Khan AA (2022) Effects of heavy metals present in sewage sludge, their impact on soil fertility, soil microbial activity, and environment. In: Shah MP, Shah N, Rodriguez-Couto S, Banerjee R (eds) Development in waste water treatment research and processes: treatment and reuse of sewage sludge: an innovative approach for wastewater treatment. Elsevier, Amsterdam, pp 197–214

Bansal P, Duhan JS, Gahlawat SK (2014) Biogenesis of nanoparticles: a review. Afr J Biotechnol 13(28):2778–2785

Bansal P, Kaur P, Kumar A, Duhan JS (2017) Microwave-assisted quick synthesis method of silver nanoparticles using citrus hybrid "Kinnow" and its potential against early blight of tomato. Res Crops 18(4):650–655

Basit F, Abbas S, Zhu M, Tanwir K, El-Keblawy A, Sheteiwy MS, Raza A, Hu J, Hu W, Guan Y (2023) Ascorbic acid and selenium nanoparticles synergistically interplay in chromium stress mitigation in rice seedlings by regulating oxidative stress indicators and antioxidant defense mechanism. Environ Sci Pollut Res 30(57):120044–120062

Bindraban PS, Dimkpa C, Nagarajan L, Roy A, Rabbinge R (2015) Revisiting fertilisers and fertilisation strategies for improved nutrient uptake by plants. Biol Fertil Soils 51(8):897–911. https://doi.org/10.1007/s00374-015-1039-7

Bot AJ, Nachtergaele FO, Young A (2000) Land resource potential and constraints at regional and country levels. World soil resources report. FAO, Rome, p 122

Brar B, Saharan BS, Seth CS, Kamboj A, Bala K, Rajput VD, Minkina T, Wong MH, Kumar D, Sadh PK, Duhan JS (2024) Nanobiochar: soil and plant interactions and their implications for sustainable agriculture. Biocatal Agric Biotechnol:103077

Cai L, Liu M, Liu Z, Yang H, Sun X, Chen J et al (2018) MgONPs can boost plant growth: evidence from increased seedling growth, morpho-physiological activities, and Mg uptake in tobacco (*Nicotiana tabacum* L.). Molecules 23(12):3375. https://doi.org/10.3390/molecules23123375

Chandrasekaran S, Dutta P, Krishnan V, Kumari M (2022) Nanotechnology in crop protection: emerging opportunities and challenges. Emerg Nanotechnol Food Agric:3–23

Chauhan N, Dilbaghi N, Gopal M, Kumar R, Kim KH, Kumar S (2017) Development of chitosan nanocapsules for the controlled release of hexaconazole. Int J Biol Macromol 97:616–624. https://doi.org/10.1016/j.ijbiomac.2016.12.059

Chen W et al (2017) Titanium dioxide nanoparticles enhance salinity tolerance in rice (*Oryza sativa* L.) through increasing antioxidant enzymatic activity and reducing lipid peroxidation. Environ Exp Bot 137:52–58

Chen J, Wu L, Lu M, Lu S, Li Z, Ding W (2020) Comparative study on the fungicidal activity of metallic MgO nanoparticles and macroscale MgO against soilborne fungal phytopathogens. Front Microbiol 11:365. https://doi.org/10.3389/fmicb.2020.00365

Choudhary R, Rajput VD, Ghodake G, Ahmad F, Meena M, Rehman RU et al (2024) Comprehensive journey from past to present to future about seed priming with hydrogen peroxide and hydrogen sulfide concerning drought, temperature, UV and ozone stresses-a review. Plant Soil:1–23. https://doi.org/10.1007/s11104-024-06499-9

Choudhary R, Ahmad F, Kaya C, Upadhyay SK, Muneer S, Kumar V, Meena M, Liu H, Upadhyay H, Seth CS (2025) Decrypting proteomics, transcriptomics, genomics, and integrated omics for augmenting the abiotic, biotic, and climate change stress resilience in plants. J Plant Physiol. https://doi.org/10.1016/j.jplph.2025.154430

Cuong HN, Pansambal S, Ghotekar S, Oza R, Thanh Hai NTT, Viet NM, Nguyen VH (2022) New frontiers in the plant extract mediated biosynthesis of copper oxide (CuO) nanoparticles and their potential applications: a review. Environ Res 203:111858. https://doi.org/10.1016/j.envres.2021.111858

Daryanto S, Wang L, Jacinthe PA (2016) Global synthesis of drought effects on maize and wheat production. PLoS One 11(5):e0156362. https://doi.org/10.1371/journal.pone.0156362

Dimkpa CO, Bindraban PS (2016) Fortification of micronutrients for efficient agronomic production: a review. Agro Sustain Develop 36(1):7

Dimkpa CO, Singh U, Bindraban PS, Elmer WH, Gardea-Torresdey JL, White JC (2019) Zinc oxide nanoparticles alleviate drought-induced alterations in sorghum performance, nutrient acquisition, and grain fortification. Sci Total Environ 688:926–934. https://doi.org/10.1016/j.scitotenv.2019.06.392

Dola DB, Mannan MA, Sarker U, Mamun MAA, Islam T, Ercisli S et al (2022) Nano-iron oxide accelerates growth, yield, and quality of *Glycine max* seed in water deficits. Front Plant Sci 13:992535. https://doi.org/10.3389/fpls.2022.992535

Duhan JS, Kumar R, Kumar N, Kaur P, Nehra K, Duhan S (2017) Nanotechnology: the new perspective in precision agriculture. Biotechnol Rep (Amst) 15:11–23. https://doi.org/10.1016/j.btre.2017.03.002

Duhan JS, Bansal P, Sadh PK, Kumar R, Kumar A (2022) Biosynthesis, characterization, toxicity assessment and bio-efficacy of silver nanoparticles synthesized by *Microbacterium mitrae* in controlling early blight in tomato (*Lycopersicon esculentum* L.). Res J Biotechnol 17:73–81

Ejaz M, Raja NI, Khan SA, Mashwani ZU, Hanif A, Iqbal M, Hussain M, Syed A, Iqbal RK, Qureshi H, Anwar T (2023) Biosynthesized silver nanoparticles ameliorate biotic stress in rice (Oryza sativa) by intricating biochemical and mineral profile. Pak J Bot 55(6):2019–2028

Elmer WH, Zuverza-Mena N, Triplett LR, Roberts EL, Silady RA, White JC (2021) Foliar application of copper oxide nanoparticles suppresses fusarium wilt development on chrysanthemum. Environ Sci Technol 55(15):10805–10810. https://doi.org/10.1021/acs.est.1c02323

El-Ramady HR, Verma KK, Rajput VD, Minkina T, Elbehiry F, Elbasiouny HY, Elsakhawy T, Omara AED, Amer MM (2022) Sources of silicon and nano-silicon in soil and plants. In: Silicon and nano-silicon in environmental stress management and crop quality improvement: progress and prospects. Academic Press, Elsevier. https://doi.org/10.1016/B978-0-323-91225-9.00003-0

Faizan M, Bhat JA, El-Serehy HA, Moustakas M, Ahmad P (2022) Magnesium oxide nanoparticles (MgO-NPs) alleviate arsenic toxicity in soybean by modulating photosynthetic function, nutrient uptake and antioxidant potential. Metals 12(12):2030. https://doi.org/10.3390/met12122030

Farhangi-Abriz S, Torabian S (2018) Nano-silicon alters antioxidant activities of soybean seedlings under salt toxicity. Protoplasma 255(3):953–962. https://doi.org/10.1007/s00709-017-1202-0

Fouda HM, Saied E, Abdelmouty ES, Osman MS (2024) Ameliorative role of copper nanoparticle in alleviating salt-induced oxidative stress in fenugreek (*Trigonella foenum-graecum* L.) plants. Biocatal Agric Biotechnol 57:103095

Gautam A, Sharma P, Ashokhan S, Yaacob JS, Kumar V, Guleria P (2023) Magnesium oxide nanoparticles improved vegetative growth and enhanced productivity, biochemical potency and storage stability of harvested mustard seeds. Environ Res 229:116023. https://doi.org/10.1016/j.envres.2023.116023

Ge Y et al (2019) Applications of nanotechnology in plant growth and crop protection: a review. Molecules 24(14):1–21

Gogos A, Voegelin A, Kaegi R (2018) Influence of organic compounds on the sulfidation of copper oxide nanoparticles. Environ Sci: Nano 5(11):2560–9

Gonzalez-Melendi P et al (2019) Silver nanoparticles as a safe alternative to mercury contamination in plants: a case study in *Echium plantagineum* L. J Nanobiotechnol 17(1):1

Grewal S, Boora R, Rani B (2024) Effect of engineered nanomaterials on the crop growth parameters under drought stress. In: Pudake RN, Tripathi RM, Singh Gill S (eds) Nanotechnology for abiotic stress tolerance and management in crop plants. Academic Press, pp 165–179. https://doi.org/10.1016/B978-0-443-18500-7.00011-9

Gupta DK et al (2016) Carbon nanotubes for drought stress management in crops. J Nanosci Nanotechnol 16(4):3365–3374

Gupta DK et al (2020) Role of nanotechnology in agriculture: a promising approach. J Appl Nat Sci 12(3):127–135

Haripriya P, Stella PM, Anusuya S (2018) Foliar spray of zinc oxide nanoparticles improves salt tolerance in finger millet crops under glasshouse condition. Sciol Biotechnol 1:20–29

Hazarika A, Yadav M, Yadav DK, Yadav HS (2022) An overview of the role of nanoparticles in sustainable agriculture. Biocatal Agric Biotechnol 43:102399. https://doi.org/10.1016/j.bcab.2022.102399

Hojjat SS, Kamyab M (2017) The effect of silver nanoparticle on fenugreek seed germination under salinity levels. Russ Agric Sci 43(1):61–65. https://doi.org/10.3103/S1068367417010189

Hussain M, Farooq S, Hasan W, Ul-Allah S, Tanveer M, Farooq M, Nawaz A (2018) Drought stress in sunflower: physiological effects and its management through breeding and agronomic alternatives. Agric Water Manag 201:152–166. https://doi.org/10.1016/j.agwat.2018.01.028

Ijaz U, Ahmed T, Rizwan M, Noman M, Shah AA, Azeem F, Alharby HF, Bamagoos AA, Alharbi BM, Ali S (2023) Rice straw based silicon nanoparticles improve morphological and nutrient profile of rice plants under salinity stress by triggering physiological and genetic repair mechanisms. Plant Physiol Biochem 201:107788

Jalil SU, Ansari MI (2019) Nanoparticles and abiotic stress tolerance in plants: synthesis, action, and signaling mechanisms. In: Plant Signal Molecules, pp 549–561

Kah M et al (2019) Nanopesticides: state of knowledge, environmental fate, and exposure modeling. Front Chem 7:1–12

Kahrilas SB, Begum NA, Rao NJ (2013) Nanoencapsulated pesticides: an emerging strategy for effective pest control. Int J Agric Biol Sci 1(4):161–165

Kalia A, Kaur H (2019) Agri-applications of nano-scale micronutrients: prospects for plant growth promotion and use-efficient micronutrient fortification. In: Nanoscale engineering in agricultural management. CRC Press, pp 81–105

Kamel SM, Elgobashy SF, Omara RI, Derbalah AS, Abdelfatah M, El-Shaer A et al (2022) Antifungal activity of copper oxide nanoparticles against root rot disease in cucumber. J Fungi (Basel) 8(9):911. https://doi.org/10.3390/jof8090911

Kashyap PL, Rai P, Sharma S, Chakdar H, Kumar S, Pandiyan K, Srivastava AK (2016) Nanotechnology for the detection and diagnosis of plant pathogens. In: Sustainable agriculture reviews, pp 253–276. https://doi.org/10.1007/978-3-319-39306-3_8

Kaur P, Duhan JS, Thakur R (2018a) Comparative pot studies of chitosan and chitosan-metal nanocomposites as nano-agrochemicals against fusarium wilt of chickpea (*Cicer arietinum* L.). Biocatal Agric Biotechnol 14:466–471. https://doi.org/10.1016/j.bcab.2018.04.014

Kaur P, Thakur R, Duhan JS, Chaudhury A (2018b) Management of wilt disease of chickpea in vivo by silver nanoparticles biosynthesized by rhizospheric microflora of chickpea (*Cicer arietinum*). J Chem Technol Biotechnol 93(11):3233–3243

Khan I, Awan SA, Rizwan M, Akram MA, Zia-ur-Rehman M, Wang X, Zhang X, Huang L (2023) Physiological and transcriptome analyses demonstrate the silver nanoparticles mediated alleviation of salt stress in pearl millet (*Pennisetum glaucum* L). Environ Pollut 318:120863

Khezerlou A, Alizadeh-Sani M, Azizi-Lalabadi M, Ehsani A (2018) Nanoparticles and their antimicrobial properties against pathogens including bacteria, fungi, parasites and viruses. Microb Pathog 123:505–526. https://doi.org/10.1016/j.micpath.2018.08.008

Kobayashi T, Nishizawa NK (2012) Iron uptake, translocation, and regulation in higher plants. Annu Rev Plant Biol 63:131–152. https://doi.org/10.1146/annurev-arplant-042811-105522

Krishnan R, Pandey S, Kumar Y, Varma A (2022) Recent advances in nanobiosensors for detection and management of plant pathogens. Sensors 22(3):843

Kumar S, Kumar D, Dilbaghi N (2017) Preparation, characterization, and bio-efficacy evaluation of controlled release carbendazim-loaded polymeric nanoparticles. Environ Sci Pollut Res 24(1):926–937

Kumar R, Ashfaq M, Verma N (2018) Synthesis of novel PVA–starch formulation-supported Cu–Zn nanoparticle carrying carbon nanofibers as a nanofertilizer: controlled release of micronutrients. J Mater Sci 53(10):7150–7164. https://doi.org/10.1007/s10853-018-2107-9

Kumar V et al (2019) Cerium oxide nanoparticles promote drought tolerance in rice (*Oryza sativa* L.) at reproductive stage. Environ Sci Pollut Res 26(15):15021–15034

Kumar R, Najda A, Duhan JS, Kumar B, Chawla P, Klepacka J et al (2021) Assessment of antifungal efficacy and release behaviour of fungicide-loaded chitosan-carrageenan nanoparticles against phytopathogenic fungi. Polymers 14(1):41. https://doi.org/10.3390/polym14010041

Kumar R, Duhan JS, Manuja A, Kaur P, Kumar B, Sadh PK (2022a) Toxicity assessment and control of early blight and stem rot of *Solanum tuberosum* l. by mancozeb-loaded chitosan-gum acacia nanocomposites. J Xenobiot 12(2):74–90. https://doi.org/10.3390/jox12020008

Kumar R, Kumar N, Rajput VDMS, Minkina T, Saharan BS et al (2022b) Advances in biopolymeric nanopesticides: a new eco-friendly/eco-protective perspective in precision agriculture. Nanomaterials (Basel) 12(22):3964. https://doi.org/10.3390/nano12223964

Kumar R, Nain V, Duhan JS (2022c) An ecological approach to control pathogens of *Lycopersicon esculentum* L. by slow release of mancozeb from biopolymeric conjugated nanoparticles. J Xenobiot 12(4):329–343

Kumar R, Poonia AK, Duhan JS (2022d) Synthesis, characterization and antimicrobial potential of mancozeb loaded chitosan nanoparticles against *Fusarium pallidoroseum*. Ann Biol 38(1):138–144

Kumar R, Nehra M, Kumar D, Saharan BS, Chawla P, Sadh PK et al (2023) Evaluation of cytotoxicity, release behavior and phytopathogens control by mancozeb-loaded guar gum nanoemulsions for sustainable agriculture. J Xenobiot 13(2):270–283. https://doi.org/10.3390/jox13020020

Kumar D, Mariyam S, Gupta KJ, Thiruvengadam M, Ghodake GS, Xing B, Seth CS (2024a) Comparative investigation on chemical and green synthesized titanium dioxide nanoparticles against chromium (VI) stress eliciting differential physiological, biochemical, and cellular attributes in Helianthus annuus L. Sci Total Environ 930:172413. https://doi.org/10.1016/j.scitotenv.2024.172413

Kumar D, Singh R, Upadhyay SK, Verma KK, Tripathi RM, Liu H et al (2024b) Review on interactions between nanomaterials and phytohormones: novel perspectives and opportunities for mitigating environmental challenges. Plant Sci 340:111964. https://doi.org/10.1016/j.plantsci.2023.111964

Kumar D, Biswas JK, Mulla SI, Singh R, Shukla R, Ahanger MA et al (2024c) Micro and nanoplastics pollution: sources, distribution, uptake in plants, toxicological effects, and innovative remediation strategies for environmental sustainability. Plant Physiol Biochem 108795. https://doi.org/10.1016/j.plaphy.2024.108795

Kumari A, Lakshmi GA, Krishna GK, Patni B, Prakash S, Bhattacharyya M, Singh SK, Verma KK (2022) Climate change and its impact on crops: a comprehensive investigation for sustainable agriculture. Agronomy 12:3008. https://doi.org/10.3390/agronomy12123008

Kumari A, Dash M, Singh SK, Jagadesh M, Mathpal B, Mishra PK, Pandey SK, Verma KK (2023) Soil microbes: a natural solution for mitigating the impact of climate change. Environ Monit Assess 195:1436. https://doi.org/10.1007/s10661-023-11988-y

Li Q et al (2017) Selenium nanoparticles for the mitigation of heavy metal toxicity. Environ Chem Lett 15(1):71–85

Liu J et al (2018) Carbon nanotubes alleviate oxidative stress in cucumber plants exposed to salinity stress. Sci Hortic 231:47–55

Liu R, Li R, Wang R, Li T, Li T, Tan W (2020) Controlled-release and targeted delivery of nutrients by nanoporous biochar for sustainable agriculture. Nanoscale 12(31):16505–16515

Lu L, Huang M, Huang Y, Corvini PF, Ji R, Zhao L (2020) Mn_3O_4 nanozymes boost endogenous antioxidant metabolites in cucumber (*Cucumis sativus*) plant and enhance resistance to salinity stress. Environ Sci Nano 7(6):1692–1703

Mariyam S, Upadhyay SK, Chakraborty K, Verma KK, Duhan JS, Muneer S et al (2024) Nanotechnology, a frontier in agricultural science, a novel approach in abiotic stress management and convergence with new age medicine-A review. Sci Total Environ 912:169097. https://doi.org/10.1016/j.scitotenv.2023.169097

Mathur S, Singh D, Ranjan R (2022) Remediation of heavy metal (loid) contaminated soil through green nanotechnology. Front Sustain Food Syst 6:932424

Mawale KS, Nandini B, Giridhar P (2024) Copper and silver nanoparticle seed priming and foliar spray modulate plant growth and thrips infestation in Capsicum spp. ACS Omega 9(3):3430–3444

Mickky B, Aldesuquy H, Elnajar M (2020) Effect of drought on yield of ten wheat cultivars linked with their flag leaf water status, fatty acid profile and shoot vigor at heading. Physiol Mol Biol Plants 26(6):1111–1117. https://doi.org/10.1007/s12298-020-00807-0

Mimmo T, Del Buono D, Terzano R, Tomasi N, Vigani G, Crecchio R et al (2014) Rhizospheric organic compounds in the soil–microorganism–plant system: their role in iron availability. Eur J Soil Sci 65:629–642. https://doi.org/10.1111/ejss.12158

Mohamed AA, Abu-Elghait M, Ahmed NE, Salem SS (2021) Eco-friendly mycogenic synthesis of ZnO and CuO nanoparticles for in vitro antibacterial, antibiofilm, and antifungal applications. Biol Trace Elem Res 199(7):2788–2799. https://doi.org/10.1007/s12011-020-02369-4

Nagore P, Ghotekar S, Mane K, Ghoti A, Bilal M, Roy A (2021) Structural properties and antimicrobial activities of *Polyalthia longifolia* leaf extract-mediated CuO nanoparticles. BioNanoScience 11(2):579–589. https://doi.org/10.1007/s12668-021-00851-4

Nehra A, Kalwan G, Gill R, Nehra K, Agarwala N, Jain PK et al (2024) Status of impact of abiotic stresses on global agriculture. In: Pudake RN, Tripathi RM, Singh Gill S (eds) Nanotechnology for abiotic stress tolerance and management in crop plants. Academic Press, pp 1–21

Nejatzadeh F (2021) Effect of silver nanoparticles on salt tolerance of *Satureja hortensis* L. During *in vitro* and *in vivo* germination tests. Heliyon 7(2):e05981. https://doi.org/10.1016/j.heliyon.2021.e05981

Noman M, Shahid M, Ahmed T, Tahir M, Naqqash T, Muhammad S et al (2020) Green copper nanoparticles from a native *Klebsiella pneumonia* strain alleviated oxidative stress impairment of wheat plants by reducing the chromium bioavailability and increasing the growth. Ecotoxicol Environ Saf 192:110303. https://doi.org/10.1016/j.ecoenv.2020.110303

Nongbet A, Panda J, Mohanta YK, Chakrabartty I, Shamim MZ, Mohanta TK (2024) Role of nanoparticles to protect plants from abiotic stress by scavenging reactive oxygen species. In: Nanotechnology for abiotic stress tolerance and management in crop plants. Academic Press, pp 95–114

Oerke E-C (2006) Crop losses to pests. J Agric Sci 144(1):31–43. https://doi.org/10.1017/S0021859605005708

Panta S, Flowers T, Lane P, Doyle R, Haros G, Shabala S (2014) Halophyte agriculture: success stories. Environ Exp Bot 107:71–83. https://doi.org/10.1016/j.envexpbot.2014.05.006

Pérez-Labrada F, López-Vargas ER, Ortega-Ortiz H, Cadenas-Pliego G, Benavides-Mendoza A, Juárez-Maldonado A (2019) Responses of tomato plants under saline stress to foliar application of copper nanoparticles. Plan Theory 8(6):151

Qin S, Liu H, Nie Z, Rengel Z, Gao W, Li C, Zhao P (2020) Toxicity of cadmium and its competition with mineral nutrients for uptake by plants: a review. Pedosphere 30(2):168–180. https://doi.org/10.1016/S1002-0160(20)60002-9

Rajiv P et al (2018) Copper nanoparticles in agriculture: synthesis and applications. Environ Chem Lett 16(3):917–929

Rajput VD, Kumari A, Upadhyay SK, Minkina T, Mandzhieva S, Ranjan A, Sushkova S, Burachevskaya M, Rajput P, Konstantinova E, Singh J, Verma KK (2023) Can nanomaterials improve the soil microbiome and crop productivity? Agriculture 13:231. https://doi.org/10.3390/agriculture13020231

Rizwan M et al (2019) Zinc oxide nanoparticles alter the wheat physiological response and reduce the cadmium uptake by plants. Environ Pollut 246:801–812

Rubina MS, Vasil'kov AY, Naumkin AV, Shtykova EV, Abramchuk SS, Alghuthaymi MA, Abd-Elsalam KA (2017) Synthesis and characterization of chitosan–copper nanocomposites and their fungicidal activity against two sclerotia-forming plant pathogenic fungi. J Nanostructure Chem 7(3):249–258. https://doi.org/10.1007/s40097-017-0235-4

Rui M, Ma C, Hao Y, Guo J, Rui Y, Tang X et al (2016) Iron oxide nanoparticles as a potential iron fertilizer for peanut (*Arachis hypogaea*). Front Plant Sci 7:815. https://doi.org/10.3389/fpls.2016.00815

Saad AM, Sitohy MZ, Sultan-Alolama MI, El-Tarabily KA, El-Saadony MT (2022) Green nanotechnology for controlling bacterial load and heavy metal accumulation in Nile tilapia fish using biological selenium nanoparticles biosynthesized by Bacillus subtilis AS12. Front Microbiol 13:1015613

Saifuddin N, Nian CY, Zhan LW, Ning KX (2011) Chitosan-silver nanoparticles composite as a point-of-use drinking water filtration system for the household to remove pesticides in water. Asian J Biochem 6(2):142–159. https://doi.org/10.3923/ajb.2011.142.159

Sánchez-Alcalá I, del Campillo MD, Barrón V, Torrent J (2014) Evaluation of preflooding effects on iron extractability and phytoavailability in highly calcareous soil in containers. J Plant Nutr Soil Sci 177(2):150–158. https://doi.org/10.1002/jpln.201200302

Sardar M, Ahmed W, Al Ayoubi S, Nisa S, Bibi Y, Sabir M et al (2022) Fungicidal synergistic effect of biogenically synthesized zinc oxide and copper oxide nanoparticles against *Alternaria citri* causing citrus black rot disease. Saudi J Biol Sci 29(1):88–95. https://doi.org/10.1016/j.sjbs.2021.08.067

Sathiyabama M, Manikandan A (2018) Application of copper-chitosan nanoparticles stimulates growth and induces resistance in finger millet (*Eleusine coracana* Gaertn.) plants against blast disease. J Agric Food Chem 66(8):1784–1790. https://doi.org/10.1021/acs.jafc.7b05921

Shahbaz M, Fatima N, Mashwani ZU, Akram A, Haq EU, Mehak A, Abasi F, Ajmal M, Yousaf T, Raja NI, UlHassan H (2022) Effect of phytosynthesized selenium and cerium oxide nanoparticles on wheat (*Triticum aestivum* L.) against stripe rust disease. Molecules 27(23):8149

Shammout MW, Awwad AM (2021) A novel route for the synthesis of copper oxide nanoparticles using Bougainvillea plant flowers extract and antifungal activity evaluation. Chem Int 7:71–78

Sharma S et al (2020) Zinc nanoparticles for agriculture: a review. J Agric Food Chem 68(7):1815–1838

Singh A, Singh NÁ, Afzal S, Singh T, Hussain I (2018a) Zinc oxide nanoparticles: a review of their biological synthesis, antimicrobial activity, uptake, translocation and biotransformation in plants. J Mater Sci 53(1):185–201

Singh P et al (2018b) Cerium oxide nanoparticles: synthesis, characterization, and applications in agriculture. J Nanosci Nanotechnol 18(9):5857–5876

Sidhu A, Bala A, Singh H, Ahuja R, Kumar A (2020) Development of MgO-sepoilite nanocomposites against phytopathogenic fungi of rice (*Oryzae sativa*): a green approach. ACS Omega 5(23):13557–13565. https://doi.org/10.1021/acsomega.0c00008

Singh B, Sharma DK, Gupta A (2009) A study towards release dynamics of thiram fungicide from starch–alginate beads to control environmental and health hazards. J Hazard Mater 161(1):208–216. https://doi.org/10.1016/j.jhazmat.2008.03.074

Song G et al (2016) Silica nanoparticles for the stabilization and phytostabilization of heavy metals in contaminated soil. Chemosphere 147:185–190

Sultana K, Ahmad B, Rauf A, Huang L, Ramadan MF (2021) Synthesis, characterization and pharmacological activities of silver nanoparticles using Bistorta affinis and Malcolmia cabulica extracts. Bioinsp Biomim Nanobiomat 9(1):7–15

Taha RS, Seleiman MF, Shami A, Alhammad BA, Mahdi AHA (2021) Integrated application of selenium and silicon enhances growth and anatomical structure, antioxidant defense system and yield of wheat grown in salt-stressed soil. Plants (Basel) 10(6):1040. https://doi.org/10.3390/plants10061040

Tantawy AS, Salama YAM, El-Nemr MA, Abdel-Mawgoud AMR (2015) Nano silicon application improves salinity tolerance of sweet pepper plants. Int J ChemTech Res 8:11–17

Thornton PK, Ericksen PJ, Herrero M, Challinor AJ (2014) Climate variability and vulnerability to climate change: a review. Glob Change Biol 20(11):3313–3328. https://doi.org/10.1111/gcb.12581

Usman M, Atiq M, Rajput NA, Sahi ST, Shad M, Lili N, Iqbal S, Arif AM, Ahmad U, Khan KS, Asif M (2024) Efficacy of green synthesized silver based nanomaterials against early blight of tomato caused by Alternaria solani. J Crop Health 76(1):105–115

Verma KK, Song XP, Joshi A, Rajput VD, Singh M, Sharma A, Singh RK, Li DM, Arora J, Minkina T, Li YR (2022a) Nanofertilizer possibilities for healthy soil, water and food in future: an overview. Front Plant Sci 13:865048. https://doi.org/10.3389/fpls.2022.865048

Verma KK, Song XP, Joshi A, Tian DD, Rajput VD, Singh M, Arora J, Minkina TM, Li YR (2022b) Recent trends in nano-fertilizer for advancing sustainable agriculture under the era of climate change to ensure food security for future generations. Nanomaterials 12:173. https://doi.org/10.3390/nano12010173

Verma KK, Song XP, Chen ZL, Tian DD, Rajput VD, Singh M, Minkina T, Li YR (2022c) Silicon and nanosilicon mitigate nutrient deficiency under stress for sustainable crop improvement. In: Silicon and nano-silicon in environmental stress management and crop quality improvement. Elsevier. https://doi.org/10.1016/B978-0-323-91225-9.00007-8

Verma KK, Song X, Degu HD, Guo D, Joshi A, Huang H, Xu L, Singh M, Huang D, Rajput VD, Li Y (2023a) Recent advances in nitrogen and nano-nitrogen fertilizers for sustainable crop production: a mini-review. Chem Biol Technol Agric 10:1–14. https://doi.org/10.1186/s40538-023-00488-3

Verma KK, Song X, Verma CL, Huang H, Singh M, Xu L, Wu J, Khan ZH, Zhang B, Li Y (2023b) Mathematical modeling of climate and fluoride effects on sugarcane photosynthesis with silicon nanoparticles. Plant Physiol Biochem 204:108089. https://doi.org/10.1016/j.plaphy.2023.108089

Verma KK, Song XP, Kumari A, Jagadesh M, Singh SK, Bhatt R, Singh M, Seth CS, Li YR (2024) Climate change adaptation: challenges for agricultural sustainability. Plant Cell Environ:1–12. https://doi.org/10.1111/pce.15078

Wang L et al (2017) Iron oxide nanoparticles increase wheat tolerance to soil water deficit stress. Front Plant Sci 8:1–13

Webber H, Ewert F, Olesen JE, Müller C, Fronzek S, Ruane AC, Bourgault M, Martre P, Ababaei B, Bindi M, Ferrise R (2018) Diverging importance of drought stress for maize and winter wheat in Europe. Nat Commun 9(1):4249

Wong SV, Li F, Li R, Tan W, Wang R (2021) Emerging opportunities and challenges for integrating nanomaterials into agricultural sensors. ACS Sensors 6(12):4254–4274

Ye L, Li L, Wang L, Wang S, Li S, Du J, Zhang S, Shou H (2015) MPK3/MPK6 are involved in iron deficiency-induced ethylene production in Arabidopsis. Front Plant Sci 6:953

Zahedi SM, Abolhassani M, Hadian-Deljou M, Feyzi H, Akbari A, Rasouli F, Koçak MZ, Kulak M, Gohari G (2023) Proline-functionalized graphene oxide nanoparticles (GO-pro NPs): a new engineered nanoparticle to ameliorate salinity stress on grape (Vitis vinifera l. cv sultana). Plant Stress 7:100128

Zhang L et al (2016) Copper nanoparticles alleviate drought stress by regulating stomatal conductance and antioxidant enzyme activity in maize (*Zea mays* L.). Plant Physiol Biochem 106:90–97

Zhang W, Long J, Geng J, Li J, Wei Z (2020) Impact of titanium dioxide nanoparticles on Cd phytotoxicity and bioaccumulation in rice (*Oryza sativa* L.). Int J Environ Res Public Health 17(9):2979

Zhao D, Fang Z, Tang Y, Tao J (2020) Graphene oxide as an effective soil water retention agent can confer drought stress tolerance to *Paeonia ostii* without toxicity. Environ Sci Technol 54(13):8269–8279

Zuo Y, Zhang FS (2011) Soil and crop management strategies to prevent iron deficiency in crops. Plant Soil 339:83–95. https://doi.org/10.1007/s11104-010-0566-0

Chapter 9
Current Scenario and Future Prospects of Nanomaterials on Plants in Response to Contaminated Soil

Haitao Liu, Qiujuan Jiao, Jingjing Zhang, and Gezi Li

9.1 Introduction

Heavy metal pollution has become a serious environmental problem worldwide. They are a class of inorganic metal and metalloid elements with an atomic density exceeding 5000 kg/m^3, generally have low chemical reactivity, and can form stable complexes (Noor et al. 2022). Metals, such as cadmium (Cd), copper (Cu), zinc (Zn), iron (Fe), chromium (Cr), vanadium (V), nickel (Ni), lead (Pb), cobalt (Co), manganese (Mn), molybdenum (Mo), titanium (Ti), cerium (Ce), lanthanum (La), etc., and metalloids include mercury (Hg), tellurium (Te), antimony (Sb), boron (B), arsenic (As), etc. (Vardhan et al. 2019; Rahman et al. 2022; Ghuge et al. 2023; Kumar et al. 2022). The sources of heavy metals in the environment are mainly divided into natural and anthropogenic factors.

Anthropogenic sources mainly include atmospheric deposition, volcanic eruptions and soil erosion, while anthropogenic sources mainly include mining and smelting, improper discharge of the three wastes, fossil fuel combustion and unreasonable application of agricultural chemicals (Kanwar et al. 2020; Yadav et al. 2023; Ghorbani et al. 2024; Azhar et al. 2022; Verma et al. 2024; Kumar et al. 2024a). Most heavy metal elements, due to their cumulative effect, can exert toxic effects on plants even under lower concentration conditions, such as inhibiting seed germination, hindering photosynthesis, impeding growth, reducing yield and quality, and can enter the human body through the food chain, endangering human health and causing cancer, skin damage, kidney disease, etc. (Vardhan et al. 2019; Ulhassan et al. 2022a; Verma et al. 2023a).

H. Liu (✉) · Q. Jiao · J. Zhang
College of Resources and Environment, Henan Agricultural University, Zhengzhou, China
e-mail: liuhaitaoky@henau.edu.cn

G. Li
National Engineering Research Center for Wheat, Henan Agricultural University, Zhengzhou, China

C. S. Seth et al. (eds.), *Nano-formulations for Sustainable Agriculture*,
Nanotechnology in the Life Sciences, https://doi.org/10.1007/978-3-031-90617-6_9

9.2 Toxicity of Heavy Metals on Plants

Heavy metals enter plant root cells from the soil via transporters and are transported into the cell interior via symplastic and apoplastic pathways, and then transported to the above-ground parts via the phloem and xylem. At the same time, heavy metals can inhibit the absorption of mineral elements, disrupt ion balance, and induce oxidative stress by causing the accumulation of toxic free radicals, and affect important cellular processes (photosynthesis and respiration), thereby damaging biological macromolecules such as proteins, lipids, and DNA, leading to DNA damage and mutations, and destroying the structure and function of cell membranes (Ghorbani et al. 2024; Kumar et al. 2024b). Most heavy metals have a high affinity for sulfur and can form stable chemical bonds with sulfhydryl groups in enzymes, thereby destroying the normal function of enzymes. Furthermore, heavy metals can also bind to carboxyl and amino groups in proteins and induce the production of reactive oxygen species, leading to excessive accumulation and destruction of cell membrane structure, causing cell damage (Sarma et al. 2019; Choudhary et al. 2025). Ultimately, heavy metals inhibit cell division and elongation, cause root browning and stunted growth, yellowing of leaves, interfere with chlorophyll synthesis, reduce photosynthesis, disrupt nutrient balance, reduce crop yield and quality, and even cause plant death (Noor et al. 2022; Ghuge et al. 2023).

9.3 Physiological and Biochemical Mechanisms of Plant Responses

Plants have evolved various defense mechanisms to enhance their tolerance to heavy metals. Various functional groups on the plant cell wall can bind to heavy metal ions, limiting the plant's uptake of heavy metal ions and acting as a physical barrier to the entry of heavy metals into the cell (Ejaz et al. 2023). In plant cells, heavy metal ions are perceived by signaling molecules, which in turn activate downstream cascade reactions. Chelating peptides can form stable complexes with heavy metal ions and transport them via transporters and sequester them in vacuoles, reducing heavy metal translocation (Ghuge et al. 2023). At the same time, regulating the content of osmoregulatory substances, maintaining cell homeostasis, maintaining normal cell structure and function, and activating enzymatic antioxidant enzyme systems and non-enzymatic antioxidant substance systems are involved in the removal of active oxygen (Yaashikaa et al. 2022; Verma et al. 2023b) (Fig. 9.1).

9.4 Detoxification of Heavy Metal by Nanomaterials

To ensure environmental safety and human health, it is urgent to eliminate or mitigate the negative effects of heavy metals. Existing methods for treating contaminated soil mainly include physical, chemical, and biological methods. However, those methods

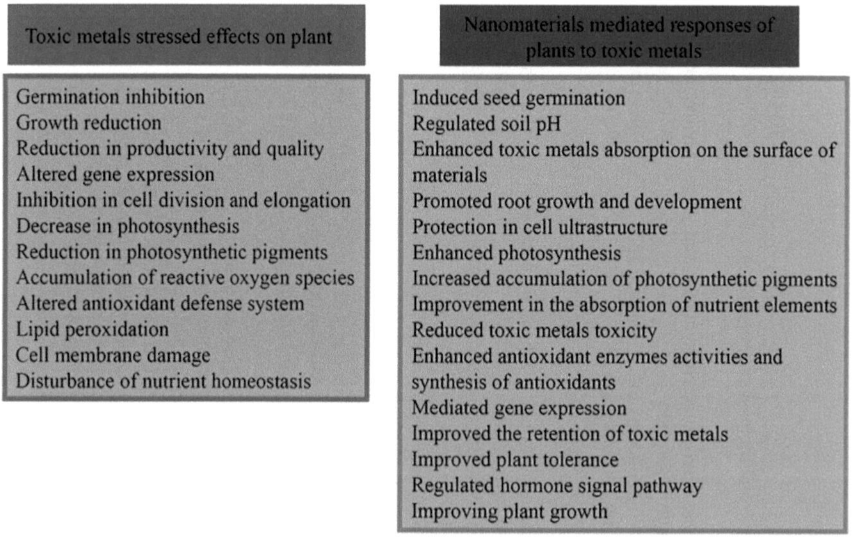

Fig. 9.1 The toxicity of toxic metals and the responses mediated by nanomaterials to toxic metals in plants

are both time-consuming and expensive and are not the best strategy for solving today's environmental problems (Rahman and Singh 2020; Kumar et al. 2024c). On the other hand, nanotechnology, as an emerging tool, has been widely used in environmental pollution remediation due to its excellent ability to immobilize and adsorb metal ions (Zhu et al. 2019; Cai et al. 2019). Nanomaterials are materials with dimensions in the range of 1–100 nm in any spatial dimension. They have the characteristics of small volume, easy penetration of plant tissues, and a large surface area (Ruotolo et al. 2018; Baby et al. 2019; Cai et al. 2019; El-Saadony et al. 2022). Studies have found that NMs can be used as slow-release fertilizers and promote plant absorption of nutrients, thereby reducing nutrient loss during the fertilization process (Rizwan et al. 2019b; Verma et al. 2023c).

NMs can improve seed germination rate, enhance photosynthesis, increase water use efficiency and transpiration rate, regulate plant hormone levels, enhance antioxidant capacity, increase the content of osmotic regulators, promote nutrient absorption, regulate gene expression, and thus promote biomass accumulation, improve resistance to adverse stress, and increase yield and quality, ultimately promoting the development of sustainable agriculture (Landa 2021; El-Saadony et al. 2022; Rajput et al. 2023). Various NMs have been widely used to enhance plant productivity, such as Zinc (Zn) NPs, Silver (Ag) NPs, Silicon dioxide (SiO_2) NPs, Copper (Cu) NPs, Iron oxide (Fe_2O_3) NPs, Titanium dioxide (TiO_2) NPs, Manganese (Mn) NPs, Carbon (C) NPs, Cerium (Ce) NPs, Potassium (K) NPs, Sulfur (S) NPs (Zulfiqar and Ashraf 2021; Landa 2021; Ulhassan et al. 2022a; El-Saadony et al. 2022) (Table 9.1).

Table 9.1 The impact of different NMs on various crops under heavy metal stresses conditions

NPs	Application concentration	NPs size	Plant species	Heavy metal stress	Effect	References
IONPs SiNPs	10 mg L^{-1} 20 mg L^{-1}	15–25 nm 40–100 nm	*Phaseolus vulgaris*	Cadmium	Promote growth and increase photosynthetic rate	Koleva et al. (2022)
FeO-NPs	25, 50, 100 mg kg^{-1}	19–40 nm	*Triticum aestivum* L.	Cadmium	Promote photosynthetic pigment synthesis and reduce cadmium absorption	Manzoor et al. (2021)
ZnO NPs Fe NPs	0, 25, 50, 75, 100 mg L^{-1} ZnO NPs 0, 5, 10, 15, 20 mg L^{-1} Fe NPs	20–30 nm 50–100 nm	*Triticum aestivum* L.	Cadmium	Promote growth, reduce oxidative damage and reduce cadmium concentration	Rizwan et al. (2019a)
ZnO NPs	25 mg L^{-1}	2–64 nm	*Leucaena leucocephala*	Lead and Cadmium	Increase the content of photosynthetic pigment and soluble protein, reduce the content of malondialdehyde and increase the antioxidant enzyme activity	Venkatachalam et al. (2017)
Si NPs	0, 300, 600, 900, 1200 mg L^{-1}		*Triticum aestivum* L.	Cadmium	Promote wheat growth and photosynthesis, alleviate oxidative stress, enhance the activity of antioxidant enzymes and reduce the cadmium content in wheat grains	Hussain et al. (2019)
Fe NPs	0, 25, 50 100 mg kg^{-1}	50–100 nm	*Triticum aestivum* L.	Cadmium	Stimulate photosynthesis, alleviate oxidative stress, reduce cadmium content in grains and increase iron content	Adrees et al. (2020)
TiO_2 NPs	0, 100, 250 mg L^{-1}	6.5 nm	*Zea mays* L.	Cadmium	Reduce cadmium content, increase antioxidant enzyme activity, upregulate energy metabolism and synthesize antioxidant metabolic pathways	Lian et al. (2020)
IONPs HGNPs	0, 25, 50, 100 mg kg^{-1}	15–31 nm 51–94 nm	*Oryza sativa* L.	Cadmium	Increase biomass accumulation, enhance antioxidant enzyme activity and photosynthesis, reduce reactive oxygen species (ROS) accumulation and cadmium content and inhibit cadmium transporter gene expression	Ahmed et al. (2021)

Si NPs	0, 300, 600, 900, 1200 mg kg^{-1}	≤ 50 nm	*Triticum aestivum* L.	Cadmium	Promote wheat growth, enhance photosynthesis, reduce oxidative stress, decrease cadmium content in wheat and increase silicon concentration	Ali et al. (2019)
ZnO NPs	0, 1, 2.5, 5, 10, 15 mg L^{-1}	20–40 nm	*Lactuca sativa* L.	Cadmium	Promote growth, enhance antioxidant capacity, alleviate oxidative stress, increase the absorption of photosynthesis-related ions and increase lignin content	Gao et al. (2022)
ZnO NPs	500 mg L^{-1}	20 nm	*Zea mays* L.	Cobalt	Increase seed zinc content, improve the antioxidant defense system, promote element absorption, and protect maize stomata and ultrastructure	Salam et al. (2022)
ZnO NPs	0, 25, 50, 75, 100 mg kg^{-1} and 0, 25, 50, 75, 100 mg L^{-1}	20–30 nm	*Triticum aestivum* L.	Cadmium	Reduce cadmium concentration, increase zinc content and improve grain yield	Hussain et al. (2018)
nZnO	0, 10, 20, 50, 100 mg L^{-1}	20–30 nm	*Oryza sativa* L.	Arsenic	Promote rice growth, reduce arsenic accumulation, and enhance arsenic and PC chelation	Yan et al. (2021)
Nano-SiO_2	4 mg kg^{-1}	20 nm	*Zea mays* L.	Aluminum	Reduce photorespiratory enzyme and NADPH oxidase activities, stimulate the antioxidant defense system, and increase organic acid accumulation	de Sousa et al. (2019)
TiO_2-NPs	40 mg L^{-1}		*Triticum aestivum* L.	Copper	Enhance nutrient uptake, increase antioxidant enzyme activity and gene expression, and reduce organic acid exudation	Alshegaihi et al. (2023)
Se-NPs IONAP	60 mg L^{-1}	10–40 nm and 15–25 nm	*Bambusoideae*	Arsenic	Enhance antioxidant performance, increase putrescine and spermidine content, protect chlorophyll integrity and reduce lipid peroxidation	Emamverdian et al. (2023)
SiO NPs	250 mg kg^{-1}	15–25 nm	*Triticum aestivum* L.	Chromium	Maintain normal cellular structure, enhance photosynthetic characteristics and increase antioxidant enzyme activity	Manzoor et al. (2022)
AstNPs	100 mg L^{-1}	12.03–30.37 nm	*Triticum aestivum* L.	Cadmium	Promote element uptake, suppress cadmium transporter gene expression, enhance antioxidant enzyme gene expression and maintain normal chloroplast structure	Zeshan et al. (2022)

(continued)

Table 9.1 (continued)

NPs	Application concentration	NPs size	Plant species	Heavy metal stress	Effect	References
TiO_2-NPs	0, 40, 80, 160 mg L^{-1}	< 100 nm	*Coriandrum sativum* L.	Cadmium	Increase soluble protein, soluble sugar, and proline content, and promote secondary metabolite synthesis	Sardar et al. (2022b)
SNPs	300 mg L^{-1}	37–61 nm	*Brassica napus* L.	Copper	Promote element uptake, enhance antioxidant capacity, and increase chlorophyll and protein content	Yuan et al. (2023)
ZnO-NPs	0, 50, 100 mg L^{-1}	20–30 nm	*portulaca oleracea* L.	Cadmium	Enhance antioxidant enzyme and glutathione reductase system activities, reduce oxidative damage, promote element uptake, and increase proline and glutathione accumulation	Pishkar et al. (2022)
SeNPs	0, 5, 10, 15 mg L^{-1}	10–45 nm	*Coriandrum sativum*	Cadmium	Promote element uptake, enhance water potential, increase antioxidant enzyme activities and reduce cadmium uptake	Sardar et al. (2022a)

9.5 The Mechanism of Nanomaterials in Alleviating Contaminated Soil by Toxic Metals

When NMs are applied to soil, they can regulate soil pH and adsorb heavy metals to the material surface, thereby reducing the mobility and bioavailability of heavy metals (Zhou et al. 2021; Ghorbani et al. 2024). After being absorbed by plants, specific NMs can regulate the expression of metal transport protein genes, enhance the plant's interception of heavy metals, and most NMs accumulate in the cell wall, bind to heavy metals to form complexes, and reduce the availability of heavy metals, thereby hindering the migration and biological activity of heavy metals in plants (Zhou et al. 2021; Rajput et al. 2022). At the same time, NMs can regulate the production of plant chelating peptides and the chelation of heavy metals in vacuoles, enhancing the retention of heavy metals and reducing their transport, thereby reducing the bioavailability of heavy metals and reducing their toxicity (Zhou et al. 2023; Ghorbani et al. 2024; Ulhassan et al. 2022a; Ahmad et al. 2023; Mariyam et al. 2024).

NMs can also upregulate the expression of antioxidant system genes to activate the antioxidant system, scavenge excessive active oxygen induced by heavy metals, promote the absorption of nutrients and the accumulation of osmotic regulators, regulate hormone homeostasis and ion balance, thereby reducing the bioavailability and toxicity of heavy metals to plants, protecting leaf ultrastructure, enhancing photosynthesis, promoting the accumulation of photosynthetic pigments, regulating the formation of cell metabolites, improving plant growth and quality, and enhancing plant resistance (Ahmad et al. 2023; Zhou et al. 2023; Ulhassan et al. 2022a, b; Zhu et al. 2019; Zou et al. 2022) (Fig. 9.1).

9.6 Application of Nanomaterials in Precision Agriculture

Currently, nanotechnology has been used to produce nano-fertilizers and nano-encapsulated nutrients, which are widely used to increase crop yields (Verma et al. 2022a, b). They can achieve targeted delivery of nutrients to plants, thereby increasing crop yield and quality through precision agriculture, and have great potential in precision agriculture (El-Saadony et al. 2022; Choudhary et al. 2024). However, the application of nanomaterials in crop improvement and sustainable agriculture is still in its early stages of research and needs further exploration. Therefore, future research should focus on the following aspects to help elucidate the mechanisms of NMs effects on ecosystems and crops:

- Develop new NMs with higher metal chelation capacity and selectivity.
- Investigate the mechanisms of metal transport and transformation in plants by nanomaterials.
- Evaluate the long-term effects of NMs on different plant species and under different pollution conditions.
- Explore the synergistic effects of NMs with other mitigation strategies, such as bioremediation and chemical chelation.

9.7 Conclusions

Summarized the recent advances in the application of NMs for the mitigation of toxic metals in plants from contaminated soil. Discussed the sources, hazards and mechanisms of heavy metal toxicity in plants, as well as the various strategies that plants employ to cope with heavy metal stress. Highlighted the unique properties of NMs that make them promising candidates for the remediation of heavy metal-contaminated soils, and the protection of plants from heavy metal toxicity. Overall, the use of NMs for the mitigation of toxic metals in plants holds great promise for the development of more sustainable and environmentally friendly agricultural practices. However, further research is needed to optimize the synthesis, characterization, and application of NMs for this purpose. In particular, more studies are needed to investigate the long-term effects of NMs on plant growth and development, as well as the synergistic effects and mechanisms of NMs with other mitigation strategies. Despite these challenges, the potential benefits of using NMs to mitigate heavy metal toxicity in plants are significant. By harnessing the unique properties of NMs, we can develop new and innovative strategies to protect plants from the harmful effects of heavy metals and ensure the sustainability of precision agriculture.

Acknowledgement This research was supported by the Key R&D and Promotion Project of Henan Provience (252102320062), Key Research Projects of Higher Education Institutions in Henan Province (24A210011) and Research and Practice Project on Teaching Reform of Higher Education at Henan Agricultural University (2024XJGLX126).

References

Adrees M, Khan ZS, Ali S, Hafeez M, Khalid S, ur Rehman MZ, Hussain A, Hussain K, Shahid Chatha SA, Rizwan M (2020) Simultaneous mitigation of cadmium and drought stress in wheat by soil application of iron nanoparticles. Chemosphere 238:124681. https://doi.org/10.1016/j.chemosphere.2019.124681

Ahmad B, Zaid A, Zulfiqar F, Bovand F, Dar TA (2023) Nanotechnology: a novel and sustainable approach towards heavy metal stress alleviation in plants. Nanotechnol Environ Eng 8(1):27–40. https://doi.org/10.1007/s41204-022-00230-8

Ahmed T, Noman M, Manzoor N, Shahid M, Abdullah M, Ali L, Wang G, Hashem A, Al-Arjani A-BF, Alqarawi AA, Abd_Allah EF, Li B (2021) Nanoparticle-based amelioration of drought stress and cadmium toxicity in rice via triggering the stress responsive genetic mechanisms and nutrient acquisition. Ecotoxicol Environ Saf 209:111829. https://doi.org/10.1016/j.ecoenv.2020.111829

Ahmed T, Masood HA, Noman M, Al-Huqail AA, Alghanem SMS, Khan MM, Muhammad S, Manzoor N, Rizwan M, Qi XJ, Abeed AHA, Li B (2023) Biogenic silicon nanoparticles mitigate cadmium (Cd) toxicity in rapeseed (*Brassica napus* L.) by modulating the cellular oxidative stress metabolism and reducing Cd translocation. J Hazard Mater 459:132070. https://doi.org/10.1016/j.jhazmat.2023.132070

Ali S, Rizwan M, Hussain A, Zia ur Rehman M, Ali B, Yousaf B, Wijaya L, Alyemeni MN, Ahmad P (2019) Silicon nanoparticles enhanced the growth and reduced the cadmium accumulation in grains of wheat (*Triticum aestivum* L.). Plant Physiol Biochem 140:1–8. https://doi.org/10.1016/j.plaphy.2019.04.041

Alshegaihi RM, Saleem MH, Saleem A, Ali B, Aziz H, Fahad S, Alataway A, Dewidar AZ, Elansary HO (2023) Silicon and titanium dioxide mitigate copper stress in wheat (*Triticum aestivum* L.) through regulating antioxidant defense mechanisms. J Plant Growth Regul 43:1519–1535. https://doi.org/10.1007/s00344-023-11204-8

Azhar U, Ahmad H, Shafqat H, Babar M, Shahzad Munir HM, Sagir M, Arif M, Hassan A, Rachmadona N, Rajendran S, Mubashir M, Khoo KS (2022) Remediation techniques for elimination of heavy metal pollutants from soil: a review. Environ Res 214:113918. https://doi.org/10.1016/j.envres.2022.113918

Baby R, Saifullah B, Hussein MZ (2019) Carbon nanomaterials for the treatment of heavy metal-contaminated water and environmental remediation. Nanoscale Res Lett 14(1):341. https://doi.org/10.1186/s11671-019-3167-8

Cai CY, Zhao MH, Yu Z, Rong HW, Zhang CS (2019) Utilization of nanomaterials for in-situ remediation of heavy metal(loid) contaminated sediments: a review. Sci Total Environ 662:205–217. https://doi.org/10.1016/j.scitotenv.2019.01.180

Choudhary R, Rajput VD, Ghodake G, Ahmad F, Meena M, Rehman RU, Prasad R, Sharma RK, Singh R, Seth CS (2024) Comprehensive journey from past to present to future about seed priming with hydrogen peroxide and hydrogen sulfide concerning drought, temperature, UV and ozone stresses-a review. Plant Soil:1–23. https://doi.org/10.1007/s11104-024-06499-9

Choudhary R, Ahmad F, Kaya C, Upadhyay SK, Muneer S, Kumar V, Meena M, Liu H, Upadhyay H, Seth CS (2025) Decrypting proteomics, transcriptomics, genomics, and integrated omics for augmenting the abiotic, biotic, and climate change stress resilience in plants. J Plant Physiol. https://doi.org/10.1016/j.jplph.2025.154430

Ejaz U, Khan SM, Khalid N, Ahmad Z, Jehangir S, Fatima Rizvi Z, Lho LH, Han H, Raposo A (2023) Detoxifying the heavy metals: a multipronged study of tolerance strategies against heavy metals toxicity in plants. Front Plant Sci 14:1154571. https://doi.org/10.3389/fpls.2023.1154571

El-Saadony MT, Saad AM, Soliman SM, Salem HM, Desoky E-SM, Babalghith AO, El-Tahan AM, Ibrahim OM, Ebrahim AAM, Abd El-Mageed TA, Elrys AS, Elbadawi AA, El-Tarabily KA, AbuQamar SF (2022) Role of nanoparticles in enhancing crop tolerance to abiotic stress: a comprehensive review. Front Plant Sci 13:946717. https://doi.org/10.3389/fpls.2022.946717

Emamverdian A, Ghorbani A, Li Y, Pehlivan N, Barker J, Ding YL, Liu GH, Zargar M (2023) Responsible mechanisms for the restriction of heavy metal toxicity in plants via the co-foliar spraying of nanoparticles. Agronomy 13(7):1748. https://doi.org/10.3390/agronomy13071748

Gao F, Zhang XD, Zhang J, Li J, Niu TH, Tang CN, Wang C, Xie JM (2022) Zinc oxide nanoparticles improve lettuce (*Lactuca sativa* L.) plant tolerance to cadmium by stimulating antioxidant defense, enhancing lignin content and reducing the metal accumulation and translocation. Front Plant Sci 13:1015745. https://doi.org/10.3389/fpls.2022.1015745

Ghorbani A, Emamverdian A, Pehlivan N, Zargar M, Razavi SM, Chen MX (2024) Nano-enabled agrochemicals: mitigating heavy metal toxicity and enhancing crop adaptability for sustainable crop production. J Nanobiotechnol 22(1):91. https://doi.org/10.1186/s12951-024-02371-1

Ghuge SA, Nikalje GC, Kadam US, Suprasanna P, Hong JC (2023) Comprehensive mechanisms of heavy metal toxicity in plants, detoxification, and remediation. J Hazard Mater 450:131039. https://doi.org/10.1016/j.jhazmat.2023.131039

Hussain A, Ali S, Rizwan M, Zia ur Rehman M, Javed MR, Imran M, SAS C, Nazir R (2018) Zinc oxide nanoparticles alter the wheat physiological response and reduce the cadmium uptake by plants. Environ Pollut 242:1518–1526. https://doi.org/10.1016/j.envpol.2018.08.036

Hussain A, Rizwan M, Ali Q, Ali S (2019) Seed priming with silicon nanoparticles improved the biomass and yield while reduced the oxidative stress and cadmium concentration in wheat grains. Environ Sci Pollut Res 26(8):7579–7588. https://doi.org/10.1007/s11356-019-04210-5

Kanwar VS, Sharma A, Srivastav AL, Rani L (2020) Phytoremediation of toxic metals present in soil and water environment: a critical review. Environ Sci Pollut Res 27(36):44835–44860. https://doi.org/10.1007/s11356-020-10713-3

Koleva L, Umar A, Yasin NA, Shah AA, Siddiqui MH, Alamri S, Riaz L, Raza A, Javed T, Shabbir Z (2022) Iron oxide and silicon nanoparticles modulate mineral nutrient homeostasis and metabolism in cadmium-stressed *Phaseolus vulgaris*. Front Plant Sci 13:806781. https://doi.org/10.3389/fpls.2022.806781

Kumar P, Gacem A, Ahmad MT, Yadav VK, Singh S, Yadav KK, Alam MM, Dawane V, Piplode S, Maurya P, Ahn Y, Jeon B-H, Cabral-Pinto MMS (2022) Environmental and human health implications of metal(loid)s: source identification, contamination, toxicity, and sustainable clean-up technologies. Front Environ Sci 10:949581. https://doi.org/10.3389/fenvs.2022.949581

Kumar D, Mariyam S, Gupta KJ, Thiruvengadam M, Ghodake GS, Xing B, Seth CS (2024a) Comparative investigation on chemical and green synthesized titanium dioxide nanoparticles against chromium (VI) stress eliciting differential physiological, biochemical, and cellular attributes in Helianthus annuus L. Sci Total Environ 930:172413. https://doi.org/10.1016/j.scitotenv.2024.172413

Kumar D, Singh R, Upadhyay SK, Verma KK, Tripathi RM, Liu HT, Dhankher OP, Tripathi RD, Sahi SV, Seth CS (2024b) Review on interactions between nanomaterials and phytohormones: novel perspectives and opportunities for mitigating environmental challenges. Plant Sci 340:111964. https://doi.org/10.1016/j.plantsci.2023.111964

Kumar D, Biswas JK, Mulla SI, Singh R, Shukla R, Ahanger MA, Shekhawat GS, Verma KK, Siddiqui MW, Seth CS (2024c) Micro and nanoplastics pollution: sources, distribution, uptake in plants, toxicological effects, and innovative remediation strategies for environmental sustainability. Plant Physiol Biochem 108795. https://doi.org/10.1016/j.plaphy.2024.108795

Landa P (2021) Positive effects of metallic nanoparticles on plants: overview of involved mechanisms. Plant Physiol Biochem 161:12–24. https://doi.org/10.1016/j.plaphy.2021.01.039

Lian JP, Zhao LF, Wu JN, Xiong HX, Bao YY, Zeb A, Tang JC, Liu WT (2020) Foliar spray of TiO_2 nanoparticles prevails over root application in reducing Cd accumulation and mitigating Cd-induced phytotoxicity in maize (*Zea mays* L.). Chemosphere 239:124794. https://doi.org/10.1016/j.chemosphere.2019.124794

Manzoor N, Ahmed T, Noman M, Shahid M, Nazir MM, Ali L, Alnusaire TS, Li B, Schulin R, Wang G (2021) Iron oxide nanoparticles ameliorated the cadmium and salinity stresses in wheat plants, facilitating photosynthetic pigments and restricting cadmium uptake. Sci Total Environ 769:145221. https://doi.org/10.1016/j.scitotenv.2021.145221

Manzoor N, Ali L, Ahmed T, Rizwan M, Ali S, Shahid MS, Schulin R, Liu Y, Wang G (2022) Silicon oxide nanoparticles alleviate chromium toxicity in wheat (*Triticum aestivum* L.). Environ Pollut 315:120391. https://doi.org/10.1016/j.envpol.2022.120391

Mariyam S, Upadhyay SK, Chakraborty K, Verma KK, Duhan JS, Muneer S, Meena M, Sharma RK, Ghodake G, Seth CS (2024) Nanotechnology, a frontier in agricultural science, a novel approach in abiotic stress management and convergence with new age medicine-A review. Sci Total Environ 912:169097. https://doi.org/10.1016/j.scitotenv.2023.169097

Noor I, Sohail H, Sun JX, Nawaz MA, Li GH, Hasanuzzaman M, Liu JW (2022) Heavy metal and metalloid toxicity in horticultural plants: tolerance mechanism and remediation strategies. Chemosphere 303:135196. https://doi.org/10.1016/j.chemosphere.2022.135196

Pishkar L, Yousefi S, Iranbakhsh A (2022) Foliar application of Zinc oxide nanoparticles alleviates cadmium toxicity in purslane by maintaining nutrients homeostasis and improving the activity of antioxidant enzymes and glyoxalase system. Ecotoxicology 31(4):667–678. https://doi.org/10.1007/s10646-022-02533-7

Rahman Z, Singh VP (2020) Bioremediation of toxic heavy metals (THMs) contaminated sites: concepts, applications and challenges. Environ Sci Pollut Res 27(22):27563–27581. https://doi.org/10.1007/s11356-020-08903-0

Rahman SU, Nawaz MF, Gul S, Yasin G, Hussain B, Li YL, Cheng HF (2022) State-of-the-art OMICS strategies against toxic effects of heavy metals in plants: a review. Ecotoxicol Environ Saf 242:113952. https://doi.org/10.1016/j.ecoenv.2022.113952

Rajput VD, Minkina T, Upadhyay SK, Kumari A, Ranjan A, Mandzhieva S, Sushkova S, Singh RK, Verma KK (2022) Nanotechnology in the restoration of polluted soil. Nanomaterials 12:769. https://doi.org/10.3390/nano12050769

Rajput VD, Kumari A, Upadhyay SK, Minkina T, Mandzhieva S, Ranjan A, Sushkova S, Burachevskaya M, Rajput P, Konstantinova E, Singh J, Verma KK (2023) Can nanomaterials improve the soil microbiome and crop productivity? Agriculture 13:231. https://doi.org/10.3390/agriculture13020231

Rizwan M, Ali S, Ali B, Adrees M, Arshad M, Hussain A, Zia ur Rehman M, Waris AA (2019a) Zinc and iron oxide nanoparticles improved the plant growth and reduced the oxidative stress and cadmium concentration in wheat. Chemosphere 214:269–277. https://doi.org/10.1016/j.chemosphere.2018.09.120

Rizwan M, Ali S, Zia ur Rehman M, Adrees M, Arshad M, Qayyum MF, Ali L, Hussain A, Chatha SAS, Imran M (2019b) Alleviation of cadmium accumulation in maize (*Zea mays* L.) by foliar spray of zinc oxide nanoparticles and biochar to contaminated soil. Environ Pollut 248:358–367. https://doi.org/10.1016/j.envpol.2019.02.031

Ruotolo R, Maestri E, Pagano L, Marmiroli M, White JC, Marmiroli N (2018) Plant response to metal-containing engineered nanomaterials: an omics-based perspective. Environ Sci Technol 52(5):2451–2467. https://doi.org/10.1021/acs.est.7b04121

Salam A, Khan AR, Liu L, Yang SQ, Azhar W, Ulhassan Z, Zeeshan M, Wu JY, Fan XM, Gan YB (2022) Seed priming with zinc oxide nanoparticles downplayed ultrastructural damage and improved photosynthetic apparatus in maize under cobalt stress. J Hazard Mater 423:127021. https://doi.org/10.1016/j.jhazmat.2021.127021

Sardar R, Ahmed S, Shah AA, Yasin NA (2022a) Selenium nanoparticles reduced cadmium uptake, regulated nutritional homeostasis and antioxidative system in Coriandrum sativum grown in cadmium toxic conditions. Chemosphere 287:132332. https://doi.org/10.1016/j.chemosphere.2021.132332

Sardar R, Ahmed S, Yasin NA (2022b) Titanium dioxide nanoparticles mitigate cadmium toxicity in *Coriandrum sativum* L. through modulating antioxidant system, stress markers and reducing cadmium uptake. Environ Pollut 292:118373. https://doi.org/10.1016/j.envpol.2021.118373

Sarma GK, Sen Gupta S, Bhattacharyya KG (2019) Nanomaterials as versatile adsorbents for heavy metal ions in water: a review. Environ Sci Pollut Res 26(7):6245–6278. https://doi.org/10.1007/s11356-018-04093-y

de Sousa A, Saleh AM, Habeeb TH, Hassan YM, Zrieq R, Wadaan MAM, Hozzein WN, Selim S, Matos M, AbdElgawad H (2019) Silicon dioxide nanoparticles ameliorate the phytotoxic hazards of aluminum in maize grown on acidic soil. Sci Total Environ 693:133636. https://doi.org/10.1016/j.scitotenv.2019.133636

Ulhassan Z, Bhat JA, Zhou WJ, Senan AM, Alam P, Ahmad P (2022a) Attenuation mechanisms of arsenic induced toxicity and its accumulation in plants by engineered nanoparticles: a review. Environ Pollut 302:119038. https://doi.org/10.1016/j.envpol.2022.119038

Ulhassan Z, Khan I, Hussain M, Khan AR, Hamid Y, Hussain S, Allakhverdiev SI, Zhou WJ (2022b) Efficacy of metallic nanoparticles in attenuating the accumulation and toxicity of chromium in plants: current knowledge and future perspectives. Environ Pollut 315:120390. https://doi.org/10.1016/j.envpol.2022.120390

Vardhan KH, Kumar PS, Panda RC (2019) A review on heavy metal pollution, toxicity and remedial measures: current trends and future perspectives. J Mol Liq 290:111197. https://doi.org/10.1016/j.molliq.2019.111197

Venkatachalam P, Jayaraj M, Manikandan R, Geetha N, Rene ER, Sharma NC, Sahi SV (2017) Zinc oxide nanoparticles (ZnONPs) alleviate heavy metal-induced toxicity in *Leucaena leucocephala* seedlings: a physiochemical analysis. Plant Physiol Biochem 110:59–69. https://doi.org/10.1016/j.plaphy.2016.08.022

Verma KK, Song XP, Joshi A, Rajput VD, Singh M, Sharma A, Singh RK, Li DM, Arora J, Minkina T, Li YR (2022a) Nanofertilizer possibilities for healthy soil, water and food in future: an overview. Front Plant Sci 13:865048. https://doi.org/10.3389/fpls.2022.865048

Verma KK, Song XP, Joshi A, Tian DD, Rajput VD, Singh M, Arora J, Minkina TM, Li YR (2022b) Recent trends in nano-fertilizer for advancing sustainable agriculture under the era of climate change to ensure food security for future generations. Nanomaterials 12:173. https://doi.org/10.3390/nano12010173

Verma KK, Song X, Verma CL, Huang H, Singh M, Xu L, Wu J, Khan ZH, Zhang B, Li Y (2023a) Mathematical modeling of climate and fluoride effects on sugarcane photosynthesis with silicon nanoparticles. Plant Physiol Biochem 204:108089. https://doi.org/10.1016/j.plaphy.2023.108089

Verma KK, Song X, Degu HD, Guo D, Joshi A, Huang H, Xu L, Singh M, Huang D, Rajput VD, Li Y (2023b) Recent advances in nitrogen and nano-nitrogen fertilizers for sustainable crop production: a mini-review. Chem Biol Technol Agric 10:1–14. https://doi.org/10.1186/s40538-023-00488-3

Verma KK, Singh M, Rajput VD, Verma CL, Burachevskaya M (2023c) Nanofertilizers: a sustainable alternative to conventional means. In: Rajput VD, Singh A, Minkina TM, Verma KK, Singh AK (eds) Nanotechnology for sustainable agriculture: an innovative and eco-friendly approach. Apple Academic Press, Inc., Taylor & Francis. https://doi.org/10.1201/9781003333128

Verma KK, Song XP, Kumari A, Jagadesh M, Singh SK, Bhatt R, Singh M, Seth CS, Li YR (2024) Climate change adaptation: challenges for agricultural sustainability. Plant Cell Environ:1–12. https://doi.org/10.1111/pce.15078

Yaashikaa PR, Kumar PS, Jeevanantham S, Saravanan R (2022) A review on bioremediation approach for heavy metal detoxification and accumulation in plants. Environ Pollut 301:119035. https://doi.org/10.1016/j.envpol.2022.119035

Yadav M, George N, Dwibedi V (2023) Emergence of toxic trace elements in plant environment: insights into potential of silica nanoparticles for mitigation of metal toxicity in plants. Environ Pollut 333:122112. https://doi.org/10.1016/j.envpol.2023.122112

Yan SW, Wu F, Zhou S, Yang JH, Tang XJ, Ye WL (2021) Zinc oxide nanoparticles alleviate the arsenic toxicity and decrease the accumulation of arsenic in rice (*Oryza sativa* L.). BMC Plant Biol 21(1):150. https://doi.org/10.1186/s12870-021-02929-3

Yuan HY, Liu QQ, Fu JH, Wang YJ, Zhang YX, Sun YM, Tong HY, Dhankher OP (2023) Co-exposure of sulfur nanoparticles and Cu alleviate Cu stress and toxicity to oilseed rape *Brassica napus* L. J Environ Sci 124:319–329. https://doi.org/10.1016/j.jes.2021.09.040

Zeshan A, Abdullah M, Adil MF, Wei DM, Noman M, Ahmed T, Sehar S, Ouyang Y, Shamsi IH (2022) Improvement of morpho-physiological, ultrastructural and nutritional profiles in wheat seedlings through astaxanthin nanoparticles alleviating the cadmium toxicity. J Hazard Mater 424:126511. https://doi.org/10.1016/j.jhazmat.2021.126511

Zhou PF, Adeel M, Shakoor N, Guo ML, Hao Y, Azeem I, Li MS, Liu MY, Rui YK (2021) Application of nanoparticles alleviates heavy metals stress and promotes plant growth: an overview. Nano 11(1):26. https://doi.org/10.3390/nano11010026

Zhou PF, Zhang P, He MK, Cao Y, Adeel M, Shakoor N, Jiang YQ, Zhao WC, Li YB, Li MS, Azeem I, Jia LL, Rui YK, Ma XM, Lynch I (2023) Iron-based nanomaterials reduce cadmium toxicity in rice (*Oryza sativa* L.) by modulating phytohormones, phytochelatin, cadmium transport genes and iron plaque formation. Environ Pollut 320:121063. https://doi.org/10.1016/j.envpol.2023.121063

Zhu Y, Xu F, Liu Q, Chen M, Liu XL, Wang YY, Sun Y, Zhang LL (2019) Nanomaterials and plants: positive effects, toxicity and the remediation of metal and metalloid pollution in soil. Sci Total Environ 662:414–421. https://doi.org/10.1016/j.scitotenv.2019.01.234

Zou CM, Lu T, Wang QRT, Xu P, Jing YF, Wang RL, Xu J, Wan JP (2022) Comparative physiological and metabolomic analyses reveal that Fe_3O_4 and ZnO nanoparticles alleviate Cd toxicity in tobacco. J Nanobiotechnol 20(1):302. https://doi.org/10.1186/s12951-022-01509-3

Zulfiqar F, Ashraf M (2021) Nanoparticles potentially mediate salt stress tolerance in plants. Plant Physiol Biochem 160:257–268. https://doi.org/10.1016/j.plaphy.2021.01.028

Chapter 10
Use of Nanomaterials as Nanofertilizers for Crop Betterment

Muazu Issifu, Mawuli K. Azameti, Abdul-Wahab Imoro, and Frederick Kankam

10.1 Introduction

In agriculture, climate change has a direct and indirect impact on the rate of fertilizer use by plants (Gangwar et al. 2023; Magnabosco et al. 2023). It alters nutrient cycling processes, affects plant access to nutrients due to changes in water availability, increases risks of leaching and runoff, and influences soil fertility management practices (Yadav et al. 2023b). Adapting fertilization strategies to the changing climate is crucial for sustainable agriculture and minimizing environmental impacts. In response to these challenges posed by climate change, efforts are being made towards developing more efficient fertilization strategies that optimize nutrient uptake while minimizing environmental impacts (Kumari et al. 2022, 2023a, b; Verma et al. 2024a). This includes Nanotechnology with its innovative technology and precision agriculture. With potential applications in agriculture, nanotechnology is considered the most rapidly emerging technology of the decade (Shang et al. 2019). Nanotools such as nanofertilizers (NFs) have revolutionized age-old agricultural techniques into precision farming (Elemike et al. 2019; Verma et al. 2022a). Additionally, the application of NFs aligns with the principles of precision

M. Issifu (✉) · F. Kankam
Department of Agricultural Biotechnology, Faculty of Agriculture Food and Consumer Sciences, University for Development Studies, Tamale, Ghana
e-mail: muazu.issifu@students.jkuat.ac.ke

M. K. Azameti
Department of Applied Biology, C. K. Tedam University of Technology and Applied Sciences, Navrongo, Ghana

A.-W. Imoro
Department of Forestry and Forest Resources Management, University for Development Studies, Tamale, Ghana

C. S. Seth et al. (eds.), *Nano-formulations for Sustainable Agriculture*,
Nanotechnology in the Life Sciences, https://doi.org/10.1007/978-3-031-90617-6_10

agriculture, which focuses on optimizing inputs like water, pesticides, and fertilizers to maximize yields while minimizing resource use and environmental impact (Gangwar et al. 2023; Kumar et al. 2024a). These advancements represent a cutting-edge approach to sustainable agriculture and have the potential to revolutionize crop production (Verma et al. 2022b).

Nanomaterials (NMs) consist of particles of substances with at least one dimension that falls within the nanoscale range, usually ranging from 1 to 100 nanometers (Mejias et al. 2021; Nongbet et al. 2022). In-depth research analysis has been performed in the area of crop improvement to unearth the potential or possible ways of using NMs in agriculture. By directly giving nutrients and agrochemicals to plants, NMs can improve nutrient absorption and alleviate the harmful consequences on the environment while also improving crop growth (An et al. 2022; Verma et al. 2024b). Soil remediation removes pollutants from soil or water, helping to improve the health of agricultural land (Elemike et al. 2019; Shang et al. 2019; Rajput et al. 2023). Nanoparticles (NPs) can be engineered to target specific plant pathogens, providing a more precise and effective method for controlling diseases in crops (Dutta et al. 2022; Worrall et al. 2018). Employing NPs as agents for seed priming could potentially boost germination rates and promote early plant growth (Nile et al. 2022). Nanosensors can sense fluctuations in temperature, humidity, nutrient levels, or disease presence at a very small scale, allowing for precise adjustments in irrigation or fertilization strategies (Miguel-Rojas and Pérez-de-Luque 2023; Yadav et al. 2023a, b, c).

To address the issue of nutrient deficiency in plants, researchers utilized nanotechnology to develop a smart delivery system that would progressively and carefully transport nutrients to the desired location in the plants. A relatively recent invention in agriculture NFs, can increase crop yield by making essential nutrients more readily available to the plant (Fincheira et al. 2023). Traditional fertilizers which are complemented by NMs consist of macro-sized particles that are applied to soil, where they dissolve and release nutrients such as nitrogen (N), phosphorus (P), and potassium (K) for plants to absorb and improve yield (Yadav et al. 2023b). Conversely, NFs are made up of considerably tiny particles that are intended to enter plant tissues and supply nutrients straight to the cells (Nongbet et al. 2022). When juxtaposed with conventional fertilizers, NFs have greater effectiveness, less wastage, and enhanced nutrient absorption by plants (Mejias et al. 2021). Nevertheless, despite NFs benefits, there are a few downsides as well to consider, including possible toxicity and expensive production costs (Yadav et al. 2023a). With the rise in availability and acceptance of NFs, farmers and home gardeners are beginning to commercialize them. At present, NFs are produced on small-scale operations and very seldom commercialized.

Nanofertilizers include NMs that provide essential nutrients like macronutrients or micronutrients, as well as NMs that transport macronutrients in fortified fertilizers (Avila-Quezada et al. 2022). Various techniques such as foliar spraying, soil treatments, root dipping, and seed coating/priming can be used to apply NFs to plants (Ding et al. 2023; Yadav et al. 2023b; Verma et al. 2023a; Haydar et al. 2024). Microscopic particles known as NFs hold the promise of supplying plants with essential nutrients without compromising their yield. By enhancing nutrient absorption and improving agricultural productivity, these tiny particles can transform the current food production systems (Haydar et al. 2024; Kumar et al. 2024b). NFs can be regarded as a "smart nutrient delivery system," which can help alleviate hunger and poverty by increasing food grain yield (Nongbet et al. 2022). Nanofertilizers have the potential to significantly improve global agriculture sustainability. As a result of their larger surface area to volume ratio and increased mobility, they can effectively promote plant growth and nutrient uptake (Verma et al. 2023b). By 2030, the use of innovative NFs could be a key factor in achieving the United Nations' sustainable development goals of eradicating world hunger and ensuring food security (Gamage et al. 2023).

Exploring the viability of NFs in sustainable farming, efforts are being made to bring these products to market and devise creative ways to enhance crop growth through their application. Essential nutrients required by plants such as N-P-K and trace minerals, are present in NFs, and their strategic application is being studied to determine their potential for sustainable use in agriculture (Haydar et al. 2024). The focus of these approaches is to optimize fertilization effectiveness and nutrient uptake, culminating in the development of novel and ameliorated products for the industry (Rajput et al. 2022). This chapter is centered on deciphering the utilization of NMs as NFs for crop betterment. The chapter further examined the composition of NFs, benefits of NFs for crop improvement, environmental impact, and safety considerations, challenges and limitations of NFs application, current research and developments in NFs and their efficacy in promoting growth and stress tolerance in crops discussed. This study highlights gaps in knowledge, potential future applications and implications, advancements in precision agriculture, and its integration with other agricultural technologies.

10.2 Nanomaterials in Agriculture: Overview

Farmers suffer large financial losses as a result of soil nutrient deficits, and human and animal consumption of grain is significantly reduced in terms of both quality and quantity. Techniques including genetic engineering, tissue culture, aeroponics, hydroponics, and others can be utilized to boost agricultural yields (El-Ramady et al. 2022). Unfortunately, these methods are too costly and unfeasible for widespread use. Therefore, to boost crop productivity by using less fertilizer and resources, inexpensive and sustainable technology is needed to give nutritional supplements (Jagadesh et al. 2024). In contrast to the nanotechnological technique, which stresses using less fertilizer, the traditional way of applying fertilizer includes using far more than is necessary (Gade et al. 2023). Due to the progress made in nanotechnology, it is now possible to create in large quantities NPs of physiologically essential metals. These may be added to fertilizer formulations to improve absorption in plant cells and save nutrients (Fig. 10.1).

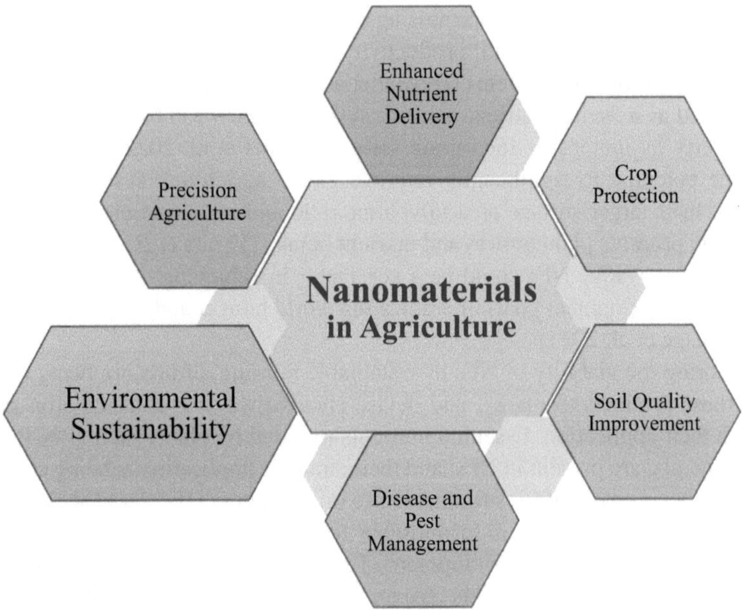

Fig. 10.1 Graphical portrayal of NMs contribution to agriculture

Nanomaterials, nanodevices, and nanoscale phenomena are all used in agricultural applications under the umbrella of nanotechnology in agriculture. It entails creating nanoscale solutions for environmental monitoring, nutrient delivery, and crop protection (Rajput et al. 2021a). This technology has the power to transform farming methods, increase crop yields, and mitigate the damaging repercussions of agricultural practices impacting the ecosystem to maximize plants' capacity to absorb essential elements and create nano-based sensors to identify pollutants (Nandini et al. 2023; Kumar et al. 2024c). Nanotechnology seeks to address agricultural issues and support efficient and sustainable food production by utilizing the special qualities of NMs (Azameti and Imoro 2023). Several research have some specific roles of nanotechnology in crop improvement including enhanced nutrient delivery (Ain et al. 2023), crop protection (Dos Santos et al. 2022; El-Saadony et al. 2022; Pérez-Labrada et al. 2019), soil quality improvement (Kalwani et al. 2022; Verma et al. 2022a, b; Singh and Singh 2021), disease and pest management (Hao et al. 2017; Saharan et al. 2015), environmental sustainability (Babu et al. 2022; Bhardwaj et al. 2022), precision agriculture (Yadav et al. 2023b).

10.3 Understanding NFs and Its Composition

Since NFs have proven to enhance productivity and eco-friendly than traditional chemical fertilizers (which played a major role in the past) they are becoming a significant part of agriculture. The key objective of introducing NFs is to enhance growth potential and accelerate production by allowing nutrients to be absorbed quickly. The soil and plant systems have a highly effective symbiotic relationship (Rajput et al. 2021b). The yield improves significantly when the same is administered gradually and effectively, allowing the plant to absorb all necessary nutrients and replenish its stored energy, which significantly boosts production (Goyal et al. 2023). This can be accomplished when fertilizers are encapsulated. Nitrogen, the primary element, is abundantly required by plants, but its absorption leads to various issues (Fig. 10.2).

Nanofertilizers are a modern approach to enhancing plant nutrition and soil fertility. They consist of macro- and micronutrient-containing NPs that are introduced into the plant rhizosphere for effective nutrient absorption (Chen et al. 2023; Mariyam et al. 2024). Nanofertilizers aid in increasing crop yield and fostering plant growth and development through the delivery of essential nutrients at the nano level (Chen et al. 2023; Deng et al. 2022; Goyal et al. 2023). According to Goyal et al. (2023), NFs are a great option for promoting plant nutrient uptake because of their notable absorption rates, enhanced usage efficiency, and less waste Their consistency, nutritional makeup and composition, and mode of action may all be used to categorize them (Mejias et al. 2021; Nongbet et al. 2022). These categories include nutrient-loaded NFs, controlled-release NFs, targeted-delivery NFs, fertilizers that stimulate plant growth, inorganic and organic NFs, hybrid NFs, fertilizers that control water and nutrient loss, and different consistency-based NFs like synthetic polymer-coated, nanocarrier-based fertilizers, surface-coated, and biological product-coated (Yadav et al. 2023b).

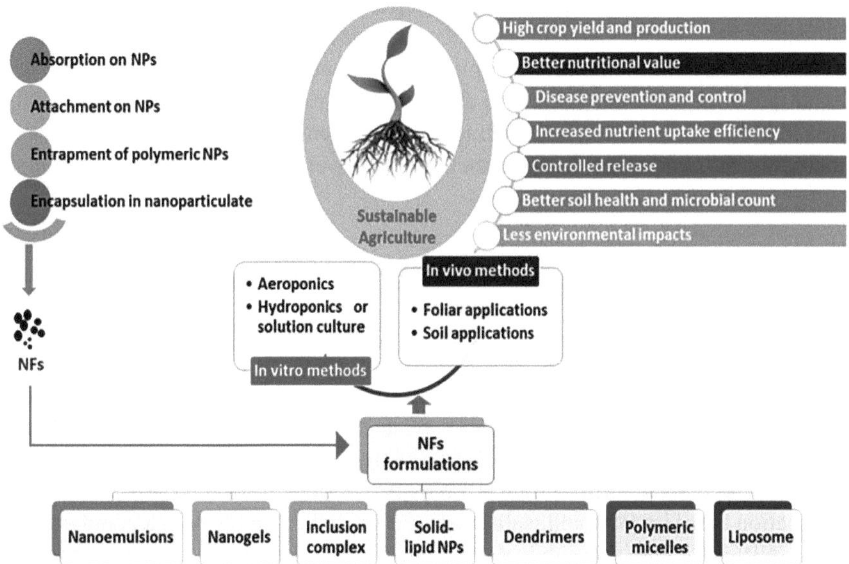

Fig. 10.2 A generalized overview of NMs in agriculture. (Adopted from Verma et al. (2022a))

Different types of NFs may exhibit varying compositions, influenced by their specific formulations and intended uses, which can be identified through both top-down and bottom-up methodologies that involve hydrocarbon polymers, dendrimers, and various nanostructured formulations (Azameti and Imoro 2023). These NPs are often developed from materials to enhance crop productivity in agriculture such as metals (examples: Gold (Au), Silver (Ag), Zinc (ZnO), Copper (Cu), Aluminum (Al), and Iron (Fe)) nanoparticles, metal oxides, like silver oxide, titanium dioxide, zinc oxide (Yaqoob et al. 2020; Altammar 2023; Balusamy et al. 2023), carbon-based particles such as carbon nanotubes, fullerenes, and graphenes (Altammar 2023), Other nanomaterials include silica (Si), polymeric, quantum dots (QDs), and liposome-based NMs or organic compounds, such as chitosan or humic acid (Balusamy et al. 2023; Yadav et al. 2023a). With their uniquely crafted design, these NPs optimize nutrient delivery to the plant, ultimately improving uptake and utilization (Yadav et al. 2023a).

Each material offers unique properties and applications in different fields. Nanofertilizers can be engineered to encapsulate or bind with macronutrients such as calcium, sulfur, potassium (K), phosphorus (P), and nitrogen (N), or micronutrients like zinc (Zn), iron (Fe), and copper (Cu), etc. This encapsulation helps to protect the nutrients from leaching or volatilization, allowing for the gradual and regulated release of nutrients, aiding in the absorption by plants, and increasing nutrient use efficiency (Pouran 2019). In addition to nutrient delivery, nanofertilizers may also incorporate other components such as biostimulants or growth-promoting substances. These additives can further enhance plant growth and development by stimulating root growth, improving stress tolerance, or enhancing nutrient absorption efficiency. With their special composition, NFs can support efficient nutrient uptake, restore soil fertility effectively, enhance absorption rates significantly, improve photosynthesis levels, increase production outputs exponentially, and reduce both soil toxicity and application frequency for healthier plants and less environmental pollution (Nongbet et al. 2022; Tyagi et al. 2022). Additionally, researchers are exploring the interactions between NPs with various materials, including soils and plants to optimize their effectiveness in sustainable agriculture.

10.3.1 Mechanisms of Action of NPs Transport into Plant Tissues

Using NFs, plants receive nutrients in a more targeted and efficient manner. The NFs small size enables them to enter plant tissues through root junctions, and wounds or upper parts and supply nutrients right to the cells, where they are most needed. The plant's ability to absorb NPs is determined by their size, shape, and interaction behavior with cell walls. Before being taken up by the plant and translocated, NPs must pass through various physiological barriers as they act as carriers. Plant cells have a size exclusion limit of 5–20 nm, which is composed of cellulose and functions as a barrier to prevent bigger particles from entering (Gade et al. 2023). A surface-functionalized NPs improves nutrient absorption by causing pore size to increase or by forming a new pore in the cell wall. The transfer of NPs to different plant tissues can occur through the symplastic and apoplastic pathways with a lipid exchange mechanism facilitating their transport into plant cells. The transport of NPs in plant cells is contingent upon their size, magnitude, and zeta potential.

Additionally, aquaporins, ion channels, complex formation with membrane transporters, root exudates, or endocytosis can all be used by NFs particles to bind to carrier proteins and enter plant cells. Nanocarriers safeguard vital nutrients in the soil by trapping them near the roots, preventing filtration (Gade et al. 2023). Through a variety of pathways, such as surface tension, molecular forces, hydrogen bonds, or viscous forces, encapsulated components can enter the soil network. By applying NFs through a foliar method, NPs can enter through stomata or the base of trichomes in leaves or cuticles. Upon entry into the plant system, NPs may traverse through stomatal openings and proceed to travel either apoplastically or symplastically via plasmodesmata within the vascular system. Notably, the plasmodesmata facilitate transport between cells (Gade et al. 2023).

Xylem vessels, neighboring cell walls, and extracellular spaces are the pathways via which apoplastic transport occurs outside of the plasma membrane. For NPs to go farther upward into the aerial portion of the plant, they must first pass via the apoplastic route, which is crucial for radial migration inside plant tissues (Larue et al. 2012; Sun et al. 2014). After entering the central cylinder, NPs can follow the transpiration stream via the xylem and travel in the direction of the aerial portion. To access the xylem through the root, the NPs must traverse the Casparian strip, which acts as a blockade to the apoplastic route, using endodermal cells in a symplastic fashion. Certain NMs can indeed collect at the Casparian strip and halt there (Pérez-de-Luque 2017).

Utilizing plasmodesmata and sieve plates, symplastic transport facilitates the movement of water and substances between the cytoplasm of neighboring cells. Additionally, symplastic transport can occur via the sieve tubes of the phloem, enabling distribution to non-photosynthetic parts of plants (Raliya et al. 2016; Wang et al. 2012). The application of nanomaterials to leaves allows them to penetrate the barrier cuticle, taking either a lipophilic or hydrophilic route. Diffusion happens through cuticular waxes in lipophilic environments, while in hydrophilic settings, diffusion occurs via polar aqueous pores located in the cuticle and/or stomata (Pérez-de-Luque 2017). Considering that cuticular pores are typically around 2 nm in diameter, the most favorable pathway for NPs penetration is through stomatal penetration, given the 20 nm size restriction.

For example, Nanoaptamers, tiny structures composed of either oligonucleotides or peptides, with the ability to alter the exterior of NFs. Nanofertilizers are an innovative type of fertilizer delivery system that is changing agricultural production. Nutrients are released from the nanostructure when triggered by signals originating from the rhizosphere as a result of the modification. Nutrients and other molecules are precisely released to the plant by Nanoaptamers. These tiny molecules can bind to plant hormones and enzymes, effectively delivering them to the plant and increasing plant nutrient uptake from the soil (Kaushal and Wani 2017; Yadav et al. 2023b). By affixing fertilizer components to NPs, such as liposome or gold NPs, nanoaptamers can transfer fertilizer to plants. This technique enables accurate distribution of the aptamer to the desired plant cells while protecting it from degradation. The NPs may then penetrate the plant cell and release the fertilizer when the aptamer binds to a particular receptor on the cell's surface. By creating a link between the root system and the soil microorganisms, nanoaptamers are used to control the quantity of fertilizer that is applied, thereby allowing for the precise supply of the appropriate amount of fertilizer (Yadav et al. 2023b).

The physiologies of different plant species vary, impacting how differently those species absorb NPs (Larue et al. 2012). Because of this, the way NMs travel throughout plants is significant since it can reveal which areas of the plant they can access and potential accumulation sites (Pérez-de-Luque 2017). varied plant species subjected to magnetic carbon-coated, titanium dioxide, or gold NPs, respectively, showed varied patterns of absorption and accumulation inside their plants (Larue et al. 2012). Nevertheless, how NMs are introduced is essential in determining a plant's ability to internalize them effectively: roots specialize in nutrient and water absorption. At the same time, leaves are adapted for gas exchange with a barrier that inhibits substance penetration (Larue et al. 2012; Choudhary et al. 2024). Applying NPs to the roots will ensure a good dispersion throughout the plant because, for instance, if the xylem rather than the phloem mostly carries them, they will probably migrate mainly from the root to the shoots and leaves rather than downward. On the other hand, foliar spraying should be used if the NPs demonstrate a good translocation via the phloem (Fig. 10.3). When trying to prevent further ingestion of NMs by humans or animals, it's important to consider the likelihood of particles traveling in the phloem and accumulating in plant organs that act as storage areas, such as fruits and grains. Nevertheless, NMs may travel laterally between the xylem

Fig. 10.3 Micrographs of $CaSO_4$ and $ZnSO_4$ used as NFs obtained using a Scanning Electron Microscope (SEM) and Transmission Electron Microscope (TEM) respectively. These images show the composition and structure of these NMs. (Source: Bhardwaj et al. 2022)

and phloem, and translocation is not always limited to a particular kind of cell (Pérez-de-Luque 2017).

10.3.2 Potential Benefits of Nanofertilizers for Crop Betterment

The utilization of NFs offers numerous benefits for augmenting crop yields. They can offer a more economical and effective means of fertilizing crops, lessen the adverse effects of fertilizer application on the environment, and improve the efficiency of nutrient absorption. Furthermore, it has been noted that using NFs can enhance plant structure, boost yields, promote seed germination, and lessen a plant's susceptibility to pests and diseases. They can help crops that are deficient in some nutrients and aid in controlling nutrition. In general, NFs can significantly increase plant health and crop productivity while lowering the amount of fertilizer required in agriculture (Fig. 10.4).

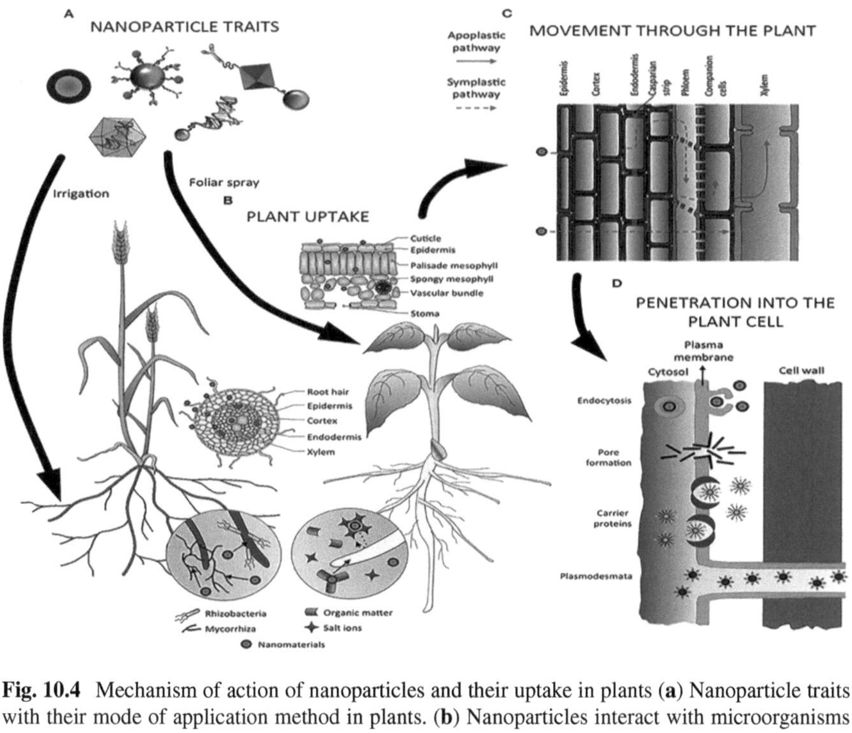

Fig. 10.4 Mechanism of action of nanoparticles and their uptake in plants (**a**) Nanoparticle traits with their mode of application method in plants. (**b**) Nanoparticles interact with microorganisms and compounds in the soil around the root zone and also vascular tissues (leaves). (**c**) Nanomaterials move within plants through the apoplastic and/or symplastic pathways and radially (**d**) Internalization of nanoparticles occurs through endocytosis, pore formation mediated by carrier proteins, and plasmodesmata. (Adopted from Pérez-de-Luque 2017)

10.3.2.1 Stimulate Seed Germination

Environmental stresses can negatively impact crop growth, productivity, and seed vigor (Akhtar and Ilyas 2022). Different agrochemical-based seed treatments improve germination, but they can also harm the ecosystem. For this reason, nano-based agrochemicals are urgently needed for a sustainable solution. The dose-dependent toxicity of seed treatment could potentially be reduced, leading to nanoagrochemicals enhancing seed viability and ensuring an appropriate release of the active substances within the product. Certain NFs, such as carbon nanotubes (CNTs), promote plant development by interacting with plant root systems and increasing hormone synthesis (Faizan et al. 2021). They raise the soil's concentration of carbon and other nutrients. Carbon nanotubes (CNTs) are made of rolled-up sheets of carbon atoms and have unique qualities that make them great fertilizers. Carbon nanotubes (CNTs) can absorb and release nutrients, stimulate plant development, and improve soil structure and water retention (Faizan et al. 2021).

10.3.2.2 Improve Plant Structure

Nanofertilizers can improve plant structure by enhancing the morpho-physiological, and biochemical attributes of plants (Akhtar and Ilyas 2022; Janmohammadi et al. 2016). They have been shown to increase nutrient uptake efficiency, optimize plant development productivity, and promote vegetative growth, resulting in healthier and more robust plants (Janmohammadi et al. 2016; Omar et al. 2023). Moreover, NFs can promote plant growth, enhance soil structure, and increase water retention, ultimately contributing to improved plant and biochemical attributes of plants. They can also improve shoot and root growth, chlorophyll contents, and overall plant performance increase water retention, lower hypoxic conditions during soil flooding, and facilitate a gradual and consistent delivery of nutrients, improving plant nutrient uptake and efficiency (Omar et al. 2023; Schaller et al. 2020; Siddiqui et al. 2023; Zahedi et al. 2020).

10.3.2.3 Increased Grain Yield and Quality

Nanofertilizer application to crops has demonstrated substantial increases in yield. By ensuring that plants receive the required nutrients precisely when they need them, these fertilizers foster optimal growth conditions, leading to more abundant harvests. The reduction in the excess usage of urea also results in stronger crops and a higher yield (Balci and Dağhan 2023; Janmohammadi et al. 2016).

10.3.2.4 Cost-Effectiveness

Although, compared to conventional fertilizers that are available, NFs may seem like an expensive investment. However, their higher nutrient absorption and reduced wastage ultimately lead to cost savings for farmers. Moreover, their long-lasting effects reduce the need for frequent applications, making them more economically viable in the long run.

10.3.2.5 Environmentally Friendly

Since NFs release nutrients gradually and directly to the plant roots, they reduce the ecological footprint of farming (Gangwar et al. 2023). Their targeted application also reduces the need for excessive fertilizer use. Through this process, the likelihood of soil and water pollution dwindles, allowing for the conservation of soil productivity and the promotion of sustainable farming practices (An et al. 2022).

10.3.2.6 Enhanced Nutrient Uptake and Efficiency

With traditional fertilizers, the majority of essential nutrients are not retained due to leaching or volatilization. On the other hand, NFs encapsulate nutrients and protect them from being destroyed (An et al. 2022). This leads to improved nutrient absorption by plants, maximizing their growth potential, and goes hand in hand with increased grain yields (Elemike et al. 2019; Shang et al. 2019).

10.3.2.7 Nano Pesticides Synergy

The ability to design NFs to release nutrients and insecticides simultaneously is one of their main benefits. This has resulted in the creation of "smart fertilizers," which shield plants against diseases and pests and provide them with nutrition (Dutta et al. 2022; Worrall et al. 2018).

10.3.2.8 Controlled Release of Nutrients

Because they are made to give nutrients to plants in a regulated way and ensure that the nutrients are released gradually over time, NFs aid in the controlled release of nutrients to plants, giving the plants a steady supply of vital nutrients. Their gradual and steady delivery of nutrients may be engineered to enhance nutrient utilization efficiency and promote ideal plant development. This prolonged release mechanism allows plants to absorb nutrients more effectively, which improves crop output and quality (Miguel-Rojas and Pérez-de-Luque 2023; Yadav et al. 2023a, b, c).

10.4 Recent Studies on Crop Response to NFs Efficacy

Smart agriculture, which acts as a connection to others, is a means to prioritize both short- and long-term growth in the face of climate change in the twenty-first century. It aims to assist nations and other relevant elements in guaranteeing essential agricultural functions. The goal of recent research has been to better understand how plants respond to NFs in the growing environment inside crop production (Kumar et al. 2022). Plant growth metrics that are influenced by NFs include root systems, photosynthetic rate, fresh and dry matter content differences, chlorophyll content, number of leaves per plant, plant height, and leaf area. As a result, careful consideration is given to an increase in yield brought about by a greater rate of photosynthetic translocation to different sections of the plants as compared to chemical fertilizer (Gade et al. 2023; Kahlel et al. 2021).

Plants by nature are subject to one or more stressors, such as biotic and abiotic, either singly or in combination, which eventually results in lower agricultural yields. These stresses include biotic stressors like diseases or the pathogens of viruses, fungi, and bacteria as well as abiotic stressors like salinity, drought, flooding, water shortages, and temperature (cold and heat stress). Throughout their life cycle, these plants may encounter one or more stressors, according to several investigations. (Chieb and Gachomo 2023; Dos Santos et al. 2022; Rivero et al. 2022; Seleiman et al. 2021). Utilizing NFs as antistress agents may enhance plant resilience to environmental and biological challenges as they have been applied to support crop growth under different stress conditions, particularly salinity stress using nano selenium (González-García et al. 2021), silicon NPs for drought (Akhtar and Ilyas 2022), salinity stress using copper NPs (González-García et al. 2021; Pérez-Labrada et al. 2019), heavy metal tolerance utilizing zinc oxide, and titanium dioxide NPs (Rizwan et al. 2019; Shah et al. 2021), heat stress tolerance using selenium NPs (Omar et al. 2023), Zirconium dioxide and silicon dioxide NPs for biotic stress tolerance (Kashyap and Siddiqui 2021; Siddiqui et al. 2020, 2023).

Cu, CuO, ZnO, Ag, Au, MgO, and TiO NPs help control fungal diseases from genera including *Fusarium, Pythium, Aspergillus, Colletotrichum*, etc. With the help of CuO, ZnO, and FeO NPs, *Ralstonia solanacearum*—a soil-borne agent that causes bacterial wilt in tomatoes—can be effectively controlled. Numerous antiviral NMs are also effective against plant viruses, such as tomato mosaic virus (ToMV), barley yellow mosaic virus (BaYMV), potato virus Y (PVY), and turnip mosaic virus (TuMV) in tobacco. Several other plant viruses may also be managed using different NPs. As a result, employing NMs to reduce stress caused by different biotic stress components is a sustainable and environmentally beneficial approach. Promising results in terms of increasing nutrient efficiency, decreasing the amount of fertilizer required in agriculture, and boosting seedling vigor have been shown in recent research on the effectiveness of NFs (Balci and Dağhan 2023; Najafi Disfani et al. 2017; Das et al. 2016). The benefits of metallic nanoparticles in supporting sustainable agriculture have been shown by research (Hao et al. 2017), and efforts are being made to create novel fertilizers that are both ecologically benign and very nutrient-efficient. Furthermore, research has indicated that NPs may be utilized to suppress plant pests and promote the growth of crops (Siddiqui et al. 2020, 2023; Kashyap and Siddiqui 2021; Hao et al. 2017) (Table 10.1).

10.5 Maximizing Nutrients for Crops Using NFs with Varying Application Methods

The most important management practice in agriculture is nutrient management, application, and utilization, which is supported by the effective use of nanotechnology, especially NFs. Nanofertilizers can penetrate crops through their roots or leaves. There are two main methods for applying NFs: *in vitro* and *in vivo*. In-vitro methods involve soil-less techniques like aeroponics and hydroponics. In-vivo methods, on the other hand, include three types of application: foliar, seed priming/coating, and soil mixing/drenching.

10.5.1 Soil Treatment

Soil mixing is the most popular method of giving plants extra nutrients. Nanofertilizer can be applied to the soil around plant roots using conventional techniques such as broadcasting, side-dressing, band placement, soaking, or fertigation. Through endocytosis and adsorption, NFs will interact with plant roots, soil particles, and microorganisms and penetrate the cells of the roots. However, various factors such as soil texture, soil salinity, plant sensitivity to salts, salt content, and pH should be taken into account when applying NFs. When using this method, the concentration of particles in a particular area is higher than when using foliar spray. This is because of the way the particles are delivered and distributed within the soil (Gade et al. 2023). Soil drenching of copper NPs increases chlorophylls, lycopene, β-carotene, and the activity of some enzymes in leaves under saline stress in bell pepper (González-García et al. 2021). Soil particles that carry a negative charge can have an impact on the way mineral nutrients are absorbed. Agricultural soils typically have a lower anion exchange capacity compared to cation exchange capacity. Among the anions, nitrate tends to remain mobile in soil solution and can easily get washed away by water passing through the soil, a process known as leaching.

Table 10.1 Response of crops to different nanofertilizer applications under control and stressed conditions

NFs	Plant(s)	Stress condition	Highlights	Reference
Abiotic stress tolerance				
Nanosilicab	Wheat	Drought stress	Enhanced germination, shoot and root length, photosynthetic pigments, membrane stability index, relative water content, osmolytes content, flavonoid, phenol, content, and antioxidant activities.	Akhtar and Ilyas (2022)
ZnO NPs + *Bacillus fortis*	Muskmelon	Alleviate Cd phytotoxicity	Bacillus fortis and ZnO-Nanoparticle modulated the plants' antioxidant system, mitigated cadmium (Cd) stress, and showed synergistic effects.	Shah et al. (2021)
TiO_2 and Si NPs	Rice	Growth, oxidative stress, and Cd accumulation in rice plants	Rice biomass increases while reducing cadmium concentration in plant tissues, improving the chlorophyll concentrations and gas exchange attributes.	Rizwan et al. (2019)
Silica NPs	Soil	Water holding capacity of soil	Substantially increase soil water retention by creating silica gels with water saturation above 700%.	Schaller et al. (2020)
Nano-iron pyrite	Vegetables, like beetroot, carrot, fenugreek, alfalfa, mustard and sesamum crops	Seed priming with iron pyrite (FeS_2) to reduce fertilizer consumption	A 12-hour seed pretreatment in an aqueous suspension of nano-iron disulfide/pyrite (FeS_2) has been found to increase vegetable crop productivity.	Das et al. (2016)

(continued)

Table 10.1 (continued)

NFs	Plant(s)	Stress condition	Highlights	Reference
Se, Si and Cu NPs	Bell pepper Fruits	Saline Stress	The NPs increased chlorophylls, enzymatic and biochemical content in fruits and leaves.	González-García et al. (2021)
Se/SiO_2 NPs	Strawberry plants	Drought stress	Photosynthetic pigments were retained when plants were treated with Se/SiO_2 (100 ppm) compared to other treated plants and exhibited higher levels of key osmolytes, such as proline and carbohydrates. RWC, membrane stability index, and WUE substantially increased. Enhanced water stress tolerance efficiency by boosting the activity of antioxidant enzymes.	Zahedi et al. (2020)
CU NPs	Tomato plants	Saline Stress	Increased the phenols in the leaves by 16% and the vitamin C content by 80%, along with an 81% increase in GSH and 7.8% increase in phenols in the fruits. The activity of phenylalanine ammonia-lyase (PAL), APX, GPX, SOD, and CAT in leaf tissues enhanced by 104, 140, 26, 8, and 93%, respectively.	Pérez-Labrada et al. (2019)

(continued)

Table 10.1 (continued)

NFs	Plant(s)	Stress condition	Highlights	Reference
Se NPs	Wheat	Drought and Heat Stress tolerance	Enhanced photosynthetic pigments, gas exchange, transpiration rates and stress tolerance capacity were achieved by increasing the activity of antioxidant enzymes and the expression level of stress-responsive genes.	Omar et al. (2023)
Biotic stress tolerance				
Si NPs + *Rhizobium leguminosarum*	Pea	Growth and bacterial blight disease complex	Increase plant growth, chlorophyll, and carotenoid in plants with pathogens. Seed priming with SiO_2 caused reduction of blight disease indices, galling, and reduced multiplication of *M. incognita*	Kashyap and Siddiqui (2021)
Si NPs	Carrot	Management of bacteria species	Plant growth, chlorophyll, and carotenoid content increased with seed priming or foliar application of SiO_2 NPs.	Siddiqui et al. (2020)
ZrO_2 NPs and AM fungus	Bottle gourd	Disease management	Increase in plant growth and photosynthetic pigments with seed priming and foliar application of ZrO_2 NPs. Reduced disease indices, galling, and population of *M. incognita*.	Siddiqui et al. (2023)
CuO NPs, Fe_2O_3 NPs, and TiO_2 NPs	Rose Petals	Antifungal roles of NPs against the fungal pathogen (*Botrytis cinerea*)	NPs inhibited the growth of *B. cinerea*	Hao et al. (2017)

(continued)

Table 10.1 (continued)

NFs	Plant(s)	Stress condition	Highlights	Reference
Plant phenology and grain yield				
ZnO NPs + Biofertilizer	Wheat	Grain yield and physicochemical activities	Increased grain yield up to 88% during water stress conditions. Upregulated plant phytochemical content including proline, soluble sugars, and enzyme activities.	Seyed Sharifi et al. (2020)
Nano-chelated Zn + Bulk NPK	Maize	Growth performance and grain yield	Significant increment in morpho-physiological traits such as ear length, ear diameter, leaf area, relative water content, and chlorophyll content	Janmohammadi et al. (2016)
Nanozeolite + PGPR	Maize	Growth development on maize plants	The nano compound treatment has been shown to improve various growth-related parameters by up to twofold	Khati et al. (2018)
Iron (III) oxide + flower waste + cow dung	Tomato	Growth and development	The growth of tomato plants showed significant improvement compared to when other fertilizers were used.	Khati et al. (2018)
ZnO NPs	Durum wheat	Growth and Zn uptake	Zinc content increased in plant organs. Chlorophyll content and plant dry weight enhanced.	Balci and Dağhan (2023)
Zn and Bo Nfs	Pomegranate	Fruit yield and quality	Enhanced pomegranate fruit yield and fruit quality. Changes in total sugars and total phenolic compounds.	Davarpanah et al. (2016)
Nano Fe/SiO_2	Barley and Maize	Germination and growth	The root lengths in plants, as well as growth, dry matter, and seedling length, increased.	Najafi Disfani et al. (2017)

10.5.2 Seed Nano-Priming

One effective method for reducing fertilizer application while achieving excellent results involves soaking seeds in powdered or liquid NFs. Using this technique, NFs penetrate seed pores, diffuse inside, and trigger growth-promoting plant hormones to function as stimulants that improve germination and development (Yadav et al. 2023b). The use of this technique may provide excellent results while reducing fertilizer use by half. When NFs are applied as seed priming, it has been found to increase seed germination by removing reactive oxygen species and regulating the hormones responsible for plant development. Furthermore, seed priming can increase resistance by inducing the expression of many genes during germination, particularly those linked to plant resilience. When maize seeds are primed with copper-chitosan (Cu-Cht) NPs, seedling development is promoted (Saharan et al. 2015). Similarly, induction and improved nitrogen metabolism have been shown to occur in non-nodulated tomato plants after seed priming with manganese nanoparticles (MnNPs). According to a study by Zheng et al. (2005) nano-TiO_2 promotes the development of spinach seeds that have naturally aged. Nickle NPs have a noticeable impact on wheat's morphophysiological characteristics. Molybdenum NPs inoculated with chickpea seeds promote the growth of "agronomically valuable" microbiota and the formation of nodules on each plant (Taran et al. 2014). Vegetable crop yield increased significantly after a 12-hour seed priming pretreatment in an aqueous solution containing nano-iron disulfide/pyrite (FeS_2) to minimize fertilizer usage (Das et al. 2016). The research suggests that different kinds of NPs can be employed as effective agents for seed priming to stimulate germination.

10.5.3 Foliar Application

During foliar application, micronutrient-containing NFs are sprayed directly onto the leaves of plants. As soon as the plants are treated, they promptly absorb the nutrients. This method is faster than soil application in replenishing deficient nutrients and providing speedy utilization of nutrients. Research has demonstrated that foliar application is more efficient for facilitating nanoscale nutrient absorption as compared to soil amendment when it comes to delivering NPs to plants. In particular, for nutrients such as iron, manganese, and copper that are immobilized in the soil, foliar application is not only more effective but also more economical compared to soil application. In wheat, seedlings, foliar treatment of Se NPs enhanced plant growth, morpho-physiological and biochemical responses, and expression of stress-responsive genes (Omar et al. 2023). Likewise, Rizwan et al. (2019) report an improvement in biomass, photosynthesis, and diminishing Cd accumulation in rice through foliar application of silicon and titanium dioxide NPs. Additionally, by activating the antioxidant mechanisms of tomatoes under salinity by foliar spraying, Cu

NPs promote stress tolerance to salt in tomatoes (Pérez-Labrada et al. 2019). Strawberry plants cultivated under normal and drought stress circumstances showed enhanced growth and yield metrics after being treated with solutions containing NPs of SiO_2, Se, and Se/SiO_2 (Zahedi et al. 2020). The method has limitations as the stomata are open only during specific times—morning and evening. In addition, failure to apply fertilizer at the proper dosage may result in damage to plants.

10.5.4 Aeroponics Treatment

Aeroponics is a technology for growing crops without soil in an air-mist environment. In this technique, plants are grown in a chamber where nutrients are delivered through oxygen mist near the floating roots in the air. Materials in the form of nutrients are continuously provided to the plants as sprays mist or dust to the roots. The plants exchange gases through their stomatal openings (Gade et al. 2023). Nanofertilizers are a better alternative to chemical fertilizers and can replace micronutrients. They increase the efficiency of nutrients by 2–20%. The size and form of the NPs determine how quickly nutrients are absorbed. While bigger particles may easily pass through non-cuticle elements including stomata, stigma, and hydathodes, smaller particles can easily penetrate the cuticle. Consequently, depending on the requirements and intended impact on the plant body, NFs may be made to be precisely the proper size and form for use in aeroponics therapy (Gade et al. 2023). Aeroponic treatments utilizing NFs may be an effective means of maximizing the intake of nutrients with the least amount of waste.

10.5.5 Hydroponics Treatment

Hydroponics uses nutrient solutions for plant cultivation. Nanofertilizers are added to help plants absorb the nutrients through their roots. Nano-encapsulated plant nutrients can enhance the effectiveness of NFs in hydroponic systems. In aeroponics and hydroponics, nanotechnology-based approaches help attain a detailed understanding of nutrient utilization and minimize resource loss. Velvet mesquite hydroponic cultures were used to test the impact of 500–4000 mg L^{-1} of 10 nm-sized ZnO NPs under stress. In response to NP exposure, the plants showed enhanced specific activity of the enzymes catalase and peroxidase, with no adverse reactions such as necrosis, stunting, wilting, or chlorosis, suggesting a high tolerance level for ZnO NPs (Hernandez-Viezcas et al. 2011). Hydroxyapatite $Ca_{10}(PO_4)_6(OH)_2$ NPs were synthesized as a kind of NFs to aid in the growth and development of *Solanum lycopersicum* L. While increased NP concentrations in hydroponic treatment greatly increased root elongation, they did not affect germination (Marchiol et al. 2019). Studies conducted in vivo with half the absolute nitrogen content of the standard urea treatment demonstrated comparable biomass development in the roots and shoots of hydroponically grown cucumber plants with N-doped ACP NPs, without any depletion of nitrogen. Due to its low cost of manufacture and excellent nitrogen utilization efficiency (up to 69%), N-doped ACP is a viable option for practical usage as NFs (Carmona et al. 2021). El-Shal et al. (2022) investigated the effects of two genotypes of alfalfa (*Medicago sativa* L.), such as Bulldog 505 and Mesa-Sirsa, at different salt concentrations (6 and 10 dSm^{-1}) in a comparable hydroponic experiment. Several plant properties, including the shoot dry weight, plant height, flower count, tiller count, root length, root fresh weight, and root dry weight, were enhanced by nano-K_2SO_4 for all salt concentrations. After adding Nano-K_2SO_4, both genotypes displayed increased electrolyte leakage, and relative water content although having differing salt concentrations.

10.6 Novel Approaches and Formulations of Nanofertilizer

Innovative methods and NFs formulations have been the subject of a recent study to improve the sustainability and effectiveness of agricultural activities. Among these advancements is the creation of innovative NMs with potent fertilizing properties, integrated NFs that include macro- and micronutritional components, and hybrid NFs that release nutrients into the soil gradually and sustainably. Furthermore, research on nano-pesticides and nano-plant protection systems is still underway and has the potential to improve the efficacy of traditional pesticides. These developments demonstrate how NFs have the power to completely transform agricultural methods by offering a more effective and environmentally friendly way to fertilize crops (Aamir Iqbal 2020).

Downsizing fertilizer inputs mechanically or chemically has been observed to improve nutrient absorption, decrease losses, and reduce nutrient toxicity. Nanoparticles have been created using a range of synthetic fertilizers, plant waste, urea, ammonia, peat, and other materials. A newly developed hybrid nanofertilizer (HNF) that releases nutrients into soil and water gradually and sustainably. Urea-modified hydroxyapatite, a calcium, phosphate, and nitrogen-rich material, was synthesized. To boost the effectiveness of the newly synthesized fertilizer, urea-modified hydroxyapatite was incorporated with NPs of copper, iron, and zinc. The Ladies' finger (*Abelmoschus esculentus*) plant was used as a model of a field experiment using the generated HNF. The results demonstrated that *A. esculentus* slow release from HNF led to a considerable increase in the absorption of Ca^{2+}, PO_4^{3-}, NO^{2-}, NO^{3-}, Cu^{2+}, Fe^{2+}, and Zn^{2+} nutrients (Tarafder et al. 2020). A novel integrated NFs with a rod-like form, 80 nm length, and a 20–30 nm width nanosize was developed. It contains both macro- and micronutritional components. Applying the novel synthetic NFs to the *Polyscias fruticosa* and *Asparagus officinalis* crops resulted in a 50% increase in total biomass and an increase in the number of branches, leaf area, and dry matter (Le et al. 2022).

However, nitrogen (N), phosphorous (P), and potassium (K) were loaded into chitosan NPs to formulate NPK NFs. Coffee seedlings were treated with the newly prepared nanofertilizer in a greenhouse setting. The findings demonstrated that the NFs enhanced the uptake of nutrients, improved photosynthesis, and increased the coffee seedlings' physiological metrics, such as leaf number, plant height, and leaf area (Ha et al. 2019). Zn and Cu ions were reduced during their exposure to basil extract to carry out the green synthesis of Zn and Cu NPs. The effects of the newly synthesized Zn and Cu NPs on the morphological and biochemical traits of the basil plant caused a significant increase in most morphological and biochemical parameters (Abbasifar et al. 2020). A urea deposition method on calcium cyanamide during the formulation process produced nanoscale nitrogen fertilizer (Wang et al. 2013). Ground urea was combined with different biofertilizers in an alternative formulation to create a specialized NFs designed to supply nutrients slowly and continuously for an extended duration (Wan et al. 2010). Similarly, fertilizers with nanoscale particles were created by combining ammonium humate, peat, and other synthetic ingredients. Such NFs are made using a mechanical cum biochemical method, in which materials are mechanically crushed into nanosized particles, and subsequently, biochemical procedures are used to create efficient nanoscale formulations. Moreover, emulsions are produced into nano-emulsions by including nanosized colloids in them (Aamir Iqbal 2020). The encapsulation of fertilizer with NPs provides many opportunities for creating plant nutrition sources that exhibit increased uptake and nutrient usage efficiency. Three potentials to encapsulate nutrients with NMs, i.e., plant-based nutrients can be encapsulated in NMs with varying chemical compositions and characteristics, nutrients can be supplied as emulsions and particles with dimensions in the range of NPs or coated with NMs like polymer film and NMs, such as polymer film can be used to coat nutrient particles.

10.7 Guidelines and Biosafety Regulations for the Use of Nanomaterials in Crop Improvement

The application of nanotechnology in agriculture is growing for several reasons, including bettering soil quality, boosting crop development, and creating more focused and effective pesticide delivery systems. However, because of worries about biosafety, creating effective and economical nano-enabled formulations for agricultural production is difficult. General public concern over the use of Agri-nanoproducts has increased significantly as a result of growing knowledge of the acute and long-term toxicity linked to NMs (Batool et al. 2022). Biosafety protocols must be developed and implemented simultaneously. This aims to specify the kind, dimensions, and configuration of NPs suitable for agricultural usage. These elements, along with solubility, stability, surface reactivity, and charges, are crucial to agricultural output. NFs reactivity in plants remains non-specific and unaffected by the type of plants being studied, probably because established protocols for their field use are unavailable (Kumari et al. 2023a, b). Inevitably, the mechanism behind the absorption of nano-based agrochemicals such as NFs in plants remains poorly understood.

Countries have different regulatory frameworks for using nanotechnology in agriculture. In Ghana, for instance, the Environmental Protection Agency (EPA) is in charge of pesticides, insecticides, weedicides, and herbicide-based products. The Food and Drugs Authority (FDA) and the Ghana Biosafety Regulation are in charge of agricultural biotechnology products. Before any agri-product containing NMs are approved, regulatory bodies require risk assessment and labeling. The process includes assessing the toxicity of the NMs being used, the likelihood of environmental release, and the potential impact on human health. This assists consumers in selecting and using items with knowledge. Regulatory bodies fund research and development efforts in this field to ensure the safety of agricultural products utilizing nanotechnology for both the environment and human health. All things considered, the responsible application of nanotechnology in agricultural goods necessitates cooperation between scientists, producers, authorities, and end users. To guarantee the security and efficacy of Agri-products based on nanotechnology, it is critical to keep an eye on and evaluate the hazards involved. Several nanotechnology-related standards have been created by the International Organization for Standardization (ISO), including one that defines and uses language specific to NPs.

10.8 Integration of NFs with Other Agricultural Technologies

To promote continuous development and technological commercialization, it is suggested to conduct an integrated analysis of smart fertilizers based on nanotechnology. If slow-release NFs can be improved to have a significant impact on the environment, energy, and economy, it would be a major achievement. Furthermore, to make NFs financially viable initiative, it is recommended to focus on research and technological interventions that optimize manufacturing processes and seek affordable, non-polluting, or biodegradable continuous matrix materials.

10.8.1 Precision Agriculture Systems

Precision agriculture (PA) is a farming method that leverages the use of state-of-the-art technology and tools for data-driven decision-making to maximize resource management and boost crop yields (Yadav et al. 2023b). It seeks to cut down on waste, boost productivity, and boost earnings. The incorporation of NFs into precision agriculture practices can involve using GPS tractors, drones, and sensor networks. The systems give constant updates on the levels of nutrients in the soil, soil pH balance, crop growth stages, and environmental conditions with unprecedented precision, enabling precise and targeted application of NFs to specific areas of the field (Yadav et al. 2023b). For example, smart NPs capable of sensing soil nutrient levels can trigger the release of NFs only when and where they are needed, minimizing over-application and reducing environmental impact. Integration with remote sensing and imaging technologies allows for the monitoring of crop health and nutrient status from a distance. Using satellite or drone imagery, farmers can pinpoint specific areas within their fields that need precision application of NFs, optimizing resource allocation and improving overall crop performance (Garg et al. 2023).

10.8.2 Smart Irrigation Systems

Integration with smart irrigation systems allows for the synchronized application of NFs with water delivery to create a more efficient irrigation system by employing NMs that regulate water flow and retention in the soil (Garg et al. 2023). By incorporating sensors with automated machinery that measure weather patterns, soil moisture levels, and plant water requirements with precision and efficiency to optimize irrigation schedules and quantities, NFs can be applied during irrigation events, ensuring optimal nutrient uptake by plants and minimizing nutrient loss through leaching and helping in conservation efforts (Yadav et al. 2023b).

10.8.3 Biological and Chemical Crop Protection

Nanofertilizers can be combined with biological control agents, such as biopesticides and biofungicides, or conventional chemical pesticides and herbicides. This integration enables the development of multifunctional agrochemical formulations that address both nutrient deficiencies and pest/disease management needs, reducing the number of separate applications and minimizing environmental impact (Yadav et al. 2023b).

10.8.4 Internet of Things (IoT) and Big Data Analytics

The Internet of Things (IoT) is a network of physically connected devices that are implanted with processors, software, sensors, antennas, and other technologies to facilitate meaningful data transmission over the network (Yadav et al. 2023b). IoT systems and nanotechnology interact in many ways, from the production of dependable sensors that comprise an IoT system to the use of nanoprocessors to compute and interpret data gathered by IoT sensors. The collection and analysis of NFs application data are made possible through the use of Internet of Things (IoT) devices and big data analytics platforms. By aggregating data on soil nutrient levels, crop responses, weather patterns, and management practices, farmers can gain valuable insights into the effectiveness of NFs and make data-driven decisions to optimize their use for future growing seasons.

10.8.5 Vertical Farming and Controlled Environment Agriculture

NFs can be seamlessly incorporated into hydroponic or aeroponic systems within indoor farming setups like vertical farms and greenhouses, providing essential nutrients directly to plant roots under regulated conditions, NFs help maximize nutrient uptake efficiency and minimize waste, supporting high-yield, year-round crop production.

10.9 Challenges and Limitations of Nanofertilizer Application

The integration of NMs is a promising development for agriculture. Because NMs can be used as both NFs and as delivery systems for insecticides and fungicides. Research has indicated that non-fungal fertilizers (NFs) are employed as soil fertilizers, growth promoters, and to augment the harvest and quality of agricultural products. NFs are a critical component of sustainable, nano-based precision farming and field applications. Despite their apparent benefits, NFs have not been able to gain traction in high-volume manufacturing or practical agricultural applications. Most research into the production and application of NFs shows benefits in terms of improved plant nutrient absorption, higher crop yields, and less environmental damage. However, at least for the time being, several problems prevent the development of the NFs market and the ensuing application of NFs in practical applications. The toxicity that nanoscale materials generally bring to plants, microorganisms, and animals is one of the main worries. The following sections address the possible limitations connected with the use of NMs as NFs.

10.9.1 Cost and Accessibility

Due to the additional expenses involved in developing and refining nanotechnology as well as the specific tools and expertise needed to administer the fertilizers, NFs are often more expensive than regular fertilizers. It is also unclear if NFs will end up being more cost-effective in the long term. The type of fertilizer and how it is applied are two things that might have an impact on the ultimate cost of NFs. The high cost of NFs makes their application in various pedoclimatic conditions across the world difficult. Because high-quality materials must be employed and precise particle dispersion is required to provide optimal efficiency, NFs can be expensive (Yadav et al. 2023b). Again, specialized equipment and expertise are required in the production process, adding to the cost. The cost of NMs has decreased, although it is still higher than that of traditional fertilizer. However, many experts are of the view that while NFs may initially cost more, their improved agricultural yields and decreased waste advantages may eventually outweigh the costs. Furthermore, NFs will probably become less expensive over time as production techniques evolve and technology advances (Yadav et al. 2023b).

10.9.2 Potential Toxicity, Ecotoxicity, and Genotoxicity

Metal and metal oxide-based nanoformulations, including CuO, CeO_2, TiO_2, Ag, NiO, and ZnO, have been thoroughly investigated and shown to promote agricultural plant development and enhance source activity. Metal oxide NPs have been found to have a range of toxicity-related effects on agricultural plants, including nanotoxicity, ecotoxicity, and genotoxicity, even though these NFs have demonstrated potential benefits in terms of boosting nutrient absorption and efficiency (Bhardwaj et al. 2022). Other nonmetal nano-metals such as fullerenes and single-walled carbon nanotubes have also been reported to exert nanotoxicity (He et al. 2021). The kind, solubility, and binding efficacy of NFs are the primary determinants of their toxicity. NFs have the potential to cause ecotoxicity because of their tiny size and high reactivity, which might have unforeseen effects on animals and ecosystems. The possibility that NPs would build up in soils and waterways and impact microbial populations, plant development, and animals is one of these concerns. Furthermore, NPs can contaminate groundwater by leaching into it (Gade et al. 2023). Decreasing NPs size correlates with a noticeable increase in toxicity (Bhardwaj et al. 2022). Genotoxicity is the term used to describe how a substance can damage genetic material, leading to potential mutations or other adverse effects. In the context of NFs, there are concerns about their potential genotoxicity as a result of their tiny dimensions and increased reactivity, NPs could interact with DNA or other cellular components, causing damage that may lead to genetic mutations or other harmful effects on plants, soil organisms, or other organisms in the ecosystem which includes, reduced DNA repair, interaction with nuclear proteins, and reactive oxygen species (ROS) generation (Bhardwaj et al. 2022; Choudhary et al. 2025).

10.9.3 Long-Term Effects on Soil Health

Continuous application of NFs can lead to imbalances in soil nutrient levels over time resulting in the accumulation of particular nutrients in the soil, leading to nutrient imbalances (Verma et al. 2022b). This can affect the availability of other essential nutrients and disrupt soil fertility. Depending on the composition of the NFs, it could either increase or decrease soil pH over time. Changes in soil pH can influence nutrient availability and microbial activity, impacting overall soil health and plant growth. Nutrient cycling and organic matter decomposition heavily rely on soil microorganisms and soil structure formation. Long-term exposure to NFs may affect the abundance and diversity of soil microbes. Some studies suggest that certain NMs can inhibit microbial activity, which can have detrimental effects on soil health and ecosystem functioning (Fatima et al. 2021).

NFs might influence soil structure and texture. Continuous use of NFs may alter soil aggregation, porosity, and water retention capacity. Changes in soil structure can affect root penetration, water infiltration, and gas exchange, ultimately impacting plant growth and soil stability. NFs can potentially accumulate in the soil over time, leading to soil contamination. The persistence of NPs in the soil may pose risks to soil biota, groundwater quality, and ecosystem health (Babu et al. 2022). Additionally, NPs may interact with soil components and undergo transformations that alter their environmental behavior and toxicity. The impact of NPs on long-term soil productivity is still uncertain (Babu et al. 2022). While NFs may initially enhance crop yields by improving nutrient uptake efficiency, their long-term effects on soil health and fertility could ultimately influence agricultural sustainability and productivity.

10.10 Potential Future Applications and Implications

To improve crop yield, researchers need to conduct extensive studies on NFs, including their use, characteristics, and mechanism of action when applied to plants. Reports suggest that excessive concentration of NFs can negatively affect plants, leading to stunted growth and reduced yield. Consequently, it is essential to maximize the dosage amount of various NFs for various plants. In-depth research is also required to ascertain the potential health effects, clearance, and safe disposal of NMs, all of which can contribute to the creation of safer and more efficient NFs. It is strongly advised that future studies concentrate on determining the proper dosage for commercial NFs field application. Technological advancements should give priority to farmer-centric suggestions for NFs. Additionally, it is anticipated that the creation and verification of NFs will supplant the manufacturing of chemical fertilizers and advance the NFs sector. Policymakers will then be better able to decide on the right time and dose for applying NFs as a result. Agriculture's use of NFs, the next generation of fertilizers, is ecologically and economically sustainable. On the other hand, their large-scale delivery poses challenges, and the supply of raw materials for their synthesis is not certain. It will be feasible to create commercially effective NFs using the right procedures when these problems are fixed. With less use of these fertilizers, agricultural yields will rise as a result. To effectively clean up polluted settings, techniques that can precisely and deliberately identify and track environmental contaminants in various biological samples are required. Governments and corporations should also encourage research groups that use a variety of animal models to investigate the safety of NMs. The basis for more effective regulation, which in some nations is now insufficient, will be laid by this research.

10.11 Conclusion on the Promising Role of Nanofertilizers for Crop Betterment

Nanofertilizers are a transformative approach to sustainable agriculture. They offer unique benefits that can improve crop productivity, resource use efficiency, and environmental sustainability. Their ability to deliver nutrients to plants at the nanoscale level ensures precise uptake for optimal growth and development. They also minimize nutrient loss and environmental pollution, promoting soil health. NFs combined with advanced agricultural technologies like precision farming, smart irrigation, remote sensing, and IoT devices can enhance their impact. By tailoring NFs applications to suit the specific requirements of crops, soil types, and surrounding conditions, farmers can enhance resource utilization while reducing unnecessary waste. The integration of NFs with other agricultural inputs offers multifunctional solutions for crop management. This approach reduces the need for separate applications and minimizes farming's environmental impact. Ongoing research is advancing NFs for modern agriculture. Researchers are refining NFs technologies by exploring novel materials, improving biocompatibility, and assessing long-lasting repercussions on soil vitality and ecosystems. Overall, NFs offer a promising solution for sustainable crop enhancement, using nanotechnology to increase yields, reduce environmental impact, and ensure food security. However, it's essential to monitor the impact on soil, ecosystems, and the health of humans to ensure safe use in agriculture.

Conflicts of interest The authors state that they do not have conflicting financial or personal interests that might skew the findings or conclusions presented in this paper.

Funding This study is not funded while it is being conducted.

References

Aamir Iqbal M (2020) Nano-fertilizers for sustainable crop production under changing climate: a global perspective. In: Hasanuzzaman M, Filho MCMT, Fujita M, Nogueira TAR (eds) Sustainable crop production. IntechOpen. https://doi.org/10.5772/intechopen.89089

Abbasifar A, Shahrabadi F, ValizadehKaji B (2020) Effects of green synthesized zinc and copper nano-fertilizers on the morphological and biochemical attributes of basil plant. J Plant Nutr 43(8):1104–1118. https://doi.org/10.1080/01904167.2020.1724305

Ain QU, Hussain HA, Zhang Q, Rasheed A, Imran A, Hussain S, Ahmad N, Bibi H, Ali KS (2023) Use of nano-fertilizers to improve the nutrient use efficiencies in plants. In: Sustainable plant nutrition. Elsevier, pp 299–321. https://doi.org/10.1016/B978-0-443-18675-2.00013-4

Akhtar N, Ilyas N (2022) Role of nanosilicab to boost the activities of metabolites in Triticum aestivum facing drought stress. Plant Soil 477(1–2):99–115. https://doi.org/10.1007/s11104-021-05285-1

Altammar KA (2023) A review on nanoparticles: characteristics, synthesis, applications, and challenges. Front Microbiol 14:1155622. https://doi.org/10.3389/fmicb.2023.1155622

An C, Sun C, Li N, Huang B, Jiang J, Shen Y, Wang C, Zhao X, Cui B, Wang C, Li X, Zhan S, Gao F, Zeng Z, Cui H, Wang Y (2022) Nanomaterials and nanotechnology for the delivery of agrochemicals: strategies towards sustainable agriculture. J Nanobiotechnol 20(1):11. https://doi.org/10.1186/s12951-021-01214-7

Avila-Quezada GD, Ingle AP, Golińska P, Rai M (2022) Strategic applications of nano-fertilizers for sustainable agriculture: benefits and bottlenecks. Nanotechnol Rev 11(1):2123–2140. https://doi.org/10.1515/ntrev-2022-0126

Azameti MK, Imoro A-WM (2023) Nanotechnology: a promising field in enhancing abiotic stress tolerance in plants. Crop Design 2(2):100037. https://doi.org/10.1016/j.cropd.2023.100037

Babu S, Singh R, Yadav D, Rathore SS, Raj R, Avasthe R, Yadav SK, Das A, Yadav V, Yadav B, Shekhawat K, Upadhyay PK, Yadav DK, Singh VK (2022) Nanofertilizers for agricultural and environmental sustainability. Chemosphere 292:133451. https://doi.org/10.1016/j.chemosphere.2021.133451

Balci GN, Dağhan H (2023) Çinko Oksit Nanopartikül Uygulamasının Makarnalık Buğdayın Büyüme ve Çinko Alımı Üzerine Etkileri. https://doi.org/10.5281/ZENODO.10003026

Balusamy SR, Joshi AS, Perumalsamy H, Mijakovic I, Singh P (2023) Advancing sustainable agriculture: a critical review of smart and eco-friendly nanomaterial applications. J Nanobiotechnol 21(1):372. https://doi.org/10.1186/s12951-023-02135-3

Batool M, Zafar MN, Nazar MF (2022) General regulations for safe handling of manufactured nanomaterials. In: Nanomaterials recycling. Elsevier, pp 61–82. https://doi.org/10.1016/B978-0-323-90982-2.00004-4

Bhardwaj AK, Arya G, Kumar R, Hamed L, Pirasteh-Anosheh H, Jasrotia P, Kashyap PL, Singh GP (2022) Switching to nanonutrients for sustaining agroecosystems and environment: the challenges and benefits in moving up from ionic to particle feeding. J Nanobiotechnol 20(1):19. https://doi.org/10.1186/s12951-021-01177-9

Carmona FJ, Dal Sasso G, Ramírez-Rodríguez GB, Pii Y, Delgado-López JM, Guagliardi A, Masciocchi N (2021) Urea-functionalized amorphous calcium phosphate nanofertilizers: optimizing the synthetic strategy towards environmental sustainability and manufacturing costs. Sci Rep 11(1):3419. https://doi.org/10.1038/s41598-021-83048-9

Chen S, Pan Z, Zhao W, Zhou Y, Rui Y, Jiang C, Wang Y, White JC, Zhao L (2023) Engineering climate-resilient rice using a nanobiostimulant-based "stress training" strategy. ACS Nano 17(11):10760–10773. https://doi.org/10.1021/acsnano.3c02215

Chieb M, Gachomo EW (2023) The role of plant growth promoting rhizobacteria in plant drought stress responses. BMC Plant Biol 23(1):407. https://doi.org/10.1186/s12870-023-04403-8

Choudhary R, Rajput VD, Ghodake G, Ahmad F, Meena M, Rehman RU et al (2024) Comprehensive journey from past to present to future about seed priming with hydrogen peroxide and hydrogen sulfide concerning drought, temperature, UV and ozone stresses-a review. Plant Soil:1–23. https://doi.org/10.1007/s11104-024-06499-9

Choudhary R, Ahmad F, Kaya C, Upadhyay SK, Muneer S, Kumar V, Meena M, Liu H, Upadhyay H, Seth CS (2025) Decrypting proteomics, transcriptomics, genomics, and integrated omics for augmenting the abiotic, biotic, and climate change stress resilience in plants. J Plant Physiol. https://doi.org/10.1016/j.jplph.2025.154430

Das CK, Srivastava G, Dubey A, Roy M, Jain S, Sethy NK, Saxena M, Harke S, Sarkar S, Misra K, Singh SK, Bhargava K, Philip D, Das M (2016) Nano-iron pyrite seed dressing: a sustainable intervention to reduce fertilizer consumption in vegetable (beetroot, carrot), spice (fenugreek), fodder (alfalfa), and oilseed (mustard, sesamum) crops. Nanotechnol Environ Eng 1(1):2. https://doi.org/10.1007/s41204-016-0002-7

Davarpanah S, Tehranifar A, Davarynejad G, Abadía J, Khorasani R (2016) Effects of foliar applications of zinc and boron nano-fertilizers on pomegranate (Punica granatum cv. Ardestani) fruit yield and quality. Sci Hortic 210:57–64. https://doi.org/10.1016/j.scienta.2016.07.003

Deng C, Wang Y, Cantu JM, Valdes C, Navarro G, Cota-Ruiz K, Hernandez-Viezcas JA, Li C, Elmer WH, Dimkpa CO, White JC, Gardea-Torresdey JL (2022) Soil and foliar exposure of soybean (Glycine max) to Cu: nanoparticle coating-dependent plant responses. NanoImpact 26:100406. https://doi.org/10.1016/j.impact.2022.100406

Ding Y, Zhao W, Zhu G, Wang Q, Zhang P, Rui Y (2023) Recent trends in foliar nanofertilizers: a review. Nanomaterials 13(21):2906. https://doi.org/10.3390/nano13212906

Dos Santos TB, Ribas AF, De Souza SGH, Budzinski IGF, Domingues DS (2022) Physiological responses to drought, salinity, and heat stress in plants: a review. Stress 2(1):113–135. https://doi.org/10.3390/stresses2010009

Dutta P, Kumari A, Mahanta M, Biswas KK, Dudkiewicz A, Thakuria D, Abdelrhim AS, Singh SB, Muthukrishnan G, Sabarinathan KG, Mandal MK, Mazumdar N (2022) Advances in nanotechnology as a potential alternative for plant viral disease management. Front Microbiol 13:935193. https://doi.org/10.3389/fmicb.2022.935193

Elemike E, Uzoh I, Onwudiwe D, Babalola O (2019) The role of nanotechnology in the fortification of plant nutrients and improvement of crop production. Appl Sci 9(3):499. https://doi.org/10.3390/app9030499

El-Ramady HR, Verma KK, Rajput VD, Minkina T, Elbehiry F, Elbasiouny HY, Elsakhawy T, Omara AED, Amer MM (2022) Sources of silicon and nano-silicon in soil and plants. In: Silicon and nano-silicon in environmental stress management and crop quality improvement: progress and prospects. Academic Press, Elsevier. https://doi.org/10.1016/B978-0-323-91225-9.00003-0

El-Saadony MT, Saad AM, Soliman SM, Salem HM, Desoky E-SM, Babalghith AO, El-Tahan AM, Ibrahim OM, Ebrahim AAM, Abd El-Mageed TA, Elrys AS, Elbadawi AA, El-Tarabily KA, AbuQamar SF (2022) Role of nanoparticles in enhancing crop tolerance to abiotic stress: a comprehensive review. Front Plant Sci 13:946717. https://doi.org/10.3389/fpls.2022.946717

El-Shal RM, El-Naggar AH, El-Beshbeshy TR, Mahmoud EK, El-Kader NIA, Missaui AM, Du D, Ghoneim AM, El-Sharkawy MS (2022) Effect of nano-fertilizers on alfalfa plants grown under different salt stresses in hydroponic system. Agriculture 12(8):1113. https://doi.org/10.3390/agriculture12081113

Faizan M, Hussain A, Mir AR, Rajput VD, Minkina T, Hayat S (2021) Effect of carbon nanotubes on abiotic stress response in plants: an overview. In: Faizan M, Hayat S, Yu F (eds) Sustainable agriculture reviews, vol 53. Springer International Publishing, pp 217–229. https://doi.org/10.1007/978-3-030-86876-5_9

Fatima F, Hashim A, Anees S (2021) Efficacy of nanoparticles as nanofertilizer production: a review. Environ Sci Pollut Res 28(2):1292–1303. https://doi.org/10.1007/s11356-020-11218-9

Fincheira P, Hoffmann N, Tortella G, Ruiz A, Cornejo P, Diez MC, Seabra AB, Benavides-Mendoza A, Rubilar O (2023) Eco-efficient systems based on nanocarriers for the controlled release of fertilizers and pesticides: toward smart agriculture. Nanomaterials (Basel, Switzerland) 13(13):1978. https://doi.org/10.3390/nano13131978

Gade A, Ingle P, Nimbalkar U, Rai M, Raut R, Vedpathak M, Jagtap P, Abd-Elsalam KA (2023) Nanofertilizers: the next generation of agrochemicals for long-term impact on sustainability in farming systems. Agrochem 2(2):257–278. https://doi.org/10.3390/agrochemicals2020017

Gamage A, Gangahagedara R, Gamage J, Jayasinghe N, Kodikara N, Suraweera P, Merah O (2023) Role of organic farming for achieving sustainability in agriculture. Farm Syst 1(1):100005. https://doi.org/10.1016/j.farsys.2023.100005

Gangwar J, Kadanthottu Sebastian J, Puthukulangara Jaison J, Kurian JT (2023) Nano-technological interventions in crop production—a review. Physiol Mol Biol Plants 29(1):93–107. https://doi.org/10.1007/s12298-022-01274-5

Garg S, Rumjit NP, Roy S (2023) Smart agriculture and nanotechnology: technology, challenges, and new perspective. Adv Agrochem S2773237123000825. https://doi.org/10.1016/j.aac.2023.11.001

González-García Y, Cárdenas-Álvarez C, Cadenas-Pliego G, Benavides-Mendoza A, Cabrera-de-la-Fuente M, Sandoval-Rangel A, Valdés-Reyna J, Juárez-Maldonado A (2021) Effect of three nanoparticles (Se, Si and Cu) on the bioactive compounds of bell pepper fruits under saline stress. Plan Theory 10(2):217. https://doi.org/10.3390/plants10020217

Goyal V, Rani D, Ritika, Mehrotra S, Deng C, Wang Y (2023) Unlocking the potential of nano-enabled precision agriculture for efficient and sustainable farming. Plan Theory 12(21):3744. https://doi.org/10.3390/plants12213744

Ha NMC, Nguyen TH, Wang S-L, Nguyen AD (2019) Preparation of NPK nanofertilizer based on chitosan nanoparticles and its effect on biophysical characteristics and growth of coffee in greenhouse. Res Chem Intermed 45(1):51–63. https://doi.org/10.1007/s11164-018-3630-7

Hao Y, Cao X, Ma C, Zhang Z, Zhao N, Ali A, Hou T, Xiang Z, Zhuang J, Wu S, Xing B, Zhang Z, Rui Y (2017) Potential applications and antifungal activities of engineered nanomaterials against Gray Mold Disease Agent Botrytis cinerea on rose petals. Front Plant Sci 8:1332. https://doi.org/10.3389/fpls.2017.01332

Haydar MS, Ghosh D, Roy S (2024) Slow and controlled release nanofertilizers as an efficient tool for sustainable agriculture: recent understanding and concerns. Plant Nano Biol 7:100058. https://doi.org/10.1016/j.plana.2024.100058

He A, Jiang J, Ding J, Sheng GD (2021) Blocking effect of fullerene nanoparticles (nC60) on the plant cell structure and its phytotoxicity. Chemosphere 278:130474. https://doi.org/10.1016/j.chemosphere.2021.130474

Hernandez-Viezcas JA, Castillo-Michel H, Servin AD, Peralta-Videa JR, Gardea-Torresdey JL (2011) Spectroscopic verification of zinc absorption and distribution in the desert plant Prosopis juliflora-velutina (velvet mesquite) treated with ZnO nanoparticles. Chem Eng J 170(2–3):346–352. https://doi.org/10.1016/j.cej.2010.12.021

Jagadesh M, Dash M, Kumari A, Singh SK, Verma KK, Kumar P, Bhatt R, Sharma SK (2024) Revealing the hidden world of soil microbes: metagenomic insights into plant, bacteria, and fungi interactions for sustainable agriculture and ecosystem restoration. Microbiol Res 285:127764. https://doi.org/10.1016/j.micres.2024.127764

Janmohammadi M, Navid A, Segherloo AE, Sabaghnia N (2016) Impact of nano-chelated micronutrients and biological fertilizers on growth performance and grain yield of maize under deficit irrigation condition. Biologija 62(2). https://doi.org/10.6001/biologija.v62i2.3339

Kahlel A-MS, Abdulla AA, Saadalla HA, Hamed MH (2021) Effect of nano fertilizers and its applying methods on growth parameters of Senna (Cassia Angustifolia) seedling plant. IOP Conference Series: Earth and Environmental Science 923(1):012003. https://doi.org/10.1088/1755-1315/923/1/012003

Kalwani M, Chakdar H, Srivastava A, Pabbi S, Shukla P (2022) Effects of nanofertilizers on soil and plant-associated microbial communities: emerging trends and perspectives. Chemosphere 287:132107. https://doi.org/10.1016/j.chemosphere.2021.132107

Kashyap D, Siddiqui ZA (2021) Effect of silicon dioxide nanoparticles and *Rhizobium leguminosarum* alone and in combination on the growth and bacterial blight disease complex of pea caused by *Meloidogyne incognita* and *Pseudomonas syringae* pv. Pisi Arch Phytopathol Plant Protect 54(9–10):499–515. https://doi.org/10.1080/03235408.2020.1843306

Kaushal M, Wani SP (2017) Nanosensors: frontiers in precision agriculture. In: Prasad R, Kumar M, Kumar V (eds) Nanotechnology. Springer Singapore, pp 279–291. https://doi.org/10.1007/978-981-10-4573-8_13

Khati P, Parul, Bhatt P, Nisha, Kumar R, Sharma A (2018) Effect of nanozeolite and plant growth promoting rhizobacteria on maize. 3 Biotech 8(3):141. https://doi.org/10.1007/s13205-018-1142-1

Kumar A, Singh K, Verma P, Singh O, Panwar A, Singh T, Kumar Y, Raliya R (2022) Effect of nitrogen and zinc nanofertilizer with the organic farming practices on cereal and oil seed crops. Sci Rep 12(1):6938. https://doi.org/10.1038/s41598-022-10843-3

Kumar D, Mariyam S, Gupta KJ, Thiruvengadam M, Ghodake GS, Xing B, Seth CS (2024a) Comparative investigation on chemical and green synthesized titanium dioxide nanoparticles against chromium (VI) stress eliciting differential physiological, biochemical, and cellular attributes in Helianthus annuus L. Sci Total Environ 930:172413. https://doi.org/10.1016/j.scitotenv.2024.172413

Kumar D, Singh R, Upadhyay SK, Verma KK, Tripathi RM, Liu H et al (2024b) Review on interactions between nanomaterials and phytohormones: novel perspectives and opportunities for mitigating environmental challenges. Plant Sci 340:111964. https://doi.org/10.1016/j.plantsci.2023.111964

Kumar D, Biswas JK, Mulla SI, Singh R, Shukla R, Ahanger MA et al (2024c) Micro and nanoplastics pollution: sources, distribution, uptake in plants, toxicological effects, and innovative remediation strategies for environmental sustainability. Plant Physiol Biochem 108795. https://doi.org/10.1016/j.plaphy.2024.108795

Kumari A, Lakshmi GA, Krishna GK, Patni B, Prakash S, Bhattacharyya M, Singh SK, Verma KK (2022) Climate change and its impact on crops: a comprehensive investigation for sustainable agriculture. Agronomy 12:3008. https://doi.org/10.3390/agronomy12123008

Kumari A, Dash M, Singh SK, Jagadesh M, Mathpal B, Mishra PK, Pandey SK, Verma KK (2023a) Soil microbes: a natural solution for mitigating the impact of climate change. Environ Monit Assess 195:1436. https://doi.org/10.1007/s10661-023-11988-y

Kumari R, Suman K, Karmakar S, Mishra V, Lakra SG, Saurav GK, Mahto BK (2023b) Regulation and safety measures for nanotechnology-based agri-products. Front Genome Editing 5:1200987. https://doi.org/10.3389/fgeed.2023.1200987

Larue C, Veronesi G, Flank A-M, Surble S, Herlin-Boime N, Carrière M (2012) Comparative uptake and impact of TiO_2 nanoparticles in wheat and rapeseed. J Toxic Environ Health A 75(13–15):722–734. https://doi.org/10.1080/15287394.2012.689800

Le TTH, Mai TTT, Phan KS, Nguyen TM, Tran TLA, Dong TN, Tran HC, Ngo TTH, Hoang PH, Ha PT (2022) Novel integrated nanofertilizers for improving the growth of Polyscias fruticosa and Asparagus officinalis. J Nanomater 2022:1–10. https://doi.org/10.1155/2022/5791922

Magnabosco P, Masi A, Shukla R, Bansal V, Carletti P (2023) Advancing the impact of plant biostimulants to sustainable agriculture through nanotechnologies. Chem Biol Technol Agric 10(1):117. https://doi.org/10.1186/s40538-023-00491-8

Marchiol L, Filippi A, Adamiano A, Degli Esposti L, Iafisco M, Mattiello A, Petrussa E, Braidot E (2019) Influence of hydroxyapatite nanoparticles on germination and plant metabolism of tomato (Solanum lycopersicum L.): preliminary evidence. Agronomy 9(4):161. https://doi.org/10.3390/agronomy9040161

Mariyam S, Upadhyay SK, Chakraborty K, Verma KK, Duhan JS, Muneer S et al (2024) Nanotechnology, a frontier in agricultural science, a novel approach in abiotic stress management and convergence with new age medicine-a review. Sci Total Environ 912:169097. https://doi.org/10.1016/j.scitotenv.2023.169097

Mejias JH, Salazar F, Pérez Amaro L, Hube S, Rodriguez M, Alfaro M (2021) Nanofertilizers: a cutting-edge approach to increase nitrogen use efficiency in grasslands. Front Environ Sci 9:635114. https://doi.org/10.3389/fenvs.2021.635114

Miguel-Rojas C, Pérez-de-Luque A (2023) Nanobiosensors and nanoformulations in agriculture: new advances and challenges for sustainable agriculture. Emerg Topics Life Sci 7(2):229–238. https://doi.org/10.1042/ETLS20230070

Najafi Disfani M, Mikhak A, Kassaee MZ, Maghari A (2017) Effects of nano Fe/SiO_2 fertilizers on germination and growth of barley and maize. Arch Agron Soil Sci 63(6):817–826. https://doi.org/10.1080/03650340.2016.1239016

Nandini B, Mawale KS, Giridhar P (2023) Nanomaterials in agriculture for plant health and food safety: a comprehensive review on the current state of agro-nanoscience. 3 Biotech 13(3):73. https://doi.org/10.1007/s13205-023-03470-w

Nile SH, Thiruvengadam M, Wang Y, Samynathan R, Shariati MA, Rebezov M, Nile A, Sun M, Venkidasamy B, Xiao J, Kai G (2022) Nano-priming as emerging seed priming technology for sustainable agriculture—Recent developments and future perspectives. J Nanobiotechnol 20(1):254. https://doi.org/10.1186/s12951-022-01423-8

Nongbet A, Mishra AK, Mohanta YK, Mahanta S, Ray MK, Khan M, Baek K-H, Chakrabartty I (2022) Nanofertilizers: a smart and sustainable attribute to modern agriculture. Plants (Basel, Switzerland) 11(19):2587. https://doi.org/10.3390/plants11192587

Omar AA, Heikal YM, Zayed EM, Shamseldin SAM, Salama YE, Amer KE, Basuoni MM, Abd Ellatif S, Mohamed AH (2023) Conferring of drought and heat stress tolerance in wheat (Triticum aestivum L.) genotypes and their response to selenium nanoparticles application. Nanomaterials 13(6):998. https://doi.org/10.3390/nano13060998

Pérez-de-Luque A (2017) Interaction of nanomaterials with plants: what do we need for real applications in agriculture? Front Environ Sci 5. https://doi.org/10.3389/fenvs.2017.00012

Pérez-Labrada F, López-Vargas ER, Ortega-Ortiz H, Cadenas-Pliego G, Benavides-Mendoza A, Juárez-Maldonado A (2019) Responses of tomato plants under saline stress to foliar application of copper nanoparticles. Plan Theory 8(6):151. https://doi.org/10.3390/plants8060151

Pouran HM (2019) Engineered nanomaterials in the environment, their potential fate and behaviour and emerging techniques to measure them. In: Hussain CM (ed) Handbook of environmental materials management. Springer International Publishing, pp 1–15. https://doi.org/10.1007/978-3-319-58538-3_95-2

Rajput VD, Singh A, Minkina TM, Shende SS, Kumar P, Verma KK, Bauer T, Gorobtsova O, Deneva S, Sindireva A (2021a) Potential applications of nanobiotechnology in plant nutrition and protection for sustainable agriculture. In: Ingle AP (ed) Nanotechnology in plant growth promotion and protection: recent advances and impacts, Ist edn. Wiley

Rajput VD, Singh A, Minkina T, Rawat S, Mandzhieva S, Sushkova S, Shuvaeva V, Nazarenko O, Rajput P, Komariah, Verma KK, Singh AK, Rao M, Upadhyay SK (2021b) Nano-enabled products: challenges and opportunities for sustainable agriculture. Plan Theory 10:2727. https://doi.org/10.3390/plants10122727

Rajput VD, Minkina T, Upadhyay SK, Kumari A, Ranjan A, Mandzhieva S, Sushkova S, Singh RK, Verma KK (2022) Nanotechnology in the restoration of polluted soil. Nanomaterials 12:769. https://doi.org/10.3390/nano12050769

Rajput VD, Kumari A, Upadhyay SK, Minkina T, Mandzhieva S, Ranjan A, Sushkova S, Burachevskaya M, Rajput P, Konstantinova E, Singh J, Verma KK (2023) Can nanomaterials improve the soil microbiome and crop productivity? Agriculture 13:231. https://doi.org/10.3390/agriculture13020231

Raliya R, Franke C, Chavalmane S, Nair R, Reed N, Biswas P (2016) Quantitative understanding of nanoparticle uptake in watermelon plants. Front Plant Sci 7. https://doi.org/10.3389/fpls.2016.01288

Rivero RM, Mittler R, Blumwald E, Zandalinas SI (2022) Developing climate-resilient crops: improving plant tolerance to stress combination. Plant J 109(2):373–389. https://doi.org/10.1111/tpj.15483

Rizwan M, Ali S, Ur Rehman MZ, Malik S, Adrees M, Qayyum MF, Alamri SA, Alyemeni MN, Ahmad P (2019) Effect of foliar applications of silicon and titanium dioxide nanoparticles on growth, oxidative stress, and cadmium accumulation by rice (Oryza sativa). Acta Physiol Plant 41(3):35. https://doi.org/10.1007/s11738-019-2828-7

Saharan V, Sharma G, Yadav M, Choudhary MK, Sharma SS, Pal A, Raliya R, Biswas P (2015) Synthesis and in vitro antifungal efficacy of Cu–chitosan nanoparticles against pathogenic fungi of tomato. Int J Biol Macromol 75:346–353. https://doi.org/10.1016/j.ijbiomac.2015.01.027

Schaller J, Cramer A, Carminati A, Zarebanadkouki M (2020) Biogenic amorphous silica as main driver for plant available water in soils. Sci Rep 10(1):2424. https://doi.org/10.1038/s41598-020-59437-x

Seleiman MF, Al-Suhaibani N, Ali N, Akmal M, Alotaibi M, Refay Y, Dindaroglu T, Abdul-Wajid HH, Battaglia ML (2021) Drought stress impacts on plants and different approaches to alleviate its adverse effects. Plants (Basel, Switzerland) 10(2):259. https://doi.org/10.3390/plants10020259

Seyed Sharifi R, Khalilzadeh R, Pirzad A, Anwar S (2020) Effects of biofertilizers and nano zinc-iron oxide on yield and physicochemical properties of wheat under water deficit conditions. Commun Soil Sci Plant Anal 51(19):2511–2524. https://doi.org/10.1080/00103624.2020.1845350

Shah AA, Aslam S, Akbar M, Ahmad A, Khan WU, Yasin NA, Ali B, Rizwan M, Ali S (2021) Combined effect of Bacillus fortis IAGS 223 and zinc oxide nanoparticles to alleviate cadmium phytotoxicity in Cucumis melo. Plant Physiol Biochem 158:1–12. https://doi.org/10.1016/j.plaphy.2020.11.011

Shang Y, Hasan MK, Ahammed GJ, Li M, Yin H, Zhou J (2019) Applications of nanotechnology in plant growth and crop protection: a review. Molecules 24(14):2558. https://doi.org/10.3390/molecules24142558

Siddiqui ZA, Hashmi A, Khan MR, Parveen A (2020) Management of bacteria *Pectobacterium carotovorum, Xanthomonas campestris* pv. *Carotae* , and fungi *Rhizoctonia solani, Fusarium solani* and *Alternaria dauci* with silicon dioxide nanoparticles on carrot. Int J Veg Sci 26(6):547–557. https://doi.org/10.1080/19315260.2019.1675843

Siddiqui ZA, Khan MR, Aqib. (2023) Comparative efficacy of zirconium dioxide nanoparticles and AM fungus against wilt disease complex of bottle gourd and upregulation of biochemical and physiological processes. Int J Veg Sci 29(3):243–260. https://doi.org/10.1080/19315260.2023.2187503

Singh P, Singh AP (2021) Nanomaterials in soil health management and crop production: potentials and limitations. In: Kharissova OV, Martínez LMT, Kharisov BI (eds) Handbook of nanomaterials and nanocomposites for energy and environmental applications. Springer International Publishing, pp 1–25. https://doi.org/10.1007/978-3-030-11155-7_35-1

Sun D, Hussain HI, Yi Z, Siegele R, Cresswell T, Kong L, Cahill DM (2014) Uptake and cellular distribution, in four plant species, of fluorescently labeled mesoporous silica nanoparticles. Plant Cell Rep 33(8):1389–1402. https://doi.org/10.1007/s00299-014-1624-5

Tarafder C, Daizy M, Alam MM, Ali MR, Islam MJ, Islam R, Ahommed MS, Aly Saad Aly M, Khan ZH (2020) Formulation of a hybrid nanofertilizer for slow and sustainable release of micronutrients. ACS Omega 5(37):23960–23966. https://doi.org/10.1021/acsomega.0c03233

Taran NY, Gonchar OM, Lopatko KG, Batsmanova LM, Patyka MV, Volkogon MV (2014) The effect of colloidal solution of molybdenum nanoparticles on the microbial composition in rhizosphere of Cicer arietinum L. Nanoscale Res Lett 9(1):289. https://doi.org/10.1186/1556-276X-9-289

Tyagi J, Ahmad S, Malik M (2022) Nitrogenous fertilizers: impact on environment sustainability, mitigation strategies, and challenges. Int J Environ Sci Technol 19(11):11649–11672. https://doi.org/10.1007/s13762-022-04027-9

Verma KK, Song XP, Joshi A, Rajput VD, Singh M, Sharma A, Singh RK, Li DM, Arora J, Minkina T, Li YR (2022a) Nanofertilizer possibilities for healthy soil, water and food in future: an overview. Front Plant Sci 13:865048. https://doi.org/10.3389/fpls.2022.865048

Verma KK, Song XP, Joshi A, Tian DD, Rajput VD, Singh M, Arora J, Minkina TM, Li YR (2022b) Recent trends in nano-fertilizer for advancing sustainable agriculture under the era of climate change to ensure food security for future generations. Nanomaterials 12:173. https://doi.org/10.3390/nano12010173

Verma KK, Song X, Degu HD, Guo D, Joshi A, Huang H, Xu L, Singh M, Huang D, Rajput VD, Li Y (2023a) Recent advances in nitrogen and nano-nitrogen fertilizers for sustainable crop production: a mini-review. Chem Biol Technol Agric 10:1–14. https://doi.org/10.1186/s40538-023-00488-3

Verma KK, Singh M, Rajput VD, Verma CL, Burachevskaya M (2023b) Nanofertilizers: a sustainable alternative to conventional means. In: Rajput VD, Singh A, Minkina TM, Verma KK, Singh AK (eds) Nanotechnology for sustainable agriculture: an innovative and eco-friendly approach. Apple Academic Press, Inc., Taylor & Francis. https://doi.org/10.1201/9781003333128

Verma KK, Song XP, Kumari A, Jagadesh M, Singh SK, Bhatt R, Singh M, Seth CS, Li YR (2024a) Climate change adaptation: challenges for agricultural sustainability. Plant Cell Environ:1–12. https://doi.org/10.1111/pce.15078

Verma KK, Joshi A, Song X-P, Singh S, Kumari A, Arora J, Singh SK, Solanki MK, Seth CS, Li Y-R (2024b) Synergistic interactions of nanoparticles and plant growth promoting rhizobacteria enhancing soil-plant systems: a multigenerational perspective. Front Plant Sci 15:1376214. https://doi.org/10.3389/fpls.2024.1376214

Wan A, Gao Q, Li H (2010) Effects of molecular weight and degree of acetylation on the release of nitric oxide from chitosan-nitric oxide adducts. J Appl Polym Sci 117(4):2183–2188. https://doi.org/10.1002/app.31130

Wang Z, Xie X, Zhao J, Liu X, Feng W, White JC, Xing B (2012) Xylem- and phloem-based transport of CuO nanoparticles in maize (*Zea mays* L.). Environ Sci Technol 46(8):4434–4441. https://doi.org/10.1021/es204212z

Wang J, Koo Y, Alexander A, Yang Y, Westerhof S, Zhang Q, Schnoor JL, Colvin VL, Braam J, Alvarez PJJ (2013) Phytostimulation of poplars and *Arabidopsis* exposed to silver nanoparticles and Ag^+ at sublethal concentrations. Environ Sci Technol 47(10):5442–5449. https://doi.org/10.1021/es4004334

Worrall E, Hamid A, Mody K, Mitter N, Pappu H (2018) Nanotechnology for plant disease management. Agronomy 8(12):285. https://doi.org/10.3390/agronomy8120285

Yadav A, Yadav K, Abd-Elsalam KA (2023a) Exploring the potential of nanofertilizers for a sustainable agriculture. Plant Nano Biol 5:100044. https://doi.org/10.1016/j.plana.2023.100044

Yadav A, Yadav K, Ahmad R, Abd-Elsalam KA (2023b) Emerging frontiers in nanotechnology for precision agriculture: advancements, hurdles and prospects. Agrochemicals 2(2):220–256. https://doi.org/10.3390/agrochemicals2020016

Yadav A, Yadav K, Abd-Elsalam KA (2023c) Nanofertilizers: types, delivery and advantages in agricultural sustainability. Agrochemicals 2(2):296–336. https://doi.org/10.3390/agrochemicals2020019

Yaqoob AA, Ahmad H, Parveen T, Ahmad A, Oves M, Ismail IMI, Qari HA, Umar K, Mohamad Ibrahim MN (2020) Recent advances in metal decorated nanomaterials and their various biological applications: a review. Front Chem 8:341. https://doi.org/10.3389/fchem.2020.00341

Zahedi SM, Moharrami F, Sarikhani S, Padervand M (2020) Selenium and silica nanostructure-based recovery of strawberry plants subjected to drought stress. Sci Rep 10(1):17672. https://doi.org/10.1038/s41598-020-74273-9

Zheng L, Hong F, Lu S, Liu C (2005) Effect of nano-TiO2 on strength of naturally aged seeds and growth of spinach. Biol Trace Elem Res 104(1):083–092. https://doi.org/10.1385/BTER:104:1:083

Chapter 11
Current Status and Future Challenges of Nanobiosensors and Nanoformulations in Agriculture

Oluwatoyin Adenike Fabiyi

11.1 Introduction

11.1.1 Agriculture, Abiotic Stress, Crop Monitoring and Nanosensor

Agriculture sector is of utmost importance because industries that are drivers of economies are directly or indirectly hinged on agriculture (FAO 2020). Annually, hunger and undernourishment are upscale largely owing to decline of cultivable hectares of land. Between the years 2000 and 2019, over 127 mha of cultivable pieces of land were no longer functional (FAO 2021). Intensive industrialization, enormous urbanization and the dynamics of weather have put so much pressure on agriculture. Yield losses incurred from incessant variabilities of climatic factors, coupled with reduction in farmlands consequently leads to food insecurity (Verma et al. 2024a). Currently, the high variability in weather is startling and has resulted in unfavourable and stressful situations for plants.

Crop stressors are conditions that impede growth, sundry climatic factors such as water deficiency that results in drought from lack of rain or insufficient rain, flooding from excessive rain, heavy metals pollution, high salinity and extremely high and low temperatures are a broad range of stressors which has resulted to influx of pests and pathogens in large numbers, eventually harming agriculture, hence productivity and quality becomes minimal (Jagadesh et al. 2024; Fabiyi and Ogundele 2025). In reflex to stressors, biochemical and morphological mechanisms are set off by the plant's complex protection system in a bid to defend plant cells (Choudhary et al. 2024). Plants are known to prompt transcriptome cells for a specialized

O. A. Fabiyi (✉)
Department of Crop Protection, Faculty of Agriculture, University of Ilorin, Ilorin, Nigeria
e-mail: fabiyi.oa@unilorin.edu.ng

C. S. Seth et al. (eds.), *Nano-formulations for Sustainable Agriculture*,
Nanotechnology in the Life Sciences, https://doi.org/10.1007/978-3-031-90617-6_11

response on discernment of stressors in the surroundings (Lamers et al. 2020). ROS generation is a primary signalling reaction, diverse stress network mechanism is activated by ROS to bring about resilience and defence of the plant system (Zhang et al. 2020). Sundry stressors are associated with distinct reaction signatures in plants (Choudhury et al. 2017).

Following this state, plant hormone levels are agitated, redox response to stressors is then orchestrated (Saxena et al. 2016). Salicylic acid, an adaptable intricate regulating hormone, augments swift and sturdy action against primary environmental stressors (Ahmad et al. 2019). Early diagnosis of plant stressors on crops through sensing of weather dynamics is a crucial part of farm management (Li et al. 2014). For fruitful high yield production, there is a great need to monitor closely the aftermath of weather variables on field crops in order to maintain sustainable yield and output (Salouti and Khadivi Derakhshan 2020). Hence, weather factor changes require monitoring for quick detection of drawbacks on crops to minimise the effect. Diverse monitoring laboratory based techniques must be manned and a lot of man hour time goes into it; thus, an immediate monitoring method is desired. The inclusion of nanomaterials into agriculture has made field crop monitoring easier and more feasible economically (Neme et al. 2021; El-Ramady et al. 2022). Through nanotechnology, very sensitive sensors have been developed encompassing combinations of computer, electronics, biology, nanoscience and optoelectronics.

Significant reduction in crop productivity is largely associated with abiotic stressors (Fabiyi et al. 2025). To guide clear of unrectifiable harm, early identification of crop stressors and timely response is vital (Verma et al. 2023). Nanosensors are unique with extreme accuracy, it can detect incidence of stressors at minimal level, while system integration is quick and seamless (Li et al. 2021). The functionality of nanosensors has enhanced great advancements in agriculture, where materials like composites, polymers, carbons and metals have been employed (Bhatia et al. 2024). Nanosensors are designed for analytical purposes which operate on the platform of biological elements fused with actuators (Shawona et al. 2020). They are coupled with bio-sensory receptors for selective probe of molecules of interest (Fig. 11.1), such as phytohormones, enzymes, amino acids, tissues, salicylic acid and H_2O_2, which are triggered in reaction to stressors. Sensors are programmed for data analysis on nanoscale in amplifying devices (Fig. 11.2). They operate with high

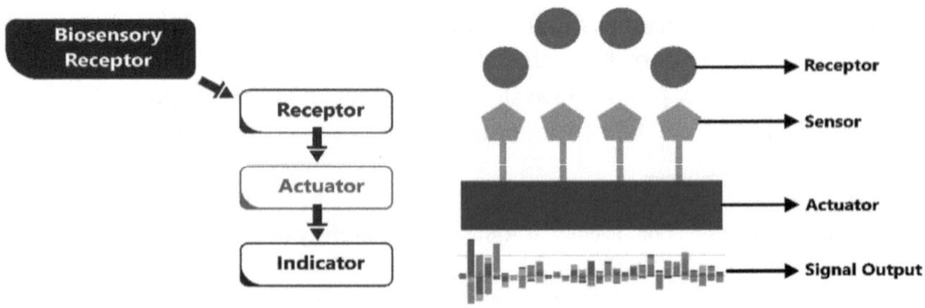

Fig. 11.1 The role and properties of nanobiosensor

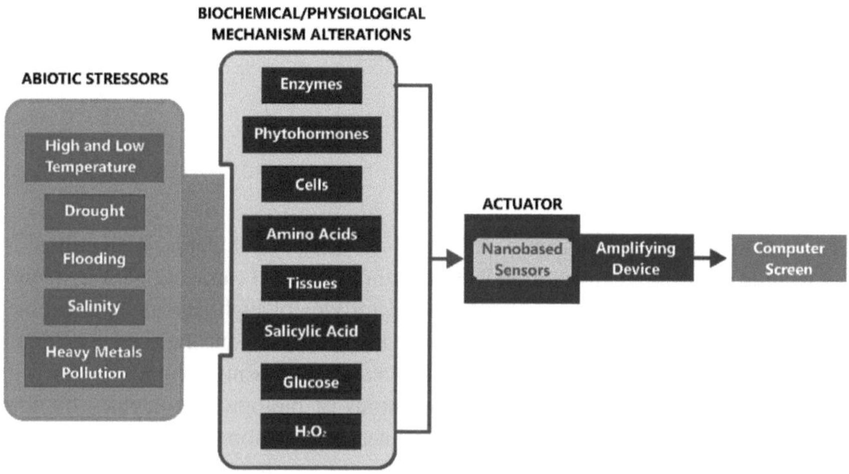

Fig. 11.2 The influence of nanobiosensor on crop plants in response to adverse agroclimatic conditions

sensitivity of detection, high volume surface ratio and are highly efficient (Shawona et al. 2020). Nanosensors are highly driven and can immediately recognize stressors and their impression on crops, excellent nanosensors have extremely high reproducibility and selective characteristics, stability and linearity are equally critical factors for consideration (Karunakaran et al. 2015; Naresh and Lee 2021).

To realize precision farming, it is crucial to fathom plant status in drastic weather conditions. In detecting crop abiotic stressors, alterations in physiology of crops could be evaluated as affected by weather and environmental variables using nanosensors (Vero et al. 2024). The use of environmentally safe and viable nanoproducts to enhance crop resilience to abiotic stress is a valuable approach, particularly in regions most impacted by extreme weather conditions (Verma et al. 2024b). This chapter presents an overview of nanosensors employed in precision farming for quick abiotic stress detection.

11.2 Development of Nanosensors

Nanosensors are harnessed in several forms in order to attain high-grade performance levels. Sensor compositions bring about precise information and groundbreaking action. Good integration of various types of nanomaterials into sensor components allows for high efficiency with inherent sensitivity (Kumar et al. 2024a). Nanosensor techniques give room to transform, convert and detect chemical signs and waves in plants which are imperceptible to electrical, optical or non-wire encoders that can be documented and interpreted using customized equipment

(Malik et al. 2023). Sensor composition and design could be hinged on wearables, FRET, genetically encoded and oxides of metals.

11.2.1 Wearable Nanosensors

These consist of nano electronics embedded into surfaces of plants. With utilization of volatile molecules and communication without wires, it can work with frequency of radio. The device consists of graphite electrodes of high sensitivity combined with channels of SWCNT placed on plant leaf surfaces for sensing (Lee et al. 2014; Oren et al. 2017). Nanosensors of SWCNT can function using complex copper with high durability and can monitor concentrations of biological molecules for long time. Aqueous medium and gas molecules can be monitored without wires using CN tube and graphene sensors. Minimal concentrations in ppb scale are reported by wearable sensors when plant volatile organic compounds are measured and health situations of crops are established beforehand (Ibrahim et al. 2022). Recovery time and sensor noise are higher when relative humidity is measured close to plant surface (Lee et al. 2014).

Wearable sensor performance depends on plant molecule type it is in contact with. Under weather variables, analyte selectivity from complex mixtures would be achieved through the use of wearable sensors of carbon nano material origin. Wearable sensors made of graphene attached to leaves of plants have been able to monitor alterations in the movement of water and alterations in relative humidity from 20% to 90%. SWCNTs have been embedded in leaves of plants for information conveyance regarding diverse stress situations, using near-infrared optical signals which transmit information to smartphones through emails and text messages. Continuous monitoring of level of water in plants and relative humidity close to leaf epidermis has been achieved by using graphene combined with carbon nanotubes printed on surface of leaf (Koman et al. 2017; Oren et al. 2017). Exchange of gas from the environment and leaf is through air gaps created by relative humidity sensors based on tape transferring of high resolution. Response to relative humidity by the graphene sensor resistance is then reported by RIC meter. On the other hand, dynamics of stomatal aperture can present water level to microfluidic SWCNT sensors on leaf epidermis once drought stress is perceived (Kumar et al. 2024b). The opening and closing time interval of stomata in times of drought passes signal to a single responsive stoma, that carries stripe and contact pads of the wearable sensor. The sensor could also detect diverse other environmental stressors which alter conductance of the stomata.

Crucial parameters which hinge on health of crops are measured by wearable sensors. In the reports of Chai et al. (2021) who monitored water and nutrient movement in watermelon plants with the aid of electronic flexible sensor device, identified shifts in distribution of water between fruits and leaves of watermelon. Rotation on day and night was detected. Similarly, Kim et al. (2019) used polymer electrode vapour-printed wearable nanosensor and identified damage in a plant's deep tissue caused by irradiation and dehydration. Further, Boonyaves et al. (2023) fabricated wearable sensors for Gibberellins, a crucial hormone in plants. The sensor was made of polymer glutenin a high molecular weight material incorporated into SWCNT. There was exactness towards gibberellins. Concentration of gibberellins in roots of lettuce plants exposed to saline stress was identified with high accuracy, interconnection was ascertained between minimal weight of fresh root and gibberellin content. Additionally, Nassar et al. (2018) observed temperature and humidity using wearable nanosensor to determine and document the microclimatic situations around the plant.

11.2.2 Encoded Sensor

These are genetically formed sensors and are used in monitoring alterations of ion and protein molecules, which have provided notable ideas in the design of smart sensors for evaluating signalling molecules in plants. Lately, Toyota et al. (2018) evaluated the concentration of calcium in a plant attacked by ruminants using genetically encoded sensors (GES). To function properly, the plant species must be amenable to generate a transgenic substance for the expression of GES to be stable. This technique has been applied in identifying signalling biomarkers in plant roots and leaves (Deuschle et al. 2006; Exposito-Rodriguez et al. 2017; Zhu et al. 2017). The mechanisms responsible for GES are dynamic and vary from covalent bond alteration to relationship between protein-protein bonds. Wavelength or fluorescence intensity is impacted and then becomes visible with electromagnetic field.

In crops like rice, sucrose and glucose are monitored and identified in reaction to abiotic stressors within radius of physiological concentration using high temporal resolution (Zhu et al. 2017). At high resolution, variation in plant calcium has been identified with application of GES, with the same technique, root glucose was detected at micromolar state (Chaudhuri et al. 2008; Keinath et al. 2015). Detection of hydrogen peroxide was achieved by GES in nuclei and protoplasts of plants through subcellular resolution of nanosensors without exceeding physiological pH levels (Kumar et al. 2024c). Thus, the GES operates within the plant's physiological level, and this allows identification of diverse plant signalling biomolecules. GES operates on the order of seconds premised on temporal resolutions, hence, biomolecule signals are recorded using this scale. Up till now GES for chloroplasts are limited to only genetically amenable crop species (Walia et al. 2018). The chloroplasts of plants exposed to abiotic stress elicit salicylic acid and ROS as signal biomolecules that stockpiles in the plant, and these can serve as identifiers of crop health status.

11.2.3 FRET Sensors

Basically, fluorescent resonance energy transfer (FRET) sensors are employed to examine levels of biomolecules in plant cells. They are premised on the consideration that if the interval between two fluorochromes is not more than a nanometer scale limit, then an exited donating fluorochrome will reverberantly convey energy to the accepting fluorochrome, thus giving rise to luminescence (Suchit et al. 2022). The efficiency of action relies upon the separating distance between the two fluorochromes and the presence of molecules is detected. Stress signalling indicators are easier to detect and monitor. They are capable of swiftly documenting the setoff of alterations in plant health. Levels of sugar, jasmonate, nitric oxide, calcium, ethylene, etc. can be checked to establish abiotic stress (Subraya et al. 2013). It is however a great challenge to determine and sense disturbances in crops without signals. Glucose is a molecule representing redox signalling reactions in plants. Regular monitoring of glucose concentration in plants is a stride in ascertaining abiotic stress impression in crops. The variation of crop cellular glucose could be evaluated using FRET nanosensors, which typically evaluate the fluctuations of glucose (Mariyam et al. 2024). For example, in examining rice under several abiotic stress using analogous resolution, the capability of glucose in roots and leaves as abiotic response signal has been noted (Zhu et al. 2017), Furthermore, salicylic acid is also a key molecule in stressor signals, it gives a significant response to stress while accumulating in chloroplasts (Chaudhuri et al. 2008, 2011).

11.3 Oxides of Metal NP Sensors

11.3.1 Copper Oxide NP Sensors

In the process of defence system activation, environmental stressors start off ROS in cells of plants in a quick wave of signals, a part of plant ROS is hydrogen peroxide, which is a significant reverse reaction stress molecule exhibited as early signal hormone in response to stressful situations (Mittler et al. 2022). The level of hydrogen peroxide coupled with its physical stability reflects through signalling pathway with a substantial interaction with other hormones to give swift response to stressors in plants. Hence, hydrogen peroxide state is often a good stress indicator (Fichman and Mittler 2020), and thus it is imperative to accurately identify hydrogen peroxide.

Oxides of copper sensors have a long standing history of sensing hydrogen peroxide continually. The production of oxides of copper is quite easy and cheap. Crystalline oxides with requisite size and shape are used in sensor production (Ji et al. 2022). Combination of oxide NPs of copper with liquid carbon ionic plate was able to detect hydrogen peroxide up to 2.5 mM uninterrupted (Ping et al. 2010). Also, hollow spheres of copper oxide based on filter paper were able to sense hydrogen peroxide. Sensing efficiency was up to 150 μM (Cheng et al. 2021). Palladium copper NPs decorated using graphene oxide identified glucose to 30 nM minimal

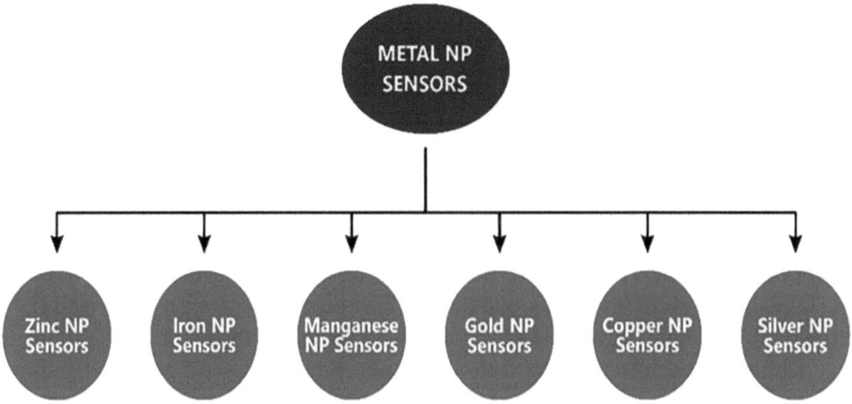

Fig. 11.3 Metallic nanoparticles biosensors

level (Dhara et al. 2015), while Zhang et al. (2015) equally sensed glucose using spheres of carbon as embellishment on copper oxide NPs, significant sensitivity was achieved, and minimal level of detection was 0.1 μM. Graphene composite copper oxide sensor prepared by Khoshhesab (2015) accurately identified ascorbic acid concurrently with caffeine and acetaminophen to the lowest level (Fig. 11.3).

11.3.2 Zinc Oxide NP Sensors

Zinc oxide is recognised as an outstanding material for sensor fabrication simply because of its stability, cheapness and most importantly eco-friendliness (Malik et al. 2023). The high rate of isoelectric point of zinc oxide gives room for quick absorption of plant proteins, enzymes and phyto-hormones for swift analysis and good ability for capturing biological molecules of focus at minute level (Choudhary et al. 2025). Compounds like uric acid, ascorbic acid and glucose have been sensed accurately by zinc oxide NP sensors, with sensing level of 2.3×10^{-7} M (Tashkhourian et al. 2014; Antonacci et al. 2018). Trisodium citrate in combination with zinc oxide was applied for glucose detection with minimal identification level of 0.02 mM. Carbon electrode NP formulation was used in sensing para nitrophenol with 0.02 μM as lower sensing limit (Bashami et al. 2015; Fang et al. 2016).

11.3.3 Iron Oxide NP Sensors

Dissimilar electrodes for sensing have been manufactured with iron oxide NPs, they are specially fabricated and tailored towards detection of organic molecules and heavy metal pollutants. Inherent sensing capabilities of oxide of iron NP sensors are premised on the morphology of sensors (Cai et al. 2019b), thus highlighting notable sensitivity of oxides of iron with exiting affinity. Fertilizers with urea hydrolyse swiftly in the environment, and thus become toxic to plantlets, owing to this, it is imperative to quantify concentration of urea. Indium-tin oxide coated glass plates combined with Fe_3O_4 NPs/chitosan were produced by Kaushik et al. (2009) for sensing residue of urea in the environment. Significant concentration of urea was detected (5–100 mg/dL) at 0.5 mg/dL minimal limit of detection. Detection of heavy metal pollutants like Zn^{2+}, Pb^{2+} and Cd^{2+} was attained by application of iron/graphene NP sensor. High range of detection was realized (1–100 μg L^{-1}) at minimal sensing levels of 0.11, 0.08, and 0.07 μg L^{-1} respectively for Zn^{2+} Cd^{2+} and Pb^{2+}. Iron oxides facilitate several redox actions with respect to the biomolecule of interest. Chitosan containers loaded with iron oxide and bromocresol green for sensing and minimal sensing level of 80 nM was obtained (Fig. 11.3).

11.3.4 Manganese Oxide NP Sensors

The distinctiveness of physicochemical qualities of oxides of manganese and interfacing capacity with biological systems has brought it into limelight in biosensor fabrication. Oxides of manganese present notable catalytic action with the capability to move electrons from molecules to surfaces of electrodes, thus creating a highly effective sensing unit (Malik et al. 2023). The oxides are hazard-free, within reach, cheap and environmentally friendly. Efficiency is high in alkaline media with a great density of energy. This has assisted in the sensing of a large number of molecules (Vukojevic et al. 2018). Manganese oxide sensor NPs are manufactured as enzyme dependent sensors with simplified designs. Heavy metals, diverse environmental pollutants, glucose and hydrogen peroxide are sensed by oxides of manganese (Han et al. 2016), bioactivity of molecules is intact owing to the adsorption capacity of manganese oxide NPs, molecules are highly immobilized, thus enhancing amplification of signal (Fig. 11.3).

11.3.5 Silver NP Sensors

Modification of electrode surfaces has been achieved with silver NPs. The compatibility of silver NPs with biological molecules preserves their potency, with combinations of atoms from sulphur, nucleotide receptors are firm silver NPs (Huang et al. 2021). The sensors have significant detection levels, very exacting and are highly stable (Fig. 11.3). Silver NPs have been employed in sensing and detection of enzymes and hydrogen peroxide. Also, NPs exhibit high conductivity of electricity which is useful in transferring electrons between electrodes and biomolecules (Wang et al. 2014).

11.3.6 Gold NP Sensors

The preparation of gold NPs biosensors entails using nitrogen or sulphur as a bonding base for the gold NPs on the electrode. The dimension for base could be two or three. The three dimensional sides hold more gold NPs. This permits high sensitivity of sensing through filling of more biomolecules. Using chitosan adsorption of gold NPs to electrode outside layer can be achieved. In electrochemical sensors, gold NPs are mostly employed for high selectivity and specificity of analytes (Hakimian and Ghourchian 2019). Also, the constancy of transfer of electrons is higher on the surface of electrode. Sensitivity and detection of gold NP sensors have been recorded to surpass that of sensors without gold NPs (Huang et al. 2021). Pesticides and heavy metal concentrations were identified in soil with gold NP, whose functionality is premised on fluorescent and colourimetric approaches (Lafleur et al. 2012). The concentration of miRNA-1886 in tomato plants subjected to drought stress was monitored by Vakilian (2019) with the application of gold NP sensors (Fig. 11.3). The specificity of the sensor was established by miRNA concentration, which was not altered by other environmental stressors except drought. Sensing of gold NPs was precise and concentration of miRNA 1886 was spotted through severe, moderate and mild levels of drought. Presence of bacteria or unhealthy plants could be sensed using gold NPs. Inhibitors of urease and urea were detected in soil, which points to the presence of released urease by a distressed plant under stress of ureolytic organism (Deng et al. 2016).

11.4 Application of Nanotechnology in Abiotic Stress Management

An array of environmental stressors is responsible for growth deficiency in crops (Fig. 11.4). Once crops are exposed to extreme weather variables, their biochemical and physiological composition changes and hormones are elicited, production of enzymes may drop, sugar may accumulate, and stomata may stop functioning in some cases (Husen 2022). This response by crops usually brings about decrease in production. Crop yield losses to abiotic stress in percentage range between 50% and 81% and this shoots up yearly (Oshunsanya et al. 2019). Recognition of stress factors is vital for immediate possible action. Diverse natural mechanisms in plants are activated to mitigate against stress. Homeostasis may be restored in cells, detoxifiers may come up and some stress hormones may likely be activated. These responses are monitored to establish that the plant is stressed.

Production and safety of food are usually affected when crops are continuously exposed to extreme weather variable factors (Singh et al. 2016; Zhu 2016). Different reports have stated that the damaging impacts of abiotic stress are minimised by sundry nanomaterials like graphene, nanoceria, carbon nanotubes and metal oxide NPs. Application of carbon nano tubes has been reported to minimise the damaging results of drought stress, and uptake of water is enhanced. Karami and Sepehri (2018) used multiwalled carbon nano tubes to boost growth of barley plants under saline stress, toxicity to arsenic in rice plants was equally minimised by nano NPs (Huang et al. 2018). Likewise, Gowayed et al. (2017) mediated salinity stress in potato plants using silicon dioxide NPs and introduction of graphene equally ameliorated saline -alkaline stress in alfalfa (Chen and Wang 2021). Furthermore, Wu et al. (2017) protected plant photosynthesis using cerium oxide NPs, while sorghum plant tolerance to drought was equally boosted by cerium oxide NPs (Djanaguiraman et al. 2018). In general, induction of tolerance in crops to abiotic stress is by raising action of antioxidant enzymes. Lead stress in carrot plants was minimised by application of manganese oxide NPs, the action of SOD and CATt was elevated by 29 and 32% individually in relation to control carrot plants. Development and growth enzyme polyamine concentration was also increased (Faiz et al. 2021).

Tomato plants under saline stress treated with copper NPs had higher levels of SOD. (Hernández-Hernández et al. 2018). Likewise, rate of photoprotection and photosynthesis in crops was increased by NPs in order for the plants to tolerate stress conditions. For example, sugarcane plants exposed to cold stress were alleviated with the introduction of graphene nanoribbons, selenium NPs and zinc oxide NPs. Photochemical efficiency became greater with foliar treatment of the sugarcane plants (Elsheery et al. 2020). Again, seeds of pearl millet treated with silver NPs before planting had increase in stomatal conductance, transpiration and photosynthesis levels (Khan et al. 2021). Plants suffering from abiotic stress had abundance of multiple protein levels from expression of genes associated with stress after NP treatments. Tolerance to drought was supported in soya beans after treatment with iron NPs, three genes related to stress were expressed in the plant leaves, while lipid and glycolysis pertaining to synthesis of protein increased on exposure of soya bean seedlings under flooding stress to aluminium oxide NPs (Mustafa et al. 2015). Similarly, iron NP treated roots of wheat undergoing drought stress presented ampleness of protein (Yasmeen et al. 2016; Linh et al. 2020). Some stress factors in crops (Fig. 11.4), monitored and alleviated with NP sensors are highlighted below.

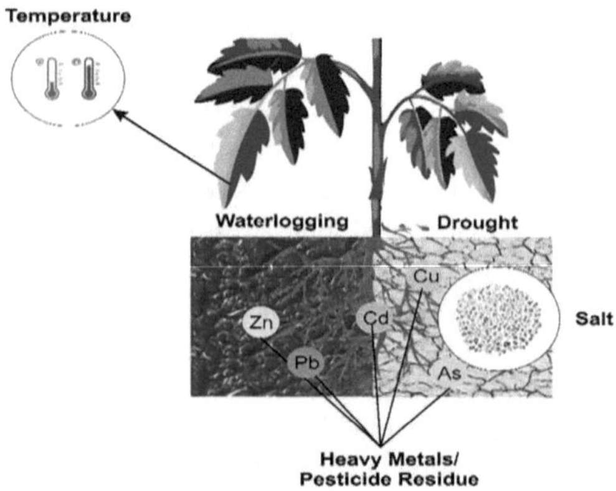

Fig. 11.4 Impact of nanobiosensors on crop plants during abiotic stress

11.4.1 Salt

The irregular, stockpiling of salt which is characteristic of semi-arid/arid regions is notably responsible for crop deaths in those parts of the globe. The damaging effect of salts is premised on the concentration and origin of the salt (Shahzad et al. 2019). Ions of sodium and chloride are concentrated in the water films of soil and becomes likely injurious, thus agitating crop water regulating activities (Munns 2005). Exposure of crops to high salt concentration results in distorted photosynthesis, minimal growth and eventual death of plant. Excessive salt concentration disturbs and alters water absorbing capacity of the plant, dehydration sets in from water deficiency and turgor pressure is lost (Jiang et al. 2020). In the same vein, essential plant functions such as movement of water and minerals from root to shoot become non-functional, hence deficiency of nutrients sets in. To minimise the outcome of high salt concentration, the crop reacts by regulating osmotic functions and some exceptional features may be developed to eliminate excess salts or atimes roots are extended to explore soil areas with less salt (Acosta-Motos et al. 2017).

Soyabean seedlings under saline stress were monitored using sensors to determine indole −3 acetic acid and salicylic acid levels, and a high level of detection was achieved with support of MWNTs (Hu et al. 2020; Mondal et al. 2022). 1.99 and 3.30 M limits of detection were attained for IAA and SA individually. Similarly, priming of seeds premised on carbon NPs notably enhanced vigour of lettuce seeds and development of seedlings by minimising elongation of primary root (Baz et al. 2020). Phosphatidic acid is a reliable molecule that may be monitored to sense crop response to stress. Phosphatidic acid has been reported to be activated by abscisic acid in reaction to salt stress. FRET controlled nanosensor with specificity for phosphatidic acid was evaluated by Li et al. (2019), it was thus concluded that phosphatidic acid accumulation is a reaction to salt stress (Fig. 11.4).

11.4.2 Drought

Exceptionally high transpiration rates from plant leaves or at times minimal supply of water to roots result in drought stress. Primarily, these occurrences are common in arid/semi-arid parts of the globe. Drought condition creates notable hindrances to crop yield (Ghaffar et al. 2020). Several complex changes in crop characteristics are induced under stress. Metabolic requirements become minimal and alteration in pathways sets in. Anthocyanin accumulates, scouting ROS enzymes pop up, osmoprotectants are released, and carotenoids are degraded (Varshney et al. 2018). Transcriptional factors like ABA, NAM, DREB and ABRE are also impacted in drought stress conditions (Nakashima and Suenaga 2017).

Sundry leaf activities are less active under drought stress. Thickness of tissue and division of cells are altered (Nelissen et al. 2018). Crop species act differently under drought stress. In some, there is improvement in holding capacity of water through leaf thickness alterations, while some minimise transpiration rate, in both cases the rate of photosynthesis becomes minimal (Rueda et al. 2017; Meng et al. 2021). Crop root zones equally suffer during drought stress, to make room for maximum water flow, stele and diameter of vessel are reduced, while there is increase in vessel number. Active roots become more in number, capacity to absorb from sheath is explored, adjustments extend to root hairs and length in order to enhance wider expansion of roots to support capacity for more water intake in drought conditions (Thangthong et al. 2019).

Water availability and plant growth are interrelated; hence water shortages will automatically lead to yield loss owing to loss of turgor pressure, minimal elongation of cells and oxidative stress (Nonami 1998). As a swift response to drought, the stomata are adjusted and transpiration rates are regulated. Thus functions of stomata could be employed to evaluate drought stress in plants. Sensor for size pore of stomata is a device that can be used to evaluate plant drought response (Koman et al. 2017). Certain optical sensors can detect hydrogen peroxide in plants as a response to drought. Sensing devices can also monitor automatic irrigation systems, activating water supply to the field during shortages (Fig. 11.4). Soil water requirement or plant water needs are estimated by nanosensors and irrigation is done when necessary while water is managed sustainably (Suchit et al. 2022).

11.4.3 Crop Health

High technology sensors that can be worn or fixed onto the plant are very efficient and accurate for monitoring. They are highly bio- compatible, cheap and very light in weight. Growth of plants, temperature, humidity, water and nutrient translocation are sensed by devices attached to plant leaf surface or stem (Nassar et al. 2018). An essential soil factor is nutrient content, crop yield and nutrient availability are interwoven. It is essential to manage and monitor soil nutrient levels with nanosensors in order to minimise waste by reducing unnecessary introduction of fertilizers (Suchit et al. 2022). Some nanosensors may be made of graphene, wearable and basically for pesticide sensing (Zhao et al. 2020) and the output of information is transmitted to a smartphone. The enormous subjection of fruits to ozone leading to fruit damage could be detected by polymer tattoos (Kim et al. 2020). Furthermore, sensing volatile compounds to determine damage to plants mechanically with the aid of graphene sensor was developed by Li et al. (2021) with high accuracy, volatile compounds were activated an hour after damage (Fig. 11.4).

11.4.4 Waterlogging

An important element for plant growth, biochemical and physiological activities is water. In waterlogged stress situations, a primary adaptation technique is aerenchyma formation. This assists root tissues submerged in water to have enough air space and absorb oxygen. For more air uptake, the epithelial cells are keratinized (Pedersen et al. 2021). In case of rice plants, more lignin is formed and the adventitious roots become bigger to enhance oxygen movement under water stress. The root modifications increase air spaces and block movement of soil toxins into root system (Yamauchi et al. 2019). Similarly, plant leaves have evolved in their reactions to water stress. There are some alterations in leaf anatomy which enhance transpiration, respiration and exchange of gas (Brodersen et al. 2017). Production of indole 3-acetic acid is recognized as an expression of water stress in maize plants. A sensor operating in amperometric mode was established to monitor water deficiency. Nanosensor utilizing optical fluorescence is applied to sense water levels in the environment (Fig. 11.4). Once the plant is under water stress, the plant's moisture content is expected to change (Chong et al. 2007).

11.4.5 Heavy Metals and Pesticide Residue

The concentration of heavy metals in the environment is of global interest and primary concern (Wawata and Fabiyi 2024). Excessive increase of metal in the environment is a fallout from poor waste management, ore extraction, emission of fuels and uncontrolled application of pesticides. Several pesticide metabolites move into water bodies, they are non-biodegradable and bioaccumulate. Environmental risk of heavy metals is very high (Fabiyi et al. 2024), oxidative stress is induced in plants, it is thus necessary to check for heavy metal pollutants in the environment (Mathivanan 2021). Arsenic, cadmium, mercury and palladium have been detected by application of nanosensors of gold NP origin. In some cases, specially made smartphones may be used in metal detection (Jin et al. 2018; Guo et al. 2020). Plants under heavy metal stress were also monitored by measuring vitronectin levels which is a signature for metal stress situations (Wang et al. 2019).

Toxicity from metals is known to destroy crop DNAs and chloroplasts, reduce enzyme activities and the crops become imbalanced to reactions. (Sperdouli 2022). However, some crop species are tolerant to metal stress and pesticides, once identified they are useful in remediation processes. An example of metal tolerant crop is seen in mustard, which was cropped together with silage maize on soil contaminated with cadmium, while peat was applied on the soil. Cadmium contaminants in soil was removed and phytoremediation was achieved (Wang et al. 2021). Likewise, nanosensors have been able to detect diverse residues of pesticides like neonicotinoids, triazines, carbamates and organophosphorus (Willner and Vikesland 2018) in the environment with the aid of fluorescent enabled sensors (Fig. 11.4).

11.4.6 Temperature

The cellular process which is mostly affected by stress from high temperatures is photosynthesis. Imbalances in crop metabolic processes lead to great reduction in all vital activities of crop. Notable dysfunction is experienced by the crop and the photosynthetic process is altered, which results in yield drop (Fahad et al. 2017). Severe plant mortalities have been recorded in extremely high temperatures. Once the photosynthetic process is slow, respiration becomes minimal and the underside of leaves receives accumulation of carbon dioxide in high concentration (Kordrostami et al. 2021). The crops are severely affected owing to their stationary nature in the sun and invariably, the crop physiology and yield are drastically affected (Lobell and Field 2007). The excessive accumulation of ROS leads to oxidative stress, which is the major consequence of fluctuations in temperature (Hasanuzzaman et al. 2013b). If there is variability in weather and the temperature turns to extreme cold, crop membranes are modified and metabolism is once again disturbed, secondary injury sets in for the crops from toxic substances that are produced by the crop (Cai et al. 2019a).

Drop in temperature minimises energy conveyance to PSII core (Su et al. 2015). Temperature stress hormones, including ethylene, abscisic acid, jasmonate, hydrogen peroxide, salicylic acid and indole- 3- acetic acid are produced in high quantities in the affected crops. Nanosensors have been developed to identify stressor elicited hormones in order to trail reflex of crops to temperature stress. FLIP-H_2O_2, a nanosensor which is very particular for hydrogen peroxide, was evaluated by Ali et al. (2020) they found it very effective for H_2O_2 identification, without recognition for other phytohormones and oxidants. Crops are stunted with increase in H_2O_2, and it is equally an important regulator in stress situations (Amjad and Serajuddin 2022). To minimise stress from heat, optimization of irrigation is necessary coupled with mulching. Circulation of air could be enhanced by giving shade to crops especially plantlets (Fig. 11.4). Heat stress impact could also be minimised using tolerant varieties and regulating and scheduling farm work time (Hasanuzzaman et al. 2013a; Nievola et al. 2017).

11.4.7 Soil

Soil is a crucial factor in the section of agriculture. It carries the charge of yield quality and productivity. There is a direct relationship between yield of crops and soil pH. It is the ultimate determinant of crop output. Soil moisture and conductivity are decided by soil. Dissimilar levels of pH are demanded by disparate crops for growth. Adequate pH measurement gives room for knowledge regarding the alkaline and base composition of soil matrix. Probe heat pulse is employed in soil measurements; it may be dual or single. The accuracy of pH is significant because it regulates the soil ecosystem and other crucial activities of the planted crops (Bristow et al. 2001; Palaparthy et al. 2013). Lately, some polymer substances are regarded as necessary and efficient for pH measurement. Nanosensors convert alterations in chemical and physical constituents of the polymer into stress. Silicon dual probe heat pulse is employed in soil moisture constituent assessment for effectivity (Valente et al. 2004). The technique relies on use of heat pulse for a specified time. Temperature rise is evaluated by keeping the probe close to source of heat. Temperature rise is employed for evaluating heat capacity and this again is used to assess constituent level of soil water. Liu et al. (2008) applied polymer premised conductive micro sensor for soil moisture evaluation (Fig. 11.4).

The polymer conducting micro sensor operates by alterations and modifications to electrical resistance sheet of the films in the polymer. The alterations translate to soil constituent moisture. Alterations in resistant polymer sheets plotted in comparison with relative humidity may be used indirectly to evaluate soil constituent moisture. Daoud et al. (2005) measured soil constituent moisture using polyamide conducting fibres fixed within PDOT-PSS aqueous dispersion. Sensitivity was established by the polymer to variabilities in temperature and humidity in soil ecosystem. The equipment is manufactured in combination with Nafion over-layer and works on the technique of alteration in an exact variable with respect to alteration in external factors. For example, when the equipment is employed for temperature sensing evaluation, alteration in conductivity of Nafion is only owing to environmental temperature (Nilsson et al. 2002). A further notable sensor device is the water flux density equipment. It is highly significant in accessing transport capabilities of soil with accuracy in times of drastic weather variabilities (Kamai et al. 2010). Exact evaluation of arsenic in soil was assessed by Lew et al. (2021) using crop as link with soil. SWCNTs were integrated into crops for observing crop change as arsenic is taken up from soil (Fig. 11.4).

11.5 Challenges in the Use of Nanosensors in Abiotic Stress Detection

A primary limitation to precision farming is data obtainment for characteristics of soil and quality of crop through a swift and cost effective method. The soil matrix is complex, so an automated sampling technique is required for soil treatment and sampling. Providing labour saving techniques for extraction is demanding and several steps are necessary owing to the low soil homogeneity. Tool for soil extraction coupled with a biosensor to give quick analysis is needed. Furthermore, sensing in plant, soil and water is to be achieved simultaneously in soil matrices, however, this is not so.

Consideration for efficient and cheap sensors is crucial for commercialization and wide use, prices of devices are to be reasonable. Energy and device power usage could be an obstacle to commercialization. It is crucial to have a supply of power standby on the farm field constantly. This is a primary drawback for automated agriculture in rural communities of developing countries. Variations in weather, relative to power usage of devices are important factors to be considered because utilization of power is usually high in bad weather. Variabilities in weather should be considered with battery system development. Battery performance should be independent of external variables and power use should be minimal. Device storage systems are deficient in space. Devices should come with large storage spaces to accommodate the massive data from agricultural fields. Management of such data should equally be of importance. Signals premised on fluorescence may be disturbed owing to minimal steadiness from phototoxicity.

Particles of nanosensors may stockpile in soil and damage balance of the soil microbial community. Contaminants may be initiated in the process of preparing nanoparticles, which may affect the functionality of the NPs, leading to unsuccessful or incompatible application with the test crop. Analyte detection at regular intervals is difficult optically, owing to plant tissues and pigments. Chemical information is often unexplainable by reason of heterogeneities and impurities, which alter signature of spectra. Hence a more complex technique is suggested for infrared and Raman spectroscopy because of their minimal ratios of noise to signal. Several factors affect nanosensor devices built for airborne sensing. Crops in high density fields and individual crops cannot be measured optimally. Shadows or overexposure to sunlight also impact monitoring, coupled with obstructing weather situations. Overall, application of nanosensors in agriculture is basically premised on congeniality, affinity and constituents of the nanomaterials in question with the species of crop.

11.6 Conclusions

There is a general downturn in agriculture by virtue of weather variability in recent years. This is again additionally intensified by human population increase, which has impacted greatly on food supplies. In order to ensure sustainable food production and eliminate food insecurity, principally in the developing nations, agriculture must be technology driven. Modern agricultural practice can be up scaled to high tech precision farming in order to accommodate the applicability of nanosensors to facilitate quick detection of abiotic stressors in crops. This would facilitate good decisions and minimise losses incurred from bad planning.

References

Acosta-Motos JR et al (2017) Plant responses to salt stress: adaptive mechanisms. Agronomy 7(1):18

Ahmad F, Singh A, Kamal A (2019) In: Iqbal M, Khan R, Reddy PS, Ferrante A, Khan NA (eds) Plant signalling molecules. Woodhead Publishing, pp 355–369, Woodhead Publishing, Elsevier, Sawston, Cambridge

Ali HM, Ahmad M, Salem MZ, Ahmad A (2020) Construction of a nanosensor for non-invasive imaging of hydrogen peroxide levels in living cells. Biology (Basel) 9(12):430

Amjad S, Serajuddin M (2022) Role of nanosensors and bio-nanosensors in crop abiotic stress. In: Augmenting crop productivity in stress environment. Springer Nature Singapore, Singapore, pp 1–12

Antonacci A, Arduini F, Moscone D, Palleschi G, Scognamiglio V (2018) Nanostructured (bio) sensors for smart agriculture. Trends Anal Chem 98:95–103. https://doi.org/10.1016/j.trac.2017.10.022

Bashami RM, Hameed A, Aslam M (2015) The suitability of ZnO film-coated glassy carbon electrode for the sensitive detection of 4-nitrophenol in aqueous medium. Anal Methods 7(2015):1794–1801

Baz H, Creech M, Chen J, Gong H, Bradford K, Huo H (2020) Water-soluble carbon nanoparticles improve seed germination and post-germination growth of lettuce under salinity stress. Plan Theory 10:1192

Bhatia D, Paul S, Acharjee T, Ramachairy SS (2024) Biosensors and their widespread impact on human health. Sens Int 5:100257

Boonyaves K, Ang MCY, Park M, Cui J, Khong DT, Singh GP, Koman VB, Gong X, Porter TK, Choi SW, Chung K, Chua NH, Urano D, Strano MS (2023) Near-infrared fluorescent carbon nanotube sensors for the plant hormone family gibberellins. Nano Lett 23(3):916924. https://doi.org/10.1021/acs.nanolett.2c04128. http://pubs.acs.org/journal/nalefd

Bristow KL, Kluitenberg GJ, Goding CJ, Fitzgerald TS (2001) A small multi-needle probe for measuring soil thermal properties, water content and electrical conductivity. Comput Electron Agric 31:265–280

Brodersen KE et al (2017) Sediment resuspension and deposition on seagrass leaves impedes internal plant aeration and promotes phytotoxic H2S intrusion. Front Plant Sci 8:657

Cai X, Magwanga RO, Xu Y (2019a) Comparative transcriptome, physiological and biochemical analyses reveal response mechanism mediated by CBF4 and ICE2 in enhancing cold stress tolerance in *Gossypium thurberi*. AoB Plants 11:plz045

Cai Z, Ye Y, Wan X, Liu J, Yang S, Xia Y, Li G, He Q (2019b) Morphology–dependent electrochemical sensing properties of iron oxide–graphene oxide nanohybrids for dopamine and uric acid. Nano 9(6):835

Chai Y, Chen C, Luo X, Zhan S, Kim J, Luo J, Wang X, Hu Z, Ying Y, Liu X (2021) Cohabiting plant-wearable sensor in situ monitors water transport in plant. Adv Sci 8:2003642

Chaudhuri B, Hörmann F, Lalonde S et al (2008) Protonophore-and pH-insensitive glucose and sucrose accumulation detected by FRET nanosensors in Arabidopsis root tips. Plant J 56:948–962

Chaudhuri B, Hörmann F, Frommer WB (2011) Dynamic imaging of glucose flux impedance using FRET sensors in wild-type Arabidopsis plants. J Exp Bot 62:2411–2417

Chen Z, Wang Q (2021) Graphene ameliorates saline-alkaline stress-induced damage and improves growth and tolerance in alfalfa (*Medicago sativa* L.). Plant Physiol Biochem 163:128–138

Cheng D, Qin J, Feng YJ, Wei J (2021) Synthesis of mesoporous CuO hollow sphere nanozyme for paper-based hydrogen peroxide sensor. Biosensors 11(8):258. https://doi.org/10.3390/bios11080258

Chong JPC, Liew OW, Li BQ, Asundi AK (2007) Optical fluorescence biosensor for plant water stress detection. In: Saratov fall meeting 2006: optical technologies in biophysics and medicine, vol III. International Society for Optics and Photonics, pp 65350–65358. https://doi.org/10.1117/12.740959

Choudhary R, Rajput VD, Ghodake G, Ahmad F, Meena M, Rehman RU, Prasad R, Sharma RK, Singh R, Seth CS (2024) Comprehensive journey from past to present to future about seed priming with hydrogen peroxide and hydrogen sulfide concerning drought, temperature, UV and ozone stresses- a review. Plant Soil. https://doi.org/10.1007/s11104-024-06499-9

Choudhary R, Ahmad F, Kaya C, Upadhyay SK, Muneer S, Kumar V, Meena M, Liu H, Upadhyay H, Seth CS (2025) Decrypting proteomics, transcriptomics, genomics, and integrated omics for

augmenting the abiotic, biotic, and climate change stress resilience in plants. J Plant Physiol. https://doi.org/10.1016/j.jplph.2025.154430
Choudhury FK, Rivero RM, Blumwald E, Mittler R (2017) Reactive oxygen species, abiotic stress and stress combination. Plant J 90:856–867
Daoud WA, Xin JH, Szeto YS (2005) Polyethylenedioxythiophene coatings for humidity, temperature and strain sensing polyamide fibers. Sensors Actuators B Chem 109:329–333
Deng HH, Hong GL, Lin FL, Liu AL, Xia XH, Chen W (2016) Colorimetric detection of urea, urease, and urease inhibitor based on the peroxidase-like activity of gold nanoparticles. Anal Chim Acta 915:74e80
Deuschle K et al (2006) Rapid metabolism of glucose detected with FRET glucose nanosensors in epidermal cells and intact roots of Arabidopsis RNA-silencing mutants. Plant Cell 18:2314–2325
Dhara K, Thiagarajan R, Nair BG, Thekkedath GSB (2015) Highly sensitive and wide-range nonenzymatic disposable glucose sensor based on a screen-printed carbon electrode modified with reduced graphene oxide and Pd-CuO nanoparticles. Microchim Acta 182(13):2183–2192
Djanaguiraman M, Nair R, Giraldo JP, Prasad PVV (2018) Cerium oxide nanoparticles decrease drought-induced oxidative damage in sorghum leading to higher photosynthesis and grain yield. ACS Omega 3(10):14406–14416
El-Ramady HR, Verma KK, Rajput VD, Minkina T, Elbehiry F, Elbasiouny HY, Elsakhawy T, Omara AED, Amer MM (2022) Sources of silicon and nano-silicon in soil and plants. In: Silicon and nano-silicon in environmental stress management and crop quality improvement: Progress and prospects. Academic, Elsevier. https://doi.org/10.1016/B978-0-323-91225-9.00003-0
Elsheery NI, Sunoj V, Wen Y, Zhu J, Muralidharan G, Cao K (2020) Foliar application of nanoparticles mitigates the chilling effect on photosynthesis and photoprotection in sugarcane. Plant Physiol Biochem 149:50–60
Exposito-Rodriguez M, Laissue PP, Yvon-Durocher G, Smirnoff N, Mullineaux PM (2017) Photosynthesis-dependent H2O2 transfer from chloroplasts to nuclei provides a high-light signalling mechanism. Nat Commun 8:49
Fabiyi OA, Ogundele AV (2025) Biosensors for monitoring climatic stressors in agricultural crops. In: Uddin I (eds) Smart nanosensors. Springer, Singapore, pp 229–245
Fabiyi OA, Wawata IG, Akinsipo OB, aogundele AV (2025) Smart nanosensors: A new approach in crop protection and pest management. In Uddin I (eds) Smart nanosensors. Springer, Singapore, pp 247–266
Fabiyi OA, Davis OA, Akanmu AO et al (2024) Biomaterials for sustainable bioremediation. In: Uddin I (eds) Sustainable nanomaterials. Sustainable materials and technology. Springer, Singapore, pp 103–123
Fahad S, Bajwa AA, Nazir U, Anjum SA, Farooq A, Zohaib A, Huang J (2017) Crop production under drought and heat stress: plant responses and management options. Front Plant Sci 8:1147
Faiz S, Yasin NA, Khan WU, Shah AA, Akram W, Ahmad A, Ali A, Naveed NH, Riaz L (2021) Role of magnesium oxide nanoparticles in the mitigation of lead-induced stress in Daucus carota: modulation in polyamines and antioxidant enzymes. Int J Phytoremediation 16:1–9
Fang L, Liu B, Liu L (2016) Direct electrochemistry of glucose oxidase immobilized on Au nanoparticles-functionalized 3D hierarchically ZnO nanostructures and its application to bio electrochemical glucose sensor. Sens Actuators B Chem 222(2016):1096–1102
Fichman Y, Mittler R (2020) Rapid systemic signalling during abiotic and biotic stresses: is the ROS wave master of all trades? Plant J 102:887–896
Food and Agriculture Organization of the United States (FAO) (2020) Statistical yearbook 2020. Food & agricultural organization. https://www.fao.org/3/cb1329en/online/cb1329en.html
Food and Agriculture Organization of the United States (FAO) (2021) Statistical yearbook 2021. Food & agricultural organization. https://www.fao.org/3/cb4477en/online/cb4477en.html#
Ghaffar N, Farrukh MA, Naz S (2020) Applications of nanobiosensors in agriculture. In: Javad S (ed) Nanoagronomy. Springer, Cham. https://doi.org/10.1007/978-3-030-41275-3_10

Gowayed S, Al-Zahrani H, Metwali E (2017) Improving the salinity tolerance in potato (Solanum tuberosum) by exogenous application of silicon dioxide nanoparticles. Int J Agric Biol 19:183–192

Guo Z, Kang Y, Liang S, Zhang J (2020) Detection of Hg (II) in adsorption experiment by a lateral flow biosensor based on streptavidin-biotinylated DNA probes modified gold nanoparticles and smartphone reader. Environ Pollut 266:115389

Hakimian F, Ghourchian H (2019) Simple and rapid method for synthesis of porous gold nanoparticles and its application in improving DNA loading capacity. Mater Sci Eng C 103:109795

Han L, Shao C, Liang B, Liu A (2016) Genetically engineered phage-templated MnO2 nanowires: synthesis and their application in electrochemical glucose biosensor operated at neutral PH condition. ACS Appl Mater Interfaces 8(22):13768–13776

Hasanuzzaman M et al (2013a) Physiological, biochemical, and molecular mechanisms of heat stress tolerance in plants. Int J Mol Sci 14(5):9643–9684

Hasanuzzaman M, Nahar K, Fujita M (2013b) Extreme temperatures, oxidative stress and antioxidant defense in plants. In: Abiotic stress-plant responses and applications in agriculture. InTech, Rijeka, pp 169–205

Hernández-Hernández H, Juárez-Maldonado A, Benavides-Mendoza A, Ortega-Ortiz H, Cadenas-Pliego G, Sánchez-Aspeytia D, González-Morales S (2018) Chitosan-PVA and copper nanoparticles improve growth and overexpress the SOD and JA genes in tomato plants under salt stress. Agronomy 8:175. https://doi.org/10.1016/j.bios.2021.113105

Hu Y, Wang X, Wang C, Hou P, Dong H, Luo B, Li A (2020) A multifunctional ratiometric electrochemical sensor for combined determination of indole-3-acetic acid and salicylic acid. RSC Adv 10:3115–3121

Huang Q, Liu Q, Lin L, Li F-J, Han Y, Song Z-G (2018) Reduction of arsenic toxicity in two rice cultivar seedlings by different nanoparticles. Ecotoxicol Environ Saf 159:261–271

Huang X, Zhu Y, Kianfar E (2021) Nano biosensors: properties, applications and electrochemical techniques. J Mater Res Technol 12:1649–1672

Husen A (2022) Environmental Pollution and Medicinal Plants. Taylor & Francis Group, LLC, Broken Sound Pkwy, NW. p. 6000

Ibrahim H, Moru S, Schnable P, Dong L (2022) Wearable plant sensor for in situ monitoring of volatile organic compound emissions from crops. ACS Sensors 7(8):22932302. https://doi.org/10.1021/acssensors.2c00834

Jagadesh M, Dash M, Kumari A, Singh SK, Verma KK, Kumar P, Bhatt R, Sharma SK (2024) Revealing the hidden world of soil microbes: metagenomic insights into plant, bacteria, and fungi interactions for sustainable agriculture and ecosystem restoration. Microbiol Res 285:127764. https://doi.org/10.1016/j.micres.2024.127764

Ji X, Yang Y, Wang A, Zhao Q (2022) One-step hydrothermal synthesis of CuO micro-crystals for non-enzymatic glucose sensors. Sci Adv Mater 14(4):638–643

Jiang T et al (2020) Simulation of plant height of winter wheat under soil water stress using modified growth functions. Agric Water Manag 232:106066

Jin B, Yang Y, He R, Park YI, Lee A, Bai D, Li F, Lu TJ, Xu F, Lin M (2018) Lateral flow aptamer assay integrated smartphone-based portable device for simultaneous detection of multiple targets using up conversion nanoparticles. Sens Actuators B Chem 276:48–56

Kamai T, Tuli A, Kluitenberg GJ, Hopmans JW (2008) Soil water flux density measurements near 1 cm d-1 using an improved heat pulse probe design. Water Resour. Res 46:W07901. https://doi.org/10.1029/2010WR009423

Karami A, Sepehri A (2018) Beneficial role of MWCNTs and SNP on growth, physiological and photosynthesis performance of barley under NaCl stress. J Soil Sci Plant Nutr 18:752–771

Karunakaran C, Rajkumar R, Bhargava K (2015) Introduction to biosensors. In: Biosensors and bioelectronics. Elsevier, pp 1–68

Kaushik A, Solanki PR, Ansari AA, Sumana G, Ahmad S, Malhotra BD (2009) Iron oxide-chitosan nanobiocomposite for urea sensor. Sens Actuator B Chem 138(2):572–580

Keinath NF et al (2015) Live cell imaging with R-GECO1 sheds light on flg22- and chitin-induced transient [Ca2+]cyt patterns in Arabidopsis. Mol Plant 8:1188–1200

Khan I, Awan SA, Raza MA, Rizwan M, Tariq R, Ali S, Huang L (2021) Silver nanoparticles improved the plant growth and reduced the sodium and chlorine accumulation in pearl millet: a life cycle study. Environ Sci Pollut Res 28:13712–13724

Khoshhesab ZM (2015) Simultaneous electrochemical determination of acetaminophen, caffeine and ascorbic acid using a new electrochemical sensor based on CuO-graphene nanocomposite. RSC Adv 5(115):95140–95148

Kim JJ, Allison LK, Andrew TL (2019) Vapor-printed polymer electrodes for long-term, on-demand health monitoring. Sci Adv 5:eaaw0463

Kim JJ, Fan R, Allison LK, Andrew TL (2020) On-site identification of ozone damage in fruiting plants using vapor-deposited conducting polymer tattoos. Sci Adv 6:eabc3296

Koman VB, Lew TTS, Wong MH, Kwak SY, Giraldob JP, Strano MS (2017) Persistent drought monitoring using a microfluidic-printed electro-mechanical sensor of stomata in planta. Lab Chip 17(23):4015–4024

Kordrostami M, Mafakheri M, Al-Khayri JM (2021) Contributions of nano biosensors in managing environmental plant stress under climatic changing era. In: Al-Khayri JM, Ansari MI, Singh AK (eds) Nanobiotechnology. https://doi.org/10.1007/978-3-030-73606-4_6

Kumar D, Mariyam S, Gupta KJ, Thiruvengadam M, Ghodake GS, Xing B, Seth CS (2024a) Comparative investigation on chemical and green synthesized titanium dioxide nanoparticles against chromium (VI) stress eliciting differential physiological, biochemical, and cellular attributes in Helianthus annuus L. Sci Total Environ 930:172413. https://doi.org/10.1016/j.scitotenv.2024.172413

Kumar D, Singh R, Upadhyay SK, Verma KK, Tripathi RM, Liu H, Dhankher OP, Tripathi RD, Sahi SV, Seth CS (2024b) Review on interactions between nanomaterials and phytohormones: novel perspectives and opportunities for mitigating environmental challenges. Plant Sci 340:111964. https://doi.org/10.1016/j.plantsci.2023.111964

Kumar D, Biswas JK, Mulla SI, Singh R, Shukla R, Ahanger MA, Shekhawat GS, Verma KK, Siddiqui MW, Seth CS (2024c) Micro and nanoplastics pollution: sources, distribution, uptake in plants, toxicological effects, and innovative remediation strategies for environmental sustainability. Plant Physiol Biochem 213:108795. https://doi.org/10.1016/j.plaphy.2024.108795

Lafleur JP, Senkbeil S, Jensen TG, Kutter JP (2012) Gold nanoparticle-based optical microfluidic sensors for analysis of environmental pollutants. Lab Chip 12(22):46514656. https://doi.org/10.1039/c2lc40543a

Lamers J, van der Meer T, Testerink C (2020) How plants sense and respond to stressful environments. Plant Physiol 182:1624–1635

Lee K et al (2014) In-situ synthesis of carbon nanotube-graphite electronic devices and their integrations onto surfaces of live plants and insects. Nano Lett 14:2647–2654

Lew TTS, Park M, Cui J, Strano MS (2021) Plant Nanobionic Sensors for Arsenic Detection. Advanced Materials 33(1):2005683. https://doi.org/10.1002/adma.202005683

Li L, Zhang Q, Huang D (2014) A review of imaging techniques for plant phenotyping. Sensors 14:20078–20111

Li W, Song T, Wallrad L et al (2019) Tissue-specific accumulation of pH-sensing phosphatidic acid determines plant stress tolerance. Nat Plants 5:1012–1021

Li Z, Zhou J, Dong T, Xu Y, Shang Y (2021) Application of electrochemical methods for the detection of abiotic stress biomarkers in plants. Biosens Bioelectron 185:113105

Linh TM, Mai NC, Hoe PT, Lien LQ, Ban NK, Hien LTT, Chau NH, Van NT (2020) Metal-based nanoparticles enhance drought tolerance in soybean. J Nanomater 2020:1–13

Liu J, Agarwal M, Varahramyan K, Berney ES, Hodo WD (2008) Polymer-based microsensor for soil moisture measurement. Sensors Actuators B Chem 129:599–604

Lobell DB, Field CB (2007) Global scale climate crop yield relationships and the impacts of recent warming. Environ Res Lett 2:5

Malik S, Singh J, Goyat R, Saharan Y, Chaudhry V, Umar A, Ibrahim AA, Akbar S, Ameen S, Baskoutas S (2023) Nanomaterials-based biosensor and their applications: a review. Heliyon 9:e19929. https://doi.org/10.1016/j.heliyon.2023.e19929

Mariyam S, Upadhyay SK, Chakraborty K, Verma KK, Duhan JS, Muneer S, Meena M, Sharma RK, Ghodake G, Seth CS (2024) Nanotechnology, a frontier in agricultural science, a novel approach in abiotic stress management and convergence with new age medicine-a review. Sci Total Environ 912:169097. https://doi.org/10.1016/j.scitotenv.2023.169097

Mathivanan S (2021) Perspectives of nano-materials and nanobiosensors in food safety and agriculture. In: Novel nanomaterials. UK. Intechopen, 197:95345. https://doi.org/10.5772/intechopen

Meng D et al (2021) The pigeon pea CcCIPK14-CcCBL1 pair positively modulates drought tolerance by enhancing flavonoid biosynthesis. Plant J 106(5):1278–1297

Mittler R, Zandalinas SI, Fichman Y, Van Breusegem F (2022) Reactive oxygen species signalling in plant stress responses. Nat Rev Mol Cell Biol 23:663–679

Mondal R, Dam P, Chakraborty J, Paret ML, Katı A, Altuntas S, Husen A (2022) Potential of nanobiosensor in sustainable agriculture: the state-of-art. Heliyon 8(12):e12207

Munns R (2005) Genes and salt tolerance: bringing them together. New Phytol 167:645–663

Mustafa G, Sakata K, Komatsu S (2015) Proteomic analysis of flooded soybean root exposed to aluminium oxide nanoparticles. J Proteome 128:280–297

Nakashima K, Suenaga K (2017) Toward the genetic improvement of drought tolerance in crops. Jpn Agric Res Q 51(1):1–10

Naresh V, Lee N (2021) A review on biosensors and recent development of nanostructured materials-enabled biosensors. Sensors 21(4):1109. https://doi.org/10.3390/s21041109

Nassar JM, Khan SM, Villalva DR, Nour MM, Almuslem AS, Hussain MM (2018) Compliant plant wearables for localized microclimate and plant growth monitoring. NPJ Flex Electron 2(1) . Available from https://doi.org/10.1038/s41528-018-0039-8

Nelissen H et al (2018) The reduction in maize leaf growth under mild drought affects the transition between cell division and cell expansion and cannot be restored by elevated gibberellic acid levels. Plant Biotechnol J 16(2):615–627

Neme K, Nafady A, Uddin S, Tola YB (2021) Application of nanotechnology in agriculture, postharvest loss reduction and food processing: food security implication and challenges. Heliyon 7:e08539

Nievola CC et al (2017) Rapid responses of plants to temperature changes. Temperature 4(4):371–405

Nilsson D, Kugler T, Svensson P-O, Berggren M (2002) An all-organic sensor-- transistor based on a novel electrochemical transducer concept printed electrochemical sensors on paper. Sensors Actuators B Chem 86:193–197

Nonami H (1998) Plant water relations and control of cell elongation at low water potentials. J Plant Res 111:373–382. https://doi.org/10.1007/BF02507801

Oren S, Ceylan H, Schnable PS, Dong L (2017) High-resolution patterning and transferring of graphene-based nanomaterials onto tape toward roll-to-roll production of tape-based wearable sensors. Adv Mater Technol 2:1700223

Oshunsanya SO, Nwosu NJ, Li Y (2019) Abiotic stress in agricultural crops under climatic conditions. In: Sustainable agriculture, forest and environmental management. Springer, pp 71–100

Palaparthy VS, Baghini MS, Singh DN (2013) Review of polymer-based sensors for agriculture-related applications. Emerg Mater Res 2:166–118

Pedersen O et al (2021) Lateral roots, in addition to adventitious roots, form a barrier to radial oxygen loss in *Zea nicaraguensis* and a chromosome segment introgression line in maize. New Phytol 229(1):94–105

Ping J, Ru S, Fan K, Wu, Ying Y (2010) Copper oxide nanoparticles and ionic liquid modified carbon electrode for the non-enzymatic electrochemical sensing of hydrogen peroxide. Microchim Acta 171(1):117–123

Rueda M, Godoy O, Hawkins BA (2017) Spatial and evolutionary parallelism between shade and drought tolerance explains the distributions of conifers in theconterminous United States. Glob. Ecol. Biogeogr. 26(1):31–42
Salouti M, Khadivi Derakhshan F (2020) Biosensors and nanobiosensors in environmental applications. In: Ghorbanpour M, Bhargava P, Varma A, Choudhary DK (eds) Biogenic nano-particles and their use in agroecosystems. Springer, Singapore, pp 515–591
Saxena I, Srikanth S, Chen Z (2016) Cross talk between H2O2 and interacting signal molecules under plant stress response. Front Plant Sci 7:570
Shahzad H, Ullah S, Iqbal M, Bilal HM, Shah GM, Ahmad S, Ahmad I (2019) Salinity types and levelbased effects on the growth, physiology and nutrient contents of maize (Zea mays). IJA 14(4):199–207
Shawona ZBZ, Hoqueb ME, Shiplu Roy Chowdhury SR (2020) Nanosensors and nanobiosensors: agricultural and food technology aspects. In: Nanofabrication for smart Nanosensor applications, pp 135–167. https://doi.org/10.1016/B978-0-12-820702-4.00006-4
Singh S, Parihar P, Singh R, Singh VP, Prasad SM (2016) Heavy metal tolerance in plants: role of transcriptomics, proteomics, metabolomics, and ionomics. Front Plant Sci 6:1143
Sperdouli I (2022) Heavy Metal Toxicity Effects on Plants. Toxics, 10(12):715. https://doi.org/10.3390/toxics10120715
Su L, Dai Z, Li S (2015) A novel system for evaluating drought cold tolerance of grapevines using chlorophyll fluorescence. BMC Plant Biol 15:82
Subraya KK, Diggs A, Porterfield DM (2013) Amperometric biosensor approaches for quantification of indole 3-acetic acid in plant stress responses. Commun Soil Sci Plant Anal 44:1749–1763
Suchit AJ, Amit C, Pramod WR, Shanthy P, Nguyen TA, Rajendran S (2022) Nanosensors for plant health monitoring. In: Nanosensors for smart agriculture. https://doi.org/10.1016/B978-0-12-824554-5.00012-4
Tashkhourian J, Hemmateenejad B, Beigizadeh H et al (2014) ZnO nanoparticles and multiwalled carbon nanotubes modified carbon paste electrode for determination of naproxen using electrochemical techniques. J Electroanal Chem 103(1):714–715
Thangthong N et al (2019) Changes in root anatomy of peanut (*Arachis hypogaea* L.) under different durations of early season drought. Agronomy 9(5):215
Toyota M et al (2018) Glutamate triggers long-distance, calcium-based plant defence signalling. Science 361:1112–1115
Vakilian KA (2019) Gold nanoparticles-based biosensor can detect drought stress in tomato by ultrasensitive and specific determination of miRNAs. Plant Physiol Biochem 145:195–204
Valente A, Morais R, Couto C, Correia JH (2004) Modelling, simulation and testing of a silicon soil moisture sensor based on the dual-probe heat-pulse method. Sens Actuators A 115:434–439
Varshney RK, Tuberosa R, Tardieu F (2018) Progress in understanding drought tolerance: from alleles to cropping systems. J Exp Bot 69(13):3175–3179
Verma KK, Song X, Verma CL, Huang H, Singh M, Xu L, Wu J, Khan ZH, Zhang B, Li Y (2023) Mathematical modeling of climate and fluoride effects on sugarcane photosynthesis with silicon nanoparticles. Plant Physiol Biochem 204:108089. https://doi.org/10.1016/j.plaphy.2023.108089
Verma KK, Song XP, Kumari A, Jagadesh M, Singh SK, Bhatt R, Singh M, Seth CS, Li YR (2024a) Climate change adaptation: challenges for agricultural sustainability. Plant Cell Environ 1–12. https://doi.org/10.1111/pce.15078
Verma KK, Joshi A, Song X-P, Singh S, Kumari A, Arora J, Singh SK, Solanki MK, Seth CS, Li Y-R (2024b) Synergistic interactions of nanoparticles and plant growth promoting rhizobacteria enhancing soil-plant systems: a multigenerational perspective. Front Plant Sci 15:1376214. https://doi.org/10.3389/fpls.2024.1376214
Vero V, Kukreja E, Thripahi RV, Pudake N (2024) Recent advances in abiotic stress management with the help of plant wearable and *in planta* nanosensors. In: Nanotechnology for abiotic stress tolerance and management in crop plants, pp 127–139. https://doi.org/10.1016/B978-0-443-18500-7.00008-9

Vukojevic V, Djurdjic S, Ognjanovic M, Fabian M, Samphao A, Kalcher K, Stankovic DM (2018) Enzymatic glucosébiosensor based on manganese dioxide nanoparticles decorated on graphene nanoribbons. J Electroanal Chem 823:610–616
Walia A, Waadt R, Jones AM (2018) Genetically encoded biosensors in plants: pathways to discovery. Annu Rev Plant Biol 69:497–524
Wang G, He X, Wang L, Gu A, Huang Y, Fang B et al (2014) Nonenzymatic electrochemical sensing of glucose. Microchim Acta 180:161e86
Wang D, Zhu J, Zhang Z, Zhang Q, Zhang W, Yu L, Jiang J, Chen X, Wang X, Li P (2019) Simultaneous lateral flow immunoassay for multi-class chemical contaminants in maize and peanut with one-stop sample preparation. Toxins (Basel) 11:56
Wang S et al (2021) Co-cropping Indian mustard and silage maize for phytoremediation of a cadmium-contaminated acid paddy soil amended with peat. Toxics 9(5):91
Wawata IG, Fabiyi OA (2024) Sustainable application of nanomaterials in the removal of heavy metals from water. In: Uddin I (eds) Sustainable nanomaterials. Sustainable materials and technology. Springer, Singapore, pp 21–44
Willner MR, Vikesland PJ (2018) Nanomaterial enabled sensors for environmental contaminants. J Nanobiotechnology 16(1):1–16. https://doi.org/10.1186/s12951-018-0419-1
Wu H, Tito N, Giraldo JP (2017) Anionic cerium oxide nanoparticles protect plant photosynthesis from abiotic stress by scavenging reactive oxygen species. ACS Nano 11:11283–11297
Yamauchi T et al (2019) Root cortex provides a venue for gas-space formation and is essential for plant adaptation to waterlogging. Front Plant Sci 10:259
Yasmeen F, Raja NI, Razzaq A, Komatsu S (2016) Gel-free/label-free proteomic analysis of wheat shoot in stress tolerant varieties under iron nanoparticles exposure. Biochim Biophys Acta 1864:1586–1598
Zhang J, Ma J, Zhang S, Wang W, Chen Z (2015) A highly sensitive non-enzymatic glucose sensor based on CuO nanoparticles decorated carbon spheres. Sens Actuator B Chem 211:385–391
Zhang H, Zhao Y, Zhu J-K (2020) Thriving under stress: how plants balance growth and the stress response. Dev Cell 55:529–543
Zhao F, He J, Li X, Bai Y, Ying Y, Ping J (2020) Smart plant-wearable biosensor for in-situ pesticide analysis. Biosens Bioelectron 170:112636
Zhu J-K (2016) Abiotic stress signalling and responses in plants. Cell 167:313–324
Zhu Q, Wang L, Dong Q et al (2017) FRET-based glucose imaging identifies glucose signalling in response to biotic and abiotic stresses in rice roots. J Plant Physiol 215:65–72

Chapter 12
Nanotechnology-Based Mutagenesis for Crop Improvement Against Abiotic Stress Tolerance

Oluwatoyin Adenike Fabiyi, David Adedayo Animasaun, and Peter Adeolu Adedibu

12.1 Introduction

Climate change and land degradation have direct impacts on agriculture, food security, and people's livelihoods (Vilakazi and Mukwada 2023). The ravaging effect of these menaces is felt across the globe. Rural communities are the most affected, with increased food, energy, shelter, and income shortages. Its impact on agriculture demands attention and concerted effort at the global stage. The need for resources, including food, land, water, and forest products, intensifies with increasing population (Verma et al. 2024a). As the world witnesses a tremendous population increase, estimated to be about 9 billion people on the planet by 2050 (UNDESA 2019), there is more and more pressure on natural resources, agricultural productivity will need to rise by more than 70% to feed the world and prevent an impending food catastrophe (World Bank 2008; van Dijk et al. 2021). To achieve this, climate change and land degradation must be addressed.

Land degradation refers to the decline or loss of the soil's ability to support current and future productivity. Numerous factors contribute to land deterioration, including harsh weather, especially drought and heat, and increased salinity. Climate significantly impacts soil formation and quality, vegetation form, and structure;

O. A. Fabiyi (✉)
Department of Crop Protection, Faculty of Agriculture, University of Ilorin, Ilorin, Nigeria
e-mail: fabiyi.oa@unilorin.edu.ng

D. A. Animasaun
Department of Plant Biology, Faculty of Life Sciences, University of Ilorin, Ilorin, Nigeria

Plant Tissue Culture Lab, Central Research Laboratories, University of Ilorin, Ilorin, Nigeria

P. A. Adedibu
School of Advanced Engineering Studies, Institute of Biotechnology, Bioengineering and Food Systems, Far Eastern Federal University, Russky Island, Vladivostok, Russia

C. S. Seth et al. (eds.), *Nano-formulations for Sustainable Agriculture*,
Nanotechnology in the Life Sciences, https://doi.org/10.1007/978-3-031-90617-6_12

high temperatures and little precipitation cause poor organic matter content and soil quality, characterizing the arid region (Kumari et al. 2022, 2023). Climate change is a primary factor propelling biodiversity loss, taking its toll on food production and security (Choudhary et al. 2024). Globally, arable land and yields have significantly reduced, owing to climate change, land degradation, and other abiotic stress factors (Gong et al. 2013). If measures are not put in place, it is estimated that global food yields will fall by an average of 10% by 2050, with certain areas experiencing reductions of up to 50%. Therefore, addressing these pressing issues is essential to increase agricultural output.

However, despite the global awakening on climate change, land degradation, and other abiotic factors, minimal improvement has been achieved; the situation defies all measures, biting harder on the ecosystem and limiting food supplies and the global economy. Approximately 33% of the global land area is prone to land degradation. According to the World Bank, an estimated 1.5 billion people whose livelihoods rely on land resources are disproportionately impacted by degradation, which affects above one-fifth of global arable land (World Bank 2022), creating an annual economic deficit of about 42 billion USD (Ioras et al. 2014). About 25% of land in Asia is vulnerable; it is twice as intense in Africa, especially in the sub-Saharan and semi-arid regions, due to their fragile soil and low-input farming. A report by the Africa Group of Negotiators Experts Support (AGNES) revealed that approximately 46% of Africa's geographical surface and about 485 million people are affected, costing 9.3 billion USD annually (Reich et al. 2001; AGNES 2020), a heavy blow on the developing nations economy. Additionally, there is over 75% degradation of the continent's farmed land, with each hectare losing an average of 30–60 kg of nutrients annually; the resultant effect is telling on agricultural yield, with a looming food crisis in the most affected regions. In Africa's arid, semi-arid, and dry sub-humid zones, only 3–30.6% of land is still unaffected, while 84% of highland regions experience moderate to severe degradation (AGNES 2020).

Abiotic stress factors are mutually related; climate change and land degradation reinforce each other. While climate change increases the pace and extent of soil degradation via heat stress and rainfall unpredictability, land degradation diminishes carbon sinks, leading to increased carbon emissions (Mariyam et al. 2024). The interplay is a crucial challenge to the world today. As climate change proves invincible despite the global efforts to curb it, it has continuously exacerbated land degradation through several factors, which include temperature rise, precipitation intensity, wind patterns, and other severe weather events. Erratic rainfall patterns alter the vegetation, initiating soil erosion in agricultural areas (Jagadesh et al. 2024). Climate change presents a significant threat to the sustainability of ecosystems and natural resources at both regional and continental levels. Moreover, scientists caution that shifting climate patterns might result in several hazards that severely harm the environment and the economy (Liu et al. 2021c).

Land degradation negatively impacts the welfare of a minimum of 3.2 billion people and propels the Earth towards a sixth major extinction event (World Bank 2022). The Intergovernmental Science-Policy Platform on Biodiversity and Ecosystem Services (IPBES) (2018) established that deforestation significantly contributes to our changing climate, accounting for approximately 10% of all anthropogenic greenhouse gas emissions. Vegetation cover loss between 2000 and 2009 has resulted in yearly worldwide emissions of up to 4.4 billion tonnes of CO_2 (Jia et al. 2019). The resultant effect is global warming, a corresponding rise in earth temperatures across strata. Studies revealed that compared to the global mean surface temperature, the land surface air temperature (LSAT) has surged significantly from the late nineteenth century to the present. A comprehensive analysis revealed that the average land surface air temperature has risen by 1.53 °C, while the average global air temperature has increased by 0.85 °C (He et al. 2019; Jia et al. 2019). Consequently, global warming has increased drastically since the 1990s, a 2 °C increase in global temperature is projected between 2026 and 2060, and up to a 3.7 °C increase before the century runs out (He et al. 2019).

In general, temperature changes, erratic rainfall patterns, reduced soil quality, and other abiotic factors interrupt plants' normal growth and development in several ways; stress-induced rise in reactive oxygen species (ROS) levels have several cytotoxicity and genotoxicity effects on plant cells and reduce yield (Fig. 12.1).

12.2 Climate Change, Land Degradation and Abiotic Stress Impact on Agricultural Productivity

Food and nutritional security are seriously threatened by abiotic pressures, especially in light of the growing global population. These intricate stressors, mostly environmental, account for approximately 50% of yield loss (Asif et al. 2023). In addition, the spread of arid climates and the emergence of hotter climates in tropical

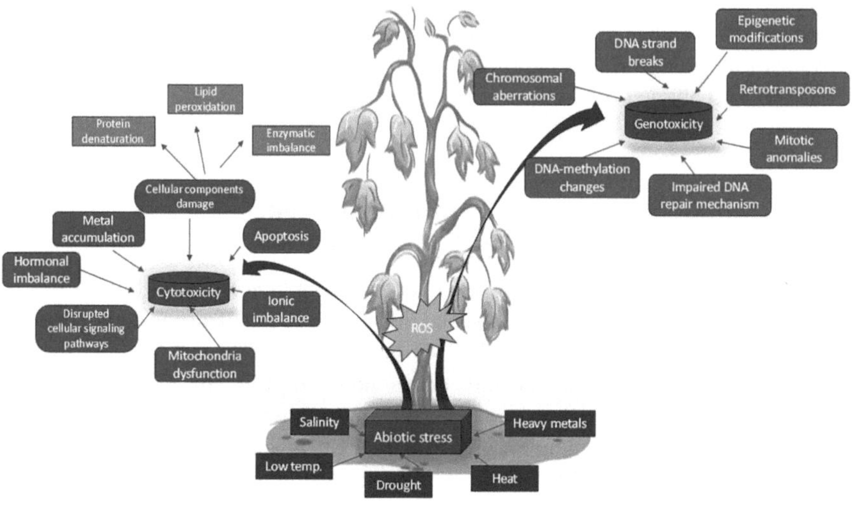

Fig. 12.1 An overview of the detrimental impact of abiotic stress on plants

areas pose more threats. Plants due to their sessile nature are continually faced with these stressors. To survive, plants use adaptation mechanisms which involve physiological, biochemical, and molecular processes. However, these mechanisms may limit plant growth and yield (Cirillo et al. 2022; Verma et al. 2024b). In all geographic regions, yields of grains have been observed to decrease by up to 7% when the mean global temperature rises by one unit (Zhao et al. 2017a). At the moment, over 35% of the world's population lives in chronic malnutrition and starvation (Chaudhry et al. 2022). The Food and Agriculture Organization projects that, at the present rate of climate change, the global output of the main cereals will drop by up to 45% by the turn of the century (FAO 2015).

Implementing innovative strategies to mitigate the yield-limiting abiotic stress factors is imperative, to prevent food crises and protect the ecosystems. This will also improve the livelihoods and provide food security in marginal rural areas. To ensure food security, many countries are intensifying agricultural efforts on the available limited resources, this may sometimes be counterproductive; resulting in depletion, and a drastic fall back to food insecurity (Vilakazi and Mukwada 2023). Consequently, climate-smart agriculture seems to be the most viable way to avoid a food crisis. Therefore, a large rise in the pace of crop production must be achieved to feed the population using techniques and approaches that are efficient and environmentally friendly.

Despite global efforts to curtail the impact of abiotic stress factors in agriculture, engaging several technologies including conventional breeding and hybridization, new planting methods, use of biostimulants and biofertilizers (Asif et al. 2023), precision agriculture, smart farming technologies, water-efficient technologies, marker-assisted selection (MAS), genetic engineering and other genomic approaches (Hasanuzzaman et al. 2019), yet, much needed to be done as its ravaging impact continues unabated. There is therefore need for new innovative technologies to complement the current efforts in fight against the yield-limiting stress factors (González Guzmán et al. 2021). To this effect, nanotechnology is a promising remedy, with its precision and multifaced applications in crop improvement (Kumar et al. 2024a; Fabiyi et al. 2025).

12.3 Nanotechnology in Crop Improvement

Nanotechnology is a field of science and engineering concerned with manufacturing, and utilization of nanostructures, and the manipulation of atoms and molecules at the nanoscale (Wawata and Fabiyi 2024). Nanomaterials have one or more dimensions on the order of 100 nanometers, or one hundred millionth of a millimeter, or less (Bayda et al. 2019; Rajput et al. 2023). Nanotechnology has great promise for increasing agriculture yield, improving crop resilience, and reducing environmental pollution (Fabiyi et al. 2020a, b). Since its introduction to agriculture at the inception of the twenty-first century, there has been a noticeable surge of research in this area (Xing et al. 2016; Bora et al. 2022). Ndlovu et al. (2020) asserted that agro-nanotechnology could help solve pressing issues in agriculture and food production, encompassing plant breeding, disease prevention, precision farming, and waste management (Fabiyi 2023a, b). Through its multifaceted applications, nanotechnology decreases yield losses and increases agricultural efficiency (Fabiyi et al. 2018; Fabiyi 2021). In modern agriculture, agrochemicals have been the farmers' aid in increasing productivity to meet the food demands of the fast-growing population. However, overuse is detrimental (Singh et al. 2021a). Overuse of pesticides and fertilizers and abiotic stress challenges in agriculture can be effectively addressed using nanotechnology (Demirer et al. 2021; Liu et al. 2021a), as it provides eco-friendly alternatives to enhance crop production to meet food demands (Fabiyi and Olatunji 2024).

Nanomaterials have been applied in farming systems, such as in nanoagrochemicals and nanosensors. Plant stress tolerance can be increased using the right nanomaterials (Zhao et al. 2020; Liu et al. 2021b). Nanosensors are very useful in monitoring and detecting changes in the environment (Shang et al. 2019; Fabiyi and Ogundele 2025). Furthermore, nanomaterials have also been found promising in the detection of pathogens, protection of plants, germination of seeds, reduction of nutrient loss, minimizing chemicals applications, and enhancement of agricultural yields (Fabiyi and Olatunji 2018; Fabiyi et al. 2024a, 2024b). In addition, nanobiocides, and nano-devices applied in animal breeding and plant genetic modification respectively are further examples of how nanotechnology has been employed in agriculture (Guleria et al. 2023). Nanotechnology can also be used in plant growth monitoring, genome modification, and plant transformation (Kwak et al. 2019). Metallic nanoparticles encapsulated with biopolymers for reduced toxicity, improved stability, and controlled release are often used as antimicrobials, nanopesticides, and nanofertilizers in modern agriculture (Kumar et al. 2022).

Nanoparticles (NPs) generally increase antioxidant enzyme activity and eliminate reactive oxygen species (ROS) to increase plant resistance to abiotic stress. An array of nanomaterials, such as ZnO NPs, TiO_2 NPs, Fe_2O_3 NPs, Si NPs, nanoceria, graphene oxides, and multi-walled carbon nanotubes have been used to minimize the adverse impacts of abiotic stress on crops including *Oryza sativa* (Huang et al. 2018), *Solanum tuberosum* (Gowayed et al. 2017), *Hordeum vulgare* (Karami and Sepehri 2018), *Medicago sativa* (Chen and Wang 2021), *Beta vulgaris* (Namjoyan et al. 2020b) *Zea mays* (Kumaraswamy et al. 2021) and *Linum usitatissimum* (Singh et al. 2021b). As human activities and climate change keep fostering abiotic stress like drought, flooding, salinity, high temperatures, and toxic metals accumulation, the development of engineered nanomaterials (ENMs) is crucial, considering their vast potential to boost plant growth and resilience towards sustainable agriculture (Ali et al. 2017).

12.4 Nanotechnology-Based Mutagenesis

12.4.1 Synthesis and Characterization of Engineered Nanomaterials

Nanomaterials could be organic, inorganic, or polymeric substances that exist naturally or are purposefully synthesized with at least one dimension ranging from 1 to 100 nm (Shang et al. 2019; Liu et al. 2021a). Scientists are able to synthesize nanoparticles with special properties, which are used extensively in several fields, such as engineering, chemistry, medicine, electronics, and agriculture (Saleh and Hassan 2023; Bora et al. 2022). Engineered nanomaterials possess unique, engineered physical, chemical, and biological attributes that help them to modify compounds and devices for targeted purposes (Bandala and Berli 2019; Viswanathan et al. 2019). The major characteristics of ENMs are their very minute size and shape, which control the kind and distribution of reactive sites in chemical reactions as well as their reactivity and delivery mechanism.

Although nanomaterials can be synthesized using inorganic compounds like sodium borohydride, sodium citrate, polyethylene glycol (PEG), polyvinylpyrrolidone (PVP), polyvinyl alcohol (PVA), bovine serum albumin (BSA), ethylene diamine tetra acetic acid (EDTA), chitosan, etc., however, green synthesis using plant extract is preferred. Plant leaves, which are the photosynthetic sites and food producers, may be combined with stems, seeds, bark, flowers, roots, and other plant parts to synthesize nanoparticles (Kumar et al. 2024b). This has been successfully carried out in several studies (Hemlata et al. 2020; Ahmad et al. 2020; Khaldari et al. 2021; Fabiyi et al. 2021; Fabiyi and Olatunji 2022; Yugay et al. 2023). Microorganisms such as bacteria, fungi, yeasts, and algae, are also utilized as biomediated pathways for ENM production. Comparing these bioreactors to physical and chemical synthesis methods, they are easier, safer, and affordable (Xu et al. 2021). The increasing attention to the green synthesis of ENM can be attributed to its economical viability and other factors; Plants are non-toxic, easily obtainable, and contain a variety of active compounds that could serve as reducing or stabilizing agents during NP synthesis (Shafey 2020; Fabiyi et al. 2024b). Furthermore, these organic molecules could help to inhibit coagulation and regulate growth (Bora et al. 2022; Fabiyi and Olatunji 2024). These compounds include flavonoids, chlorophyll, amino acids, citric acid, carotenoids, polyphenols, vitamins (B, C, D, and K), polysaccharides, etc. (Fabiyi 2021; Antunes Filho et al. 2023).

ENMs may be very reactive and most times exhibit heterogeneity due to their ability to adsorb species from their surroundings (such as the creation of a protein corona around inorganic nanoparticles). The surrounding matrix often affects the physicochemical characteristics of ENMs, which might aggregate or have their surface modified after some time due to slight environmental disturbances (Labuda et al. 2023). Based on their composition, nanomaterials are sometimes categorized as metals and metal oxides, non-metal inorganics, nanocarbon, and organic nanomaterials. More than half of the nanoparticles found in products are either titanium dioxide or silver, making metals and metal oxides the most prevalent materials. Metal nanoparticles (such as gold) may be manufactured in well-defined forms and with minimal polydispersity, however, metal and metal oxide ENMs usually contain spherical or irregularly shaped particles of different sizes (Labuda et al. 2023).

Numerous techniques have been used to characterize NPs; however, these techniques are used to examine defined characteristics or a combination of numerous properties. Techniques based on microscopy, such as Transmission electron microscopy (TEM), Atomic force microscopy (AFM), and High-resolution transmission electron microscopy (HRTEM), may provide details on the size, shape, and crystal structure of nanomaterials. Methods, like the magnetic approaches, are tailored to certain groups of materials. These methods include superconducting quantum interference device (SQUID), X-ray magnetic circular dichroism (XMCD), Ferromagnetic resonance (FMR), and Vibrating sample magnetometer (VSM), as examples. Fourier-transform infrared spectroscopy (FTIR) is more adaptable for NP surface composition and ligand binding properties. The ligand density may also be examined using optical microscopy and bioanalytical techniques (Jayawardena et al. 2021). UV-Vis for elucidating optical properties, agglomeration state, NP shape, size, and concentration. More details about the general and more specific physical features of nanoparticle samples, such as their optical qualities, elemental content, and structure can be obtained using a variety of different approaches, which include, spectroscopy, and X-ray methods (Mourdikoudis et al. 2018; Jayawardena et al. 2021).

12.4.2 Plant NP Uptake and Mutagenesis Mechanism

The variations in the morphology and physiology of plant species and the characteristics of nanoparticles substantially influence their uptake in plants. Plants mostly absorb NPs via leaves and roots (Wang et al. 2023). Plant leaves are often embedded with a cuticle, which is a protective layer of soluble wax covering the epidermal membrane. Therefore, only NPs of defined sizes (<10 nm) pass the stomata into the cells. After passing through the cuticle, two pathways -hydrophilic and lipophilic- mediate the NPs' route of transport. Hydrophilic routes involve movement via polar aqueous pores found in cuticles or stomata, while lipophilic pathways facilitate the transport through cuticular waxes (Ahmad et al. 2022).

NPs may also be absorbed by roots, however, the endodermal Casparian strip and root epidermal cells limit their absorption. NPs can also enter through passage cells, secondary root birth sites, and injured roots into the cells. Nanomaterials enter cells and membranes by a variety of mechanisms, such as diffusion, endocytosis, plasmodesmata, or chemical and physical degradation. NPs must traverse the mesophyll cells after penetrating the cuticle to get into the vascular system, this is indispensable for NP transport in plants. According to Banerjee et al. (2019), certain NPs are capable of overcoming entry barriers on their own. In mesophyll cells, NPs are mostly transported via the symplastic and apoplastic tracks (Kumar et al. 2024c).

Symplastic movement occurs via plasmodesmata, while apoplastic movement occurs via extracellular space. The apoplastic routes allow NPs to enter the central cylinder of the root, and through the xylem to the aerial portion of the plant (Ahmad et al. 2022). It is hypothesized that the apoplast might be the plants' translocation mechanism for NPs as recent studies indicate that plants limit fluid flow in the apoplast after the introduction of NPs (Wang et al. 2021b, 2023). By symplastic movement, NPs are channeled through phloem sieve tubes (Ahmad et al. 2022).

Nanomaterials often find their way to the nucleus. Nuclear pores allow nanomaterials of very small size to enter the nucleus. Nevertheless, larger NMs have been found in the nucleus, suggesting alternative pathways for nuclear absorption. For instance, NMs could easily get into the nucleus during membrane disintegration in mitosis, then, directly interacts with the DNA and interfere with the transcription and replication processes. It has also been proven that during cell division, it interferes with spindle fibers and centrioles (Kisin et al. 2011; Kohl et al. 2020).

Numerous biological systems have been used to study the mutagenic and genotoxic impact of synthesized nanoparticles (Wang et al. 2007; Girgis et al. 2012; Mei et al. 2012; Yin et al. 2012). The mutagenic mechanism has been the subject of much research lately. Studies have attributed this effect to the generation of ROS by NMs or by toxic ions discharged during their breakdown. The chemical makeup of the ENMs determines how much ROS is produced. Their tiny size, high specific surface area, and high surface reactivity cause increased amounts of ROS generation (Oberdörster et al. 2005), which in turn intensifies oxidative stress in tissues either directly or indirectly (Fu et al. 2014). Their mutagenetic effect is linked to oxidative DNA damage. ROS in cells targets the DNA. Free radicals interacts with DNA and other biomolecules in cells. ROS causes strand breaks and lesions from purine or pyrimidine oxidation. This can further lead to more severe chromosomal anomalies and gene mutations such as DNA-protein crosslinks, abasic sites, and strand breakages (Valko et al. 2006; Gonzalez et al. 2008' Kohl et al. 2020). Moreover, NMs can also affect the DNA through other molecules that play roles in DNA replication or have interactions with the DNA. For instance, protein kinases that control cell cycle processes like DNA replication and cell division are affected by NMs (Magdolenova et al. 2013; Kohl et al. 2020).

12.5 Applications of Engineered Nanomaterials in Plant Stress Control

12.5.1 Nano-based Mutagenesis in Plant Abiotic Stress Tolerance

The variability and incessant fluctuations in natural environment has subjected plants to hard situations. A single abiotic stress factor may affect a particular plant, while in some cases combination of stress factors may affect a single plant, which badly devastates the crop and yield is excessively low (Chaudhry et al. 2022). Salt accumulation in soil leads to ionic and osmotic problems in crops. Na^+ and mineral ions build up results in toxicity which distrupts the cellular processes of crops (Melino and Tester 2023). The complexities of extreme temperature results from lack of rain and very high temperatures, water content of leaves are affected, rate of transpiration, frequency of photosynthesis and stomatal conductance are all disoriented. This graduates to meiosis, poor cell division and finally necrosis of the crop, occurring most especially in semi arid and arid zones of the world (Hammond et al. 2022). Response of crops to environmental stressors depends largely on the crop developmental stage, constituents of crop gene and intensity of stress. The episode progreses into severe losses of crop up to 50%, culminating into food insecurity which is a global concern.

Efforts are directed towards sustainable production of crops that are tolerant to unprecedented weather variables. Positive alterations in plant genomic information gives crop the necessary required strength for adaption to sundary abiotic stressors. Genes responsible for conferment of resistance are edited for crop performance boosting to withstand harsh weather variables. Under stressed conditions for plants, application of nanomaterials is a reliable technique that has contributed significantly to plant support systems for improved productivity (Yusefi-Tanha et al. 2020) Several biochemical and physiochemical processes of growth regulation in crops are altered with nanomaterials via diverse route (Fig. 12.2) for positive response to weather variability. Genetic methods are used in conjunction with nanocarriers to reduce ROS and stimulate plant tissues to release antioxidants, enhancing

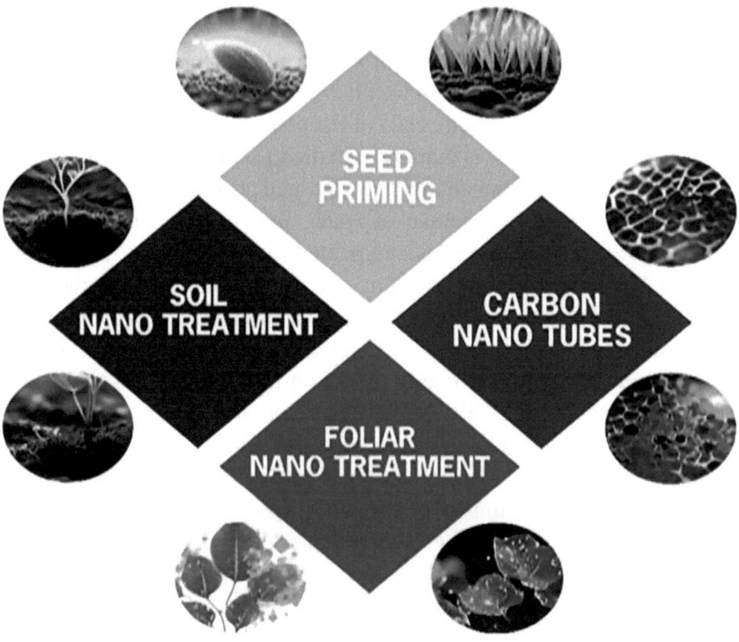

Fig. 12.2 Overview of nano-based methods for abiotic stress management in crops

self-defense and promoting accelerated growth. (Mazhar et al. 2022). The efficiency of nanomaterials relies primarily on age of crop development, crop type and dosage of nanomaterials applied. Diverse nano based methods for crop tolerance to abiotic stressors are highlighted below.

12.5.1.1 Seed Priming

The effectiveness of seed priming using nanomaterials have been documented to supercede that of other substances used in priming. Priming facilitates quick emergence and development of seedlings. Crops at the emergence stage are highly at risk of environmental variabilities, while they struggle with some metabolic adjustments. Nanomaterials primed seeds are not dormant, all necessary enzymes are activated and defence systems are equally stimulated as they emerge into the environment (Szőllősi et al. 2020). Silver NPs primed seeds of *Oryza sativa* exhibited high rate of propagation. The activity of amylase and soluble carbohydrates in the seeds were high. This assisted a faster rate of development. Likewise, priming of *O. sativa* with zinc oxide NPs improved proline, antioxidant levels thus assisting the plants to withstand stress (Mahakham et al. 2017; Waqas Mazhar et al. 2022). Zinc oxide and selenium NPs, at varied concentrations minimised the resultant effect of salt stress in *Brassica napus* seeds, rate of germination was significantly higher than untreated seeds (El-Badri et al. 2021), high factor of resistance to drought was also exhibited by the plants. The levels of osmoprotectants and ROS scavengers was high, these were deemed responsible for dought tolerance stress exhibited.

In *Zea mays* plants exposed to salt stress vigor of seedlings increased tremendously, with rate of germination and antioxidants after priming with titanium oxide nanoparticles at 60 ppm. Under harsh weather variability, priming has been reported to reduce damage to photosynthetic cells of the plant. The antioxidant system of the plant is turned on while H_2O_2 is equally signalled (Shah et al. 2021). Germination enhancement through environmentally friendly seed priming techniques is achieved by antioxidant boosting, high intake of water and increased hydrolysis of seed starch content. Copper nanoparticles augmented *Zea mays* crops using priming of seed method had improved anthocyanin and carotenoid content, the ROS was greatly reduced and this supported yield of maize grains (Van Nguyen et al. 2022). Colloidal bimetals of zinc and copper NPs applied as primers for *Triticum aestivum* seeds under drought stress gave rise to reduction in thiobarbituric acid contents of the seeds. Furthermore, iron NPs primed *Triticum aestivum* seeds had increased proline and beta carotene content (Taran et al. 2017; Noor et al. 2022). Sulphur NPs applied as primers at 0.01 ppm on salt stressed *Triticum aestivum* supported the seedlings and alleviated the impact of stress of the growing seedlings by increasing metabolism of nitrogen and level of antioxidants. Primed seeds of *Gossypium hirsutism* using coated cerium oxide-NPs facilitated vitality in plant roots and minimised production of ROS thus enabling the plants to withstand salinity stress. Enhanced sprouting and quick development was observed in *Lactuca sativa* seedlings whose seeds were primed with carbon NPs (An et al. 2020; Baz et al. 2020).

12.5.1.2 Carbon Nanotubes

Improved tolerance to salinity in *Vitis vinifera* seeds and *Brassica oleracea* seedlings were reported by Li et al. (2022) and Martínez-Ballesta et al. (2016) in their research on MW carbon nano tubes. Germination rate was very fast in V. vinifera seeds, while malondialdehyde (MDA) content of seed was minimised. High rate of aquaporin transduction was seen in *B. oleracea* seedlings, thus supporting high water intake of the plant. The introduction of NPs during seedling growth has alleviated several stress conditions in crops. Single walled carbon nanos alleviated salt stress in *Sophora alopecuroides* seedlings through adjustment of nitrogen and carbon pathways, with the involvement of citric and glycolysis process for crop growth energy enhancement (Wan et al. 2020).

Carbon nanotube priming of *Zea mays* seeds facilitated germination speed and increased carotenoid percentage in drought stressed *Zea mays*. Further, *Glycine max* subjected to critical water shortage and augmented with SW carbon nano tubes presented good germination and increased yield. In the same vein, *Triticum aestivum* seedlings treated with SW carbon nano tubes of silicon oxide facilitated drought tolerance of the seedlings, there was high efficiency of use by the seedlings with increased activity of amylase (Shahriari et al. 2019; Sun et al. 2020a, b; Akhtar et al. 2021a).

12.5.1.3 Foliar Nano Treatments

Reports by Ghani et al. (2022) established that stress in plants could be treated with NPs through foliar route. Zinc oxide NPs was applied on *Cucumis sativus* seedlings to ease drought stress. The non enzymatic and enzymatic processes of the plant was boosted, generation of ROS was decreased, while the plant biomass improved. Similarly, yield and seed weight of *Hordeum vulgare* plants was increased significantly while under drought stress by foliar application of silicon and titanium oxide NPs. Foliar treatments of *Mangifera indica* trees which are in bloosom stage with silicon and titanium oxides under salt stress afforded minimal malformation of flowers, reduced abortion of seeds, good uptake of nutrients and assimilation of carbon, culminating in high production of fruits annually (Ghorbanian et al. 2017; Elsheery et al. 2020). The foliar treatment of *Solanum melongena* against drought stress at intervals of 30 and 45 days post transplanting assisted photosynthesis efficiency in the plants, arrangements of stem anatomy and stability index were greatly increased (Semida et al. 2021).

Foroutan et al. (2019) in their research on seedlings of *Moringa peregrine* was able to induce tolerance against drought with foliar application of zinc oxide NPs at varying concentrations. Degradation of Chi was notably decreased while phenolic and antioxidants were enhanced in the plant (Foroutan et al. 2019). Likewise, Ali et al. (2021) actuated biosynthesis of alkaloid through gene modification in *Catharanthus roseus* seedlings using foliar technique of application of 1% chitosan NPs. Tolerance to drought was achieved with reduction in hydrogen peroxide content of the crop. Further, leaf aging of *Beta vulgaris* seedlings subjected to drought stress was reduced significantly with foliar treatment of silicon NPs introduced at 10 leaf stage, while soluble sugar content and weight was increased at harvest (Namjoyan et al. 2020a, b).

Shortage of water damage in *Hordeum vulgare* plants was corrected with manganese and iron NPs applied as foliar spray. Nutrient uptake of *Glycine max* seedlings was equally increased with foliar treatment of nano chelates of manganese. Effect of drought was reduced and high grain yield was achieved irrespective of water shortage. *Vigna radiate* seedlings had high level of carotenoid and significant reduction of water stress effect after foliar treatment with nano chelates of manganese (Sanavi and Mohammad 2018; Vaghar et al. 2021; Raza et al. 2023). Lipid peroxidation was minimal in *Cucumis sativus* seedlings treated with zinc NPs via foliar route, despite drought, amino acid, sugar and proline contents were high (Ghani et al. 2022).

The foliar treatment of *Catharanthus roseus* seedlings at 2 weeks post planting facilitated tolerance of seedlings affected with salinity stress, tolerance genes were activated and there was increase in alkaloid content of plant (Hassan et al. 2021). Reaction of *Linum usitatissimum* to salt stress was minimal with foliar treatment of seedlings with silver NPs. Carbohydrate, proline and protein content increased significantly and pigments for photosynthesis were notably improved (Khalofah et al. 2021). Salt toxicity in *Triticum aestivum* was corrected through foliar use of iron NPs introduced during flowering and tillering periods of the plants, again zinc oxide NPs applied on the leaves of *Solanum lycopersicum* boosted growth and good fruit production under salt toxicity (El-Saber et al. 2021; Faizan et al. 2021a, b).

12.5.1.4 Soil Treatments

At times nanomaterials may be incorporated into the soil for crop health boost. B_2O_3, CuO and ZnO NP_S introduced in the soil as micronutrients gave upto 36% grain yield to wheat *T. aestivum* plants under stress of drought. Effect of drought on *Sorghum bicolor* was suppressed when zinc oxide was used as an admix in the soil. Enhancement of grain yield was upto 183%, while nutrients like nitrogen, zinc and potassium were greatly absorbed into the sorghum grain from the soil (Dimkpa et al. 2017, 2019). *T. aestivum* subjected to drought and cadmium as stress factors had minimal oxidative stress with high grain out put in iron NP amended soil (Adrees et al. 2020). Pre sowing introduction of iron NPs into the soil medium at 7 days before sowing assisted *T. aestivum* to withstand drought and cadmium stress. Uptake of cadmium in the *Triticum aestivum* plants was significantly minimal, while high concentration of iron was noted in the grains (Adrees et al. 2020). Enhanced carotenoid was noted in *Hordeum vulgare* seedlings planted in soil modified with silicon NPs. Effect of drought stress on the seedlings was suppressed, functions of antioxidant enzymes was increased with full shoot biomass.

Further, the adverse effect of drought on *Zea mays* plants was reduced with soil treatment of potassium NPs, there was accelerated nutrient uptake with high concentration of essential minerals in the *Zea mays* seeds (Aqaei et al. 2020; Ghorbanpour et al. 2020). Soil treatment with Titanium oxide NPs influenced tolerance to *T. aestivum* plants subdued to drought stress, stability of membrane and osmolyte content increased. Similarly, titanium NPs incorporated into the soil improved proline content in *Zea mays* leaves, with enhancement of antioxidant activities and higher yield in the absence of water (Mustafa et al. 2021; Karvar et al. 2022). Enzyme activities were upregulated to reduce salt stress in *Eucalyptus tereticornis* planted in soil supplemented with iron NPs while Cucumis sativus seedlings under multiple stress of drought and salt grown in silicon NPs irrigated soil had good nutrient uptake with boosted yield (Alsaeedi et al. 2019; Singh et al. 2021). *Vicia faba* subjected to salt toxicity had high proline and soluble sugars after irrigation with calcium phosphate NPs at 30 days after emergence (Nasrallah et al. 2021).

12.5.1.5 Nano-Based Genome Editing

The dynamics of weather has resulted in frequent crop failures and reduced yield. The natural system of plants can be modified to augment tolerance of sundry crop species to weather variability for yield improvement. Several method of genetic breeding are effective, but arduous and only applicable to a limited group of crops. Targeted techniques like TALENs, ZFNs, and CRISPR-Cas9 alleviated some bottlenecks with short-term option of appearance. However, there is the need for unification of gene firmness. Nanomaterials have been applied approvingly in the field of breeding and alteration of genes. Biomolecules are conveyed smartly with porous and polymetric nanomaterials (Sanzari et al. 2019). This has facilitated accuracy, speed and dependability. Guarded navigation amid the complexity of cell cytoplasm for DNA conveyance is assured with nano mechanism. There is acceleration and hopeful outcomes in the fundamentals of plant genetic techniques and gene alteration using nanomaterials. Some available nano techniques are analysed (Fig. 12.3).

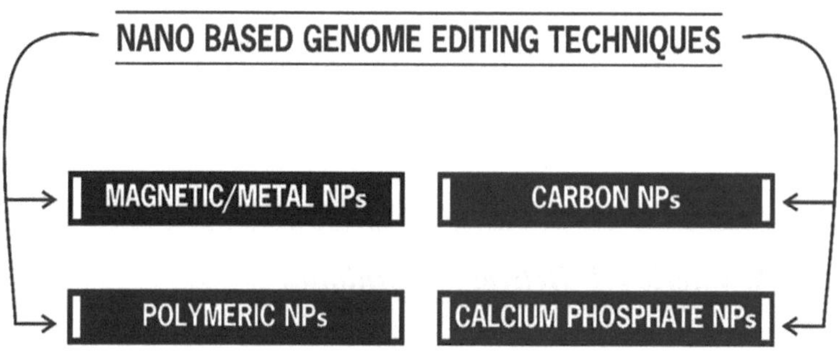

Fig. 12.3 Overview of nano techniques for genome editing

12.6 Magnetic/Metal NP Delivery Techniques

Metal based NPs are engineered from zinc, silver, copper, iron, gold, cerium, silicon, aluminium, nickle and titanium. Gold NPs are versatile and have been employed in DNA delivery of non-viral vectors. It has exhibited minimal toxicity and highly stable in cells (Sangeetha et al. 2019). Carriage of siRNA for non-viral/toxic into plants with gold NPs altered using polyethylenimine was found to be very effective as a gene expression redundancy vehicle with 77% success. The alteration of plant genes has been greatly achieved with gold and iron NPs. The bio-conjugation attributes of gold NPs is responsible for their repeated applications as gene carrier vehicles. Biolistic procedure was employed by Martin-Ortigosa et al. (2012) to deliver protein and plasmid into epidermal tissues of *Allium sativum* with gold NP as the vehicle. Pollen grains of *Gossypium hirsutum* were outstandingly transfigured with DNA complex of magnetic NP origin using magnetic and gold iron NPs, delivery went through pollen open apertures. The transferred DNA integrated successfully and self pollination afforded offspring expression (Zhao et al. 2017). Nanoclusters of gold were employed for siRNA movement into *N. benthamiana* leaves with carrier internalization. Minimal damage to plant tissues was incurred with the use of biolistic gold NPs and efficiency of more than 40% has been recorded (Zhang et al. 2021).

12.6.1 Carbon NP Delivery Techniques

These include materials made of nanotubes, nanodiamonds, graphene and fullerenes. Nano tubes of carbon can permeate cell membranes and are therefore good nano-vehicles in plants. Carbon nano cargo systems have distinct properties, it is a rapid and simple technique of gene movement. DNA cargoes are shielded from nucleases while mutation of genes are restricted (Chandrasekaran et al. 2020). Carbon tubes of S&MW were used for plasmid delivery in *Nicotiana tabacum*, a stable transformation was achieved. siRNA delivery in *T. aestivum* and *Gossypium hirsutum* DNA plasmids were loaded with grafting, mature plant leaves were then pervaded and gene expression was clearly achieved using carbon tubes (Demirer et al. 2019). Compatibility of carbon cargo carriers is largely based on choice of polymer. Dots of carbon are equally good delivery vehicles, they function specifically with regards to species, efficiency relies greatly on dot siize. Minimal siRNA deposition is achieved with large sized dots (Demirer et al. 2019).

12.6.2 Polymeric NP Delivery Techniques

Cellulose and starch are used in the preparation of nanoparticles and have been reported as an effective carrier of RNA, DNA and proteins (Fabiyi, 2025). They are biodegradable with minimal toxicity. Polymer based enantiomers can function via vesicular dissociation and membrane diffusion. A well studied polymeric nanomaterials that can degrade quickly in cells is chitosan. Successful gene redundancy in *Caenorhabditis elegans* was achieved with chitosan RNA NPs. Quatum dots chitosan DNA composite were unified successfully into *Jatropha curcas* for cell transformation, while starch composite NPs have been applied as carriers for genes in *Dioscorea* spp. Also, polyamidoamine NPs served as vehicle with plasmid content in gene alteration of *Medicago sativa* (Lichtenberg et al. 2020; Liu et al. 2008; Wang et al. 2013; Amani et al. 2018). Polymer dendrimer NPs were put in the alteration of *Agrostis stolonifera* genes, the plant cells took up the polymer DNA successfully.

12.6.3 Calcium Phosphate NPs Delivery Techniques

In some procedures, DNAs are mixed with calcium phosphate NPs, this technique of plant gene altration has been demonstrated to be very effective. Evolving cells of plants are kept in complex solutions of calcium phosphate NP-DNA. 80.7% alteration efficiency was achieved by Naqvi et al. (2012) when plasmids was delivered to hypocotyls of *Brassica juncea* with the aid of calcium phosphate NP as vehicle. Rafsanjani et al. (2016) equally applied CaP NPs in the movement of genes to *Cichorium intybus,* increase in plant soluble protein was realized.

12.7 Applications and Impacts of Nanotechnology-Based Mutagenesis in Fortifying Crops Against Abiotic Stress

Abiotic stress is a critical challenge to agricultural productivity. Plants are often faced with multiple stresses with detrimental physiological effects, hindering their normal growth and development (Bellard et al. 2012). Nanotechnology is an efficient and promising approach that applies ENMs to solve real-life problems across fields. Nanotechnology can effectively address agricultural challenges, especially, regarding stress tolerance and yield enhancement. It can also help to address issues associated with excessive fertilizer and pesticide usage and their negative impacts on the ecosystem (Khot et al. 2012; Ahmad et al. 2022). Thus, the adoption of nanotechnology in agriculture is increasing, enhancing food quality and global production in an eco-friendly approach. NPs play a vital role during stress conditions, they alter the expression of genes involved in vital physiological processes, such as biosynthesis, cell structure, electron transport, and energy storage. They lessen oxidative and osmotic stress, safeguard photosynthetic equipment, and speed up the scavenging of ROS (Gohari et al. 2020; Choudhary et al. 2025). In recent years, a variety of NPs with their limitless potential to address abiotic stress in agriculture have been employed (Table 12.1).

Table 12.1 Application of nanoparticles to plant species under abiotic stress conditions

Stress condition	Plant species	Nanoparticle	Observation	References
High Temperature (Heat) stress	Maize (*Zea mays* L.)	Au NPs	Enhanced germination and vigor of seedlings	Tovar et al. (2020)
	Maize (*Zea mays* L.)	ZnO NPs	Enhanced vigor, germination, root and shoot development	Itroutwar et al. (2020)
	Wheat (*Triticum aestivum* L.)	Se NPs	Enhanced plant growth and yield and significantly lowered the severity of crown root rot	El-Saadony et al. (2021)
	Wheat (*Triticum aestivum* L.)	Si NPs	Restored the deformed nucleus and chloroplast, and increased the photosystem II's photochemical efficiency	Younis et al. (2020)
	Wheat (*Triticum aestivum* L.)	ZnO NPs	Significantly improved tolerance, and growth.	Azmat et al. (2022)
	Barley (*Hordeum vulgare* L.)	Fe_3O_4 NPs	Enhanced photosynthesis and vigor	Tombuloglu et al. (2019)
	Sorghum	Se NPs	Enhanced membrane integrity and antioxidant defense system	Djanaguiraman et al. (2018b)
	Wheat (*Triticum aestivum* L.)	ZnO and TiO_2 NPs	Enhanced antioxidant enzyme activity, MDA levels, and hydrogen peroxide.	Thakur et al. (2021)
Salt stress	Maize (*Zea Mays* L.)	TiO_2 NPs	Improved antioxidant enzyme activity and seed vigor	Shah et al. (2021)
	Lily (*Lilium* ssp.)	Ag NPs	Enhanced growth, flowering, and photosynthetic pigment and biomass.	Salachna et al. (2019)
	Cotton (*Gossypium* spp. L.)	CeO_2 NPs	Improved growth, vigor, photosynthesis, leaf K^+ retention, and Na^+ leaf exclusion	Liu et al. (2021b)
	Trigonella foenum-graecum	ZnO NPs	Increased quantities of proline and protein, improved antioxidant activity, and decreased levels of MDA and H_2O_2	Noohpisheh et al. (2020)
	Tomato (*Solanum lycopersicon* L.)	CuO NPs	Enhanced growth and oxidative stress modulation via activation of JA and SOD	Hernández-Hernández et al. (2018)
	Abelmoschus esculentus L.	ZnO NPs	Improvements to the total soluble sugar and proline accumulation, the antioxidant enzymes activity, and photosynthetic pigments	Alabdallah and Alzahrani (2020)
	Lycopersicon esculentum	SiO_2 NPs	Increased germination and root vigor	Haghighi et al. (2012)
	Rice (*Oryza sativa* L.)	Mn NPs	Increased enzymatic activity, recovery from chlorosis, concentration of phenolic compounds, and detoxification of ROS	Rahman et al. (2016)
	Rice (*Oryza sativa* L.)	CeO_2 NPs	Improved photosynthesis, growth, and yield.	Zhou et al. (2021)
	Mung bean (*Vigna radiata*) (L.)	Mn NPs	Increased nitrate reductase activity, chlorophyll concentration, and membrane stability.	Shahi and Srivastava (2018)

Drought stress	Wheat (*Triticum aestivum* L.)	TiO_2 NPs	Enhanced photosynthesis, growth, and ROS elimination	Faraji and Sepehri (2020)
	Sorghum (*Sorghum bicolor* L.)	ZnO NPs	Higher uptake of nitrogen, potassium, and other vital nutrients; faster growth of plants; higher yield	Dimkpa et al. (2019)
	Wheat (*Triticum aestivum* L.)	Se NPs	Increased growth and yield	El-Saadony et al. (2021)
	Maize (*Zea mays* L.)	CuO NPs	Increased photosynthetic pigments	Van Nguyen et al. (2021)
	Maize (*Zea mays* L.)	ZnO NPs	Enhanced production of melatonin and antioxidant enzyme activities. Upregulation of SOD, CAT, APX, CAT, TDC, and SNAT relative transcript abundance	Sun et al. (2020a)
	Ginger (*Zingiber officinale*)	Fe_2O_3 NPs	Increased photosynthetic pigments (chlorophyll, carotenoid) and POD content	Noor et al. (2022)
	Solanum melongena L.	ZnO NPs	Enhanced growth, photosynthesis, and yield	Semida et al. (2021)
	Ocimum basilicum L.	TiO_2 NPs	Reduced toxicity and increased biomass	Kiapour et al. 2015
	Cucumber (*Cucumis sativus* L.)	ZnO NPs	Enhanced uptake of nutrients, osmolyte buildup, and antioxidant defense system	Ghani et al. (2022)
	Soybean (*Glycine max*) (L.)	ZnO, Fe, Cu, and Co NPs	Enhanced drought tolerance, and relative water content. Drought-responsive gene expression is upregulated in both roots and shoots, notably *GmERD1*.	Linh et al. (2020)
Cold stress	Rice (*Oryza sativa* L.)	ZnO NPs	Growth in biomass, root length, chlorophyll accumulation, and plant height. Lower MDA and H_2O_2 levels to lessen oxidative damage. Expression of genes and transcriptional factors triggered by chilling.	Song et al. (2021)
	Wheat (*Triticum aestivum* L.)	Au NPs	Enhanced the plants' resilience to cold stress by maintaining their photosynthetic activity.	Venzhik et al. (2022)
	Rocket (*Eruca sativa* Mill.)	Fe_3O_4 NPs	Enhanced growth, chlorophyll a, root-shoot elongation, and seed germination	Plaksenkova et al. (2019)
	Chickpea (*Cicer arietinum* L.)	TiO_2 NPs	Increased levels of carotenoids and chlorophyll, a stronger immune system.	Mohammadi et al. (2014)
	Banana (*Musa acuminata* Var. Baxi)	Chitosan NPs	An increase in osmoprotectants, yield, antioxidant enzymes, and photosynthetic pigments. Reduced oxidative stress.	Wang et al. (2021a)

(continued)

Table 12.1 (continued)

Stress condition	Plant species	Nanoparticle	Observation	References
Metal stress	Wheat (*Triticum aestivum* L.)	SiO_2 NPs	Enhanced Cd tolerance by reducing the absorption, buildup, and translocation of Cd while increasing the consumption of nutrients and reducing oxidative stress.	Ur Rahman et al. (2021)
	Wheat (*Triticum aestivum* L.)	ZnO NPs	Growth, photosynthesis, and zinc concentration in the roots, shoots, and grains all increased. Elevated levels of POD and SOD in leaves. Decrease in the toxicity of lead in roots and shoots	Hussain et al. (2018)
	Rice (*Oryza sativa* L.)	ZnO NPs	A significant fall in levels of arsenic and cadmium in the roots and shoots	Ma et al. (2020)
	Wheat (*Triticum aestivum* L.)	Fe_2O_3 NPs	Increased in photosynthesis, dry weight of grains, and antioxidant enzyme activity.	Hussain et al. (2019a)
	Soybean (*Glycine max* (L.)	TiO_2 NPs	Enhanced physiology, vigor, and photosynthesis	Singh and Lee (2016)
	Tomato (*Solanum lycopersicum* L.)	Chitosan NPs	Increased biomass, photosynthesis, and decrease in H_2O_2	Faizan et al. (2021a, b)
	Rice (*Oryza sativa* L.)	ZnO NPs	Enhanced plant growth, and tolerance to heavy metals	Akhtar et al. (2021b)

12.7.1 Salinity

Agricultural productivity, crop yield, and useable land area are significantly limited by salinity and sodicity which affects approximately 1 billion hectares of farmland globally (Mukhopadhyay et al. 2021). Salt stress in plants induces ionic imbalances and osmotic stress. Nanomaterials have been applied to mitigate their growth and yield limiting effects on different plants, with significant success (Liu et al. 2021a). Under salt stress, chlorophyll becomes less active, and consequently, the rate of photosynthesis declines. Salt-stressed tomato seedlings were treated with SeNPs, chloroplast's antioxidant defense system was reinforced, and the seedlings showed increased photosynthesis (Diao et al. 2014). A similar result was observed in salt-stressed cotton plants when treated with CeO_2 NPs, photosynthetic rate, plant growth, K^+ retention, and Na^+ exclusion in the leaves were significantly improved (Liu et al. 2021b). Photosynthetic pigments and vigor also increased in ZnONPs-treated saline-grown *Abelmoschus esculentus* L. (okra) along with the antioxidant enzymes activities (Alabdallah and Alzahrani 2020).

AgNps have also been used in this regard. Rice seeds were treated with Ag NPs (100–150 nm). It was observed that under salt stress, the roots and shoots had higher levels of antioxidant enzyme activity (Thuesombat et al. 2016). Likewise in rice plants, MnNPs helped the plants recover, eliminate ROS, and increase phenolic compounds and enzymatic activities (Rahman et al. 2016). In mung bean and bell peppers, MnNPs improved plant vigor, chlorophyll content, and enzymatic activities to resuscitate the plants from salt stress (Shahi and Srivastava 2018; Ye et al. 2020).

12.7.2 Drought

Drought has been identified to be the most prevalent abiotic stressor that affects agricultural yield, particularly in arid regions of the world (Ilyas et al. 2020). Due to inadequate water resource management and climate change, drought is becoming more frequent globally. Each year, drought stress limits agricultural output, costing billions of dollars. Drought stress generates ROS, leading to lipid oxidative damage and disruptions in the membranes of affected plants.

Plant tolerance to drought stress has also been enhanced with the application of NPs. Treating corn with yttrium-doped Fe_2O_3 NPs significantly improved its growth and yield under drought stress (Van Nguyen et al. 2021). Jute seedlings treated with hydroxylapatite nanoparticle (HANP) which triggered the proline biosynthesis pathway showed notable resilience to drought stress (Das et al. 2016). SiO_2 and SeNPs were also used to enhance photosynthesis in drought-stressed strawberries (Zahedi et al. 2020). Furthermore, CeO_2NPs proved effective in inducing drought resilience in sorghum by enhancing the photosynthetic rate and grain yield (Djanaguiraman et al. 2018a). Chitosan nanoparticles (CSNPs) increase drought tolerance in *C. roseus* leaves by significantly lowering the H_2O_2 and malondialdehyde levels under drought stress. It also preserves the integrity of the membrane and the cell's activities. Chitosan modifies osmotic pressure and mitigates the effects of ROS on cells (Hidangmayum et al. 2019). In related studies, drought stress in maize and wheat was successfully mitigated by applying only CuNPs (Van Nguyen et al. 2021), combined with Zn (Taran et al. 2017) and Ag NPs (Ahmed et al. 2021). It has also been established that exposure to TiO_2NPs significantly reduces the number of root tips and activates enzymatic defense mechanisms and secondary metabolites to better adapt plants to drought (Rajpal et al. 2023). There is a need for more studies to fully explore NPs' potential to enhance plant tolerance to drought.

12.7.3 Temperature

Temperature is critical in plant growth. An alteration in the optimum temperature has detrimental impacts on plants. Today, crops are faced with extremes and unprecedented changes in temperature due to climate change, which causes ion imbalance in the cells of plants, limiting their performance and yield (Ahmad et al. 2022; Markarian et al. 2023). Plants exposed to temperatures above the optimum (heat stress) experience additional osmotic and oxidative stress, disrupting their regular cellular processes (Rajpal et al. 2023). Selenium nanoparticles enhanced *Lycopersicum esculentum* resilience to both high and low temperatures (Haghighi et al. 2014); Au NPs and Ag NPs improved wheat tolerance to cold and heat stress respectively (Iqbal et al. 2017). Also, Fe_3O_4 NPs when applied to barley under heat stress, significantly enhanced its vigor and chlorophyll content (Tombuloglu et al. 2019), Likewise, ZnO NPs when applied to maize mitigate extreme temperature effects. Furthermore, selenium nanoparticles enhanced sorghum resilience to high temperatures by activating the antioxidant system to scavenge ROS caused by heat (Djanaguiraman et al. 2018b). TiO_2 NPs applied under cold stress conditions increased the concentration of glycyrrhizin in *Glycyrrhiza glabra* (Kardavan Ghabel and Karamian 2020), whilst chitosan NPs decreased ROS and improved osmoprotectants in bananas (Wang et al. 2021a).

12.7.4 Heavy Metal Stress

The limiting effect of heavy metals on soil and agricultural yield has made it a prominent issue. Its build up in plant tissues is detrimental to the growth and development of plants. Stressed plants show a range of symptoms which include reduced seed germination, transpiration, photosynthesis, declined growth, chlorosis, necrosis, and shortened roots. These occur as a result of intrinsic biochemical responses, such as increased ROS, enzyme denaturation, and disrupted protein structures (Feng et al. 2023). Nanoparticles have been widely employed to enhance plant growth and ameliorate metal-induced cytotoxicity. (Hussain et al. 2019a, b; Zhou et al. 2020; Akhtar et al. 2021a, b). Singh et al. reported that adding TiO_2NPs to the soil enhanced photosynthesis in soybean plants, and also significantly improved its vigor, hence reducing the toxicity of cadmium (Cd) in the plant tissues (Singh et al. 2016). Likewise, maize planted in soil polluted with Cd was treated with 100 and 250 ppm of TiO_2 NPs. Results showed that the foliar application of TiO_2NPs considerably improved biomass and impeded the absorption of Cd (Lian et al. 2020).

Furthermore, the application of Fe_2O_3 NPs to wheat under Cd stress showed a considerable increase in dry weight and the activity of antioxidant enzymes, as well as a reduction in the rate of electrolyte leakage in the leaves (Hussain et al. 2019a). CeO_2 NPs have also been applied to reduce metal stress in plants primarily due to Ce^{3+} and Ce^{4+} on their surface, which mitigate oxidative stress and mop off ROS (Wu et al. 2017). ZnO NPs alleviate plant stress by enhancing the physiological and biochemical activities of the plant and activating the oxidative stress system. It was demonstrated that applying ZnO NPs to rice (Ma et al. 2020) and *T. aestivum* plants (Hussain et al. 2018) under Cd stress enhanced growth, photosynthesis, and grain production (Table 12.1).

12.8 Challenges in the Development and Use of Nanotechnology for Crop Improvement

Remodification of tough varieties of crop could be achieved through nano assisted techniques to include traits that are fascinating, especially tolerance to environmental variability to meet food request of world population. However, most delivery techniques come up with poor efficiency thus creating a lag in the technology, while in some cases integration sites are not specific, some expression are insignificant and tissues are equally damaged (Altpeter et al. 2016). Furthermore, the techniques are host specific, tissues are regenerated which atimes cause fertility issues in modified plants. Plant cell wall physicochemical attributes creates delivery bottlenecks for biomolecules, where cell pore sizes are narrow, cargo delivery becomes impossible. The emergence of nanomaterials biomolecule delivery brought about unmatched success in the procedure, but, toxicity of the materials inhibit applications and in some cases DNA structural damage has been reported (Rastogi et al. 2017), hence there is the need to address this to come up with non-toxic plants. The oxidative properties of some NPs are very high, thus metabolism of normal plant cells are disturbed which often times lead to oxidative rupture of the modified cells (Du et al. 2017). Bonding of biomolecules to NPs is a difficult task, the affinity to bind varies in sundry molecules, and the discharge of the bond complex in cells of plants could be a herculean task as well, consequently, the inherent attributes of biomolecules and NPs are to be well understood to enhance adaptability for delivery (Dasgupta et al. 2014). Nano carrier preparations are very complex and comes with very high cost, this has impeded large scale use of the cargo carriers and this highlights that there is great need to develop cost effective carriers.

12.9 Conclusions

The benefits of nanomaterials cannot be amplified, however, the behaviour of NPs in soil, plants, seeds and leaves needs to be monitored to understand the pathways in order to determine the most appropriate mode of NP use to avoid environmental pollution given the toxic nature of NPs. Close monitoring would minimise NP contact with cells of plants, enzyme and plant cell damaging will be eliminated and thus improving NP delivery methods. Advancements in gene editing and mutagenesis of crops are very slow largely because of adoption, protection of culture and the desire for indigenous crops are factors working against adoption of modified crops. There is variation between regulations of the government and perception of public, this hurdle has to be overcomed in order to have abiotic tolerant crops.

References

Adrees M, Khan ZS, Ali S, Hafeez M, Khalid S et al (2020) Simultaneous mitigation of cadmium and drought stress in wheat by soil application of iron nanoparticles. Chemosphere 238:124681

AGNES (2020) Land degradation and climate change in Africa. Africa Group of Negotiators Experts Support. https://production-new-commonwealth-files.s3.eu-west-2.amazonaws.com/s3fs-public/documents/Policy-brief-2_Land-Degradation_Final_09032020.pdf?VersionId=YY9IxFeMAXNuaz6U66Kr6muz_SMCIe6Y

Ahmad W, Kumar Jaiswal K, Amjad M (2020) *Euphorbia herita* leaf extract as a reducing agent in a facile green synthesis of iron oxide nanoparticles and antimicrobial activity evaluation. Inorg Nano-Met Chem:1–8. https://doi.org/10.1080/24701556.2020.1815062

Ahmad Z, Tahseen S, Wasi A, Ganie IB, Shahzad A, Emamverdian A, Ramakrishnan M, Ding Y (2022) Nanotechnological interventions in agriculture. Nano 12(15):2667. https://doi.org/10.3390/nano12152667

Ahmed F, Javed B, Razzaq A, Mashwani Z (2021) Applications of copper and silver nanoparticles on wheat plants to induce drought tolerance and increase yield. IET Nanobiotechnol 15(1):68–78. https://doi.org/10.1049/nbt2.12002

Akhtar N, Ilyas N, Mashwani ZR, Hayat R, Yasmin H, Noureldeen A et al (2021a) Synergistic effects of plant growth promoting rhizobacteria and silicon dioxide nano-particles for amelioration of drought stress in wheat. Plant Physiol Biochem 166:160–176

Akhtar N, Khan S, Rehman SU, Rehman ZU, Khatoon A, Rha ES, Jamil M (2021b) Synergistic effects of zinc oxide nanoparticles and bacteria reduce heavy metals toxicity in rice (Oryza sativa L.) plant. Toxics 9(5):113. https://doi.org/10.3390/toxics9050113

Alabdallah NM, Alzahrani HS (2020) The potential mitigation effect of ZnO nanoparticles on [Abelmoschus esculentus L. Moench] metabolism under salt stress conditions. Saudi J Biol Sci 27(11):3132–3137. https://doi.org/10.1016/j.sjbs.2020.08.005

Ali S, Liu Y, Ishaq M, Shah T, Abdullah IA, Din I (2017) Climate change and its impact on the yield of major food crops: evidence from Pakistan. Food Secur 6(6):39. https://doi.org/10.3390/foods6060039

Ali EF, El-Shehawi AM, Ibrahim OHM, Abdul-Hafeez EY, Moussa MM, Hassan FAS (2021) A vital role of chitosan nanoparticles in improvisation the drought stress tolerance in Catharanthus roseus (L.) through biochemical and gene expression modulation. Plant Physiol Biochem 161:166–175

Alsaeedi A, El-Ramady H, Alshaal T, El-Garawany M, Elhawat N, Al-Otaibi A (2019) Silica nanoparticles boost growth and productivity of cucumber under water deficit and salinity stresses by balancing nutrients uptake. Plant Physiol Biochem 139:1–10

Altpeter F, Springer NM, Bartley LE, Blechl AE, Brutnell TP, Citovsky V et al (2016) Advancing crop transformation in the era of genome editing. Plant Cell 28:1510–1520. https://doi.org/10.1105/tpc.16.00196

Amani A, Zare N, Asadi A, Zakaria RA (2018) Ultrasound-enhanced gene delivery to alfalfa cells by hPAMAM dendrimer nanoparticles. Turk J Biol 42(1):63–75

An J, Hu P, Li F, Wu H, Shen Y, White JC et al (2020) Emerging investigator series: molecular mechanisms of plant salinity stress tolerance improvement by seed priming with cerium oxide nanoparticles. Environ Sci Nano 7:2214–2228

Antunes Filho S, dos Santos MS, dos Santos OAL, Backx BP, Soran M-L, Opriş O, Lung I, Stegarescu A, Bououdina M (2023) Biosynthesis of nanoparticles using plant extracts and essential oils. Molecules 28(7):3060. https://doi.org/10.3390/molecules28073060

Aqaei P, Weisany W, Diyanat M, Razmi J, Struik PC (2020) Response of maize (*Zea mays* L.) to potassium nano-silica application under drought stress. J Plant Nutr 43:1205–1216

Asif A, Ali M, Qadir M, Karthikeyan R, Singh Z, Khangura R, Di Gioia F, Ahmed ZFR (2023) Enhancing crop resilience by harnessing the synergistic effects of biostimulants against abiotic stress. Front Plant Sci 14. https://doi.org/10.3389/fpls.2023.1276117

Azmat A, Tanveer Y, Yasmin H, Hassan MN, Shahzad A, Reddy M, Ahmad A (2022) Coactive role of zinc oxide nanoparticles and plant growth promoting rhizobacteria for mitigation of synchronized effects of heat and drought stress in wheat plants. Chemosphere 297:133982. https://doi.org/10.1016/j.chemosphere.2022.133982

Bandala ER, Berli M (2019) Engineered nanomaterials (ENMs) and their role at the nexus of food, energy, and water. Mater Sci Energy Technol 2(1):29–40. https://doi.org/10.1016/j.mset.2018.09.004

Banerjee K, Pramanik P, Maity A, Joshi DC, Wani SH, Krishnan P (2019) Methods of using nanomaterials to plant systems and their delivery to plants (mode of entry, uptake, translocation, accumulation, biotransformation and barriers). In: Advances in phytonanotechnology. Elsevier, pp 123–152. https://doi.org/10.1016/b978-0-12-815322-2.00005-5

Bayda S, Adeel M, Tuccinardi T, Cordani M, Rizzolio F (2019) The history of nanoscience and nanotechnology: from chemical–physical applications to nanomedicine. Molecules 25(1):112. https://doi.org/10.3390/molecules25010112

Baz H, Creech M, Chen J, Gong H, Bradford K, Huo H (2020) Water-soluble carbon nanoparticles improve seed germination and post-germination growth of lettuce under salinity stress. Agronomy 2020(10):1192

Bellard C, Bertelsmeier C, Leadley P, Thuiller W, Courchamp F (2012) Impacts of climate change on the future of biodiversity. Ecol Lett 15(4):365–377. https://doi.org/10.1111/j.1461-0248.2011.01736.x

Bora KA, Hashmi S, Zulfiqar F, Abideen Z, Ali H, Siddiqui ZS, Siddique KHM (2022) Recent progress in bio-mediated synthesis and applications of engineered nanomaterials for sustainable agriculture. Front Plant Sci 13. https://doi.org/10.3389/fpls.2022.999505

Chandrasekaran R, Rajiv P, Abd-Elsalam KA (2020) Carbon nanotubes: plant gene delivery and genome editing. In: AbdElsalam K (ed) Carbon nanomaterials for agri-food and environmental applications. Elsevier, Amsterdam, pp 279–296

Chaudhry UK, Gökçe ZNÖ, Gökçe AF (2022) Salt stress and plant molecular responses. IntechOpen, London, UK

Chen Z, Wang Q (2021) Graphene ameliorates saline-alkaline stress-induced damage and improves growth and tolerance in alfalfa (Medicago sativa L.). Plant Physiol Biochem 163:128–138. https://doi.org/10.1016/j.plaphy.2021.03.039

Choudhary R, Rajput VD, Ghodake G, Ahmad F, Meena M, Rehman RU, Prasad R, Sharma RK, Singh R, Seth CS (2024) Comprehensive journey from past to present to future about seed priming with hydrogen peroxide and hydrogen sulfide concerning drought, temperature, UV and ozone stresses- a review. Plant Soil. https://doi.org/10.1007/s11104-024-06499-9

Choudhary R, Ahmad F, Kaya C, Upadhyay SK, Muneer S, Kumar V, Meena M, Liu H, Upadhyay H, Seth CS (2025) Decrypting proteomics, transcriptomics, genomics, and integrated omics for augmenting the abiotic, biotic, and climate change stress resilience in plants. J Plant Physiol. https://doi.org/10.1016/j.jplph.2025.154430

Cirillo V, Molisso D, Aprile AM, Maggio A, Rao R (2022) Systemin peptide application improves tomato salt stress tolerance and reveals common adaptation mechanisms to biotic and abiotic stress in plants. Environ Exp Bot 199:104865. https://doi.org/10.1016/j.envexpbot.2022.104865

Das A, Ray R, Mandal N, Chakrabarti K (2016) An analysis of transcripts and enzyme profiles in drought stressed jute (*Corchorus capsularis*) and rice (Oryza sativa) seedlings treated with CaCl2, hydroxyapatite nano-particle and β-amino butyric acid. J Plant Growth Regul 79(3):401–412. https://doi.org/10.1007/s10725-015-0144-9

Dasgupta S, Auth T, Gompper G (2014) Shape and orientation matter for the cellular uptake of nonspherical particles. Nano Lett 14:687–693. https://doi.org/10.1021/nl403949h

Demirer GS, Zhang H, Matos JL, Goh NS, Cunningham FJ, Sung Y, Chang R, Aditham AJ, Chio L, Cho MJ (2019) High aspect ratio nanomaterials enable delivery of functional genetic material without DNA integration in mature plants. Nat Nanotechnol 14:456–464

Demirer GS, Silva TN, Jackson CT, Thomas JB, Ehrhardt DW, Rhee SY, Mortimer JC, Landry MP (2021) Nanotechnology to advance CRISPR–Cas genetic engineering of plants. Nat Nanotechnol 16(3):243–250. https://doi.org/10.1038/s41565-021-00854-y

Diao M, Ma L, Wang J, Cui J, Fu A, Liu H (2014) Selenium promotes the growth and photosynthesis of tomato seedlings under salt stress by enhancing chloroplast antioxidant defense system. J Plant Growt Regul 33(3):671–682. https://doi.org/10.1007/s00344-014-9416-2

Dimkpa CO, Bindraban PS, Fugice J, Agyin-Birikorang S, Singh U, Hellums D (2017) Composite micronutrient nanoparticles and salts decrease drought stress in soybean. Agron Sustain Dev 37:5

Dimkpa CO, Singh U, Bindraban PS, Elmer WH, Gardea-Torresdey JL, White JC (2019) Zinc oxide nanoparticles alleviate drought-induced alterations in sorghum performance, nutrient acquisition, and grain fortification. Sci Total Environ 688:926–934

Djanaguiraman M, Nair R, Giraldo JP, Prasad PVV (2018a) Cerium oxide nanoparticles decrease drought-induced oxidative damage in sorghum leading to higher photosynthesis and grain yield. ACS Omega 3(10):14406–14416. https://doi.org/10.1021/acsomega.8b01894

Djanaguiraman M, Belliraj N, Bossmann SH, Prasad PVV (2018b) High-temperature stress alleviation by selenium nanoparticle treatment in grain sorghum. ACS Omega 3(3):2479–2491. https://doi.org/10.1021/acsomega.7b01934

Du W, Tan W, Peralta-Videa JR, Gardea-Torresdey JL, Ji R, Yin Y et al (2017) Interaction of metal oxide nanoparticles with higher terrestrial plants: physiological and biochemical aspects. Plant Physiol Biochem 110:210–225. https://doi.org/10.1016/j.plaphy.2016.04.024

El-Badri AM, Batool M, Mohamed IA, Khatab A, Sherif A, Wang Z et al (2021) Modulation of salinity impact on early seedling stage via nano-priming application of zinc oxide on rapeseed (Brassica napus L.). Plant Physiol Biochem 166:376–392

El-Saadony MT, Saad AM, Najjar AA, Alzahrani SO, Alkhatib FM, Shafi ME, Selem E, Desoky E-SM, Fouda SEE, El-Tahan AM, Hassan MAA (2021) The use of biological selenium nanoparticles to suppress *Triticum aestivum* L. crown and root rot diseases induced by Fusarium species and improve yield under drought and heat stress. Saudi J Biol Sci. https://doi.org/10.1016/j.sjbs.2021.04.043

El-Saber MM, Mahdi AA, Hassan AH, Farroh KY, Osman A (2021) Effects of magnetite nanoparticles on physiological processes to alleviate salinity induced oxidative damage in wheat. J Sci Food Agric 2021(101):5550–5562

Elsheery NI, Helaly MN, El-Hoseiny HM, Alam-Eldein SM (2020) Zinc oxide and silicone nanoparticles to improve the resistance mechanism and annual productivity of salt-stressed mango trees. Agronomy 10:558

Fabiyi OA (2021) Sustainable management of *Meloidogyne incognita* infecting carrot: green synthesis of silver nanoparticles with *Cnidoscolus aconitifolius*: (*Daucus carota*). Vegetos 34(2):277–285

Fabiyi OA (2023a) Nano clay materials and their role in plant disease management. In: Avila-Quezada GD, Rai M (eds) Nanophytopathology, pp 145–156. https://doi.org/10.1201/9781003344513-11

Fabiyi OA (2023b) The perspective of nanotechnology in plant protection. In: Avila-Quezada GD, Rai M (eds) Nanophytopathology, pp 23–38. https://doi.org/10.1201/9781003344513-3

Fabiyi OA, Olatunji GA (2018) Application of green synthesis in nano particles preparation: *Ficus mucoso* extracts in the management of *Meloidogyne incognita* parasitizing groundnut *Arachis hypogea*. Indian J Nematol 48(1):13–17

Fabiyi OA, Olatunji GA (2022) Application of green synthesis in silver nanoparticles for root knot nematode management on tomato. Indian J Nematol 52(2):164–173. https://doi.org/10.5958/0974-4444.2022.00020.8

Fabiyi OA, Olatunji GA (2024) Nematicidal potency of silver nanoparticles and other extracts from *Borreria verticillata* (L.) Mey on *Meloidogyne incognita* infected tomato plants. Indian Phytopathol 77(2):311–321

Fabiyi OA, Olatunji GA, Saadu AO (2018) Suppression of *Heterodera sacchari* in rice with agricultural waste-silver nano particles. J Solid Waste Tech Manag 44(2):87–91

Fabiyi OA, Olatunji GA, Atolani O, Olawuyi RO (2020a) Preparation of bio-nematicidal nanoparticles of *Eucalyptus officinalis* for the control of cyst nematode (*Heterodera sacchari*). J Anim Plant Sci 30(5):1172–1177

Fabiyi OA, Alabi RO, Ansari RA (2020b) Nanoparticles' synthesis and their application in the management of phytonematodes: an overview. In: Ansari RA, Rizvi R, Mahmood I (eds) Management of phytonematodes: recent advances and future challenges, Springer, Singapore, pp 125–140

Fabiyi OA, Claudius-Cole AO, Olatunji GA, Abubakar DO, Adejumo OA (2021) Evaluation of the *in vitro* response of *Meloidogyne incognita* to silver nano particle liquid from agricultural wastes. Agrivita J Agric Sci 43(3):524–534

Fabiyi OA, Lateef A, Gueguim-Kana EB, Beukes LS, Matyumza N, Bello TT, Olatunji GA (2024a) Correction characterization and nematicidal potential of copper, iron and zinc nanoparticles synthesized from *Tridax procumbens* L. extract on *Meloidogyne incognita* infected cabbage plants. Eur J Plant Pathol 168(4):683–695. https://doi.org/10.1007/s10658-023-02792-y

Fabiyi OA, Lateef A, Gueguim-Kana EB, Beukes LS, Matyumza N, Bello TT, Olatunji GA (2024b) Evaluation of *Borreria verticillata* (L.) synthesized copper, iron and zinc nanoparticles against *Meloidogyne incognita* on cowpea plants and the effect on cowpea seed quality. J Crop Health 76:1507–1518

Faizan M, Bhat JA, Chen C, Alyemeni MN, Wijaya L, Ahmad P et al (2021a) Zinc oxide nanoparticles (ZnO-NPs) induce salt tolerance by improving the antioxidant system and photosynthetic machinery in tomato. Plant Physiol Biochem 2021(161):122–130

Faizan M, Rajput VD, Al-Khuraif AA, Arshad M, Minkina T, Sushkova S, Yu F (2021b) Effect of foliar fertigation of chitosan nanoparticles on cadmium accumulation and toxicity in *Solanum lycopersicum*. Biol 10(7):666. https://doi.org/10.3390/biology10070666

FAO (2015) Climate change and food security: risks and responses. Food And Agriculture Organization of the United Nations. isbn:978-92-5-108998-9. https://www.fao.org/3/i5188e/I5188E.pdf

Fabiyi OA (2025) Nanomaterials of cellulose origin: Novel strategies for plant disease control. In: Rajan R, Ahmad F, Koshariya A, Seth CS (eds). Sustainable era of nanomaterials. Sustainable development and biodiversity. Springer, Singapore. 379–397 pp

Fabiyi OA, Ogundele AV (2025) biosensors for monitoring climatic stressors in agricultural crops. In: Uddin I (eds). Smart nanosensors. Springer, Singapore. 229–245 pp

Fabiyi OA, Wawata IG, Akinsipo OB, Ogundele AV (2025) Smart nanosensors: A new approach in crop protection and pest management. In: Uddin I (eds) Smart nanosensors. Springer, Singapore. 247–266 pp

Faraji J, Sepehri A (2020) Exogenous nitric oxide improves the protective effects of tio2 nanoparticles on growth, antioxidant system, and photosynthetic performance of wheat seedlings under drought stress. J Soil Sci Plant Nutr 20(2):703–714. https://doi.org/10.1007/s42729-019-00158-0

Feng D, Wang R, Sun X, Liu L, Liu P, Tang J, Zhang C, Liu H (2023) Heavy metal stress in plants: ways to alleviate with exogenous substances. Sci Total Environ 897:165397. https://doi.org/10.1016/j.scitotenv.2023.165397

Foroutan L, Solouki M, Abdossi V, Fakheri BA, Mahdinezhad N, Gholamipourfard K et al (2019) The effects of zinc oxide nanoparticles on drought stress in *Moringa peregrina* populations. Int J Basic Sci Med 4:119–127

Fu PP, Xia Q, Hwang H-M, Ray PC, Yu H (2014) Mechanisms of nanotoxicity: generation of reactive oxygen species. J Food Drug Anal 22(1):64–75. https://doi.org/10.1016/j.jfda.2014.01.005

Ghani MI, Saleem S, Rather SA, Rehmani MS, Alamri S, Rajput VD, Kalaji HM, Saleem N, Sial TA, Liu M (2022) Foliar application of zinc oxide nanoparticles: an effective strategy to mitigate drought stress in cucumber seedling by modulating antioxidant defense system and osmolytes accumulation. Chemosphere 289:133202. https://doi.org/10.1016/j.chemosphere.2021.133202

Ghorbanian H, Janmohammadi M, Ebadi-Segherloo A, Sabaghnia N (2017) Genotypic response of barley to exogenous application of nanoparticles under water stress condition. Ann Univ Mariae Curie-Sklodowska Sect. C- Biologia 272(2) https://cejsh.icm.edu.pl/cejsh/element/bwmeta1.element.ojs-doi10_17951_c_2017_72_2_15-27

Ghorbanpour M, Mohammadi H, Kariman K (2020) Nanosilicon-based recovery of barley (Hordeum vulgare) plants subjected to drought stress. Environ Sci Nano 7:443–461

Girgis E, Khalil WKB, Emam AN, Mohamed MB, Rao KV (2012) Nanotoxicity of gold and gold–cobalt nanoalloy. Chem Res Toxicol 25(5):1086–1098. https://doi.org/10.1021/tx300053h

Gohari G, Safai F, Panahirad S, Akbari A, Rasouli F, Dadpour MR, Fotopoulos V (2020) Modified multiwall carbon nanotubes display either phytotoxic or growth promoting and stress protecting activity in *Ocimum basilicum* L. in a concentration-dependent manner. Chemosphere 249:126171. https://doi.org/10.1016/j.chemosphere.2020.126171

Gong H, Meng D, Li X, Zhu F (2013) Soil degradation and food security coupled with global climate change in northeastern China. Chin Geogr Sci 23(5):562–573. https://doi.org/10.1007/s11769-013-0626-5

González Guzmán M, Cellini F, Fotopoulos V, Balestrini R, Arbona V (2021) New approaches to improve crop tolerance to biotic and abiotic stresses. Physiol Plant 174(1). https://doi.org/10.1111/ppl.13547

Gonzalez L, Lison D, Kirsch-Volders M (2008) Genotoxicity of engineered nanomaterials: a critical review. Nanotoxicol 2(4):252–273. https://doi.org/10.1080/17435390802464986

Gowayed S, Al-Zahrani HSM, Metwali EMR (2017) Improving the salinity tolerance in potato (Solanum tuberosum) by exogenous application of silicon dioxide nanoparticles. Int J Agric Biol 19(01):183–192. https://doi.org/10.17957/ijab/15.0262

Guleria G, Shweta T, Shandilya M, Sharma S, Sapna T, Kalia S (2023) Nanotechnology for sustainable agro-food systems: the need and role of nanoparticles in protecting plants and improving crop productivity. Plant Physiol Biochem 194:533–549. https://doi.org/10.1016/j.plaphy.2022.12.004

Haghighi M, Afifipour Z, Mozafarian M (2012) The alleviation effect of silicon on seed germination and seedling growth of tomato under salinity stress. J Fruit Ornam Plant Res 76(1):119–126. https://doi.org/10.2478/v10032-012-0008-z

Haghighi M, Abolghasemi R, Teixeira da Silva JA (2014) Low and high temperature stress affect the growth characteristics of tomato in hydroponic culture with Se and nano-Se amendment. Sci Hortic 178:231–240. https://doi.org/10.1016/j.scienta.2014.09.006

Hammond WM, Williams AP, Abatzoglou JT, Adams HD, Klein T, López R et al (2022) Global field observations of tree die-off reveal hotter-drought fingerprint for earth's forests. Nat Commun 13:1–11

Hasanuzzaman M, Hakeem KR, Nahar K, Alharby HF (2019) Plant abiotic stress tolerance: agronomic, molecular and biotechnological approaches, 1st edn. Springer, New York City. https://doi.org/10.1007/978-3-030-06118-0

Hassan FAS, Ali E, Gaber A, Fetouh MI, Mazrou R (2021) Chitosan nanoparticles effectively combat salinity stress by enhancing antioxidant activity and alkaloid biosynthesis in *Catharanthus roseus* (L.) G. Don. Plant Physiol Biochem 162:291–300

He K, Gutiérrez EE, Heming NM, Koepfli K, Wan T, He S, Jin W, Liu S, Jiang X (2019) Cryptic phylogeographic history sheds light on the generation of species diversity in sky-island mountains. J Biogeogr 46(10):2232–2247. https://doi.org/10.1111/jbi.13664

Hemlata MPR, Singh AP, Tejavath KK (2020) Biosynthesis of silver nanoparticles using Cucumis prophetarum aqueous leaf extract and their antibacterial and antiproliferative activity against cancer cell lines. ACS Omega 5(10):5520–5528. https://doi.org/10.1021/acsomega.0c00155

Hernández-Hernández H, González-Morales S, Benavides-Mendoza A, Ortega-Ortiz H, Cadenas-Pliego G, Juárez-Maldonado A (2018) Effects of chitosan–PVA and Cu nanoparticles on the growth and antioxidant capacity of tomato under saline stress. Molecules 23(1):178. https://doi.org/10.3390/molecules23010178

Hidangmayum A, Dwivedi P, Katiyar D, Hemantaranjan A (2019) Application of chitosan on plant responses with special reference to abiotic stress. Physiol Mol Biol Plants 25(2):313–326. https://doi.org/10.1007/s12298-018-0633-1

Huang Q, Liu Q, Lin L, Li F-J, Han Y, Song Z-G (2018) Reduction of arsenic toxicity in two rice cultivar seedlings by different nanoparticles. Ecotoxicol Environ Saf 159:261–271. https://doi.org/10.1016/j.ecoenv.2018.05.008

Hussain A, Ali S, Rizwan M, Zia ur Rehman M, Javed MR, Imran M, Chatha SAS, Nazir R (2018) Zinc oxide nanoparticles alter the wheat physiological response and reduce the cadmium uptake by plants. Environ Pollut 242:1518–1526. https://doi.org/10.1016/j.envpol.2018.08.036

Hussain A, Ali S, Rizwan M, ur Rehman MZ, Qayyum MF, Wang H, Rinklebe J (2019a) Responses of wheat (Triticum aestivum) plants grown in a Cd contaminated soil to the application of iron oxide nanoparticles. Ecotoxicol Environ Saf 173:156–164. https://doi.org/10.1016/j.ecoenv.2019.01.118

Hussain A, Rizwan M, Ali Q, Ali S (2019b) Seed priming with silicon nanoparticles improved the biomass and yield while reduced the oxidative stress and cadmium concentration in wheat grains. Environ Sci Pollut Res 26(8):7579–7588. https://doi.org/10.1007/s11356-019-04210-5

Ilyas M, Nisar M, Khan N, Hazrat A, Khan AH, Hayat K, Fahad S, Khan A, Ullah A (2020) Drought tolerance strategies in plants: a mechanistic approach. J Plant Growth Regul 40(3):926–944. https://doi.org/10.1007/s00344-020-10174-5

Ioras F, Bandara I, Kemp C (2014) Introduction to climate change and land degradation. In: Arraiza MP, Santamarta JC, Ioras F, García-Rodríguez JL, Abrudan IV, Korjus H, Borála G (eds) Climate change and restoration of degraded land. Colegio de Ingenieros de Montes, Madrid, pp 15–52. isbn:978-84-617-1377-6

IPBES (2018) Summary for policymakers of the assessment report on land degradation and restoration of the Intergovernmental Science-Policy Platform on Biodiversity and Ecosystem Services. In: Scholes R, Montanarella L, Brainich A, Barger N, ten Brink B, Cantele M, Erasmus B, Fisher J, Gardner T, Holland TG, Kohler F, Kotiaho JS, Von Maltitz G, Nangendo G, Pandit R, Parrotta J, Potts MD, Prince S, Sankaran M, Willemen L (eds) . IPBES Secretariat, Bonn, p 44

Iqbal M, Raja NI, Mashwani Z-U-R, Hussain M, Ejaz M, Yasmeen F (2017) Effect of silver nanoparticles on growth of wheat under heat stress. Iran J Sci Technol Trans A Sci 43(2):387–395. https://doi.org/10.1007/s40995-017-0417-4

Itroutwar PD, Kasivelu G, Raguraman V, Malaichamy K, Sevathapandian SK (2020) Effects of biogenic zinc oxide nanoparticles on seed germination and seedling vigor of maize (Zea mays). Biocatal Agric Biotechnol 29:101778. https://doi.org/10.1016/j.bcab.2020.101778

Jagadesh M, Dash M, Kumari A, Singh SK, Verma KK, Kumar P, Bhatt R, Sharma SK (2024) Revealing the hidden world of soil microbes: metagenomic insights into plant, bacteria, and fungi interactions for sustainable agriculture and ecosystem restoration. Microbiol Res 285:127764. https://doi.org/10.1016/j.micres.2024.127764

Jayawardena HSN, Liyanage SH, Rathnayake K, Patel U, Yan M (2021) Analytical methods for characterization of nanomaterial surfaces. Anal Chem 93(4):1889–1911. https://doi.org/10.1021/acs.analchem.0c05208

Jia G, Shevliakova E, Artaxo P, De Noblet-Ducoudré N, Houghton R, House J, Kitajima K, Lennard C, Popp A, Sirin A, Sukumar R, Verchot L (2019) Land–climate interactions. In: Shukla PR, Skea J, Calvo Buendia E, Masson-Delmotte V, Pörtner HO, Roberts DC, Zhai P, Slade R, Connors S, van Diemen R, Ferrat M, Haughey E, Luz S, Neogi S, Pathak M, Petzold J, Portugal Pereira J, Vyas P, Huntley E, Kissick K, Belkacemi M, Malley J (eds) Climate change and land: an IPCC special report on climate change, desertification, land degradation, sustainable land management, food security, and greenhouse gas fluxes in terrestrial ecosystems. https://doi.org/10.1017/9781009157988.004

Karami A, Sepehri A (2018) Beneficial role of MWCNTs and SNP on growth, physiological and photosynthesis performance of barley under NaCl stress. J Soil Sci Plant Nutr. https://doi.org/10.4067/s0718-95162018005002202

Kardavan Ghabel V, Karamian R (2020) Effects of TiO_2 nanoparticles and spermine on antioxidant responses of Glycyrrhiza glabra L. to cold stress. Acta Bot Croat 79(2):137–147. https://doi.org/10.37427/botcro-2020-025

Karvar M, Azari A, Rahimi A, Maddah-Hosseini S, Ahmadi-Lahijani MJ (2022) Titanium dioxide nanoparticles (TiO2-NPs) enhance drought tolerance and grain yield of sweet corn (Zea mays L.) under deficit irrigation regimes. Acta Physiol Plant 44:14

Khaldari I, Naghavi MR, Motamedi E (2021) Synthesis of green and pure copper oxide nanoparticles using two plant resources via solid-state route and their phytotoxicity assessment. RSC Adv 11(6):3346–3353. https://doi.org/10.1039/d0ra09924d

Khalofah A, Kilany M, Migdadi H (2021) Phytostimulatory influence of comamonas testosteroni and silver nanoparticles on *Linum usitatissimum* L. under salinity stress. Plan Theory 10:790

Khot LR, Sankaran S, Maja JM, Ehsani R, Schuster EW (2012) Applications of nanomaterials in agricultural production and crop protection: a review. J Crop Prot 35:64–70. https://doi.org/10.1016/j.cropro.2012.01.007

Kiapour H, Moaveni P, Habibi D, Sani B (2015) Evaluation of the application of gibbrellic acid and titanium dioxide nanoparticles under drought stress on some traits of basil (*Ocimum basilicum* L.). Int J Agron Agric Res 6:138–150

Kisin ER, Murray AR, Sargent L, Lowry D, Chirila M, Siegrist KJ, Schwegler-Berry D, Leonard S, Castranova V, Fadeel B et al (2011) Genotoxicity of carbon nanofibers: are they potentially more or less dangerous than carbon nanotubes or asbestos? Toxicol Appl Pharmacol 252(1):1–10. https://doi.org/10.1016/j.taap.2011.02.001

Kohl Y, Rundén-Pran E, Mariussen E, Hesler M, El Yamani N, Longhin EM, Dusinska M (2020) Genotoxicity of nanomaterials: advanced in vitro models and high throughput methods for human hazard assessment—a review. Nano 10(10):1911. https://doi.org/10.3390/nano10101911

Kumar S, Shandilya M, Uniyal P, Thakur S, Parihar N (2022) Efficacy of polymeric nanofibrous membranes for proficient wastewater treatment. Polym Bull 80(7):7145–7200. https://doi.org/10.1007/s00289-022-04417-6

Kumar D, Mariyam S, Gupta KJ, Thiruvengadam M, Ghodake GS, Xing B, Seth CS (2024a) Comparative investigation on chemical and green synthesized titanium dioxide nanoparticles against chromium (VI) stress eliciting differential physiological, biochemical, and cellular attributes in Helianthus annuus L. Sci Total Environ 930:172413. https://doi.org/10.1016/j.scitotenv.2024.172413

Kumar D, Singh R, Upadhyay SK, Verma KK, Tripathi RM, Liu H, Dhankher OP, Tripathi RD, Sahi SV, Seth CS (2024b) Review on interactions between nanomaterials and phytohormones: novel perspectives and opportunities for mitigating environmental challenges. Plant Sci 340:111964. https://doi.org/10.1016/j.plantsci.2023.111964

Kumar D, Biswas JK, Mulla SI, Singh R, Shukla R, Ahanger MA, Shekhawat GS, Verma KK, Siddiqui MW, Seth CS (2024c) Micro and nanoplastics pollution: sources, distribution, uptake in plants, toxicological effects, and innovative remediation strategies for environmental sustainability. Plant Physiol Biochem 213:108795. https://doi.org/10.1016/j.plaphy.2024.108795

Kumaraswamy RV, Saharan V, Kumari S, Chandra Choudhary R, Pal A, Sharma SS, Rakshit S, Raliya R, Biswas P (2021) Chitosan-silicon nanofertilizer to enhance plant growth and yield in maize (*Zea mays* L.). Plant Physiol Biochem 159:53–66. https://doi.org/10.1016/j.plaphy.2020.11.054

Kumari A, Lakshmi GA, Krishna GK, Patni B, Prakash S, Bhattacharyya M, Singh SK, Verma KK (2022) Climate change and its impact on crops: a comprehensive investigation for sustainable agriculture. Agronomy 12:3008. https://doi.org/10.3390/agronomy12123008

Kumari A, Dash M, Singh SK, Jagadesh M, Mathpal B, Mishra PK, Pandey SK, Verma KK (2023) Soil microbes: a natural solution for mitigating the impact of climate change. Environ Monit Assess 195:1436. https://doi.org/10.1007/s10661-023-11988-y

Kwak S-Y, Lew TTS, Sweeney CJ, Koman VB, Wong MH, Bohmert-Tatarev K, Snell KD, Seo JS, Chua N-H, Strano MS (2019) Chloroplast-selective gene delivery and expression in planta using chitosan-complexed single-walled carbon nanotube carriers. Nat Nanotechnol 14(5):447–455. https://doi.org/10.1038/s41565-019-0375-4

Labuda J, Barek J, Gajdosechova Z, Goenaga-Infante H, Johnston LJ, Mester Z, Shtykov S (2023) Analytical chemistry of engineered nanomaterials: part 1. Scope, regulation, legislation, and metrology (IUPAC Technical Report). Pure Appl Chem 95(2):133–163. https://doi.org/10.1515/pac-2021-1001

Li S, Chen S, Zhang Z, Huang Y, Li G, Li Y, Li J (2022) Short-term exposure to silver nanoparticles alters the physiology and induces stress-related gene expression in Nelumbo nucifera. Plant Physiology and Biochemistry 177:38–45. https://doi.org/10.1016/j.plaphy.2022.02.021

Lian J, Zhao L, Wu J, Xiong H, Bao Y, Zeb A, Tang J, Liu W (2020) Foliar spray of TiO_2 nanoparticles prevails over root application in reducing Cd accumulation and mitigating Cd-induced phytotoxicity in maize (*Zea mays* L.). Chemosphere 239:124794. https://doi.org/10.1016/j.chemosphere.2019.124794

Lichtenberg SS, Nuti K, DeRouchey JE, Tsyusko O, Unrine J (2020) Efficacy of chitosan/double-stranded RNA polyplex nanoparticles for gene silencing under variable environmental conditions. Environ Sci Nano 7(5):1582–1592

Linh TM, Mai NC, Hoe PT, Lien LQ, Ban NK, Hien LTT, Chau NH, Van NT (2020) Metal-based nanoparticles enhance drought tolerance in soybean. J Nanomater 2020:1–13. https://doi.org/10.1155/2020/4056563

Liu J, Agarwal M, Varahramyan K, Berney ES, Hodo WD (2008) Polymer-based microsensor for soil moisture measurement. Sens. Actuators B 129:599–604

Liu C, Zhou H, Zhou J (2021a) The applications of nanotechnology in crop production. Molecules 26(23):7070. https://doi.org/10.3390/molecules26237070

Liu J, Li G, Chen L, Gu J, Wu H, Li Z (2021b) Cerium oxide nanoparticles improve cotton salt tolerance by enabling better ability to maintain cytosolic K^+/Na^+ ratio. Research Square Platform LLC. https://doi.org/10.21203/rs.3.rs-210835/v1

Liu Z, Herman JD, Huang G, Kadir T, Dahlke HE (2021c) Identifying climate change impacts on surface water supply in the southern Central Valley, California. Sci Total Environ 759:143429. https://doi.org/10.1016/j.scitotenv.2020.143429

Ma X, Sharifan H, Dou F, Sun W (2020) Simultaneous reduction of arsenic (As) and cadmium (Cd) accumulation in rice by zinc oxide nanoparticles. Chem Eng J 384:123802. https://doi.org/10.1016/j.cej.2019.123802

Magdolenova Z, Drlickova M, Henjum K, Rundén-Pran E, Tulinska J, Bilanicova D, Pojana G, Kazimirova A, Barancokova M, Kuricova M et al (2013) Coating-dependent induction of cytotoxicity and genotoxicity of iron oxide nanoparticles. Nanotoxicology 9(Suppl 1):44–56. https://doi.org/10.3109/17435390.2013.847505

Mahakham W, Sarmah AK, Maensiri S, Theerakulpisut P (2017) Nanopriming technology for enhancing germination and starch metabolism of aged rice seeds using phytosynthesized silver nanoparticles. Sci Rep 7:8263

Mariyam S, Upadhyay SK, Chakraborty K, Verma KK, Duhan JS, Muneer S, Meena M, Sharma RK, Ghodake G, Seth CS (2024) Nanotechnology, a frontier in agricultural science, a novel approach in abiotic stress management and convergence with new age medicine-a review. Sci Total Environ *912*:169097. https://doi.org/10.1016/j.scitotenv.2023.169097

Markarian S, Shariatmadari H, Shirvani M, Mirmohammady Maibody SAM (2023) Impacts of ZnO nanoparticles and dissolved zinc ($ZnSO_4$) on low temperature induced responses of wheat. J Plant Nutr 46(14):3435–3449. https://doi.org/10.1080/01904167.2023.2206421

Martin-Ortigosa S, Valenstein JS, Lin VSY, Trewyn BG, Wang K (2012) Gold functionalized mesoporous silica nanoparticle mediated protein and DNA codelivery to plant cells via the biolistic method. Adv. Funct Mater 22:3576–3582. https://doi.org/10.1002/adfm.201200359

Martínez-Ballesta C, Zapata L, Chalbi N, Carvajal M (2016) Multiwalled carbon nanotubes enter broccoli cells enhancing growth and water uptake of plants exposed to salinity. J Nanobiotechnology 14:42

Mazhar WM, Ishtiaq M, Hussain I, Parveen A, Hayat Bhatti K, Azeem M et al (2022) Seed nanopriming with zinc oxide nanoparticles in rice mitigates drought and enhances agronomic profile. PLoS One 17:e0264967

Mei N, Zhang Y, Chen Y, Guo X, Ding W, Ali SF, Biris AS, Rice P, Moore MM, Chen T (2012) Silver nanoparticle-induced mutations and oxidative stress in mouse lymphoma cells. Environ Mol Mutagen 53(6):409–419. https://doi.org/10.1002/em.21698

Melino V, Tester M (2023) Salt-tolerant crops: time to deliver. Annu Rev Plant Biol 74:671–696

Mohammadi R, Maali-Amiri R, Mantri NL (2014) Effect of TiO_2 nanoparticles on oxidative damage and antioxidant defense systems in chickpea seedlings during cold stress. Russ J Plant Physiol 61(6):768–775. https://doi.org/10.1134/s1021443714050124

Mourdikoudis S, Pallares RM, Thanh NTK (2018) Characterization techniques for nanoparticles: comparison and complementarity upon studying nanoparticle properties. Nanoscale 10(27):12871–12934. https://doi.org/10.1039/c8nr02278j

Mukhopadhyay R, Sarkar B, Jat HS, Sharma PC, Bolan NS (2021) Soil salinity under climate change: challenges for sustainable agriculture and food security. J Environ Manag 280:111736. https://doi.org/10.1016/j.jenvman.2020.111736

Mustafa H, Ilyas N, Akhtar N, Raja NI, Zainab T, Shah T et al (2021) Biosynthesis and characterization of titanium dioxide nanoparticles and its effects along with calcium phosphate on physicochemical attributes of wheat under drought stress. Ecotoxicol Environ Safety 2021(223):112519

Namjoyan S, Rajabi A, Sorooshzadeh A, Agha Alikhani M (2020a) Impact of nanosilicon and tebuconazole foliar application on some physiological traits, growth and white sugar yield of sugar beet under drought stress. J Sugar Beet 35:157–173

Namjoyan S, Sorooshzadeh A, Rajabi A, Aghaalikhani M (2020b) Nano-silicon protects sugar beet plants against water deficit stress by improving the antioxidant systems and compatible solutes. Acta Physiol Plant 42(10). https://doi.org/10.1007/s11738-020-03137-6

Naqvi S, Maitra AN, Abdin MZ, Akmal MD, Arora I, Samim MD (2012) Calcium phosphate nanoparticle mediated genetic transformation in plants. J Mater Chem 22(8):3500–3507

Nasrallah AK, Kheder AA, Kord MA, Fouad AS, El-Mogy MM, Atia MAM (2021) Mitigation of salinity stress effects on broad bean productivity using calcium phosphate nanoparticles application. Horticulturae 2022(8):75

Ndlovu N, Mayaya T, Muitire C, Munyengwa N (2020) Nanotechnology applications in crop production and food systems. Int J Plant Breed Crop Sci 7(1):603–613

Noohpisheh Z, Amiri H, Mohammadi A, Farhadi S (2020) Effect of the foliar application of zinc oxide nanoparticles on some biochemical and physiological parameters of Trigonella foenum-graecum under salinity stress. Plant Biosyst 155(2):267–280. https://doi.org/10.1080/11263504.2020.1739160

Noor R, Yasmin H, Ilyas N, Nosheen A, Hassan MN, Mumtaz S et al (2022) Comparative analysis of iron oxide nanoparticles synthesized from ginger (*Zingiber officinale*) and cumin seeds (*Cuminum cyminum*) to induce resistance in wheat against drought stress. Chemosphere 292:133201

Oberdörster G, Oberdörster E, Oberdörster J (2005) Nanotoxicology: an emerging discipline evolving from studies of ultrafine particles. Environ Health Perspect 113(7):823–839. https://doi.org/10.1289/ehp.7339

Plaksenkova I, Jermaļonoka M, Bankovska L, Gavarāne I, Gerbreders V, Sledevskis E, Sniķeris J, Kokina I (2019) Effects of Fe_3O_4 nanoparticle stress on the growth and development of rocket Eruca sativa. J Nanomater 2019:1–10. https://doi.org/10.1155/2019/2678247

Rafsanjani MSO, Kiran U, Ali A, Abdin MZ (2016) Transformation efficiency of calcium phosphate nanoparticles for genetic fluorescently labeled mesoporous silica nanoparticles. Plant Cell Rep 33:1389–1402

Rahman A, Hossain MS, Mahmud J-A, Nahar K, Hasanuzzaman M, Fujita M (2016) Manganese-induced salt stress tolerance in rice seedlings: regulation of ion homeostasis, antioxidant defense and glyoxalase systems. Physiol Mol Biol Plants 22(3):291–306. https://doi.org/10.1007/s12298-016-0371-1

Rajpal VR, Prakash S, Mehta S, Minkina T, Rajput VD, Deswal R (2023) A comprehensive review on mitigating abiotic stresses in plants by metallic nanomaterials: prospects and concerns. Clean Techn Environ Policy. https://doi.org/10.1007/s10098-023-02561-9

Rajput VD, Kumari A, Upadhyay SK, Minkina T, Mandzhieva S, Ranjan A, Sushkova S, Burachevskaya M, Rajput P, Konstantinova E, Singh J, Verma KK (2023) Can nanomaterials improve the soil microbiome and crop productivity? Agriculture 13:231. https://doi.org/10.3390/agriculture13020231

Rastogi A, Zivcak M, Sytar O, Kalaji HM, He X, Mbarki S et al (2017) Impact of metal and metal oxide nanoparticles on plant: a critical review. *Front Chem* 5:78. https://doi.org/10.3389/fchem.2017.00078

Raza A, Charagh S, Salehi H, Abbas S, Saeed F, Poinern GEJ et al (2023) Nano-enabled stress-smart agriculture: can nanotechnology deliver drought and salinity-smart crops? J Sustain Agric Environ 2023(2):189–214. https://doi.org/10.1002/sae2.12061

Reich PF, Numben ST, Almaraz RA, Eswaran H (2001) Land resources stresses and desertification in Africa. In: Bridges EM, Hannam ID, Penning de Vries FWT, Scherr SJ, Sombatpanit S (eds) Response to land degradation. Science Publishers Inc., Enfield, pp 101–114

Salachna P, Byczyńska A, Zawadzińska A, Piechocki R, Mizielińska M (2019) Stimulatory effect of silver nanoparticles on the growth and flowering of potted oriental lilies. Agronomy 9(10):610. https://doi.org/10.3390/agronomy9100610

Saleh HM, Hassan AI (2023) Synthesis and characterization of nanomaterials for application in cost-effective electrochemical devices. Sustain For 15(14):10891. https://doi.org/10.3390/su151410891

Sanavi M, Mohammad SA (2018) Effects of iron and manganese nano-chelate foliar application on some qualitative traits of (Vigna Radiata L.Wilczek Parto variety) under water deficit condition. Iran J Field Crop Sci 49:61–70

Sangeetha J, Sarim KM, Thangadurai D, Gupta A, Mundaragi A, Sheth BP, Wani SA, Baqual MF, Habib H (2019) Nanoparticle-mediated plant gene transfer for precision farming and sustainable agriculture. In: Panpatte D, Jhala Y (eds) Nanotechnology for agriculture. Springer, Singapore

Sanzari I, Leone A, Ambrosone A (2019) Nanotechnology in plant science: to make a long story short. Front Bioeng Biotechnol 7:120

Semida WM, Abdelkhalik A, Mohamed Gamal F, Abd El-Mageed TA, Abd El-Mageed SA, Rady MM, Ali EF (2021) Foliar application of zinc oxide nanoparticles promotes drought stress tolerance in eggplant (Solanum melongena L.). Plan Theory 10(2):421. https://doi.org/10.3390/plants10020421

Shafey AME (2020) Green synthesis of metal and metal oxide nanoparticles from plant leaf extracts and their applications: a review. Green Process Synth 9(1):304–339. https://doi.org/10.1515/gps-2020-0031

Shah T, Latif S, Saeed F, Ali I, Ullah S, Abdullah Alsahli A et al (2021) Seed priming with titanium dioxide nanoparticles enhances seed vigor, leaf water status, and antioxidant enzyme activities in maize (*Zea mays* L.) under salinity stress. J King Saud Univ Sci 33(1):101207. https://doi.org/10.1016/j.jksus.2020.10.004

Shahi S, Srivastava M (2018) Influence of foliar application of manganese on growth, pigment content, and nitrate reductase activity of *Vigna radiata*(L.) R. Wilczek under salinity. J Plant Nutr 41(11):1397–1404. https://doi.org/10.1080/01904167.2018.1454470

Shahriari A, Omidi H, Mohammadi H, Mohammadi A, Ahmadi K (2019) The effect of carbon nanotubes seed priming on germination and photosynthetic pigmentation of maize hybrids under drought stress. Agroecol J 15:1–12

Shang Y, Hasan MK, Ahammed GJ, Li M, Yin H, Zhou J (2019) Applications of nanotechnology in plant growth and crop protection: a review. Molecules 24(14):2558. https://doi.org/10.3390/molecules24142558

Singh J, Lee B-K (2016) Influence of nano-TiO_2 particles on the bioaccumulation of Cd in soybean plants (Glycine max): a possible mechanism for the removal of Cd from the contaminated soil. J Environ Manag 170:88–96. https://doi.org/10.1016/j.jenvman.2016.01.015

Singh D, Sillu D, Kumar A, Agnihotri S (2021) Dual nanozyme characteristics of iron oxide nanoparticles alleviate salinity stress and promote the growth of an agroforestry tree, *Eucalyptus tereticornis* Sm. Environ Sci Nano 2021(8):1308–1325

Singh H, Sharma A, Bhardwaj SK, Arya SK, Bhardwaj N, Khatri M (2021a) Recent advances in the applications of nano-agrochemicals for sustainable agricultural development. Environ Sci Process Impacts 23(2):213–239. https://doi.org/10.1039/d0em00404a

Singh P, Arif Y, Siddiqui H, Sami F, Zaidi R, Azam A, Alam P, Hayat S (2021b) Nanoparticles enhances the salinity toxicity tolerance in Linum usitatissimum L. by modulating the antioxidative enzymes, photosynthetic efficiency, redox status and cellular damage. Ecotoxicol Environ Saf 213:112020. https://doi.org/10.1016/j.ecoenv.2021.112020

Song Y, Jiang M, Zhang H, Li R (2021) Zinc oxide nanoparticles alleviate chilling stress in rice (oryza sativa L.) by regulating antioxidative system and chilling response transcription factors. Molecules 26(8):2196. https://doi.org/10.3390/molecules26082196

Sun L, Song F, Guo J, Zhu X, Liu S, Liu F, Li X (2020a) Nano-ZnO-induced drought tolerance is associated with melatonin synthesis and metabolism in maize. Int J Mol Sci 21(3):782. https://doi.org/10.3390/ijms21030782

Sun W, Shahrajabian MH, Huang Q (2020b) Soybean seeds treated with single walled carbon nanotubes (SwCNTs) showed enhanced drought tolerance during germination. Int J Adv Biol Biomed Res 8:9–16

Szőllősi R, Molnár Á, Kondak S, Kolbert Z (2020) Dual effect of nanomaterials on germination and seedling growth: stimulation vs. phytotoxicity. Plan Theory 9:1745

Taran N, Storozhenko V, Svietlova N, Batsmanova L, Shvartau V, Kovalenko M (2017) Effect of zinc and copper nanoparticles on drought resistance of wheat seedlings. Nanoscale Res Lett 12(1):60. https://doi.org/10.1186/s11671-017-1839-9

Thakur S, Asthir B, Kaur G, Kalia A, Sharma A (2021) Zinc oxide and titanium dioxide nanoparticles influence heat stress tolerance mediated by antioxidant defense system in wheat. Cereal Res Commun 50(3):385–396. https://doi.org/10.1007/s42976-021-00190-w

Thuesombat P, Hannongbua S, Ekgasit S, Chadchawan S (2016) Effects of silver nanoparticles on hydrogen peroxide generation and antioxidant enzyme responses in rice. J Nanosci Nanotechnol 16(8):8030–8043. https://doi.org/10.1166/jnn.2016.12754

Tombuloglu H, Slimani Y, Tombuloglu G, Almessiere M, Baykal A (2019) Uptake and translocation of magnetite (Fe_3O_4) nanoparticles and its impact on photosynthetic genes in barley (Hordeum vulgare L.). Chemosphere 226:110–122. https://doi.org/10.1016/j.chemosphere.2019.03.075

Tovar GI, Briceño S, Suarez J, Flores S, González G (2020) Biogenic synthesis of iron oxide nanoparticles using Moringa oleifera and chitosan and its evaluation on corn germination. Environ Nanotechnol Monit Manag 14:100350. https://doi.org/10.1016/j.enmm.2020.100350

UNDESA (2019) United Nations Department of Economic and Social Affairs (Population Division). https://www.worldometers.info/population/world/

Ur Rahman S, Xuebin Q, Kamran M, Yasin G, Cheng H, Rehim A, Riaz L, Rizwan M, Ali S, Alsahli AA, Alyemeni MN (2021) Silicon elevated cadmium tolerance in wheat (Triticum aestivum L.) by endorsing nutrients uptake and antioxidative defense mechanisms in the leaves. Plant Physiol Biochem 166:148–159. https://doi.org/10.1016/j.plaphy.2021.05.038

Vaghar MS, Sayfzadeh S, Zakerin HR, Kobraee S, Valadabady AR (2021) Effect offoliar application of iron, zinc and manganese nano-chelate on some quantitative and qualitative characteristics of soybean (Glycine max L.) under water deficit stress. Environ Stress Crop Sci 14:703–718

Valko M, Rhodes CJ, Moncol J, Izakovic M, Mazur M (2006) Free radicals, metals and antioxidants in oxidative stress-induced cancer. Chem Biol Interact 160(1):1–40. https://doi.org/10.1016/j.cbi.2005.12.009

van Dijk M, Morley T, Rau ML, Saghai Y (2021) A meta-analysis of projected global food demand and population at risk of hunger for the period 2010–2050. Nat Food 2(7):494–501. https://doi.org/10.1038/s43016-021-00322-9

Van Nguyen D, Nguyen HM, Le NT, Nguyen KH, Nguyen HT, Le HM, Nguyen AT, Dinh NTT, Hoang SA, Van Ha C (2021) Copper nanoparticle application enhances plant growth and grain yield in maize under drought stress conditions. J Plant Growth Regul 41(1):364–375. https://doi.org/10.1007/s00344-021-10301-w

Van Nguyen D, Nguyen HM, Le NT, Nguyen KH, Nguyen HT, Le HM et al (2022) Copper nanoparticle application enhances plant growth and grain yield in maize under drought stress conditions. J Plant Growth Regul 41(364):75

Venzhik YV, Deryabin AN, Popov VN, Dykman LA, Titov AF, Moshkov IE (2022) Influence of gold nanoparticles on the tolerance of wheat to low temperature. Doklady Biochem Biophys 502(1):5–9. https://doi.org/10.1134/s1607672922010100

Verma KK, Song XP, Kumari A, Jagadesh M, Singh SK, Bhatt R, Singh M, Seth CS, Li YR (2024a) Climate change adaptation: challenges for agricultural sustainability. Plant Cell Environ:1–12. https://doi.org/10.1111/pce.15078

Verma KK, Joshi A, Song X-P, Singh S, Kumari A, Arora J, Singh SK, Solanki MK, Seth CS, Li Y-R (2024b) Synergistic interactions of nanoparticles and plant growth promoting rhizobacteria enhancing soil-plant systems: a multigenerational perspective. Front Plant Sci 15:1376214. https://doi.org/10.3389/fpls.2024.1376214

Vilakazi BS, Mukwada G (2023) Curbing land degradation and mitigating climate change in mountainous regions: a systemic review. Environ Monit Assess 195(2). https://doi.org/10.1007/s10661-022-10906-y

Viswanathan VK, Rajaram Manoharan SR, Subramanian S, Moon A (2019) Nanotechnology in spine surgery: a current update and critical review of the literature. World Neurosurg 123:142–155. https://doi.org/10.1016/j.wneu.2018.11.035

Wan J, Wang R, Bai H, Wang Y, Xu J (2020) Comparative physiological and metabolomics analysis reveals that single-walled carbon nanohorns and ZnO nanoparticles affect salt tolerance in Sophora alopecuroides. Environ Sci Nano 2020(7):2968–2981

Wang JJ, Sanderson BJS, Wang H (2007) Cyto- and genotoxicity of ultrafine TiO_2 particles in cultured human lymphoblastoid cells. Mutat Res 628(2):99–106. https://doi.org/10.1016/j.mrgentox.2006.12.003

Wang Q, Chen JN, Zhan P, Zhang L, Kong QQ (2013) Establishment of a suspension cell system for transformation of *Jatropha curcas* using nanoparticles. Adv Mater Res 608–609:314–319

Wang A, Li J, AL-Huqail AA, AL-Harbi MS, Ali EF, Wang J, Ding Z, Rekaby SA, Ghoneim AM, Eissa MA (2021a) Mechanisms of chitosan nanoparticles in the regulation of cold stress resistance in banana plants. Nano 11(10):2670. https://doi.org/10.3390/nano11102670
Wang JW, Cunningham FJ, Goh NS, Boozarpour NN, Pham M, Landry MP (2021b) Nanoparticles for protein delivery in planta. Curr Opin Plant Biol 60:102052. https://doi.org/10.1016/j.pbi.2021.102052
Wang X, Xie H, Wang P, Yin H (2023) Nanoparticles in plants: uptake, transport and physiological activity in leaf and root. Materials (Basel) 16(8):3097. https://doi.org/10.3390/ma16083097
Waqas Mazhar M, Ishtiaq M, Hussain I, Parveen A, Hayat Bhatti K, Azeem M et al (2022) Seed nano-priming with zinc oxide nanoparticles in rice mitigates drought and enhances agronomic profile. PLoS One 17:e0264967
Wawata IG, Fabiyi OA (2024) Sustainable application of nanomaterials in the removal of heavy metals from water. In: Uddin I (eds) Sustainable nanomaterials. sustainable materials and technology. Springer, Singapore.
World Bank (2008) World development report 2008: agriculture for development. World Bank, Washington, DC
World Bank (2022) Land degradation overview. World Bank, Washington, DC. https://www.worldbank.org/en/topic/land-degradation
Wu H, Tito N, Giraldo JP (2017) Anionic cerium oxide nanoparticles protect plant photosynthesis from abiotic stress by scavenging reactive oxygen species. ACS Nano 11(11):11283–11297. https://doi.org/10.1021/acsnano.7b05723
Xing K, Shen X, Zhu X, Ju X, Miao X, Tian J, Feng Z, Peng X, Jiang J, Qin S (2016) Synthesis and in vitro antifungal efficacy of oleoyl-chitosan nanoparticles against plant pathogenic fungi. Int J Biol Macromol 82:830–836. https://doi.org/10.1016/j.ijbiomac.2015.09.074
Xu L, Zhu Z, Sun D-W (2021) Bioinspired nanomodification strategies: moving from chemical-based agrosystems to sustainable agriculture. ACS Nano 15(8):12655–12686. https://doi.org/10.1021/acsnano.1c03948
Ye Y, Cota-Ruiz K, Hernández-Viezcas JA, Valdés C, Medina-Velo IA, Turley RS, Peralta-Videa JR, Gardea-Torresdey JL (2020) Manganese nanoparticles control salinity-modulated molecular responses in Capsicum annuum L. through priming: a sustainable approach for agriculture. ACS Sustain Chem Eng 8(3):1427–1436. https://doi.org/10.1021/acssuschemeng.9b05615
Yin J-J, Liu J, Ehrenshaft M, Roberts JE, Fu PP, Mason RP, Zhao B (2012) Phototoxicity of nano titanium dioxides in HaCaT keratinocytes—generation of reactive oxygen species and cell damage. Toxicol Appl Pharmacol 263(1):81–88. https://doi.org/10.1016/j.taap.2012.06.001
Younis AA, Khattab H, Emam MM (2020) Impacts of silicon and silicon nanoparticles on leaf ultrastructure and TaPIP1 and TaNIP2 gene expressions in heat stressed wheat seedlings. Biol Plant 64:343–352. https://doi.org/10.32615/bp.2020.030
Yugay YA, Sorokina MR, Grigorchuk VP, Rusapetova TV, Silant'ev VE, Egorova AE, Adedibu PA, Kudinova OD, Vasyutkina EA, Ivanov VV et al (2023) Biosynthesis of functional silver nanoparticles using callus and hairy root cultures of *Aristolochia manshuriensis*. J Funct Biomater 14(9). https://doi.org/10.3390/jfb14090451
Yusefi-Tanha E, Fallah S, Rostamnejadi A, Pokhrel LR (2020) Zinc oxide nanoparticles (ZnONPs) as a novel nanofertilizer: influence on seed yield and antioxidant defense system in soil grown soybean (Glycine max cv. Kowsar). Sci Total Environ 738:140240
Zahedi SM, Moharrami F, Sarikhani S, Padervand M (2020) Selenium and silica nanostructure-based recovery of strawberry plants subjected to drought stress. Sci Rep 10(1). https://doi.org/10.1038/s41598-020-74273-9
Zhang H et al (2021) Gold-nanocluster-mediated delivery of SiRNA to intact plant cells for efficient gene knockdown. Nano Lett 21:5859–5866
Zhao X, Meng Z, Wang Y, Chen W, Sun C, Cui B, Cui J, Yu M, Zeng Z, Guo S, Luo D (2017) Pollen magnetofection for geneticmodification with magnetic nanoparticles as gene carriers. Nat Plants 3(12):956–964

Zhao C, Liu B, Piao S, Wang X, Lobell DB, Huang Y, Huang M, Yao Y, Bassu S, Ciais P et al (2017a) Temperature increase reduces global yields of major crops in four independent estimates. Proc Natl Acad Sci 114(35):9326–9331. https://doi.org/10.1073/pnas.1701762114

Zhao L, Lu L, Wang A, Zhang H, Huang M, Wu H, Xing B, Wang Z, Ji R (2020) Nano-biotechnology in agriculture: use of nanomaterials to promote plant growth and stress tolerance. J Agric Food Chem 68(7):1935–1947. https://doi.org/10.1021/acs.jafc.9b06615

Zhou P, Adeel M, Shakoor N, Guo M, Hao Y, Azeem I, Li M, Liu M, Rui Y (2020) Application of nanoparticles alleviates heavy metals stress and promotes plant growth: an overview. Nano 11(1):26. https://doi.org/10.3390/nano11010026

Zhou H, Wu H, Zhang F, Su Y, Guan W, Xie Y, Giraldo JP, Shen W (2021) Molecular basis of cerium oxide nanoparticle enhancement of rice salt tolerance and yield. Environ Sci Nano 8(11):3294–3311. https://doi.org/10.1039/d1en00390a

Chapter 13
Nanotechnology-Based Biosensors for Early Detection of Abiotic Stress in Crops

Safal Kumar Paikray, Nigam Sekhar Tripathy, Liza Sahoo, Amarendra N. Misra, and Fahima Dilnawaz

13.1 Introduction

Nanotechnology is an emerging technology in research and innovation that utilizes the practical application of nanomaterials (NMs). Nanotechnology deals with building and manipulating things on the scale of atoms and molecules (McNeil 2005). NMs are microscopic particles with a size one thousand millionth of a meter (10^{-9} m). NMs have a large surface area-to-volume ratio. This means that they have a large amount of exposed surface area for their size, which makes them very reactive (Theivasanthi and Alagar 2011). Nanotechnology is becoming the most significant and advanced technology of the twenty-first century. Nanotechnology was first introduced in 1959 by Nobel Prize Laurate Richard Feynman, and the word 'nanotechnology' was coined by Prof. Norio Taniguchi, Tokyo Science University, in 1974 (Bayda et al. 2019; Parveen et al. 2016). The world of nanotechnology is constantly evolving and has applications across all the science fields, such as chemistry, biology, physics, materials science, and engineering (Choudhary et al. 2024). Seeing the potential of nanotechnology, numerous governments and investors are investing in this leading-edge technology. It can revolutionize industries such as electronics, healthcare, agriculture, food, cosmetics, and manufacturing by developing innovative materials, devices, and systems at the atomic or nanoscale (Malik et al. 2023).

S. K. Paikray · N. S. Tripathy · L. Sahoo · F. Dilnawaz (✉)
School of Biotechnology, Centurion University of Technology and Management, Bhubaneswar, Odisha, India

A. N. Misra
Office of the Vice Chancellor, Jagadguru Krupalu University, Banara, Odisha, India

Faculty of Sciences, Department of Biology, Sri Sri University, Godisahi, Odisha, India

C. S. Seth et al. (eds.), *Nano-formulations for Sustainable Agriculture*, Nanotechnology in the Life Sciences, https://doi.org/10.1007/978-3-031-90617-6_13

Agriculture is considered the backbone of a country's economy. The world's increasing population is driving up demand for food supply. To meet the demand, agriculture scientists and agriculture engineers have been exploring and adopting new approaches that can increase agricultural production. In agriculture, the adaptation of nanotechnology is increasing due to its potential to enhance harvesting efficacy, promote soil health, and respond to environmental challenges (Neme et al. 2021; Usman et al. 2020).

Agricultural nanotechnology has several applications, such as stimulating plant growth, using nanopesticides, nanoinsecticides, and nanofertilizers to enhance productivity. Nanoclay, nanozeolites, and hydrogels help improve soil quality and provide smart monitoring (Verma et al. 2023). It facilitates the determination of plant traits, enhances genome sequencing, and has been used to reduce after-harvesting losses by developing multifunctional packaging ingredients. Nanoscale edible coating is applied to food to manage after-harvest quality and nanosensors are used in grain quality control (Rajput et al. 2021; Choudhary et al. 2025). NMs are also used in food processing for additives, anti-cracking agents, and food nano-sensing (Mali et al. 2020; Sharon et al. 2010; Zambrano-Zaragoza et al. 2018; Bouwmeester et al. 2009; Rashidi and Khosravi-Darani 2011; Ezhilarasi et al. 2013). Biosensors are tiny sensing or analytical devices for sensing specific biological and chemical substances. Biosensors are precise detection tools used in many fields, such as food safety testing, disease detection, and environmental monitoring (Kumar et al. 2024a). The integration of biotechnology with sensing technology gives rise to biosensor technology, which involves the utilization of physiochemical transducer systems to transform the bio-recognition activity into a quantifiable signal (Velasco-Garcia and Mottram 2003; Glynne-Jones et al. 2004).

A biosensor comprises three parts: a sensitive biological element called a bioreceptor that interacts with the analyte (biomolecules), a transducer that converts the analyte signal into more easily measured and quantified signals, and a detector system or biosensor that uses electronic or signal processors to display results (Arora and Saini 2013). Biosensors use bioreceptors such as DNA, enzymes, antibodies, whole cells, and tissue. Leland C. Clark conducted the first biosensor experiment by using platinum electrodes to detect oxygen and glucose oxidase enzymes to trap against the electrodes using a dialysis membrane (Fracchiolla et al. 2012). There are generally four types of biosensors: electrochemical biosensors, ICS biosensors (ion channel switch biosensors), DNA biosensors, and microbial biosensors Arora and Saini 2013). Biosensors are electronic devices that can accurately measure agricultural samples, thereby preventing damage and ensuring the quality of crops and livestock (Kumar et al. 2024b). Biosensors are cost-effective and real-time monitoring devices for food authentication and monitoring over the traditional methods for quality safety checking of food processing industries (Neethirajan et al. 2018). Biosensors can do detection of pathogens like *E. coli* in food (Pebdeni et al. 2022). Biosensors can be used to monitor pesticide toxicity on agricultural lands, reduce farmer health risks and prevent harmful chemicals or pesticides from being absorbed by crops and fruits, thereby preventing environmental pollution (Rapini and Marrazza 2016).

13.2 Types of Abiotic Stress

Plant life cycle is greatly influenced by a various of biotic and abiotic stresses, ultimately affecting their growth and reproduction. Abiotic stresses are responsible for nearly 50% of all yield losses. Among these, high temperatures account for 20%, salinity for 10%, drought for 9%, low temperatures for 7%, and other abiotic stresses for 4% (Ningombam et al. 2021). Non-living factors such as temperatures, drought, flooding, high salinity, and heavy metal toxicity, nutrient deficiency known as abiotic stresses. Abiotic stress has an adverse effect on growth, development, reproduction, and yield, which leads to reduced productivity and ultimately the death of the crop causing economical loss (He et al. 2018; Dresselhaus and Hückelhoven 2018; Verma et al. 2024) (Figs. 13.1 and 13.2).

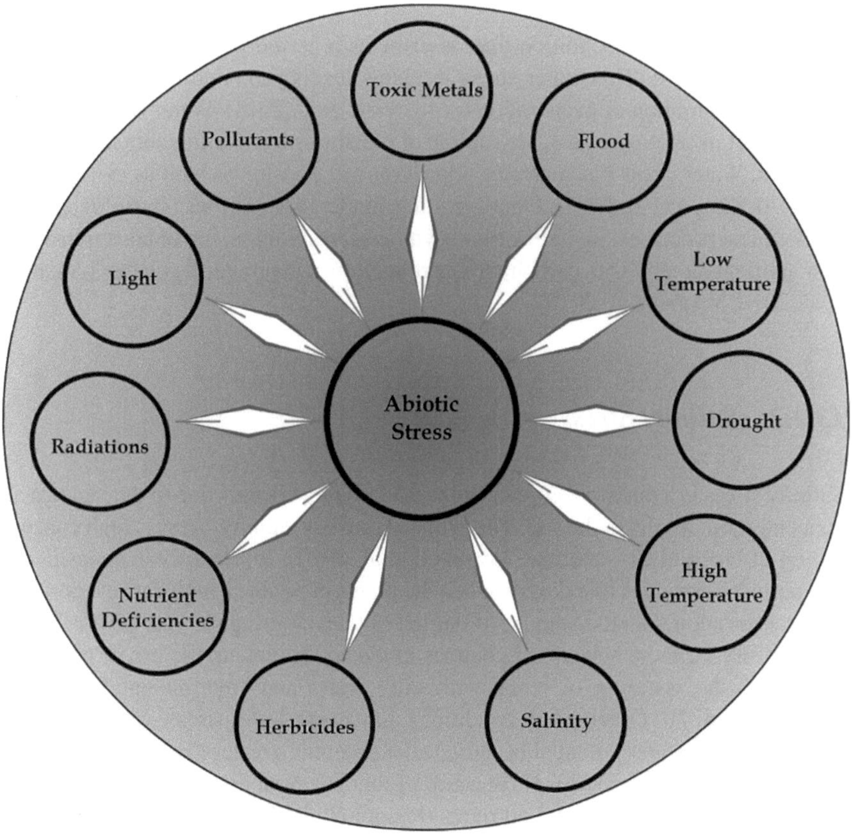

Fig. 13.1 Impact of common abiotic stress on plants growth and development

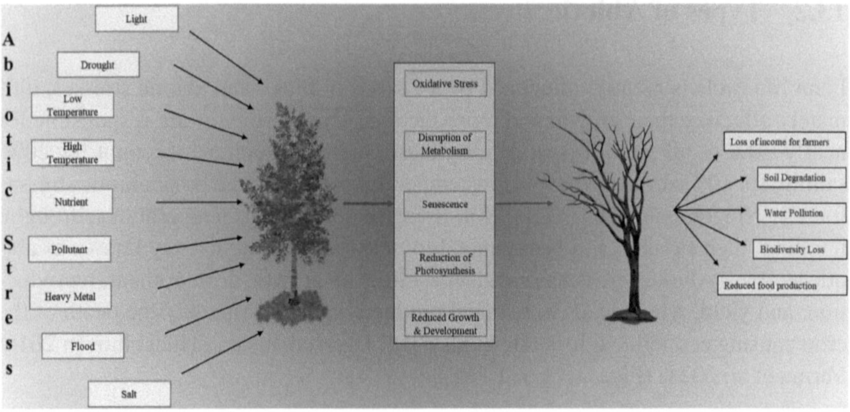

Fig. 13.2 Graphical representations of effects on trees of abiotic stress

13.2.1 Water Stress

Drought is a climatic condition that describe as a period of less rainfall resulting water deficit stress. This water stress in plants leading to continues loss of water through transpiration or evaporation (Ashkavand et al. 2018). Water stress can lead to the arrest of photosynthesis, disruption of metabolism and ultimately the death of the plant. Water stress has a greater impact on cell division as well as cell enlargement (Al-Khayri et al. 2023). Plant growth is hindered by various physiological and biochemical processes, including photosynthesis, respiration, ion uptake, translocation, nutrient metabolism, carbohydrate production, and growth regulators (Ahanger et al. 2016; Verma et al. 2020).

13.2.2 Salinity Stress

Salinity stress is a threat to agricultural production that occurs due to the increase in concentration of salts in the soil. The primary cause of salinity stress is high concentration of Na^+ and Cl^- solution (Tavakkoli et al. 2010). It primarily happens in dry and partially dry areas like deserts when the plants experience high-rate evaporation and transpiration than the amount of rainfall obtained during the year (Okon 2019). Salinity stress can be because of changes in the soil's content, the use of more fertilisers, or the watering of crops with salty water and environmental pollution (Gabrijel et al. 2011). High salt conditions may cause water stress, which causes a decrease in soil water potential because water potential greater than the salt osmotic potential (Mariyam et al. 2024). Because of this potential disturbance it is hard for the plants to absorb water and nutrients (Rasool et al. 2013). Salt stress may cause increase in the concentration of abscisic acid, which is a growth inhibitor hormone. It may cause senescence and ultimately death of the plant (Jouyban 2012).

13.2.3 Heavy Metal Stress

Heavy metal stress is caused by different anthropogenic activities like mining, industrial waste and sewage disposal, fertilizer applications etc. Heavy metals are non-biodegradable metallic elements having an atomic mass of more than 20 and a density of more than 5 g cm^{-3} (Raychaudhuri et al. 2021). Some heavy metals are micronutrients that play a crucial role in plant growth, like Fe, Mg, Mo, Zn, Mn, Cu, and Ni and on the other hand, there are some non-essential heavy metals, including Ag, Cr, Cd, Co, As, Sb, Pb, Se, and Hg (Alves et al. 2016). These micronutrients are essential for plant metabolism, growth (Tripathi et al. 2015), and development in very small amounts, but the execs amount of these metals inhibits the growth of the plant and may affect different metabolic reactions and physiological pathways (Ghori et al. 2019). This stress leads to the inactivation or denaturation of crucial enzymes and proteins (Ghori et al. 2019).

13.2.4 Temperature Stress

Temperature is an essential abiotic factor that affects plant growth and development, as it has a role in determining the rate of photosynthesis, respiration, and other physiological process (Atkin et al. 2005). When the temperature is too high or too low than the optimum temperature negatively affects plant's physiological and metabolic pathways (Toscano et al. 2019). Exposure to high temperatures can reduce the rate of photosynthesis, leading to dehydration. Heat stress in plants stimulates the generation of ROS and creates oxidative stress as a result, cell damage occurs (Farmer and Davoine 2007). Cold stress can adversely affect the cells by altering membrane structure, decreasing cytoplasmic streaming, and leakage of electrolytes and plasmolysis, which can lead to cellular death (Kumar et al. 2018). It causes abnormal anaerobic respiration, which leads to an increase in the synthesis of abnormal metabolites, which can result in metabolic abnormalities (Yadav et al. 2020). Cold stress disrupts the normal plant growth cycle, leading to abnormalities in development, fruit ripening, and decay, which can ultimately result in death (Zhang et al. 2020).

13.3 Emergence of Nanotechnology-Based Biosensors

A biosensor is an apparatus that uses signals proportionate to the concentration of an analyte in a reaction to quantify biological or chemical responses. Nanobiosensors are tiny analytical devices composed of NMs with biological elements such as proteins, enzymes, and cell organelles like mitochondria, etc as transducers (Arora and Saini 2013; Choudhary et al. 2025). Nano sensors are very useful for specific

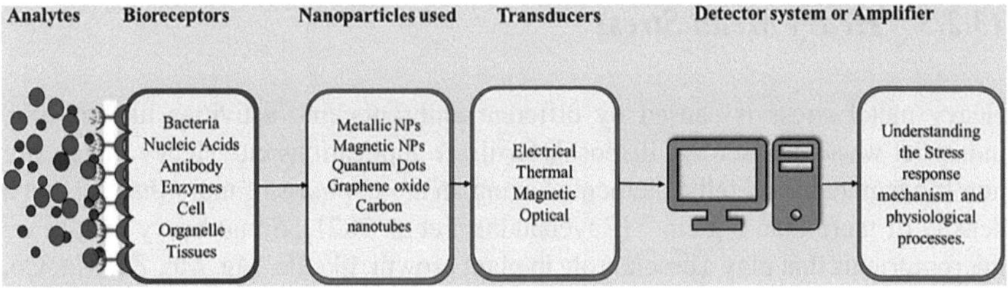

Fig. 13.3 Schematic representations of compounds of biosensors

applications because of their small size. They help to make biosensors more miniaturized, allowing them to work at a lower level on the nanoscale. Due to their small size, these sensors are highly effective in agriculture and other specific applications (Sharma et al. 2021) (Fig. 13.3).

13.3.1 Types of Nanomaterials Used in Biosensor Development

Numerous NMs such as metal NPs, magnetic NPs, carbon NPs graphene, porous carbon, silica NPs, and quantum dots are being used for biosensors. Metallic NPs, such as gold, platinum, and silver, are commonly used as biosensors due to their unique characteristics, such as ease of functionalization, catalysis of electrochemical reactions, and improved electron transfer (Jamkhande et al. 2019). Magnetic nanoparticles are used to make biosensors as they can be easily manipulated by an external magnetic field (Abedini-Nassab et al. 2021). Carbon NMs such as nanotubes, nanofibers, graphene, and fullerene are utilized in the production of agricultural biosensors (Regiart et al. 2016).

13.4 Types of Nanobiosensors

13.4.1 Electrochemical Nanobiosensors

Electrochemical biosensors use receptors linked to a transducer surface to produce an electrochemical signal when interacting with analytes. Electrochemical nanosensors employ chemical reactions to generate or consume ions or electrons, which are then quantified as voltage, current, or impedance (Huang et al. 2021). The signals produced by these sensors reflect the amount of analyte present in the sample, which is then analyzed to determine the quantity of analyte present (Kimmel et al. 2012). These are highly sensitive and can be used in various fields, including clinical, environmental, food, and agriculture sectors (Baranwal et al. 2022). They can be categorized into amperometry, voltammetry, and potentiometry. Amperometric sensors quantify the electrical current of an electroactive species using redox reaction (Kumar et al. 2024c). Voltammetry sensors measure electrical current by detecting the potential generated by a chemical reaction. Potentiometric sensors quantify the voltage difference between the working and reference electrodes (Srivastava et al. 2018).

13.4.2 Optical Nanobiosensor

Optical biosensors detect changes in the optical signal, such as wavelength, intensity, and polarity, using various spectroscopic measurements like refraction, Raman scattering, fluorescence, and absorption (Wang and Wolfbeis 2019). Many optical methods have been used in biosensors based on interferometry, fluorescence spectroscopy, surface Plasmon resonance and spectroscopy (Ligler 2008). These optical biosensors are useful tools for detecting and analyzing biological molecules and processes (Ligler 2008). Optical biosensors work by using a closed path for light to travel through. When an analyte connects to a resonator, it causes a change in the resonant frequency (Mehrotra et al. 2019). This change is measured to detect the analyte. An optical resonator is a device that allows light to circulate in a closed path by using optical components arranged in a specific way (Hodgson and Weber 2013). Optical biosensors can be classified into two types based on optical methods: robbing and reacting biosensors. Probing biosensors rely on differences in interactions between analyte and recognition elements governed by affinities (Tatiana Duque et al. 2013). Reacting biosensors can detect chemical processes through optical responses. These processes include chemisorptions, catalytic reactions, and the formation of new chemical bonds, which produce detectable optical changes (Tatiana Duque et al. 2013). Tiny optical fibers are essential for sensing in nano biosensors, allowing for intercellular and intracellular monitoring of biological and physiological parameters (Li et al. 2020).

13.4.3 Mechanical Nanobiosensors

Nanoscale mechanical forces between molecules aid in constructing small, label-free, and highly sensitive biosensors for measuring biomolecular interactions (Tamayo et al. 2013). Nanoscale mechanical forces between biomolecules make biomolecular interaction measurement exciting. Microscale cantilever beams can identify biomolecules by deflecting upon interaction with specific biomolecules. The quantification and interpretation of a substance's concentration in a solution can be achieved by measuring the deflection of cantilever beams upon interaction with molecules. Three mechanisms convert an analyte into micromechanical bending of a cantilever: surface stress, mass loading, and temperature change (Ziegler 2004).

13.4.4 Magnetic Nanobiosensors

Magnetic biosensors utilize specially developed magnetic NPs, typically ferrite-based materials, either separately or collectively (Sobczak-Kupiec et al. 2016). These sensors detect hazardous metal ions and poisonous chemical molecules such as polycyclic aromatic hydrocarbons (PAHs), antibiotics, and fungicides (Dong and Zhang 2017; Du and Jing 2011; Zamora-Gálvez et al. 2016; Kumar et al. 2016). Magnetic sensors operate based on the principle that magnetic materials experience a change in their magnetic moment when exposed to a magnetic field (Sobczak-Kupiec et al. 2016). Magnetic compounds, including iron and other transition metals like alloys, are used in screening. Adding magnetic NPs has increased the sensitivity and power of detection devices (Chen et al. 2017). Superconducting quantum interference devices (SQUIDs) utilize magnetic NPs superparamagnetic nature to detect biological targets quickly. By utilizing antibodies attached to magnetic NPs, these devices detect and separate specific antigens from mixtures, taking advantage of the superparamagnetic effect of magnetic materials, which is particularly evident in nanoscale particles (Pretzell 2012).

13.5 Advantages of Nanotechnology-Based Biosensors

Nanobiosensors are highly sensitive and precise, capable of detecting even the slightest disturbances or changes in the system (Malik et al. 2013). This technology is capable of monitoring various essential elements with great efficiency, including pH levels, hormones, proteins, genes, enzymes, soil moisture, plant biotic and abiotic stress levels, and micro and macronutrients (Ali et al. 2021). It is capable of detecting pathogens and microbes present in crops or plants with precision and high sensitivity, enabling early detection of diseases, pathogens, and abiotic stress and therefore early prevention can be made (Ijaz et al. 2023). The nanosensor can manage the irrigation and fertigation system by monitoring and regulating the water and fertilizer levels, enabling precision agriculture. Its precise nature reduces water wastage and prevents the excess use of fertilizers and pesticides, which can minimize environmental pollution (Abioye et al. 2020).

Nanosensors are emerging as the best tool for monitoring agricultural fields as they are low-cost and portable. Real-time monitoring of crop progression, crops, or agricultural parameters like moisture content, soil fertility, temperature, crop nutrient capacity, pathogens, plant diseases, etc can be done (Kundu et al. 2019). Nanobiosensors can monitor soil quality and crop health in real time by incorporating the global positioning system (GPS). Real-time monitoring plays a vital role in enhancing agricultural productivity by allowing farmers to make decisions on time (Dar et al. 2020) (Fig. 13.4).

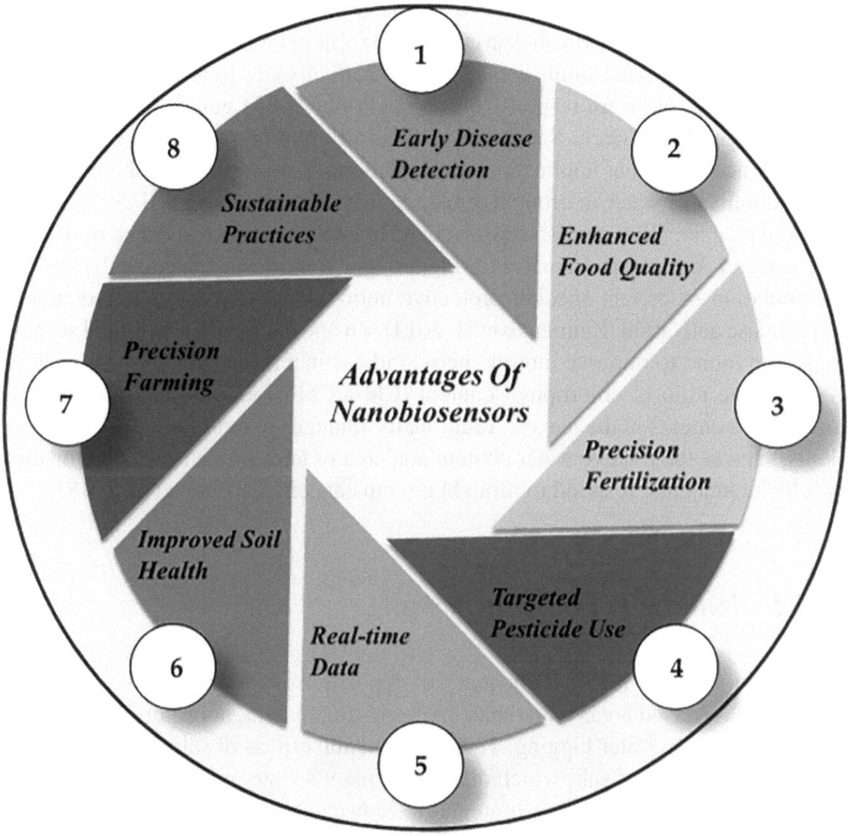

Fig. 13.4 Advantages of nanobiosensors in sustainable agriculture

13.6 Nanotechnology-Based Biosensors to Monitor Abiotic Stress in Plants

Specific examples of nanotechnology-based biosensors for detecting various abiotic stresses in crops.

13.6.1 Detection of Drought Stress

Cerium oxide NPs application improved the drought tolerance ability in sorghum (Djanaguiraman et al. 2018). Carbon nanotubes proposed to improve water uptake and reduce the negative effects of drought stress by interacting with aquaporins (Martínez-Ballesta et al. 2016). A recent study conducted by Chandrashekar et al. (2023) indicated that seed priming with Single-walled carbon nanotube (SWCNT) was shown to alleviate the oxidative damage resulting from drought stress in *Hyoscyamus niger* (Chandrashekar et al. 2023). García-Sánchez et al. (2015) used TiO_2 NPs, Ag NPs, and multi-walled carbon nanotubes to treat *A. thaliana*. They found 16 genes that were responsive to drought out of 351 genes that showed differential regulation (García-Sánchez et al. 2015). Furthermore, Borišev et al. (2016) demonstrated that foliar application of nanofullerenol may influence the intracellular metabolism of water in drought-stressed plants (Borišev et al. 2016). The application of analcite NPs was also characterized to encourage the resistance of drought in wheat and corn plants. In maize during drought conditions, analcite enhanced the accumulation of certain specific protective antioxidants (flavonoids, carotenoids) and catalase activation (Zaimenko et al. 2014). Zn and Cu NPs in a colloidal suspension were found to improve morphometric indices in a recent study which resulted in a variable ratio of chlorophyll content (Chl a/Chl b) as well as an increase in carotenoid contents in the leaves. Additionally, changes in morphometric indices of plants, such as the relative water content and area of leaves, result from the method by which plants first respond to drought circumstances (Jacobson et al. 2018).

13.6.2 Detection of Salinity Stress

In arid and semiarid regions, unequal salt deposition is the major reason for crop failure. Salt-affected areas experience high sodium content, poor porosity, and loss of nutrients due to water logging. These detrimental effects of salt stress depend on the source and level of salt, which disrupt the plant's water relation (Shahzad et al. 2019). Several bionanosensing strategies have been applied to improve the cultivation and growth rates of resistant crops in saline environments. In a study by Hu et al. (2020), an electrochemical biosensor was designed to detect the existence of IAA at 1.99 M and SA 3.30 M in soybean seedlings subjected to salt stress (Hu et al. 2020). Mondal et al. (2022) enhanced this by utilizing multiwall carbon nanotubes (MWNT) and carbon black. The application of carbon NPs in seed priming was found to have a significant effect on lettuce seed strength (Mondal et al. 2022). The growth of seedlings under high salinity stress is achieved by reducing primary root elongation (Baz et al. 2020). Engineers from the Massachusetts Institute of Technology (MIT) have successfully designed and implanted single-wall carbon nanotube (SWCNT) nanosensors within the leaves of plants for the purpose of detecting and transmitting data regarding various kinds of stress through near-infrared (nIR) optical signals. Afterward, the data gathered can be acquired through electronic mail and smartphone text messages (Amjad and Serajuddin 2022). The effects of three different concentrations of titanium dioxide NPs ($nTiO_2$) (0.01–0.03%) on the growth of broad bean plants in order to mitigate soil salinity and enhance plant growth under stress. Under standard conditions, applying $nTiO_2$ resulted in notable enhancements in antioxidant activity, shoot length, leaf area, and root dry weight (Abdel Latef et al. 2018).

13.6.3 Detection of Heavy Metal Stress

Optical chemical sensors are commonly used for the detection of heavy metals. These chemical sensors use electromagnetic radiation to create a signal in the transducer. An optical chemical sensor comprises a receptor linked to a transduction element for detecting chemical compounds (Li et al. 2013). The interaction between the sample and radiation changes optical properties, enabling measurement of chemicals or analytes concentration (Ray 2010). Optical sensors use different principles and regions of the spectra to measure light intensity and related properties such as lifetime, refractive index, scattering, diffraction, and polarization (Lobnik et al. 2012).

In recent times, optical techniques based on graphene-based NMs have emerged as a promising approach for detecting heavy metal ions. Due to their simple construction and high sensitivity towards certain metal ions, these techniques have gained attention (Chang et al. 2014). Noble NPs such as Ag, Au, and Pd when combined with graphene enhance their strength and improve their catalytic performance (Sharma et al. 2021). A nano biosensor-based QD-DNAzyme labeled with a quencher can detect the presence of heavy metal ions like lead and copper. When heavy metal ions are present, the DNA molecule cleaves and the QDs reflect ion-specific fluorescence signals, indicating the presence of the metal ion. This new technology is faster and more accurate than existing methods (Kordrostami et al. 2021).

13.6.4 Detection Temperature Stress

In the year Nietzel et al. (2019) successfully devised biosensors capable of detecting the existence of hydrogen peroxide (H_2O_2) within Arabidopsis. The primary application of these biosensors is to investigate the dynamics of hydrogen peroxide (H_2O_2) during oxidative bursts induced by elicitors, with a temporal resolution of 15 seconds. Changes in pH do not have an impact on the biosensors (Nietzel et al. 2019). Meanwhile, Exposito-Rodriguez and co-workers (2017) employed H_2O_2 signaling to allow interaction between chloroplasts and nuclei when cells are exposed to variable light intensity. Temperature stress promotes the rise in the concentration of plant hormones which include abscisic acid (ABA) and indole-3-acetic acid (IAA), salicylic acid (SA), ethylene (ET), and jasmonates (JAs) play an essential part in ensuring that plants survive with adverse environments (Siddiqi and Husen 2019). To track a plant's response to abiotic stress, an electrochemical biosensor has been developed to detect the phytotronmones IAA and SA (Hu et al. 2020). Amreen (2020) created a pH-stable nanosensor (FLIP-H_2O_2) that is extremely specific for H_2O_2 but resistant to various oxidizing agents such superoxide anions, peroxynitrite, and nitric oxide by the use of genetic engineering. FLIP-H_2O_2 can be utilized to monitor the subcellular targeting signals in organisms by focusing on organelles like mitochondria and chloroplasts (Amreen et al. 2020). The effect of silver nanoparticles (Ag NPs) on wheat development under high temperature (30–40 °C for 3 days, 3 hr/day) stress was investigated. The results revealed that Ag NPs have the ability to promote wheat growth and can shield wheat against the damaging effects of high temperatures (Iqbal et al. 2019).

13.7 Future Perspectives and Conclusion

The utilization of nanosensors in contemporary agriculture is progressively growing at a consistent rate. Advancements in biosensor technology will soon allow for real-time quality and safety assessments on portable smart devices, benefiting farmers and industries. However, existing sensors have limitations in detection, sensitivity, specificity, electrode design, matrix fabrication, and cost. There is still room for improvement to address the rising problems in agriculture (Pretzell 2012). It is widely accepted that nanotechnology has the potential to transform the agriculture industry (Pretzell 2012). Through the application of nanotechnology, it is possible to develop sustainable agricultural practices with greater precision and efficiency. With extensive research and targeted efforts, using nano biosensor can improve crop yield, reduce the need for harmful pesticides, and enhance soil health, among other benefits. Nanobiosensor has emerging biosensing tool for the detection and monitoring abiotic stress. Nowadays, advancements in nanotechnology-based (bio)sensors are steadily increasing to deliver speedy, sensitive, and cost-effective analysis focused on farmer requirements (Salouti and Khadivi Derakhshan 2020).

Nano biosensors can detect various elements and monitor crop health, soil quality, and irrigation and fertigation systems with precision and efficiency. They are low-cost and portable, making them a valuable tool for real-time monitoring of agricultural fields (Dar et al. 2020). This technology can improve agricultural productivity by enabling early detection and prevention of diseases and abiotic stress. GPS-enabled nano biosensors can monitor soil and crop health in real-time, improving agricultural productivity for farmers and efforts can be made to reduce crop losses in agroecosystems (Dar et al. 2020). Despite recent advancements in biosensors, limitations remain in the field, including low storage and working stability of biological components (Naresh and Lee 2021). Long-term stability and biocompatibility should be ensured. Biotechnology, biomimetic chemistry, nanotechnology, and material science are being used to design more robust biological recognition elements, such as aptamers, molecular imprinting polymers, peptide nucleic acids, and synthetic peptides (Scognamiglio et al. 2015). Efficient data management and analysis tools are needed for processing and analyzing large amounts of data.

Acknowledgements SKP, NST, LS gratefully acknowledge Centurion University of Technology and Management for research fellowship.

References

Abdel Latef AAH, Srivastava AK, Abd El-sadek MS, Kordrostami M, Tran L-SP (2018) Titanium dioxide nanoparticles improve growth and enhance tolerance of broad bean plants under saline soil conditions. Land Degrad Dev 29(4):1065–1073

Abedini-Nassab R, Pouryosef Miandoab M, Şaşmaz M (2021) Microfluidic synthesis, control, and sensing of magnetic nanoparticles: a review. Micromachines (Basel) 12(7):768

Abioye EA, Abidin MSZ, Mahmud MSA, Buyamin S, Ishak MHI, Rahman MKIA, Otuoze AO, Onotu P, Ramli MSA (2020) A review on monitoring and advanced control strategies for precision irrigation. Comput Electron Agric 173:105441. https://doi.org/10.1016/j.compag.2020.105441

Ahanger MA, Morad-Talab N, Abd-Allah EF, Ahmad P, Hajiboland R (2016) Plant growth under drought stress. In: Water stress and crop plants: a sustainable approach, pp 649–668. https://doi.org/10.1002/9781119054450.ch37

Ali SS, Al-Tohamy R, Koutra E, Moawad MS, Kornaros M, Mustafa AM, Mahmoud YAG, Badr A, Osman ME, Elsamahy T, Jiao H, Sun J (2021) Nanobiotechnological advancements in agriculture and food industry: applications, nanotoxicity, and future perspectives. Sci Total Environ 792:148359. https://doi.org/10.1016/j.scitotenv.2021.148359

Al-Khayri JM, Rashmi R, Surya Ulhas R, Sudheer WN, Banadka A, Nagella P, Aldaej MI, Rezk AA, Shehata WF, Almaghasla MI (2023) The role of nanoparticles in response of plants to abiotic stress at physiological, biochemical, and molecular levels. Plants (Basel) 12(2):292

Alves LR, Reis AR, Gratao PL (2016) Heavy metals in agricultural soils: from plants to our daily life (a review). Científica 44(3):346–361

Amjad S, Serajuddin M (2022) Role of nanosensors and bionanosensors in crop abiotic stress. In: Augmenting crop productivity in stress environment. Springer, Singapore, pp 1–12

Amreen AHM, Ahmad M, Salem MZM, Ahmad A (2020) Construction of a nanosensor for non-invasive imaging of hydrogen peroxide levels in living cells. Biology (Basel) 9(12):430, PMID: 33260458; PMCID: PMC7760702. https://doi.org/10.3390/biology9120430

Arora RK, Saini RP (2013) Biosensors: way of diagnosis. Int J Pharm Sci Res 4(7):2517–2527

Ashkavand P et al (2018) Application of SiO_2 nanoparticles as pretreatment alleviates the impact of drought on the physiological performance of Prunus mahaleb (Rosaceae). Bol Soc Argent Bot 53(2):1–10

Atkin OK et al (2005) The hot and the cold: unravelling the variable response of plant respiration to temperature. Funct Plant Biol 32(2):87–105

Baranwal J et al (2022) Electrochemical sensors and their applications: a review. Chemosensors 10(9):363

Bayda S, Adeel M, Tuccinardi T, Cordani M, Rizzolio F (2019) The history of nanoscience and nanotechnology: from chemical-physical applications to nanomedicine. Molecules 25(1):112

Baz H et al (2020) Water-soluble carbon nanoparticles improve seed germination and post-germination growth of lettuce under salinity stress. Agronomy 10(8):1192

Borišev M et al (2016) Drought impact is alleviated in sugar beets (Beta vulgaris L.) by foliar application of fullerenol nanoparticles. PLoS One 11(11):e0166248

Bouwmeester H, Dekkers S, Noordam MY, Hagens WI, Bulder AS, de Heer C, ten Voorde SE, Wijnhoven SW, Marvin HJ, Sips AJ (2009) Review of health safety aspects of nanotechnologies in food production. Regul Toxicol Pharmacol 53(1):52–62

Chandrashekar HK et al (2023) Nanoparticle-mediated amelioration of drought stress in plants: a systematic review. 3 Biotech 13(10):336

Chang J et al (2014) Graphene-based sensors for detection of heavy metals in water: a review. Anal Bioanal Chem 406(16):3957–3975

Chen Y-T et al (2017) Biosensing using magnetic particle detection techniques. Sensors 17(10):2300

Choudhary R, Rajput VD, Ghodake G, Ahmad F, Meena M, Rehman RU, Prasad R, Sharma RK, Singh R, Seth CS (2024) Comprehensive journey from past to present to future about seed priming with hydrogen peroxide and hydrogen sulfide concerning drought, temperature, UV and ozone stresses- a review. Plant Soil. https://doi.org/10.1007/s11104-024-06499-9

Choudhary R, Ahmad F, Kaya C, Upadhyay SK, Muneer S, Kumar V, Meena M, Liu H, Upadhyay H, Seth CS (2025) Decrypting proteomics, transcriptomics, genomics, and integrated omics for augmenting the abiotic, biotic, and climate change stress resilience in plants. J Plant Physiol. https://doi.org/10.1016/j.jplph.2025.154430

Dar FA, Qazi G, Pirzadah TB (2020) Nano-Biosensors: NextGen Diagnostic Tools in Agriculture. In: Hakeem, K., Pirzadah, T. (eds) Nanobiotechnology in Agriculture. Nanotechnology in the Life Sciences. Springer, Cham. https://doi.org/10.1007/978-3-030-39978-8_7

Djanaguiraman M et al (2018) Cerium oxide nanoparticles decrease drought-induced oxidative damage in sorghum leading to higher photosynthesis and grain yield. ACS Omega 3(10):14406–14416

Dong Y, Zhang L (2017) Constructed ILs coated porous magnetic nickel cobaltate hexagonal nanoplates sensing materials for the simultaneous detection of cumulative toxic metals. J Hazard Mater 333:23–31

Dresselhaus T, Hückelhoven R (2018) Biotic and abiotic stress responses in crop plants. Agronomy 8(11):267

Du J, Jing C (2011) Preparation of thiol modified Fe3O4@ Ag magnetic SERS probe for PAHs detection and identification. J Phys Chem C 115(36):17829–17835

Exposito-Rodriguez M et al (2017) Photosynthesis-dependent H_2O_2 transfer from chloroplasts to nuclei provides a high-light signalling mechanism. Nat Commun 8(1):49

Ezhilarasi PN, Karthik P, Chhanwal N, Anandharamakrishnan C (2013) Nanoencapsulation techniques for food bioactive components: a review. Food Bioprocess Technol 6:628–647

Farmer EE, Davoine C (2007) Reactive electrophile species. Curr Opin Plant Biol 10(4):380–386

Fracchiolla NS, Artuso S, Cortelezzi A (2012) Biosensors in clinical practice: focus on oncohematology. Sensors (Basel) 13(5):6423–6447

Gabrijel O, Zed R, Szilvia V (2011) Chapter 8: Soil salinisation and salt stress in crop production. In: Arun S, Venkateswarlu B (eds) Abiotic stress in plants. IntechOpen, Rijeka

García-Sánchez S, Bernales I, Cristobal S (2015) Early response to nanoparticles in the Arabidopsis transcriptome compromises plant defence and root-hair development through salicylic acid signalling. BMC Genomics 16:1–17

Ghori NH et al (2019) Heavy metal stress and responses in plants. Int J Environ Sci Technol 16(3):1807–1828

Glynne-Jones P, Tudor MJ, Beeby SP, White NM (2004) An electromagnetic, vibration-powered generator for intelligent sensor systems. Sensors Actuators A Phys 110(1-3):344–349

He M, He CQ, Ding NZ (2018) Abiotic stresses: general defenses of land plants and chances for engineering multistress tolerance. Front Plant Sci 9:1771

Hodgson N, Weber H (2013) Optical resonators: fundamentals, advanced concepts and applications. Springer Science & Business Media

Hu Y et al (2020) A multifunctional ratiometric electrochemical sensor for combined determination of indole-3-acetic acid and salicylic acid. RSC Adv 10(6):3115–3121

Huang X, Zhu Y, Kianfar E (2021) Nano biosensors: properties, applications and electrochemical techniques. J Mater Res Technol 12:1649–1672

Ijaz M et al (2023) Nanobiotechnology to advance stress resilience in plants: current opportunities and challenges. Mater Today Bio 22:100759

Iqbal M et al (2019) Effect of silver nanoparticles on growth of wheat under heat stress. Iran J Sci Technol Trans A: Sci 43:387–395

Jacobson A et al (2018) Interactions between a plant probiotic and nanoparticles on plant responses related to drought tolerance. Ind Biotechnol 14(3):148–156

Jamkhande PG et al (2019) Metal nanoparticles synthesis: an overview on methods of preparation, advantages and disadvantages, and applications. J Drug Deliv Sci Technol 53:101174

Jouyban Z (2012) The effects of salt stress on plant growth. Tech J Eng Appl Sci 2(1):7–10

Kimmel DW et al (2012) Electrochemical sensors and biosensors. Anal Chem 84(2):685–707

Kordrostami M, Mafakheri M, Al-Khayri JM (2021) Contributions of nano biosensors in managing environmental plant stress under climatic changing era. In: Al-Khayri JM, Ansari MI, Singh AK (eds) Nanobiotechnology: mitigation of abiotic stress in plants. Springer International Publishing, Cham, pp 117–137
Kumar S et al (2016) Molecularly imprinted star polymer-modified superparamagnetic iron oxide nanoparticle for trace level sensing and separation of mancozeb. RSC Adv 6(43):36751–36760
Kumar R et al (2018) A review report: low temperature stress for crop production. Int J Pure Appl Biosci 6(2):575–598
Kumar D, Mariyam S, Gupta KJ, Thiruvengadam M, Ghodake GS, Xing B, Seth CS (2024a) Comparative investigation on chemical and green synthesized titanium dioxide nanoparticles against chromium (VI) stress eliciting differential physiological, biochemical, and cellular attributes in Helianthus annuus L. Sci Total Environ 930:172413. https://doi.org/10.1016/j.scitotenv.2024.172413
Kumar D, Singh R, Upadhyay SK, Verma KK, Tripathi RM, Liu H, Dhankher OP, Tripathi RD, Sahi SV, Seth CS (2024b) Review on interactions between nanomaterials and phytohormones: novel perspectives and opportunities for mitigating environmental challenges. Plant Sci 340:111964. https://doi.org/10.1016/j.plantsci.2023.111964
Kumar D, Biswas JK, Mulla SI, Singh R, Shukla R, Ahanger MA, Shekhawat GS, Verma KK, Siddiqui MW, Seth CS (2024c) Micro and nanoplastics pollution: sources, distribution, uptake in plants, toxicological effects, and innovative remediation strategies for environmental sustainability. Plant Physiol Biochem 213:108795. https://doi.org/10.1016/j.plaphy.2024.108795
Kundu M et al (2019) Recent developments in biosensors to combat agricultural challenges and their future prospects. Trends Food Sci Technol 88:157–178
Li M et al (2013) Nanostructured sensors for detection of heavy metals: a review. ACS Sustain Chem Eng 1(7):713–723
Li Y et al (2020) Optical fiber technologies for nanomanipulation and biodetection: a review. J Lightwave Technol 39(1):251–262
Ligler FS (2008) Fluorescence-based optical biosensors. In: Biophotonics. Springer, Berlin, Heidelberg, pp 199–215
Lobnik A, Turel M, Urek SK (2012) Optical chemical sensors: design and applications. In: Advances in chemical sensors, pp 3–28, Optical Chemical Sensors: Design and Applications, In book: Advances in Chemical Sensors, https://doi.org/10.5772/31534, https://www.intechopen.com/chapters/26343
Mali SC et al (2020) Nanotechnology a novel approach to enhance crop productivity. Biochem Biophys Rep 24:100821
Malik P et al (2013) Nanobiosensors: concepts and variations. Int Sch Res Notices 2013, pp 1–9
Malik S, Muhammad K, Waheed Y (2023) Nanotechnology: a revolution in modern industry. Molecules 28(2):661
Mariyam S, Upadhyay SK, Chakraborty K, Verma KK, Duhan JS, Muneer S, Meena M, Sharma RK, Ghodake G, Seth CS (2024) Nanotechnology, a frontier in agricultural science, a novel approach in abiotic stress management and convergence with new age medicine-a review. Sci Total Environ 912:169097. https://doi.org/10.1016/j.scitotenv.2023.169097
Martínez-Ballesta MC et al (2016) Multiwalled carbon nanotubes enter broccoli cells enhancing growth and water uptake of plants exposed to salinity. J Nanobiotechnol 14:1–14
McNeil SE (2005) Nanotechnology for the biologist. J Leukoc Biol 78(3):585–594
Mehrotra P, Chatterjee B, Sen S (2019) EM-wave biosensors: a review of RF, microwave, mm-wave and optical sensing. Sensors 19(5):1013
Mondal R et al (2022) Potential of nanobiosensor in sustainable agriculture: the state-of-art. Heliyon 8(12):e12207
Naresh V, Lee N (2021) A review on biosensors and recent development of nanostructured materials-enabled biosensors. Sensors 21(4):1109
Neethirajan S, Ragavan V, Weng X, Chand R (2018) Biosensors for sustainable food engineering: challenges and perspectives. Biosensors 8(1):23

Neme K, Nafady A, Uddin S, Tola YB (2021) Application of nanotechnology in agriculture, post-harvest loss reduction and food processing: food security implication and challenges. Heliyon 7(12):e08539

Nietzel T et al (2019) The fluorescent protein sensor roGFP 2-Orp1 monitors in vivo H2O2 and thiol redox integration and elucidates intracellular H2O2 dynamics during elicitor-induced oxidative burst in Arabidopsis. New Phytol 221(3):1649–1664

Ningombam B, Devi AS, Darvhankar M (2021) Effects of abiotic stresses on crop yield: a review. Pharm Innov 10(5):418–422

Okon OG (2019) Effect of salinity on physiological processes in plants. In: Giri B, Varma A (eds) Microorganisms in saline environments: strategies and functions. Springer International Publishing, Cham, pp 237–262

Parveen K, Banse V, Ledwani L (2016) Green synthesis of nanoparticles: their advantages and disadvantages. In: AIP conference proceedings. AIP Publishing, Melville

Pebdeni AB et al (2022) Recent advances in optical biosensors for specific detection of E. coli bacteria in food and water. Food Control 135:108822

Pretzell A (2012) High temperature radio-frequency superconducting quantum interference device system for detection of magnetic nanoparticles, vol 48. Forschungszentrum Jülich

Rajput VD, Minkina T, Kumari A, Harish SVK, Verma KK, Mandzhieva S, Sushkova S, Srivastava S, Keswani C (2021) Coping with the challenges of abiotic stress in plants: new dimensions in the field application of nanoparticles. Plan Theory 10:1221. https://doi.org/10.3390/plants10061221

Rapini R, Marrazza G (2016) Biosensor potential in pesticide monitoring. In: Comprehensive analytical chemistry. Elsevier, Amsterdam, pp 3–31

Rashidi L, Khosravi-Darani K (2011) The applications of nanotechnology in food industry. Crit Rev Food Sci Nutr 51(8):723–730

Rasool S et al (2013) Salt stress: causes, types and responses of plants. In: Ahmad P, Azooz MM, Prasad MNV (eds) Ecophysiology and responses of plants under salt stress. Springer New York, New York, pp 1–24

Ray PC (2010) Size and shape dependent second order nonlinear optical properties of nanomaterials and their application in biological and chemical sensing. Chem Rev 110(9):5332–5365

Raychaudhuri SS et al (2021) Chapter 6: Polyamines, metallothioneins, and phytochelatins—natural defense of plants to mitigate heavy metals. In: Attaur R (ed) Studies in natural products chemistry. Elsevier, Amsterdam, pp 227–261

Regiart M et al (2016) An ordered mesoporous carbon modified electrochemical sensor for solid-phase microextraction and determination of triclosan in environmental samples. Sensors Actuators B Chem 232:765–772

Salouti M, Khadivi Derakhshan F (2020) Biosensors and Nanobiosensors in Environmental Applications. In: Ghorbanpour, M., Bhargava, P., Varma, A., Choudhary, D. (eds) Biogenic Nano-Particles and their Use in Agroecosystems. Springer, Singapore. https://doi.org/10.1007/978-981-15-2985-6_26

Scognamiglio V et al (2015) Synthetic biology and biomimetic chemistry as converging technologies fostering a new generation of smart biosensors. Biosens Bioelectron 74:1076–1086

Shahzad H et al (2019) Salinity types and level-based effects on the growth, physiology and nutrient contents of maize (Zea mays). Ital J Agron 14(4):199–207

Sharma P et al (2021) A review on biosensors and nanosensors application in agroecosystems. Nanoscale Res Lett 16(1):136

Sharon M, Choudhary AK, Kumar R (2010) Nanotechnology in agricultural diseases and food safety. J Phytology 2(4):83–92

Siddiqi KS, Husen A (2019) Plant response to jasmonates: current developments and their role in changing environment. Bull Natl Res Cent 43(1):1–11

Sobczak-Kupiec A et al (2016) Magnetic nanomaterials and sensors for biological detection. Nanomedicine 12(8):2459–2473

Srivastava AK, Dev A, Karmakar S (2018) Nanosensors and nanobiosensors in food and agriculture. Environ Chem Lett 16(1):161–182

Tamayo J et al (2013) Biosensors based on nanomechanical systems. Chem Soc Rev 42(3):1287–1311

Tatiana Duque M et al (2013) Chapter 5: New insights on optical biosensors: techniques, construction and application. In: Toonika R (ed) State of the art in biosensors. IntechOpen, Rijeka

Tavakkoli E, Rengasamy P, McDonald GK (2010) High concentrations of Na+ and Cl- ions in soil solution have simultaneous detrimental effects on growth of faba bean under salinity stress. J Exp Bot 61(15):4449–4459

Theivasanthi T, Alagar M (2011) Studies of copper nanoparticles effects on micro-organisms. ArXiv:1–6. https://arxiv.org/abs/1110.1372

Toscano S et al (2019) Effect of preharvest abiotic stresses on the accumulation of bioactive compounds in horticultural produce. Front Plant Sci 10:1212

Tripathi DK et al (2015) Micronutrients and their diverse role in agricultural crops: advances and future prospective. Acta Physiol Plant 37(7):139

Usman M et al (2020) Nanotechnology in agriculture: current status, challenges and future opportunities. Sci Total Environ 721:137778

Velasco-Garcia MN, Mottram T (2003) Biosensor technology addressing agricultural problems. Biosyst Eng 84(1):1–12

Verma KK, Singh P, Song X-P, Malviya MK, Singh RK, Chen G-L, Solomon S, Li YR (2020) Mitigating climate change for sugarcane improvement: role of silicon in alleviating abiotic stresses. Sugar Tech 22(5):741–749. https://doi.org/10.1007/s12355-020-00831-0

Verma KK, Song X, Degu HD, Guo D, Joshi A, Huang H, Xu L, Singh M, Huang D, Rajput VD, Li Y (2023) Recent advances in nitrogen and nano-nitrogen fertilizers for sustainable crop production: a mini-review. Chem Biol Technol Agric 10:1–14. https://doi.org/10.1186/s40538-023-00488-3

Verma KK, Song XP, Kumari A, Jagadesh M, Singh SK, Bhatt R, Singh M, Seth CS, Li YR (2024) Climate change adaptation: challenges for agricultural sustainability. Plant Cell Environ:1–12. https://doi.org/10.1111/pce.15078

Wang X-d, Wolfbeis OS (2019) Fiber-optic chemical sensors and biosensors (2015–2019). Anal Chem 92(1):397–430

Yadav S et al (2020) Effect of abiotic stress on crops. Sustain Crop Prod 1–21

Zaimenko NV et al (2014) Enhancement of drought resistance in wheat and corn by nanoparticles of natural mineral analcite. Ecol Balk 6(1):1–10

Zambrano-Zaragoza ML, González-Reza R, Mendoza-Muñoz N, Miranda-Linares V, Bernal-Couoh TF, Mendoza-Elvira S, Quintanar-Guerrero D (2018) Nanosystems in edible coatings: a novel strategy for food preservation. Int J Mol Sci 19(3):705

Zamora-Gálvez A et al (2016) Molecularly imprinted polymer-decorated magnetite nanoparticles for selective sulfonamide detection. Anal Chem 88(7):3578–3584

Zhang F et al (2020) Transcriptome analysis of postharvest blueberries (Vaccinium corymbosum 'Duke') in response to cold stress. BMC Plant Biol 20:1–20

Ziegler C (2004) Cantilever-based biosensors. Anal Bioanal Chem 379:946–959

Chapter 14
Green Nanotechnology-Based Formulations for Abiotic Stress Tolerance for Sustainable Agriculture

Roohi Sharma, Priyanka Saini, Raman Tikoria, Nitika Kapoor, and Puja Ohri

14.1 Introduction

Agricultural production obstacles include a lack of food, scarce land and water, and a lack of funding for sustainability (Calicioglu et al. 2019). Abiotic stressors that impair crop development and productivity, such as salt, drought, heat, and nutritional deficits, cost farmers money (Minhas et al. 2017; Kaur et al. 2022). These problems are made worse by climate change, which further reduces yields and stresses resources. The farming industry must maximize resources, implement cutting-edge methods, and encourage ethical resource usage in order to meet these difficulties (Verma et al. 2024a). The effects of climate change can be lessened by incorporating cutting-edge materials into farming. It will take a variety of efforts, financial backing, and broad adoption of sustainable techniques to meet the growing demand for food (Usman et al. 2020; Kumari et al. 2022a).

In the recent 10 years, scientists have been examining novel technologies, like nanotechnology, to address global agricultural challenges and fulfill the demands of an expanding population. Innovative products and equipment designed to enhance

R. Sharma · P. Ohri
Department of Zoology, Guru Nanak Dev University, Amritsar, Punjab, India

P. Saini
Department of Zoology, Deshbandhu College, University of Delhi, New Delhi, India

R. Tikoria (✉)
Department of Zoology, School of Basic Sciences, Central University of Punjab, Bathinda, Punjab, India
e-mail: raman.tikoria@cup.edu.in

N. Kapoor
Department of Botanical and Environmental Sciences, Guru Nanak Dev University, Amritsar, Punjab, India

C. S. Seth et al. (eds.), *Nano-formulations for Sustainable Agriculture*, Nanotechnology in the Life Sciences, https://doi.org/10.1007/978-3-031-90617-6_14

crop plant features have been created by integrating nanotechnology especially green or agri-nanotechnology with several scientific domains (Zulfiqar and Ashraf 2021; Verma et al. 2024b). Green nanotechnology minimizes negative effects, lowers energy consumption, generates less waste, and poses fewer health hazards in order to create sustainable solutions for societal and environmental problems (Khan 2020; Verma et al. 2023). It encourages ecologically benign technologies and eco-friendly activities by incorporating sustainability throughout the life cycle of nano-products. Using effectively regulated inputs, such as fertilizers, herbicides, and agrochemicals, mediated by real-time monitoring systems, agri-nanotechnology seeks to increase crop output (Shende et al. 2023). Crop productivity and quality will benefit from a range of nanotechnology applications at every stage of agriculture, from germination of seeds to post-harvesting. Numerous everyday applications exist for nanotechnology. In the year 2003, the food and farming sectors began implementing nanotechnology in various forms (Xing et al. 2016). Among the many benefits of nanotechnology is increased food safety and quality. With the use of nanotechnology, scientists may create particles with special, adaptable properties for use in fuel cells, electronics, medicines, water and gas purification, and agriculture, among other industries (Verma et al. 2022).

In the Agri-tech revolution, engineered nanomaterials (ENMs), such as silvers (Ag), gold (Au), copper (Cu), zinc oxide (ZnO), titanium dioxide (TiO_2), Silica (SiO_2) nanoparticles etc. are the major participants (Dikshit et al. 2021). These materials <100 nm nanoscale size allows them to effectively distribute nutrients and insecticides through foliar or root treatments by overcoming biological barriers (Hong et al. 2021). Green NPs derive from natural sources or eco-friendly synthesis methods, such as plant extracts (aloe vera, lemon, green tea) (Ying et al. 2022), microbes (bacteria and fungi) (Priya et al. 2021), biofuels (Reddy Yadav et al. 2018), biopolymers (starch, cellulose, chitosan) (Mallakpour et al. 2021), waste materials, such as fruit peels, agricultural residues (Bhardwaj and Singh 2023), and green chemistry techniques (Kaur et al. 2021).

Nanomaterial-based optical and electrochemical sensors can also be used to identify environmental pollutants and heavy metals (HMs) in soil media (Thakur and Kumar 2022). Because ENMs may function as sensors in plants, they are perfectly equipped to spearhead this transformation and allow for extensive crop health monitoring at both high spatial resolution and temporal resolution (Rajput et al. 2021; Tikoria et al. 2023a). By enabling tailored delivery of nutrients and additives that result in more effective resource use and better stress and disease control, nanotechnology also helps to reduce preharvest food loss. Because of the aforementioned qualities, ENMs could be a viable option for sustainable agriculture. This chapter explores the diverse applications of nano-enabled methods such as smart nanosensors, precise genome editing, and targeted nutrient delivery in enhancing resilience to abiotic stresses in agriculture. It addresses challenges and highlights the potential of agro-based ENMs, offering insights into their benefits and limitations for sustainability.

14.2 Effect of Abiotic Stress in Sustainable Agriculture

In agriculture, plants are greatly impacted by abiotic factors such as drought, waterlogging, heavy metals, salinity, UV radiation, pH variations and temperature (cold or heat) stress. Water supply is decreased by drought, which results in stunted growth and yield losses (Minhas et al. 2017; Tikoria et al. 2022, 2023b). Flooding damages roots and makes them more prone to illness by interfering with the oxygen flow to the roots. Contaminated soil with HMs such as cadmium (Cd), arsenic (As), chromium (Cr), lead (Pb), mercury (Hg), nickel (Ni), copper (Cu) and zinc (Zn) can lead to poisoning symptoms and poor nutrient uptake (Calanca 2017). Osmotic equilibrium is upset by salinity, which prevents growth and yield. UV light interferes with photosynthesis and degrades (deoxyribonucleic acid) DNA. Changes in pH affect the availability of nutrients and the activity of enzymes. Thermal stress causes cellular metabolism to be disturbed, which results in wilting and decreased output (Abdelrahman et al. 2020). In order to guarantee food security and agricultural sustainability, it is imperative to mitigate these stresses (Fig. 14.1).

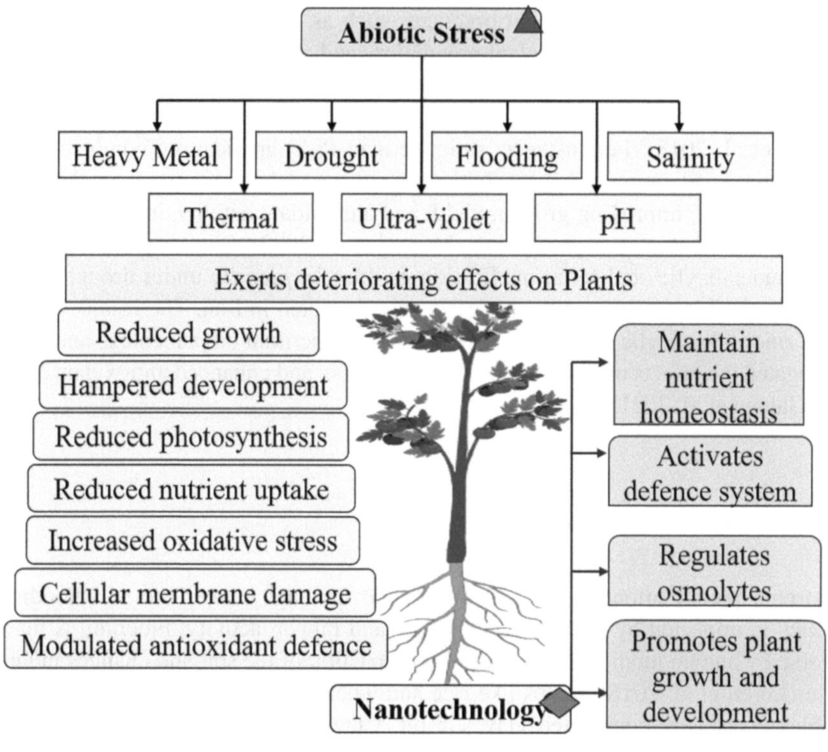

Fig. 14.1 Implications of abiotic stress in plants and its bioremediation by nanobiotechnology

14.2.1 Drought

Rising air temperatures and CO_2 levels from global climate change disrupt rainfall, worsening drought and salinity stress in plants. Glacial melting reduces water reserves, impacting crop irrigation (Chaudhry and Sidhu 2022). Erratic rainfall due to human activities intensifies water stress. Monsoon shifts reduce predictability, affecting plant productivity. Sustainable water management and crop planning are essential for agricultural resilience (Ondrasek 2013). Many factors affect a plant's ability to grow and develop, but in agriculture, drought is the most restrictive factor that reduces grain yield and quality (Pamungkas and Farid 2022). It is still essential to comprehend how resilient plants are to water constraint, particularly in arid areas. Grain yield and quality are impacted by variations in water availability that go beyond ideal thresholds, and problems with water scarcity are made worse by global warming (Geng et al. 2015).

Plants undergo adaptive responses to drought stress, such as withering and rolling of the leaves, as well as growth inhibition and osmotic adjustment (Choudhary et al. 2024). Conditions, such as the length of the drought and the properties of the soil affect how severe the symptoms are (Iqbal et al. 2020). Sustainable agriculture requires an understanding of these stress reactions and how to manage them. Implementing best management practices, such as optimum sowing time, phytohomones, bioremediation on plant population, and soil management, coupled with using transgenic plants and contemporary irrigation methods, mulching, and microbial inoculations, are effective solutions for mitigating drought stress in field crops (Solis et al. 2018). For instance, salicylic acid (SA) application, combined with potassium fertilization, enhances *Triticum aestivum* yield and nutrient uptake under drought stress, improving growth, yield, and antioxidant activity, indicating potential for sustainable crop production (Munsif et al. 2022). Similarly, the effects of zinc and salicylic acid foliar application on *Mentha piperita* under drought stress were studied in a two-year field experiment conducted in Iran. The results showed that zinc and salicylic acid treatment increased peppermint oil percentage and yield, mitigated the effects of drought on oil composition, and enhanced antioxidant activity (Jahani et al. 2021).

14.2.2 Flooding Stress

Extreme precipitation events and inadequate drainage exacerbate soil flooding, which is impacted by both climate change and human activity. Flooding is made worse by human modifications, such as compaction of the soil and changes in land use (Loreti et al. 2016). Crops like rice and wheat are impacted by perched water tables caused by compacted clay subsoil. Claypan layers and heavy clay soils impede drainage, which aggravate generalized problems with soil flooding (Singh et al. 2021). To reduce soil flooding and its detrimental consequences on ecosystems and agriculture, effective management techniques are essential. Soil flooding causes large losses in crop productivity; in the US, flooding and high soil moisture content account for more than 70% of yield decline (Rosenzweig et al. 2002). Prolonged flooding causes an increase in yield losses; at the V6 stage, each day of soil flooding reduces maize yields by 0.57 Mg ha^{-1} (Kaur et al. 2017). Furthermore, crop damage is intensified by the frequency and length of flooding; longer floods result in larger yield reductions, underscoring the significance of prompt management actions. The prolonged wetness of winter *Brassica napus* seedlings results in elevated levels of methane and nitrous oxide emissions, reduced leaf gas exchange parameters, and an increased risk of global warming. These findings underscore the significance of employing drainage strategies for sustainable crop cultivation (Li et al. 2023).

Plants facing soil flooding can enhance aeration with root aeration tubes or pneumatophores, adjust root morphology for oxygen uptake, regulate metabolic activities toward anaerobic pathways, develop morphological adaptations like stilt roots, and form symbiotic relationships with microbes for nutrient uptake, ensuring survival by maintaining oxygen availability and metabolic balance (Mariyam et al. 2024). Dripping techniques, raised beds, cover crops and using plant tolerant varieties is other specialized strategies employed by farmers (Kaur et al. 2020). For example, the investigation of *Triarrhena sacchariflora* physiological, biochemical, and transcriptome responses to flooding stress reveals notable alterations in the photosynthetic system, antioxidative enzyme activity, and gene expression linked to defense and stress response pathways (Wang et al. 2019a).

14.2.3 Salinity

Excess ions like Na^+ and Cl^- disrupt vital nutrient absorption and membrane integrity, which reduces photosynthesis and growth in a variety of plant species. Soil salinity also reduces water uptake and induces cell injury, which affects germination rates, growth, and nutrient uptake (Okon 2019). By reducing the water potential and uptaking toxic ions such as Na^+ and Cl^- in chloroplasts, salt stress negatively impacts plant photosynthesis. This results in a decrease in chlorophyll content and photosynthetic efficiency (Hameed et al. 2021). Oxidative stress, which is created by the accumulation of reactive oxygen species (ROS), intensifies cellular damage and ultimately lowers crop yield by influencing reproductive growth and yield components (Sachdev et al. 2021; Choudhary et al. 2025).

Plants' salt tolerance relies on stress-responsive genes like SOS pathway genes for salt regulation, along with osmotic protectants and antioxidant enzymes (Hao et al. 2021). Growth-related genes likely engage common signaling pathways across various stressors. Plant metabolism is altered by salt stress, leading to the accumulation of osmolytes, such as proline and glycine betaine as well as the production of osmoprotective proteins (Parihar et al. 2015). Changes in lipid metabolism and phytohormone levels, especially ABA, ethylene, and JA, are indicative of adaptations to preserve membrane integrity and function under stressful environments (Rahman et al. 2023). As exemplary 24 sugarcane genotypes were evaluated using additive main effects and multiplicative interaction (AMMI) and genotype by environment interaction (GGE) models in various induced salinity stress environments. The results showed significant effects of both genotype and environment on cane yield and related traits, with some genotypes demonstrating productivity and stability in saline conditions (Kumar et al. 2023). Also, the potential of 28-homoBL as a protective agent against salt stress in maize seedlings was suggested by the study's investigation of the effects of 28-homobrassinolide (28-homoBL) on *Zea mays* L. seedlings under salt stress (Kumar et al. 2024a). Results showed that 28-homoBL treatments mitigated salt-induced growth inhibition, reduced lipid peroxidation, and enhanced antioxidative enzyme activities (Arora et al. 2008).

14.2.4 Heavy Metal

The toxic effects of HMs on plants include growth inhibition, chlorosis, interference with vital metabolic pathways Zn, Cd, Cu, Hg, Cr, As, Ni and Pb. These disruptions of physiological processes ultimately lower crop productivity and cause oxidative stress (Ghori et al. 2019). Cu′s involvement in oxidative stress and ion leakage (Giannakoula et al. 2021), Pb disruption of membrane permeability and photosynthetic pathways (Yildirim et al. 2019), Cd restriction of chlorophyll synthesis and water balance (Shanmugaraj et al. 2019) and Zn interference with enzymatic reactions and chlorosis are a few examples. Research emphasizes the negative impacts on photosynthesis, growth, and enzyme activities, highlighting the necessity of efficient mitigation techniques in contaminated environments.

Plant roots take up, accumulate, and detoxify HMs from soil or water as part of the process known as HM remediation. This procedure frequently uses rhizofiltration, phytoextraction, and phytostabilization as methods to reduce pollutants in the environment and enhance ecosystem health (Chandra et al. 2015). As applying silicon improves antioxidation, chelates metal ions, and decreases the uptake of HMs by plants (Adrees et al. 2015). Mycorrhizal fungi that live in symbiosis reduce metal intake. Metal tolerance is increased by genetically modifying cells to optimize ethylene levels (Emamverdian et al. 2015). In another study, application of 24-epibrassinolide inhibits the build-up of chromium and stimulates antioxidative enzymes, compartmentalization and phenolic compounds help lessen the effects of HM stress efficient methods of mitigation in polluted areas (Shahzad et al. 2018).

14.2.5 Others (Thermal, UV Radiation, pH)

Thermal stress on crops, induced by extreme temperatures, disrupts plant metabolism, impairs photosynthesis, and reduces crop yield (Abdelrahman et al. 2020). The natural tolerance mechanisms and strategic techniques at the molecular level are studied to relieve heat stress and improve output. The studies focused on the negative effects of rising temperatures on crop yield, especially wheat (Kayastha et al. 2023). UV radiation exposure causes DNA damage, inhibits growth, and leads to decreased chlorophyll content in crops. As in case of *Rosmarinus officinalis* shoots, melatonin application mitigated the UV-B stress by enhancing antioxidant enzyme activities, increasing secondary metabolite contents, and boosting bioactivity, highlighting its potential role in plant defense and stress response mechanisms (Zhao et al. 2023). Similarly, pH stress alters nutrient availability, disrupts root function, and impairs nutrient uptake, ultimately affecting crop growth and development (Odoh et al. 2020) (Table 14.1).

14.3 Green Nanotechnology for Sustainable Agriculture

Green nanotechnology presents a viable avenue for sustainable agriculture by fusing the concepts of nanoscience with eco-friendly methods. Nanotechnology is the exact manipulation of materials at the nanoscale, providing unique control over their various properties and functions (Nasrollahzadeh et al. 2019). The creation of nanoscale pesticide and fertilizer delivery systems is one use for green nanotechnology. These inputs can have their release regulated by encasing them in nanocarriers, which improves their effectiveness and lowers total consumption while minimizing leaking into soil and water bodies (Sahai et al. 2019). Furthermore, real-time monitoring of pest infestations, nutrient levels, and soil health can be accomplished through the use of nanoscale sensors, allowing for focused and accurate actions. Additionally, NPs show promise as soil additives, enhancing soil structure, water retention, and nutrient availability. Examples of these are nanosilicates and nanoclays (Papatzani 2016). In addition to decreasing the demand for synthetic fertilizers and slowing down soil degradation, these minerals can increase crop productivity.

Furthermore, green nanotechnology makes it easier to create biodegradable, low-risk biopesticides and fertilizers based on NMs that won't harm non-target organisms or ecosystems. This promotes the development of sustainable crop protection measures. A viable method for sustainably reducing abiotic stress in plants is provided by green nanotechnology (Saxena et al. 2016). It reduces the influence on the environment by utilizing NPs made from natural sources, such as plant extracts and biopolymers (Banerjee et al. 2021). Plants with nano-enabled delivery systems are more resilient to ecological strains such as salinity, flooding, drought and heavy metals because they have improved nutrient absorption, water retention, and stress tolerance (Ijaz et al. 2023). Furthermore, precise action is made possible by real-time plant health monitoring made possible by nanoscale sensors (Kumar et al. 2024b). Green nanotechnology offers an integrated strategy to improving agricultural sustainability while reducing negative effects on ecosystems and human health through environmentally friendly synthesis techniques and focused applications.

Table 14.1 Effect of various abiotic stresses in crop plants

Abiotic stress	Scientific name	Common name	Effect	References
UV radiation	*Coffea* sp.	Coffee	UV triggered acclimation responses, which also resulted in a decrease in overall biomass.	de Paula et al. (2024)
Flooding	*Oryza sativa*	Rice	Decreased oxygen supply in plant cells that led to yield loss.	Haque et al. (2023)
Flooding	*Brassica napus*	Rapeseed	Reduced weight of roots and shoots, seed production and fewer main branches.	Li et al. (2023)
Salinity	*Saccharum officinarum*	Sugarcane	Decreased the crop yield via increased ion toxicity.	Kumar et al. (2023)
Heat	*Triticum aestivum*	Wheat	Poor productivity of crop.	Kayastha et al. (2023)
UV radiation	*Rosmarinus officinalis*	Rosemary	Reduced biomass, altered phenolic compounds	Zhao et al. (2023)
Drought	*Triticum aestivum*	Wheat	Reduced yield components and overall yield, and grain nutrient content and absorption diminished productivity and nutrient uptake.	Munsif et al. (2022)
Drought	*Glycine max*	Soybean	Elevated potassium and sodium concentrations in grains, leaves, and stems while reduced quantities of manganese, copper, zinc, and phosphorus in the grain, stem, and leaves.	Ezzati Lotfabadi et al. (2022)
Heavy metal (As)	*Betula celtiberica*	Silverbirch	Decreased the plant biomass and thereby yield.	Alves et al. (2022)
Drought	*Mentha piperita*	Peppermint	Decreased oil percentage and levels of chlorophyll and elevated amounts of total flavonoids and phenolics.	Jahani et al. (2021)
Salinity	*Oryza sativa*	Rice	Reduction in rice grains due to low germination and seedling growth.	Singh et al. (2021)
Heavy metal (Cr)	*Medicago sativa*	Alfalfa	Accumulation of Cr decreased plant biomass, root and shoot growth.	Tirry et al. (2021)
Heavy metal (Cu and Cd)	*Lolium perenne*	Perennial ryegrass	Increased the oxidative stress and increased the antioxidative enzymes.	Ke et al. (2021)
Heat	*Hordeum vulgare*	Barley	The variable temperatures during growth stage highly influenced the yield rate.	Otero et al. (2021)

(continued)

Table 14.1 (continued)

Abiotic stress	Scientific name	Common name	Effect	References
Drought	*Beta vulgaris*	Sugarbeet	Decreased Leaf area index (LAI), sugar content, photosynthetic pigments, relative water content (RWC) and root production. Increased osmolytes (proline, glycine betaine) and antioxidative enzyme activities such as catalase (CAT), superoxide dismutase (SOD) and peroxidase (PPO).	Khodadadi et al. (2020)
Salinity	*Triticum aestivum*	Wheat	Reduced plant height, RWC and antioxidative enzymes are enhanced due to stress.	Yadav et al. (2020)
Flooding	*Triarrhena sacchariflora*	Nakai	Anti-oxidative enzymes SOD and POD first increased and then declined their expression in roots.	Wang et al. (2019a, b)
Flooding	*Arabidopsis thaliana*	Arabidopsis	Decreased starch concentration and glucose content.	Loreti et al. (2018)
Salinity	*Arachis hypogaea*	Groundnut	Significant reduction in plant growth parameters along with reduced oilseed yield.	
Heavy metal (Cd)	*Leucaena leucocephala*	Acacia	Diminished the plant growth and plant growth parameters like RWC and LAI.	Kang et al. (2018)
Heavy metal (Cu)	*Brassica nigra*	Black mustard	Reduced shoot growth and root growth of overall plant.	Hansda et al. (2017)

14.4 Synthesis of Bio-nanomaterials

14.4.1 Synthesis of Engineered Nanomaterials

For the production of ENMs, bio-mediated synthesis appears to be a cost-effective, environmentally friendly, and time-efficient method (Albalawi et al. 2021). When compared to traditional procedures, this green technology guarantees the manufacturing of ENMs that are more stable and have less contamination hazards (Bandala and Berli 2019). Utilizing plant extracts especially those derived from different plant components including leaves, stems, seeds, bark, flowers, and roots are more effective method than using chemical agents like polyethylene glycol or sodium borohydride (Shafey 2020). Plants provide a plentiful, non-toxic source of active chemicals that aid in the reduction and stabilizing processes that occur during the production of NPs. Plant extracts contain organic compounds that function as potent ligand-capping agents and reducing agents to maintain colloidal stability and prevent agglomeration. Furthermore, effective bioreactors for ENM synthesis are provided by microorganisms such bacteria, fungus, yeasts, and algae, providing affordable and environmentally beneficial pathways (Dikshit et al. 2019) (Fig. 14.2).

These bio-mediated techniques highlight sustainability by outperforming physical and chemical synthesis methods. Given the variety of effects different synthesis pathways have on the end products, it is imperative to comprehend the properties and applications of ENMs produced through bio-mediated mechanisms (Shreyash et al. 2021). Green techniques are used to show general procedures for ENM synthesis, and bio-mediated synthesis approaches are provided with citation data.

14.4.1.1 Synthesis from Microbes

Unicellular and multicellular microbes are important components of ecosystems, affecting the cycles of greenhouse gases. They are employed in industrial operations that produce various chemicals, including antibiotics. Some improve soil quality, while others induce illnesses in plants. Among these, fungi are essential in many industries because they produce foods high in protein and help plants absorb nutrients. Yeasts, commonly used in fermentation, simplify the synthesis of ENMs, offering cost-effectiveness and environmental friendliness. Bao et al. (2010) synthesized cadmium telluride (CdTe) quantum dots using yeast cells grown in a nutrient supplied updated medium, resulting in well-dispersed CdTe ENMs with a cubic zinc-blended shape (3.6 nm) after purification steps. Also, using the fungus Verticillium, a novel biological method is introduced to its biomass by exposing it to aqueous Ag^+ ions. This results in intracellular diminution and the formation of

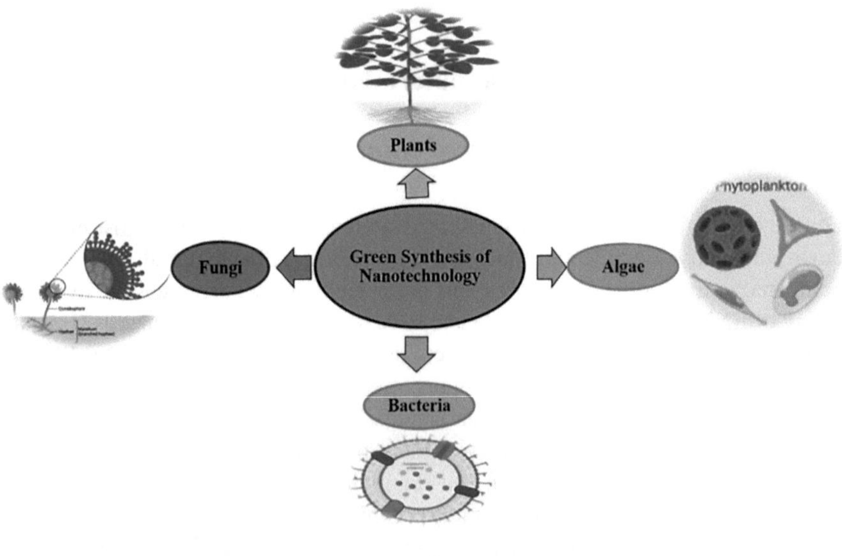

Fig. 14.2 Synthesis of green nanoparticles by different organisms

silver nanoparticles (25 ± 12 nm) beneath the cell wall surface through enzyme-mediated processes. The fungal cells are not harmed, so they can continue to multiply (Mukherjee et al. 2001).

Similarly, yeasts isolated from the gut of *Cornitermes cumulans* termites were used to synthesized silver and silver chloride nanoparticles. Colonies were cultured and then incubated in liquid medium with silver nitrate, yielding circular silver chloride nanoparticles (2–10 nm). These nanoparticles showed dose-dependent antibacterial effects against S. aureus and K. pneumoniae, with up to 85% inhibition (Eugenio et al. 2016). Similarly, silver nanoparticles (ENMs) were synthesized using bacteria *Lactobacillus rhamnosus* GG, with exopolysaccharide (EPS) acting as a reducing and stabilizing agent. These NPs exhibited antimicrobial activity against *E. coli*, *L. monocytogenes*, *K. pneumoniae*, *P. aeruginosa*, *Aspergillus* spp. and *Penicillium* spp. with dose-dependent effects observed against both bacterial and fungal pathogens (Kanmani and Lim 2013). Also, AuNPs were synthesized using a novel bacterial strain, identified as *Stenotrophomonas malophilia* (AuRed02), isolated from a site near Indian gold mines. After 8 hours of incubation, monodisperse GNPs (~40 nm) were obtained, stabilized with negatively charged phosphate (PO_4^{2-}) groups from NADP (Nicotinamide adenine dinucleotide phosphate), rendering them the stability in aqueous medium (Nangia et al. 2009). Therefore, the production of ENMs using microbial species proves to be a time-efficient, environmentally friendly and straightforward process, yielding particles with consistent shape and size (Fig. 14.2).

14.4.1.2 Synthesis from Plants

The production of ENMs through plant mediation occurs at low temperatures, swiftly, and in an environmentally sustainable manner, utilizing benign solvents (Ying et al. 2022). Various plants like *Aloe vera* (Medda et al. 2015), *Citrus lemon* (Nabi et al. 2022), *Cocos nucifera* (Govarthanan et al. 2016), *Musa paradisiaca* (Deena et al. 2015) etc. have been utilized to obtain the ENMs. AgNPs were synthesized using the aqueous leaf extract of *Decaschistia crotonifolia* (Dc), a rare medicinal plant which showed potential antimicrobial activity, antioxidative and catalytic property (Palithya et al. 2021). Similarly, Ag NPs were synthesized using *Cucumis prophetarum* leaf extract and $AgNO_3$ solution at a 1:9 ratio, heated at 80 °C. Brown Ag-ENMs were obtained post-centrifugation and drying. They exhibited stronger antibacterial activity against Gram-negative *S. typhi* than gram-positive *Staphylococcus aureus*. Additionally, these green-synthesized Ag-ENMs demonstrated potential in inhibiting cancer cell line growth, with concentration-dependent reductions in cell viability observed (Meena et al. 2020).

14.5 Role of Engineered Nanomaterials in Abiotic Stress Resilience

Engineered NMs offer diverse applications in agriculture, including as biosensors, fertilizers, pesticides, and carriers for genetic engineering. These innovations enhance stress resilience in crops by improving nutrient delivery, enabling predictive stress detection, and optimizing crop management, fostering sustainable and resilient agricultural practices with increased yield and reduced environmental impact (Pal et al. 2022). Nanotechnology application in mitigating biological and environmental stress in plants is particularly noteworthy. Diverse nanosensors tailored to detect various environmental stresses underscore the necessity for continuous innovation in nanotechnology design and implementation (Thakur et al. 2022). Nanocapsules function as a protective shield; ensuring encapsulated components remain contained and do not seep into the environment (Kumar et al. 2024c). When integrated with other organic substances, they can potentially bolster plants' resilience to non-living factors such as severe temperatures and insufficient water levels (Shang et al. 2019). Also, NMs in seed priming boosts germination rates and counters seed aging. This study contrasts heavy zinc acetate compound with algae-based synthesis of ZnO NMs on maize germination. ZnO-treated seeds show marked early germination enhancement, with 100 mg/L ZnO achieving the highest 87% rate after a week, along with superior root and shoot growth (Itroutwar et al. 2020) (Table 14.2).

14.5.1 *Heavy Metal Stress*

Globalization leads to increased environmental contamination from higher levels of toxic metals, adversely affecting soil dynamics, microbial organization, soil fertility, and crop efficiency, but phytogenic NPs show promise in initiating resistance mechanisms and regulating the plant's growth and physiological programming under HM stress (Gill 2014; Ali et al. 2023). Various green NPs, including metal oxides, such as TiO_2, MgO and ZnO, magnetic NPs and nanostructured materials, are combined with nanobiosensors for target monitoring, utilizing surface chemistry to link biomolecules such as antibodies, enzymes, ssDNA, or cells, establishing suitable bionanointerfaces capable of detecting minute concentrations of specific target molecules (Malhotra and Ali 2018; Hameed et al. 2023). NPs can transform or absorb heavy metals (HMs) in soil, reducing their bioavailability and mobility; for instance, the application of Fe_3O_4 NPs decreased Cd mobility in *Hevea brasiliensis* Muell (Sebastian et al. 2019). Plant growth is inhibited by several physiological and biochemical processes, such as photosynthesis, respiration, ion uptake, translocation, nutrient metabolism, carbohydrate synthesis, and growth regulators. A living plant cell-based biosensor was designed to detect damage caused by Cd(II) and Pb(II). It involves modifying a glassy carbon electrode with L-cysteine, followed by anti-IgG-Au antibody attachment. Then, living plant cells, treated with

Table 14.2 Role of NPs in abiotic stress alleviation

NPs used	Abiotic stress	Plant studied	Impact of NPs	References
TiO_2 NPs	Cold	Licorice	Reduced lipid peroxidation and H_2O_2 and elevated phenolics, total protein and osmolytes.	Ghabel and Karamian (2020)
Ag NPs	Flooding	Soyabean	Enhanced plant growth and control over the quality of protein.	Hshimoto et al. (2020)
SiO_2 NPs	Hg	Soybean	Disintegrated and immobilized Hg.	Li et al. (2020)
Zno NPs	Salinity	Fenugreek	Elevated levels of proline and trigonelline content were observed.	Noohpish et al. (2021)
Fe NPs	Drought	Foxtail millet	Enhanced enzymatic antioxidant activities, osmolyte concentrations, photosynthetic efficiency, growth traits, and biochemical responses.	Sreelakshmi et al. (2020)
SiO_2 NPs	Cd	Wheat	Enhanced antioxidative enzymes activity.	Ali et al. (2019)
Si NPs	Drought	Cucumber	Enhanced productivity and growth via signaling the intake of nutrients, such as higher potassium and nitrogen levels.	Alsaeedi et al. (2019)
ZnO NPs	Cd	Wheat	Reduced oxidative stress, boosted antioxidant enzymes, Si-NPs lowered Cd in wheat grains and boosted Si in plants.	Hussain et al. (2019)
Ag NPs	Heat	Wheat	Enhanced fresh weight, dry weight, root number, shoot length, and root length.	Iqbal et al. (2019)
ZnO and Fe_3O_4 NPs	Heat	Wheat	The activities of antioxidant enzymes (GST, SOD, POD, CAT) were elevated, while lipid peroxidation decreased.	Hassan et al. (2018)
CeO_2 NPs	Salinity	Arabidopsis	Enhanced the retention of K+ in leaf mesophyll, elevated chlorophyll content, boostedbiomass production and improved photosynthetic efficiency.	Wu et al. (2018)
SiO_2 NPs	Salinity	Common bean	Increased seed germination metrics such as percentage, vigor, rate, lengths, and masses.	Alsaeedi et al. (2017)
Anatase TiO_2 NPs	Drought	Flaxseed	Enhanced chlorophyll and carotenoids, promoted flax growth and yield, reducing H_2O_2 and MDA.	Aghdam et al. (2016)
Al_2O_3 NPs	Flooding	Soyabean	Controlled the ascorbate/glutathione process, membrane permeability and activity of the tricarboxylic acid (TCA) cycle.	Mustafa and Komatsu (2016)
TiO_2 and SiO_2 NPs	Drought	Cotton	The levels of total phenolic compounds, total soluble proteins, total free amino acids, proline content, total antioxidant activity and antioxidant capacity were all elevated.	Shallan et al. (2016)

anti-vitronectin, were affixed onto the electrode, enabling detection via changes in electrochemical impedance, effectively identifying Cd(II) and Pb(II) (Wang et al. 2019b).

For instance, investigated the impact of ZnO-NPs (25 ppm) synthesized from *Ulva lactuca* on *Leucaena leucocephala* under heavy metal stress induced by lead (Pb, 100 ppm) and cadmium (Cd, 50 ppm) was investigated. These environmentally-friendly ZnO-NPs significantly boosted plant parameters and photosynthetic pigment levels in HM-stressed *L. leucocephala* (Venkatachalam et al. 2017). Similarly, treatment with Fe_3O_4 NPs (0.5 ppm) derived from husk extract of *Cocos nucifera* improved plant yield, photosystem II (PSII) quantum efficiency, chlorophyll pigment content and crop productivity in Cd-stressed *Oryza sativa* (Sebastian et al. 2018). Furthermore, the external application of TiO_2 NPs (0.1%) produced from leaf extract of *Musa paradisica* on As stressed *Vigna radiata* resulted in significant reduction of As^{3+} ion accumulation. Moreover, it elevated amino acid content and enhanced activities of enzymatic antioxidant such as SOD, CAT, and APX facilitating detoxification of ROS (Katiyar et al. 2020; Tikoria et al. 2024) (Table 14.2).

14.5.2 Salt Stress

Green synthesized NPs play a pivotal role in alleviating salinity stress-induced adversities in plants by maintaining ion balance and osmotic potential, pivotal for promoting salinity tolerance outcomes (Munns and Tester 2008; Dhiman et al. 2021, 2023). Selenium nanoparticles (Se NPs) at a concentration of 100 ppm, synthesized using *H. vulgare* leaves, effectively mitigated the harmful impacts of salinity stress. This mitigation was observed through enhancements in root and shoot parameters, such as length, fresh weight FW), and dry weight along with enhancement in phenolic compounds and photosynthetic pigment contents while simultaneously reduced the stress markers, like MDA and H_2O_2 (Habibi and Aleyasin 2020). The alleviation of salinity stress through enhanced growth and physiological performance with the use of plant-derived nanoparticles has been substantiated by various studies. As exemplary, externally applied Zn-NPs synthesized from *Sorghum bicolor* leaf extract enhanced growth and photosynthetic parameters in salinity-stressed *Abelmoschus esculentus* at a range of 0.3%, by up-regulating the antioxidant defense system (Zafar et al. 2021).

An organic electrochemical transistor (OECT) biosensor measures plant response to salt stress via impedance spectroscopy. Impedance data from the biosensor in the plant stem helps derive equivalent circuit models, revealing changes in ionic concentration under salt stress (Sneha et al. 2024). Utilizing an electrochemical sensor, precise measurement of H_2O_2 release from *Arabidopsis* leaves during varying periods of salt stress highlights the efficacy of electrochemical sensing for assessing plant response to salt stress through H_2O_2 detection (Zhang et al. 2022). The utilization of plant extract-based synthesis for NPs has demonstrated effectiveness in stimulating stress-responsive signaling (Table 14.2). Calcium oxide (CaO NPs,

1.5 ppm), synthesized from *P. granatum* fruit extract, showed remarkable potential in alleviating salinity stress-induced challenges in *Triticale callus* (Yazıcılar et al. 2021). Additionally, these NPs have been noted for their role in regulating nutrient balance and protecting plants from ion imbalance during salt stress (Wahid et al. 2022).

14.5.3 Drought Stress

Drought, a major abiotic stressor, diminishes crop yield and quality through temperature fluctuations, decreased light intensity, and insufficient rainfall, prompting various plant responses such as changes in growth characteristics and biochemical profiles involving enzymatic antioxidant activities, protein, and metabolite levels (Kumari et al., 2022b; Tikoria and Ohri 2023). For example, *Chaetomorpha antennina* was employed in the production of iron nanoparticles (FeNPs), with some being uncoated while others were coated with citrate compounds (Sreelakshmi et al. 2020). Reportedly, these NPs were effective in improving drought stress tolerance in *Setaria italica* by positively influencing enzymatic antioxidant activities including catalase (CAT), superoxide dismutase (SOD), and peroxidase (POX), as well as osmolyte concentrations. Moreover, FeNPs also enhanced photosynthetic functioning, growth attributes and various biochemical results in *S. italica* at concentrations ranging from 50 to 120 mg L^{-1}. Additionally, SeNPs (30 ppm), derived from *A. sativum* bud extract, were administered to *T. aestivum* under water deficit conditions (Table 14.2). These Se NPs notably improved growth parameters such as plant length, biomass accumulation, leaf area, number while diminishing lipid peroxidation, thereby assisting in mitigating cellular toxicity induced by drought stress (Ikram et al. 2020).

14.5.4 Other Abiotic Stress

Recent years have seen heat and cold stress worsen due to climate change, harming global crop yields. Yet, green-synthesized NPs show promise in mitigating heat stress effects on plants. Limited evidence depicts their role in acclimatizing plants to high temperatures. Iqbal et al. (2019) utilized Ag NPs from *Moringa oleifera* leaf extract, noting their efficacy at 50 and 75 ppm concentrations in reducing MDA and H_2O_2 levels while enhancing antioxidant defense in *T. aestivum* under heat stress (Iqbal et al. 2019). Similarly, a study investigated the impact of TiO_2 NPs on cold tolerance (CT) in chickpea genotypes. Results revealed differential gene expression related to metabolism, defense, and signaling pathways (Table 14.2). TiO_2 NPs enhanced CT, possibly by reducing oxidative stress, suggesting their potential for field applications in crop productivity (Amini et al. 2017).

14.6 Nanoformulations: Environmental and Toxicity Issues

The utilization of ENMs in agriculture offers several benefits for enhancing crop production, but it also raises concerns regarding potential long-term effects on plants (Rajput et al. 2020). The impact of ENMs depends on factors such as exposure level, concentration, size, shape, and mode of contact. Research suggests that ENMs can influence various aspects of plant physiology, including metabolism, nutrient transport, and root morphology (Zuverza-Mena et al. 2017). Inside cells, ENMs can disrupt protein function, induce oxidative stress, and alter antioxidant patterns, potentially leading to oxidative damage. Additionally, ENMs effects vary based on material type, synthesis method, particle characteristics, and application techniques, adding complexity to their interactions at the cellular level (Vardakas et al. 2022).

Studies have investigated that certain ENMs, such as CeO_2, can significantly alter plant biochemical composition, with potential implications for human health, such as reducing cancer risk (Zhao et al. 2023b). However, in-depth toxicological studies are necessary to ensure the safety of ENM use in agriculture. A major concern is the potential accumulation of ENMs in plant tissues, which could impact plant growth promotion (Verma et al. 2018). In a recent research, it was found that ENMs, specifically SiO_2 and CeO_2 NPs, at a concentration of 100 mg L^{-1}, significantly increased the accumulation of polycyclic aromatic hydrocarbons (PAHs), such as Pyrene (Pyr), by 18.4–34.8% in *Glycine max* tissues over a 15-day hydroponic cultivation period, indicating potential risks and specific molecular mechanisms associated with co-exposure of ENMs and PAHs in agriculture (Cai et al. 2023). Research has demonstrated adverse effects of ENMs on various growth parameters, including root elongation, biomass, and chlorophyll levels, highlighting the need for careful consideration of their application and potential risks in agricultural practices.

14.7 Conclusions

This chapter explores eco-friendly methods for synthesizing novel NMs in agriculture, focusing on utilizing natural plant extracts, microbes and algae to exploit their chemical and physical properties. Engineered NMs offer potential applications in agriculture, including as NFs for enhancing crop growth, in seed priming and foliar spray to combat various environmental strains and diseases. Nanoformulations have shown promising effect in alleviating abiotic stress in plants, offering potential solutions for improving crop resilience and productivity. Challenges include understanding ENMs complete life cycle in the food chain, identifying potential toxicity mechanisms, scaling up green synthesis methods, developing multifunctional biomediated ENMs, and assessing their impact on plant physiology and nutritional quality. Addressing these challenges is crucial for wider adoption of ENMs in agriculture amidst increasing food demand and changing climate conditions.

References

Abdelrahman M, Burritt DJ, Gupta A, Tsujimoto H, Tran LS (2020) Heat stress effects on source–sink relationships and metabolome dynamics in wheat. J Exp Bot 71(2):543–554

Adrees M, Ali S, Rizwan M, Zia-ur-Rehman M, Ibrahim M, Abbas F, Farid M, Qayyum MF, Irshad MK (2015) Mechanisms of silicon-mediated alleviation of heavy metal toxicity in plants: a review. Ecotoxicol Environ Saf 119:186–197

Aghdam MT, Mohammadi H, Ghorbanpour M (2016) Effects of nanoparticulate anatase titanium dioxide on physiological and biochemical performance of *Linum usitatissimum* (Linaceae) under well-watered and drought stress conditions. Brazil J Bot 39:139–146

Albalawi F, Hussein MZ, Fakurazi S, Masarudin MJ (2021) Engineered nanomaterials: the challenges and opportunities for nanomedicines. Int J Nanomedicine 16:161–184

Ali S, Rizwan M, Hussain A, ur Rehman MZ, Ali B, Yousaf B, Wijaya L, Alyemeni MN, Ahmad P (2019) Silicon nanoparticles enhanced the growth and reduced the cadmium accumulation in grains of wheat (*Triticum aestivum* L.). Plant Physiol Biochem 140:1–8

Ali M, Kumar D, Tikoria R, Sharma R, Parkirti P, Vikram V, Kaushal K, Ohri P (2023) Exploring the potential role of hydrogen sulfide and jasmonic acid in plants during heavy metal stress. Nitric Oxide 140–141:16–29

Alsaeedi AH, El-Ramady H, Alshaal T, El-Garawani M, Elhawat N, Almohsen M (2017) Engineered silica nanoparticles alleviate the detrimental effects of Na+ stress on germination and growth of common bean (*Phaseolus vulgaris*). Environ Sci Pollut Res 24:21917–21928

Alsaeedi A, El-Ramady H, Alshaal T, El-Garawany M, Elhawat N, Al-Otaibi A (2019) Silica nanoparticles boost growth and productivity of cucumber under water deficit and salinity stresses by balancing nutrients uptake. Plant Physiol Biochem 139:1–10

Alves A, Boiu-Sicuia O, Popescu A (2022) Use of endophytic and rhizosphere bacteria to improve phytoremediation of arsenic-contaminated industrial soils by autochthonous Betula celtiberica. Oltenia Stud Comun Stiint Nat 38:79–84

Amini S, Maali-Amiri R, Mohammadi R, Kazemi-Shahandashti SS (2017) cDNA-AFLP analysis of transcripts induced in chickpea plants by TiO2 nanoparticles during cold stress. Plant Physiol Biochem 111:39–49

Arora N, Bhardwaj R, Sharma P, Arora HK (2008) Effects of 28-homobrassinolide on growth, lipid peroxidation and antioxidative enzyme activities in seedlings of *Zea mays* L. under salinity stress. Acta Physiol Plant 30:833–839

Bandala ER, Berli M (2019) Engineered nanomaterials (ENMs) and their role at the nexus of food, energy, and water. Mater Sci Energy Technol 2(1):29–40

Banerjee A, Sarkar A, Acharya K, Chakraborty N (2021) Nanotechnology: an emerging hope in crop improvement. Lett Appl Nano Bio Sci 10(4):2784–2803

Bao H, Hao N, Yang Y, Zhao D (2010) Biosynthesis of biocompatible cadmium telluride quantum dots using yeast cells. Nano Res 3:481–489

Bhardwaj K, Singh AK (2023) Bio-waste and natural resource mediated eco-friendly synthesis of zinc oxide nanoparticles and their photocatalytic application against dyes contaminated water. Chem Eng J Adv 16:100536

Cai Y, Ma X, Yuan B, Fang G, Ullah H, Zhou D, Gao J (2023) Metabolomics of soybean (Glycine max L.) response to co-exposure of pyrene and three metal oxide engineered nanomaterials. J Hazard Mater 441:129985

Calanca PP (2017) Effects of abiotic stress in crop production. In: Ahmed M, Stockle C (eds) Quantification of climate variability, adaptation and mitigation for agricultural sustainability. Springer, Cham, pp 165–180
Calicioglu O, Flammini A, Bracco S, Bellù L, Sims R (2019) The future challenges of food and agriculture: an integrated analysis of trends and solutions. Sustain For 11(1):222
Chandra R, Saxena G, Kumar V (2015) Phytoremediation of environmental pollutants: an eco-sustainable green technology to environmental management. In: Chandra R, Saxena G, Vineet Kumar V (eds) Advances in biodegradation and bioremediation of industrial waste. CRC Press, Boca Raton, pp 1–29
Chaudhry S, Sidhu GP (2022) Climate change regulated abiotic stress mechanisms in plants: a comprehensive review. Plant Cell Rep 41(1):1–31
Choudhary R, Rajput VD, Ghodake G, Ahmad F, Meena M, Rehman RU, Prasad R, Sharma RK, Singh R, Seth CS (2024) Comprehensive journey from past to present to future about seed priming with hydrogen peroxide and hydrogen sulfide concerning drought, temperature, UV and ozone stresses- a review. Plant Soil. https://doi.org/10.1007/s11104-024-06499-9
Choudhary R, Ahmad F, Kaya C, Upadhyay SK, Muneer S, Kumar V, Meena M, Liu H, Upadhyay H, Seth CS (2025) Decrypting proteomics, transcriptomics, genomics, and integrated omics for augmenting the abiotic, biotic, and climate change stress resilience in plants. J Plant Physiol. https://doi.org/10.1016/j.jplph.2025.154430
de Paula W, Santos AR, Vale EM, Pireda S, Correia LZ, Desouza GA, de Abreu DP, Carvalho LK, Almeida FA, Baroni DF, Partelli FL (2024) UV-B reduction and excess: management strategies regarding *Coffea* sp. crop. Sci Hortic 323:112499
Deena S, Dakshinamurthy A, Mosae Selvakumar P (2015) Green synthesis of silver nanoparticle using banana (Musa) sap. Adv Mater Res 1086:7–10
Dhiman S, Ibrahim M, Devi K, Sharma N, Kapoor N, Kaur R, Sharma N, Tikoria R, Ohri P, Mir BA, Bhardwaj R (2021) Biochar assisted remediation of toxic metals and metalloids. In: Prasad MNV (ed) Handbook of assisted and amendment: enhanced sustainable remediation technology. Wiley, USA, pp 131–162
Dhiman S, Singh AD, Kour J, Bhardwaj T, Devi K, Tikoria R, Ohri P, Bhardwaj R, Sharma P (2023) Redox homeostasis in response to abiotic stresses in plants. In: Singh A, Bhardwaj R, Tripathi DK (eds) The role of growth regulators and phytohormones in overcoming environmental stress. Academic, London, pp 15–37
Dikshit PK, Kumar J, Das P, Sinha S, Pandey A (2019) Green synthesis of metallic nanoparticles: applications and limitations. Environ Chem Lett 17:1103–1120
Dikshit PK, Kumar J, Das AK, Sadhu S, Sharma S, Singh S, Gupta PK, Kim BS (2021) Green synthesis of metallic nanoparticles: applications and limitations. Catalysts 11:902
Emamverdian A, Ding Y, Mokhberdoran F, Xie Y (2015) Heavy metal stress and some mechanisms of plant defense response. ScientificWorldJournal 2015:756120
Eugenio M, Müller N, Frasés S, Almeida-Paes R, Lima LM, Lemgruber L, Farina M, de Souza W, Sant'Anna C (2016) Yeast-derived biosynthesis of silver/silver chloride nanoparticles and their antiproliferative activity against bacteria. RSC Adv 6(12):9893–9904
Ezzati Lotfabadi Z, Weisany W, Abdul-razzak Tahir N, Mohammadi Torkashvand A (2022) Arbuscular mycorrhizal fungi species improve the fatty acids profile and nutrients status of soybean cultivars grown under drought stress. J Appl Microbiol 132(3):2177–2188
Geng SM, Yan DH, Zhang TX, Weng BS, Zhang ZB, Qin TL (2015) Effects of drought stress on agriculture soil. Nat Hazards 75:1997–2011
Ghabel VK, Karamian R (2020) Effects of TiO2 nanoparticles and spermine on antioxidant responses of Glycyrrhiza glabra L. to cold stress. Acta Bot Croat 79(2):137–147
Ghori NH, Ghori T, Hayat MQ, Imadi SR, Gul A, Altay V, Ozturk M (2019) Heavy metal stress and responses in plants. Int J Environ Sci Technol 16:1807–1828
Giannakoula A, Therios I, Chatzissavvidis C (2021) Effect of lead and copper on photosynthetic apparatus in citrus (*Citrus aurantium* L.) plants. The role of antioxidants in oxidative damage as a response to heavy metal stress. Plan Theory 10(1):155

Gill M (2014) Heavy metal stress in plants: a review. Int J Adv Res 2(6):1043–1055

Govarthanan M, Seo YS, Lee KJ, Jung IB, Ju HJ, Kim JS, Cho M, Kamala-Kannan S, Oh BT (2016) Low-cost and eco-friendly synthesis of silver nanoparticles using coconut (*Cocos nucifera*) oil cake extract and its antibacterial activity. Artif Cells Nanomed Biotechnol 44(8):1878–1882

Habibi G, Aleyasin Y (2020) Green synthesis of Se nanoparticles and its effect on salt tolerance of barley plants. Int J Nano Dimens 11(2):145–157

Hameed A, Ahmed MZ, Hussain T, Aziz I, Ahmad N, Gul B, Nielsen BL (2021) Effects of salinity stress on chloroplast structure and function. Cells 10(8):2023

Hameed A, Saif MJ, Qayyum MA, Khalid T, Farooq T (2023) Nanomaterial-based sensors for real-time monitoring of crop plants growth, development, production, and protection. In: Husen A (ed) Engineered nanomaterials for sustainable agricultural production, Soil improvement and stress management. Academic, London, pp 357–385

Hansda A, Kumar V, Anshumali (2017) Cu-resistant Kocuria sp. CRB15: a potential PGPR isolated from the dry tailing of Rakha copper mine. Biotech 7:1–11

Hao S, Wang Y, Yan Y, Liu Y, Wang J, Chen S (2021) A review on plant responses to salt stress and their mechanisms of salt resistance. Horticulture 7(6):132

Haque MA, Rafii MY, Yusoff MM, Ali NS, Yusuff O, Arolu F, Anisuzzaman M (2023) Flooding tolerance in rice: adaptive mechanism and marker-assisted selection breeding approaches. Mol Biol Rep 50(3):2795–2812

Hashimoto T, Mustafa G, Nishiuchi T, Komatsu S (2020) Comparative analysis of the effect of inorganic and organic chemicals with silver nanoparticles on soybean under flooding stress. Int J Mol Sci 21(4):1300

Hassan NS, Salah El Din TA, Hendawey MH, Borai IH, Mahdi AA (2018) Magnetite and zinc oxide nanoparticles alleviated heat stress in wheat plants. Curr Nanomater 3(1):32–43

Hong J, Wang C, Wagner DC, Gardea-Torresdey JL, He F, Rico CM (2021) Foliar application of nanoparticles: mechanisms of absorption, transfer, and multiple impacts. Environ Sci Nano 8(5):1196–1210

Hussain A, Oves M, Alajmi MF, Hussain I, Amir S, Ahmed J, Rehman MT, El-Seedi HR, Ali I (2019) Biogenesis of ZnO nanoparticles using Pandanus odorifer leaf extract: anticancer and antimicrobial activities. RSC Adv 9(27):15357–15369

Ijaz M, Khan F, Ahmed T, Noman M, Zulfiqar F, Rizwan M, Chen J, Siddique KH, Li B (2023) Nanobiotechnology to advance stress resilience in plants: current opportunities and challenges. Mater Today Bio 22:100759

Ikram M, Raja NI, Javed B, Mashwani ZU, Hussain M, Hussain M, Ehsan M, Rafique N, Malik K, Sultana T, Akram A (2020) Foliar applications of bio-fabricated selenium nanoparticles to improve the growth of wheat plants under drought stress. Green Process Synth 9(1):706–714

Iqbal M, Raja NI, Mashwani ZU, Hussain M, Ejaz M, Yasmeen F (2019) Effect of silver nanoparticles on growth of wheat under heat stress. Iran J Sci Technol A 43:387–395

Iqbal A, Fahad S, Iqbal M, Alamzeb M, Ahmad A, Anwar S, Khan AA, Amanullah AM, Inamullah S (2020) Special adaptive features of plant species in response to drought. In: Hasanuzzaman M, Tanveer M (eds) Salt and drought stress tolerance in plants: signaling networks and adaptive mechanisms. Springer, Cham, pp 77–118

Itroutwar PD, Kasivelu G, Raguraman V, Malaichamy K, Sevathapandian SK (2020) Effects of biogenic zinc oxide nanoparticles on seed germination and seedling vigor of maize (Zea mays). Biocatal Agric Biotechnol 29:101778

Jahani F, Tohidi-Moghadam HR, Larijani HR, Ghooshchi F, Oveysi M (2021) Influence of zinc and salicylic acid foliar application on total chlorophyll, phenolic components, yield and essential oil composition of peppermint (*Mentha piperita* L.) under drought stress condition. Arab J Geosci 14:1–12

Kanmani P, Lim ST (2013) Synthesis and structural characterization of silver nanoparticles using bacterial exopolysaccharide and its antimicrobial activity against food and multidrug-resistant pathogens. Process Biochem 48:1099–1106

Kang X, Yu X, Zhang Y, Cui Y, Tu W, Wang Q, Li Y, Hu L, Gu Y, Zhao K, Xiang Q (2018) Inoculation of *Sinorhizobium saheli* YH1 leads to reduced metal uptake for Leucaena leucocephala grown in mine tailings and metal-polluted soils. Front Microbiol 9:379917

Katiyar P, Yadu B, Korram J, Satnami ML, Kumar M, Keshavkant S (2020) Titanium nanoparticles attenuates arsenic toxicity by up-regulating expressions of defensive genes in *Vigna radiata* L. J Environ Sci 92:18–27

Kaur G, Zurweller BA, Nelson KA, Motavalli PP, Dudenhoeffer CJ (2017) Soil waterlogging and nitrogen fertilizer management effects on corn and soybean yields. Agron J 109(1):97–106

Kaur G, Singh G, Motavalli PP, Nelson KA, Orlowski JM, Golden BR (2020) Impacts and management strategies for crop production in waterlogged or flooded soils: a review. Agron J 112(3):1475–1501

Kaur H, Kaur S, Kumar S, Singh J, Rawat M (2021) Eco-friendly approach: synthesis of novel green TiO_2 nanoparticles for degradation of reactive green 19 dye and replacement of chemical synthesized TiO 2. J Clust Sci 32:1191–1204

Kaur R, Sharma N, Tikoria R, Ali M, Kour S, Kumar D, Ohri P (2022) Insights into biosynthesis and signaling of cytokinins during plant growth, development and stress tolerance. In: Aftab T (ed) Auxins, cytokinins and gibberellins signaling in plants. Springer International Publishing, Cham, pp 153–187

Kayastha P, Barsha KC, Pandey B, Magar BR, Chand H, Bhandari J, Lamichhane P, Baduwal P, Poudel MR (2023) Molecular basis of heat stress tolerance in wheat. J Agric Appl Biol 4(1):11–19

Ke T, Guo G, Liu J, Zhang C, Tao Y, Wang P, Xu Y, Chen L (2021) Improvement of the Cu and Cd phytostabilization efficiency of perennial ryegrass through the inoculation of three metal-resistant PGPR strains. Environ Pollut 271:116314

Khan SH (2020) Green nanotechnology for the environment and sustainable development. In: Naushad M, Lichtfouse E (eds) Green materials for wastewater treatment. Springer, Cham, pp 13–46

Khodadadi S, Chegini MA, Soltani A, Ajam Norouzi H, Sadeghzadeh Hemayati S (2020) Influence of foliar-applied humic acid and some key growth regulators on sugar beet (*Beta vulgaris* L.) under drought stress: antioxidant defense system, photosynthetic characteristics and sugar yield. Sugar Tech 22(5):765–772

Kumar R, Dhansu P, Kulshreshtha N, Meena MR, Kumaraswamy MH, Appunu C, Chhabra ML, Pandey SK (2023) Identification of salinity tolerant stable sugarcane cultivars using AMMI, GGE and some other stability parameters under multi environments of salinity stress. Sustain For 15(2):1119

Kumar D, Mariyam S, Gupta KJ, Thiruvengadam M, Ghodake GS, Xing B, Seth CS (2024a) Comparative investigation on chemical and green synthesized titanium dioxide nanoparticles against chromium (VI) stress eliciting differential physiological, biochemical, and cellular attributes in Helianthus annuus L. Sci Total Environ 930:172413. https://doi.org/10.1016/j.scitotenv.2024.172413

Kumar D, Singh R, Upadhyay SK, Verma KK, Tripathi RM, Liu H, Dhankher OP, Tripathi RD, Sahi SV, Seth CS (2024b) Review on interactions between nanomaterials and phytohormones: novel perspectives and opportunities for mitigating environmental challenges. Plant Sci 340:111964. https://doi.org/10.1016/j.plantsci.2023.111964

Kumar D, Biswas JK, Mulla SI, Singh R, Shukla R, Ahanger MA, Shekhawat GS, Verma KK, Siddiqui MW, Seth CS (2024c) Micro and nanoplastics pollution: sources, distribution, uptake in plants, toxicological effects, and innovative remediation strategies for environmental sustainability. Plant Physiol Biochem 213:108795. https://doi.org/10.1016/j.plaphy.2024.108795

Kumari S, Khanna RR, Nazir F, Albaqami M, Chhillar H, Wahid I, Khan MI (2022a) Biosynthesized nanoparticles in developing plant abiotic stress resilience: a new boon for sustainable approach. Int J Mol Sci 23(8):4452

Kumari A, Lakshmi GA, Krishna GK, Patni B, Prakash S, Bhattacharyya M, Singh SK, Verma KK (2022b) Climate change and its impact on crops: a comprehensive investigation for sustainable agriculture. Agronomy 12:3008. https://doi.org/10.3390/agronomy12123008

Li Y, Zhu N, Liang X, Bai X, Zheng L, Zhao J, Li YF, Zhang Z, Gao Y (2020) Silica nanoparticles alleviate mercury toxicity via immobilization and inactivation of Hg (ii) in soybean (Glycine max). Environ Sci Nano 7(6):1807–1817

Li L, Zhang L, Tang J, Xing H, Zhao L, Jie H, Jie Y (2023) Waterlogging increases greenhouse gas release and decreases yield in winter rapeseed (*Brassica napus* L.) seedlings. Sci Rep 13(1):18673

Loreti E, van Veen H, Perata P (2016) Plant responses to flooding stress. Curr Opin Plant Biol 33:64–71

Loreti E, Valeri MC, Novi G, Perata P (2018) Gene regulation and survival under hypoxia requires starch availability and metabolism. Plant Physiol 176(2):1286–1298

Malhotra BD, Ali MA (2018) Nanomaterials in biosensors: fundamentals and applications. Nanomat Biosen:1–74

Mallakpour S, Sirous F, Hussain CM (2021) A journey to the world of fascinating ZnO nanocomposites made of chitosan, starch, cellulose, and other biopolymers: Progress in recent achievements in eco-friendly food packaging, biomedical, and water remediation technologies. Int J Biol Macromol 170:701–716

Mariyam S, Upadhyay SK, Chakraborty K, Verma KK, Duhan JS, Muneer S, Meena M, Sharma RK, Ghodake G, Seth CS (2024) Nanotechnology, a frontier in agricultural science, a novel approach in abiotic stress management and convergence with new age medicine-a review. Sci Total Environ 912:169097. https://doi.org/10.1016/j.scitotenv.2023.169097

Medda S, Hajra A, Dey U, Bose P, Mondal NK (2015) Biosynthesis of silver nanoparticles from Aloe vera leaf extract and antifungal activity against Rhizopus sp. and Aspergillus sp. Appl Nanosci 5:875–880

Meena PR, Singh AP, Tejavath KK (2020) Biosynthesis of silver nanoparticles using *Cucumis prophetarum* aqueous leaf extract and their antibacterial and antiproliferative activity against cancer cell lines. ACS Omega 5(10):5520

Minhas PS, Rane J, Pasala RK (2017) Abiotic stresses in agriculture: an overview. In: Minhas P, Rane J, Pasala R (eds) Abiotic stress management for resilient agriculture. Springer, Singapore, pp 3–8

Mukherjee P, Ahmad A, Mandal D, Senapati S, Sainkar SR, Khan MI, Parishcha R, Ajaykumar PV, Alam M, Kumar R, Sastry M (2001) Fungus-mediated synthesis of silver nanoparticles and their immobilization in the mycelial matrix: a novel biological approach to nanoparticle synthesis. Nano Lett 1(10):515–519

Munns R, Tester M (2008) Mechanisms of salinity tolerance. Annu Rev Plant Biol 59:651–681

Munsif F, Farooq U, Arif M, Shah T, Jehangir M, Zaheer S, Akhtar K, Khan MS, Ahmad I, Ahmad W, Ali S (2022) Potassium and salicylic acid function synergistically to promote the drought resilience through upregulation of antioxidant profile for enhancing potassium use efficiency and wheat yield. Ann Appl Biol 180(2):273–282

Mustafa G, Komatsu S (2016) Insights into the response of soybean mitochondrial proteins to various sizes of aluminum oxide nanoparticles under flooding stress. J Proteome Res 15(12):4464–4475

Nabi G, Ain QU, Tahir MB, Nadeem Riaz K, Iqbal T, Rafique M, Hussain S, Raza W, Aslam I, Rizwan M (2022) Green synthesis of TiO2 nanoparticles using lemon peel extract: their optical and photocatalytic properties. Int J Environ Anal Chem 102(2):434–442

Nangia Y, Wangoo N, Goyal N, Shekhawat G, Suri CR (2009) A novel bacterial isolate Stenotrophomonas maltophilia as living factory for synthesis of gold nanoparticles. Microb Cell Factories 8:1–7

Nasrollahzadeh M, Sajjadi M, Sajadi SM, Issaabadi Z (2019) Green nanotechnology. In: de Farias RB (ed) Interface science and technology. Elsevier, Amsterdam, pp 145–198

Noohpisheh Z, Amiri H, Mohammadi A, Farhadi S (2021) Effect of the foliar application of zinc oxide nanoparticles on some biochemical and physiological parameters of *Trigonella foenum-graecum* under salinity stress. Plant Biosyst 155(2):267–280

Odoh CK, Sam K, Zabbey N, Eze CN, Nwankwegu AS, Laku C, Dumpe BB (2020) Microbial consortium as biofertilizers for crops growing under the extreme habitats. In: Yadav A, Singh J, Rastegari A, Yadav N (eds) Plant microbiomes for sustainable agriculture. Springer, Cham, pp 381–424

Okon OG (2019) Effect of salinity on physiological processes in plants. In: Giri B, Varma A (eds) Microorganisms in saline environments: strategies and functions. Springer, Cham, pp 237–262

Ondrasek G (2013) Water scarcity and water stress in agriculture. In: Ahmad P, Wani M (eds) Physiological mechanisms and adaptation strategies in plants under changing environment. Springer, New York, pp 75–96

Otero EA, Miralles DJ, Benech-Arnold RL (2021) Development of a precise thermal time model for grain filling in barley: a critical assessment of base temperature estimation methods from field-collected data. Field Crop Res 260:108003

Pal A, Kaur P, Dwivedi N, Rookes J, Bohidar HB, Yang W, Cahill DM, Manna PK (2022) Clay-Nanocomposite based smart delivery systems: a promising tool for sustainable farming. ACS Agric Sci Technol 3(1):3–16

Palithya S, Gaddam SA, Kotakadi VS, Penchalaneni J, Challagundla VN (2021) Biosynthesis of silver nanoparticles using leaf extract of Decaschistia crotonifolia and its antibacterial, antioxidant, and catalytic applications. Green Chem Lett Rev 14:137–152

Pamungkas SS, Farid N (2022) Drought stress: responses and mechanism in plants. Rev Agric Sci 10:168–185

Papatzani S (2016) Effect of nanosilica and montmorillonite nanoclay particles on cement hydration and microstructure. Mater Sci Technol 32(2):138–153

Parihar P, Singh S, Singh R, Singh VP, Prasad SM (2015) Effect of salinity stress on plants and its tolerance strategies: a review. Environ Sci Pollut Res 22:4056–4075

Priya N, Kaur K, Sidhu AK (2021) Green synthesis: an eco-friendly route for the synthesis of iron oxide nanoparticles. Front Nanotechnol 3:655062

Rahman SU, Li Y, Hussain S, Hussain B, Riaz L, Ashraf MN, Khaliq MA, Du Z, Cheng H (2023) Role of phytohormones in heavy metal tolerance in plants: a review. Ecol Indic 146:109844

Rajput V, Minkina T, Mazarji M, Shende S, Sushkova S, Mandzhieva S, Burachevskaya M, Chaplygin V, Singh A, Jatav H (2020) Accumulation of nanoparticles in the soil-plant systems and their effects on human health. Ann Agric Sci 65(2):137–143

Rajput VD, Singh A, Minkina T, Rawat S, Mandzhieva S, Sushkova S, Shuvaeva V, Nazarenko O, Rajput P, Komariah VKK, Singh AK, Rao M, Upadhyay SK (2021) Nano-enabled products: challenges and opportunities for sustainable agriculture. Plan Theory 10:2727. https://doi.org/10.3390/plants10122727

Reddy Yadav LS, Raghavendra M, Udayabhanu MK, Nagaraju G (2018) Photocatalytic, biodiesel, electrochemical sensing properties and formylation reactions of ZnO nanoparticles synthesized via eco-friendly green synthesis method. J Mater Sci Mater Electron 29:8747–8759

Rosenzweig C, Tubiello FN, Goldberg R, Mills E, Bloomfield J (2002) Increased crop damage in the US from excess precipitation under climate change. Glob Environ Change 12(3):197–202

Sachdev S, Ansari SA, Ansari MI, Fujita M, Hasanuzzaman M (2021) Abiotic stress and reactive oxygen species: generation, signaling, and defense mechanisms. Antioxidants (Basel) 10(2):277

Sahai P, Sinha VB, Dutta R (2019) Bioformulation and nanotechnology in pesticide and fertilizer delivery system for eco-friendly agriculture: a review. Sci Agric 3:2–10

Saxena R, Tomar RS, Kumar M (2016) Exploring nanobiotechnology to mitigate abiotic stress in crop plants. J Pharm Sci 8(9):974

Sebastian A, Nangia A, Prasad MN (2018) A green synthetic route to phenolics fabricated magnetite nanoparticles from coconut husk extract: implications to treat metal contaminated water and heavy metal stress in *Oryza sativa* L. J Clean Prod 174:355–366

Sebastian A, Nangia A, Prasad MN (2019) Cadmium and sodium adsorption properties of magnetite nanoparticles synthesized from Hevea brasiliensis Muell. Arg. bark: relevance in amelioration of metal stress in rice. J Hazard Mater 371:261–272
Shafey AME (2020) Green synthesis of metal and metal oxide nanoparticles from plant leaf extracts and their applications: a review. Green Process Synth 9(1):304–339
Shahzad B, Tanveer M, Che Z, Rehman A, Cheema SA, Sharma A, Song H, ur Rehman S, Zhaorong D (2018) Role of 24-epibrassinolide (EBL) in mediating heavy metal and pesticide induced oxidative stress in plants: a review. Ecotoxicol Environ Saf 147:935–944
Shallan MA, Hassan H, Namich AA, Ibrahim AA (2016) Biochemical and physiological effects of TiO2 and SiO2 nanoparticles on cotton plant under drought stress. Res J Pharm Biol Chem Sci 7(4):1540–1551
Shang Y, Hasan MK, Ahammed GJ, Li M, Yin H, Zhou J (2019) Applications of nanotechnology in plant growth and crop protection: a review. Molecules 24(14):2558
Shanmugaraj BM, Malla A, Ramalingam S (2019) Cadmium stress and toxicity in plants: an overview. In: Mirza Hasanuzzaman M, Prasad MNV, Fujita M (eds) Cadmium toxicity and tolerance in plants. Academic Press, London, pp 1–17
Shende S, Rajput V, Minkina T, Sushkova S, Mandzhieva S (2023) Strategic role of nanotechnology in plant growth improvement and crop production. In: Ingle AP (ed) Nanotechnology in agriculture and agroecosystems. Elsevier, pp 25–49
Shreyash N, Bajpai S, Khan MA, Vijay Y, Tiwary SK, Sonker M (2021) Green synthesis of nanoparticles and their biomedical applications: a review. ACS Appl Nano Mater 4(11):11428–11457
Singh RK, Kota S, Flowers TJ (2021) Salt tolerance in rice: seedling and reproductive stage QTL mapping come of age. Theor Appl Genet 134:3495–3533
Sneha M, Ravindranath NA, Jayaraman V (2024) Electrochemical investigations of plant response to salt stress: insights through impedance studies of OECT-based biosensor. ChemistrySelect 9(1):e202302928
Solis J, Gutierrez A, Mangu V, Sanchez E, Bedre R, Linscombe S, Baisakh N (2018) Genetic mapping of quantitative trait loci for grain yield under drought in rice under controlled greenhouse conditions. Front Chem 5:129
Sreelakshmi B, Induja S, Adarsh P, Rahul H, Arya S, Aswana S, Haripriya R, Aswathy B, Manoj P, Vishnudasan D (2020) Drought stress amelioration in plants using green synthesised iron oxide nanoparticles. Mater Today Proc 41:7
Thakur A, Kumar A (2022) Recent advances on rapid detection and remediation of environmental pollutants utilizing nanomaterials-based (bio) sensors. Sci Total Environ 834:155219
Thakur M, Wang B, Verma ML (2022) Development and applications of nanobiosensors for sustainable agricultural and food industries: recent developments, challenges and perspectives. Environ Technol Innov 26:102371
Tikoria R, Ohri P (2023) Application of neem waste vermicompost in compensating nematode induced stress and upregulating physiological markers of tomato plants under glass house conditions after 10 days of exposure. Environ Sci Pollut Res. https://doi.org/10.1007/s11356-023-30324-y#citeas
Tikoria R, Sharma N, Kour S, Kumar D, Ohri P (2022) Vermicomposting: an effective alternative in integrated pest management. In: Vig AP, Singh J, Suthar S (eds) Earthworm engineering and applications. Nova Science Publishers, New York, pp 103–118
Tikoria R, Kaur A, Ohri P (2023a) Amelioration of oxidative stress and growth enhancement by application of vermicompost via modulating phyto-constituents in tomato plants during nematode stress. J Soil Sci Plant Nutr 23(3):3944–3960
Tikoria R, Kaur A, Ohri P (2023b) Physiological, biochemical and structural changes in tomato plants by vermicompost application in different exposure periods under glass house conditions. Plant Physiol Biochem 197:107656
Tikoria R, Kumar D, Ali M, Ohri P (2024) Boosting of free radical scavenging capacity and physiological markers of tomato plants by vermicompost application during nematode stress. J Soil Sci Plant Nutr 24:1507–1518

Tirry N, Kouchou A, El Omari B, Ferioun M, El Ghachtouli N (2021) Improved chromium tolerance of Medicago sativa by plant growth-promoting rhizobacteria (PGPR). J Genet Eng Biotechnol 19:1–14

Usman M, Farooq M, Wakeel A, Nawaz A, Cheema SA, ur Rehman H, Sanaullah M (2020) Nanotechnology in agriculture: current status, challenges and future opportunities. Sci Total Environ 721:137778

Vardakas P, Kyriazis ID, Kourti M, Skaperda Z, Tekos F, Kouretas D (2022) Oxidative stress–mediated nanotoxicity: mechanisms, adverse effects, and oxidative potential of engineered nanomaterials. In: Liu JL, Basshir S (eds) Advanced nanomaterials and their applications in renewable energy. Elsevier, Amsterdam, pp 179–218

Venkatachalam P, Jayaraj M, Manikandan R, Geetha N, Rene ER, Sharma NC, Sahi SV (2017) Zinc oxide nanoparticles (ZnONPs) alleviate heavy metal-induced toxicity in Leucaena leucocephala seedlings: a physiochemical analysis. Plant Physiol Biochem 110:59–69

Verma SK, Das AK, Patel MK, Shah A, Kumar V, Gantait S (2018) Engineered nanomaterials for plant growth and development: A perspective analysis. Sci Total Environ 630:1413–1435

Verma KK, Song XP, Joshi A, Rajput VD, Singh M, Sharma A, Singh RK, Li DM, Arora J, Minkina T, Li YR (2022) Nanofertilizer possibilities for healthy soil, water and food in future: an overview. Front Plant Sci 13:865048. https://doi.org/10.3389/fpls.2022.865048

Verma KK, Singh M, Rajput VD, Verma CL, Burachevskaya M (2023) Nanofertilizers: a sustainable alternative to conventional means. In: Rajput VD, Singh A, Minkina TM, Verma KK, Singh AK (eds) Nanotechnology for sustainable agriculture: an innovative and eco-friendly approach. Apple Academic Press, Inc., Taylor & Francis, New York. https://doi.org/10.1201/9781003333128

Verma KK, Song XP, Kumari A, Jagadesh M, Singh SK, Bhatt R, Singh M, Seth CS, Li YR (2024a) Climate change adaptation: challenges for agricultural sustainability. Plant Cell Environ:1–12. https://doi.org/10.1111/pce.15078

Verma KK, Song X-P, Singh S, Kumari A, Arora J, Singh SK, Solanki MK, Seth CS, Li Y-R (2024b) Synergistic interactions of nanoparticles and plant growth promoting rhizobacteria enhancing soil-plant systems: a multigenerational perspective. Front Plant Sci 15:1376214. https://doi.org/10.3389/fpls.2024.1376214

Wahid I, Rani P, Kumari S, Ahmad R, Hussain SJ, Alamri S, Tripathy N, Khan MI (2022) Biosynthesized gold nanoparticles maintained nitrogen metabolism, nitric oxide synthesis, ions balance, and stabilizes the defense systems to improve salt stress tolerance in wheat. Chemosphere 287:132142

Wang X, Cheng M, Yang Q, Wei H, Xia A, Wang L, Ben Y, Zhou Q, Yang Z, Huang X (2019a) A living plant cell-based biosensor for real-time monitoring invisible damage of plant cells under heavy metal stress. Sci Total Environ 697:134097

Wang J, Sun H, Sheng J, Jin S, Zhou F, Hu Z, Diao Y (2019b) Transcriptome, physiological and biochemical analysis of *Triarrhena sacchariflora* in response to flooding stress. BMC Genet 20:1–15

Wu H, Shabala L, Shabala S, Giraldo JP (2018) Hydroxyl radical scavenging by cerium oxide nanoparticles improves Arabidopsis salinity tolerance by enhancing leaf mesophyll potassium retention. Environ Sci Nano 5(7):1567–1583

Xing K, Shen X, Zhu X, Ju X, Miao X, Tian J, Feng Z, Peng X, Jiang J, Qin S (2016) Synthesis and in vitro antifungal efficacy of oleoyl-chitosan nanoparticles against plant pathogenic fungi. Int J Biol Macromol 82:830–836

Yadav T, Kumar A, Yadav RK, Yadav G, Kumar R, Kushwaha M (2020) Salicylic acid and thiourea mitigate the salinity and drought stress on physiological traits governing yield in pearl millet-wheat. Saudi J Biol Sci 27(8):2010–2017

Yazıcılar B, Böke F, Alaylı A, Nadaroglu H, Gedikli S, Bezirganoglu I (2021) In vitro effects of CaO nanoparticles on *Triticale callus* exposed to short and long-term salt stress. Plant Cell Rep 40:29–42

Yildirim E, Ekinci M, Turan M, Güleray AG, Selda ÖR, Dursun A, Raziye KU, Balci T (2019) Impact of cadmium and lead heavy metal stress on plant growth and physiology of rocket (*Eruca sativa* L.). Kahramanmaraş Sütçü İmam Üniversitesi Tarım ve Doğa Dergisi 22(6):843–850

Ying S, Guan Z, Ofoegbu PC, Clubb P, Rico C, He F, Hong J (2022) Green synthesis of nanoparticles: current developments and limitations. Environ Technol Innov 26:102336

Zafar S, Hasnain Z, Aslam N, Mumtaz S, Jaafar HZ, Wahab PE, Qayum M, Ormenisan AN (2021) Impact of Zn nanoparticles synthesized via green and chemical approach on okra (*Abelmoschus esculentus* L.) growth under salt stress. Sustain For 13(7):3694

Zhang J, Lu M, Zhou H, Du X, Du X (2022) Assessment of salt stress to arabidopsis based on the detection of hydrogen peroxide released by leaves using an electrochemical sensor. Int J Mol Sci 23(20):12502

Zhao Z, Yun C, Gu L, Liu J, Yao L, Wang W, Wang H (2023a) Melatonin enhances biomass, phenolic accumulation, and bioactivities of rosemary (Rosmarinus officinalis) in vitro shoots under UV-B stress. Physiol Plant 175(4):e13956

Zhao L, Liu W, Zhang Y, Wang X, Li Y (2023b) Effects of CeO2 nanoparticles on the nutritional quality of corn and soybean. Molecules 28:1798

Zorb C, Geilfus CM, Dietz KJ (2019) Salinity and crop yield. Plant Biol 21:31–8

Zulfiqar F, Ashraf M (2021) Nanoparticles potentially mediate salt stress tolerance in plants. Plant Physiol Biochem 160:257–268

Zuverza-Mena N, Martínez-Fernández D, Du W, Hernandez-Viezcas JA, Bonilla-Bird N, López-Moreno ML, Komárek M, Peralta-Videa JR, Gardea-Torresdey JL (2017) Exposure of engineered nanomaterials to plants: insights into the physiological and biochemical responses-a review. Plant Physiol Biochem 110:236–264

Chapter 15
Potential of Nano-based Biosensors on Plant Growth, Development and Stress Adaptation Strategies: An Overview

Bharat M. Bhalerao and Kiran D. Pawar

15.1 Introduction

The global agricultural sector faces immense challenges in meeting the food demands of a rapidly growing population. One of the most pressing issues is the impact of abiotic stress on crop health and productivity. Abiotic stress factors, including drought, salinity, extreme temperatures, and heavy metal toxicity, can significantly impair plant growth and yield. Timely detection and management of these stressors are crucial for sustaining agricultural productivity and ensuring food security (Choudhary et al. 2025). Recent advancements in nanotechnology offer promising solutions to this challenge, particularly through the development of nanotechnology-based biosensors for the early detection of abiotic stress in crops. Abiotic stress is a major constraint in agriculture, reducing crop yields by more than 50% globally (Mahajan and Tuteja 2005; Verma et al. 2022). Traditional methods of stress detection, such as visual inspection and biochemical assays, are often time-consuming and lack the sensitivity needed for early intervention. Nanotechnology-based biosensors, however, provide a highly sensitive, specific, and rapid means of detecting abiotic stress markers at the molecular level. These sensors can detect minute changes in the physiological and biochemical status of plants, allowing for timely and precise stress management (Mishra et al. 2014).

Nanomaterials, such as carbon nanotubes (CNTs), quantum dots, and metallic nanoparticles (NPs), play a critical role in the functionality of these biosensors. Their unique properties, including high surface area, enhanced electrical

B. M. Bhalerao (✉)
Department of Biochemistry, Post Graduate Institute, Mahatma Phule Krishi Vidyapeeth, Rahuri, Maharashtra, India

K. D. Pawar
School of Nanoscience and Biotechnology, Shivaji University, Kolhapur, Maharashtra, India

C. S. Seth et al. (eds.), *Nano-formulations for Sustainable Agriculture*,
Nanotechnology in the Life Sciences, https://doi.org/10.1007/978-3-031-90617-6_15

conductivity, and customizable optical characteristics, significantly improve the sensitivity and accuracy of biosensors (Alharbi et al. 2018). For instance, carbon nanotube-based sensors have been shown to detect stress-related phytohormones and reactive oxygen species with high precision, enabling early detection of drought and salinity stress in crops (Khodakovskaya et al. 2012). The integration of nanotechnology-based biosensors with wireless communication networks and data analytics platforms further enhances their utility in precision agriculture. These integrated systems enable real-time monitoring of crop health and environmental conditions, facilitating data-driven decision-making for farmers. This approach not only improves the efficiency of resource use but also minimizes the environmental impact of agricultural practices (Ray et al. 2012). Despite their potential, the deployment of nanotechnology-based biosensors in agriculture faces several challenges, including flexibility, cost, and potential environmental risks. Addressing these challenges requires interdisciplinary research and collaboration among scientists, engineers, and policymakers (Wang et al. 2016). The development of cost-effective and environmentally friendly NMs, along with standardized protocols for sensor fabrication and deployment, is essential for the widespread adoption of this technology in agriculture (Cinteza and Olaru 2017).

Present chapter explores the applications of nanotechnology-based biosensors for the early detection of abiotic stress in crops and discusses various types of NMs used in biosensor construction, their mechanisms of detecting stress markers, and their practical applications in agricultural settings (Choudhary et al. 2024). It aims to provide insights into how these biosensors can contribute to enhancing crop resilience, optimizing resource management, and supporting sustainable agricultural practices. Additionally, the chapter will address current challenges and future perspectives in the field, emphasizing the potential impact of nanotechnology on improving food security and agricultural sustainability.

15.2 Importance of Abiotic Stress Detection in Crops

The detection of abiotic stress in crops is critically important due to its significant impact on agricultural productivity and food security. Abiotic stress factors such as drought, salinity, extreme temperatures, and heavy metal toxicity can severely hinder crop growth and yield, leading to substantial economic losses (Verma et al. 2020). Early detection of these stresses is essential for timely intervention, which can mitigate their adverse effects and improve crop resilience. Abiotic stresses are major contributors to the variability in crop yield and quality. For instance, drought and extreme heat have been identified as responsible for approximately 40% of global crop yield variability (Lesk et al. 2016). Timely detection allows for interventions such as optimized irrigation and the application of stress-relief agents, preventing significant yield losses. Furthermore, real-time monitoring of abiotic stress facilitates precision agriculture, where resources are utilized efficiently to manage stress conditions.

Sensor technologies in agriculture can enhance water use efficiency and reduce the overuse of fertilizers and pesticides (Jones et al. 2010). In addition to improving crop management practices, early detection of abiotic stress is invaluable for crop breeding programs. Understanding plant responses to abiotic stress at the molecular level is crucial for identifying and developing stress-tolerant genotypes. Zandalinas et al. (2018) underscored the importance of early stress detection in facilitating the selection of resilient crop varieties. Moreover, effective management of abiotic stress through early detection contributes to sustainable agricultural practices. Fleury et al. (2010) noted that mitigating the impact of abiotic stress on crops reduces the need for extensive land use and conserves natural resources.

The economic benefits of early abiotic stress detection are also significant. Early detection can result in significant cost savings by preventing crop loss and optimizing the use of inputs. Lobell et al. (2011) highlighted the economic advantages of adapting agricultural practices based on early stress detection, which include improved crop resilience and optimized input usage. Additionally, with the increasing frequency of extreme weather events due to climate change, early detection of abiotic stress is crucial for adapting agricultural practices (Verma et al. 2024a). Bailey-Serres et al. (2019) discussed how climate change exacerbates abiotic stress, emphasizing the need for advanced monitoring systems to support climate-resilient agriculture. Other studies further corroborate the importance of abiotic stress detection. For example, Mittler (2006) described the critical role of reactive oxygen species in plant stress responses, highlighting the need for early detection mechanisms. According to Bita and Gerats (2013), the combination of multiple abiotic stresses can compound their effects, making early detection even more crucial for effective management. Additionally, Pareek et al. (2010) discussed the potential of genetic engineering and molecular breeding in developing crops with enhanced stress tolerance, which is heavily reliant on early stress detection for success.

15.3 Role of Nanotechnology in Modern Agriculture

Nanotechnology holds immense promise for revolutionizing modern agriculture by offering innovative solutions to longstanding challenges in crop production, sustainability, and food security. This burgeoning field leverages nanoscale materials and devices to enhance various aspects of agricultural practices, from crop protection and nutrient management to environmental monitoring and precision farming. One significant application of nanotechnology in agriculture is in the development of nanoscale delivery systems for agrochemicals. NPs can encapsulate pesticides, herbicides, and fertilizers, improving their efficiency and targeted delivery to plants while minimizing environmental impact (Khot et al. 2012; Verma et al. 2024b). This targeted delivery ensures that agrochemicals reach their intended targets with minimal waste, reducing the quantity needed and mitigating adverse effects on non-target organisms and ecosystems. Moreover, NMs such as carbon nanotubes, nanoclays, and metallic NPs are utilized in soil remediation and enhancement. They can improve soil structure, water retention, and nutrient availability, thereby promoting healthier root systems and enhancing overall crop productivity (Khodakovskaya et al. 2009; Dimkpa et al. 2012).

Nanotechnology also plays a crucial role in plant genetics and breeding. Nanoscale tools enable precise manipulation of genetic material and facilitate the development of stress-tolerant crop varieties. NPs can deliver genes or RNA molecules into plant cells, enhancing traits such as drought tolerance, disease resistance, and nutrient uptake (Liu et al. 2019). Furthermore, nano sensors and nanodevices enable real-time monitoring of environmental parameters and plant health (Kumar et al. 2024a). These devices detect early signs of stress, such as nutrient deficiencies or pathogen infections, allowing farmers to implement timely interventions and optimize resource use (Tuteja et al. 2013). Nano sensors can also monitor soil quality, moisture levels, and nutrient content, providing valuable data for precision agriculture and decision-making. Despite these advancements, the widespread adoption of nanotechnology in agriculture faces challenges related to safety, regulation, and public acceptance. Concerns about NPs toxicity, environmental impact, and potential long-term effects on ecosystems and human health necessitate rigorous testing and regulatory frameworks (Miralles et al. 2012; Rajput et al. 2021). Addressing these concerns requires collaborative efforts among scientists, policymakers, and stakeholders to ensure the responsible development and deployment of nanotechnology in agriculture (Fig. 15.1).

15.4 Overview of Abiotic Stress in Crops

Abiotic stress refers to environmental factors that can adversely affect plant growth, development, and productivity, independent of biological factors (Kumari et al. 2022). These stresses are significant challenges in agriculture and can severely limit crop yield and quality.

15.4.1 Limited Water Supply and Saline Soil

Drought is characterized by inadequate water availability for plants, leading to water stress. It is one of the most prevalent and detrimental abiotic stresses in agriculture. Drought stress disrupts physiological processes in plants, such as photosynthesis and nutrient uptake, ultimately impairing growth and reducing yield (Chaves et al. 2003). Research indicates that drought can reduce crop yields globally by more than 50% (Lesk et al. 2016). Salinity stress arises from high concentrations of soluble salts in soil and irrigation water, particularly sodium chloride (NaCl). Excess salts disrupt osmotic balance in plants, impairing water uptake and leading to ion toxicity and oxidative stress (Munns and Tester 2008). Salinity stress limits crop

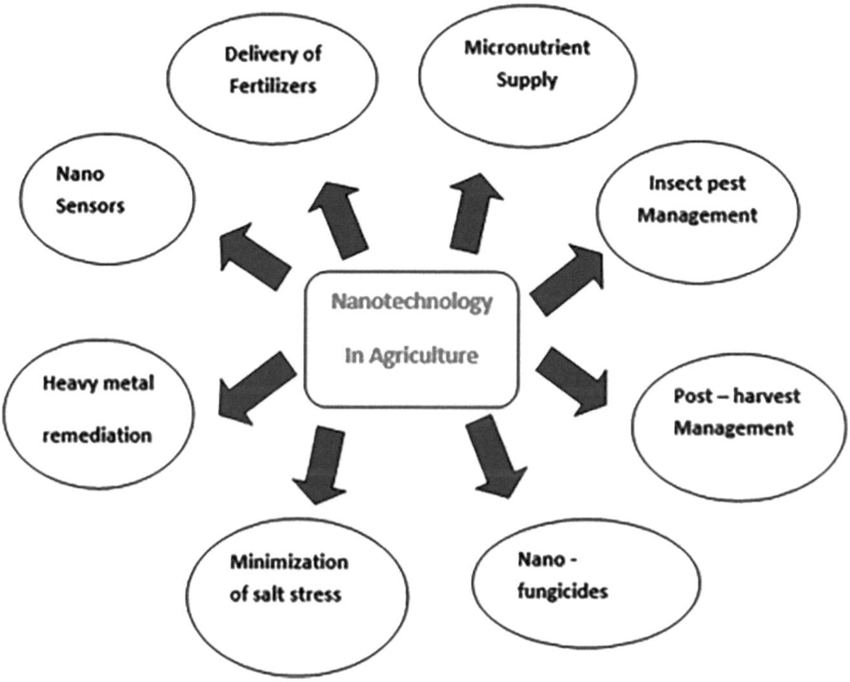

Fig. 15.1 Application of nanotechnology in modern agriculture

productivity in arid and semi-arid regions and is exacerbated by poor irrigation practices and climate change.

15.4.2 Extreme Temperature and Heavy Metals

Temperature extremes include both heat and cold stress. Heat stress occurs when temperatures exceed optimal ranges for plant growth, causing cellular damage, protein denaturation, and accelerated senescence (Bita and Gerats 2013). Cold stress, on the other hand, affects plants during germination, growth, and reproduction, leading to reduced enzymatic activity and membrane rigidification. Both types of temperature extremes can significantly impact crop yields. Heavy metal stress results from the accumulation of toxic metals such as cadmium (Cd), lead (Pb), and mercury (Hg) in soil or water sources. These metals can enter plants through roots and disrupt physiological processes, leading to oxidative stress, enzyme inhibition, and impaired nutrient uptake (Clemens 2006). Heavy metal contamination poses serious health risks and limits crop cultivation in contaminated areas.

15.5 Current Methods for Detection of Abiotic Stress

Detecting abiotic stress in crops is crucial for timely intervention and effective management to mitigate its detrimental effects on yield and quality. Several advanced methods have been developed to detect and quantify various types of abiotic stressors such as drought, salinity, temperature extremes, and heavy metals (Mariyam et al. 2024). Physiological and biochemical assays remain fundamental for assessing plant responses to abiotic stress. These methods include measurements of chlorophyll content, photosynthetic efficiency, leaf water potential, and enzymatic activities (Mittler 2006). For example, stomatal conductance and leaf relative water content are commonly used indicators of drought stress severity (Flexas et al. 2006). Spectroscopic techniques such as infrared thermography and chlorophyll fluorescence imaging provide non-destructive means to assess plant stress responses.

Thermal imaging detects changes in leaf temperature associated with water or heat stress, while chlorophyll fluorescence imaging assesses photosynthetic efficiency and stress-induced damage (Baker 2008). Advancements in molecular biology have facilitated the identification of stress-responsive genes and molecular markers associated with abiotic stress tolerance in crops. Techniques like quantitative PCR (qPCR) and RNA sequencing (RNA-seq) quantify gene expression levels and identify candidate genes involved in stress responses (Fahad et al. 2015). Remote sensing technologies, including satellite-based sensors and unmanned aerial vehicles (UAVs), provide spatial and temporal information on crop health and stress conditions. These technologies monitor vegetation indices such as normalized difference vegetation index (NDVI) to detect stress-induced changes in plant canopy structure and physiology (Mahlein 2016). Emerging electrochemical and nanotechnology-based sensors offer sensitive detection of stress-related biomarkers in plants. Theses sensors helps in detecting various changes such as, ion fluxes, reactive oxygen species (ROS), and metabolites with high specificity and fast turnaround (Zhou et al. 2017; Pandey et al. 2020).

15.6 Principles of Biosensors

15.6.1 Types and Mechanism of Biosensors

Biosensors are analytical devices that integrate biological components with physicochemical transducers to detect and quantify analytes of interest. They play a crucial role in various fields including medicine, environmental monitoring, and agriculture. Biosensors can be categorized into different types based on their bioreceptor component (Kumar et al. 2024b). Enzyme-based biosensors utilize enzymes as bioreceptors to catalyze reactions with specific analytes, producing detectable signals. Antibody-based biosensors employ antibodies for selective binding to antigens, enabling sensitive detection of pathogens or biomarkers (Wei et al. 2014).

DNA-based biosensors utilize DNA strands as recognition elements, detecting complementary sequences or mutations with high specificity. The fundamental mechanism of biosensing involves the interaction between the bioreceptor and the target analyte, leading to a measurable signal. Upon binding with the analyte, the bioreceptor undergoes a specific biochemical reaction or conformational change. This change is transduced into a measurable signal by the transducer component of the biosensor (Heller et al. 2009).

15.7 Role of Nanotechnology in Biosensing

Nanotechnology has revolutionized biosensing by leveraging nanoscale materials and structures to enhance the performance and capabilities of biosensors across diverse applications. Key advancements include improved sensitivity, specificity, and the miniaturization of detection platforms. Nanomaterials such as NPs, nanowires, and nanotubes play a pivotal role in biosensing by amplifying signal transduction mechanisms. These materials offer unique optical, electrical, and catalytic properties that enhance the detection limits and response times of biosensors (Cao et al. 2002). For instance, NPs provide a large surface area for efficient bioreceptor immobilization, facilitating enhanced interaction with target analytes and thereby improving sensitivity (Patolsky et al. 2004). Functionalized NMs also enable precise control over bioreceptor orientation and immobilization, ensuring selective recognition of specific analytes. This capability is crucial in antibody-based biosensors for detecting pathogens or biomarkers with high specificity (Wei et al. 2014).

Additionally, nanotechnology-driven biosensors integrate nanostructured transducers such as nanoelectrodes and nanopores, which enable direct electrical detection of biomolecular interactions at a scale previously unachievable (Heller et al. 2009). The miniaturization enabled by nanotechnology allows for the development of portable and point-of-care biosensors. These devices are essential for applications in healthcare diagnostics, environmental monitoring, and personalized medicine. Nano-enabled biosensors integrated into wearable devices or implantable sensors offer continuous monitoring of biomarkers, enabling early disease detection and personalized treatment strategies (Gao et al. 2016). In environmental biosensing, nanotechnology facilitates the development of sensors capable of detecting pollutants, toxins, and pathogens in air, water, and soil. Functionalized NPs enhance the capture and concentration of target molecules from complex environmental samples, improving detection efficiency and reliability (Kumar et al. 2012).

15.8 Common Nanomaterials Used in Biosensors and Their Advantages Over Traditional Methods

Nanotechnology has introduced various NMs into biosensing, enhancing their performance and capabilities compared to traditional methods. Key NMs include NPs, i.e., gold, silver, quantum dots, nanowires, nanotubes, and nanocomposites. NPs, particularly gold and silver, offer better optical properties and high surface area-to-volume ratio. They are easily functionalized with biomolecules, enabling sensitive detection due to enhanced signal amplification and specific binding capacity (Sperling et al. 2008). Quantum dots are semiconductor NPs with tunable emission spectra and superior photostability. They allow for multiplexed detection and imaging, surpassing traditional organic dyes in fluorescence-based assays (Medintz et al. 2005). Nanowires and nanotubes exhibit high electrical conductivity and mechanical strength, enabling direct electrical detection of biomolecules. They facilitate label-free detection, providing actual time monitoring of biomolecular interactions with high sensitivity (Heller et al. 2009).

Nanocomposites combine NPs or nanowires with polymers or carbon-based materials, offering synergistic properties such as enhanced stability and biocompatibility. These hybrids are advantageous for robust biosensing platforms capable of detecting a wide range of analytes in complex samples (Zhu et al. 2015). Advantages of NMs in biosensing include enhanced sensitivity, selectivity, miniaturization, and the ability to perform multiplexed and real-time monitoring (Kumar et al. 2024c). These advancements are crucial for applications in healthcare diagnostics, environmental monitoring, and personalized medicine, offering efficient and reliable detection solutions. Nanotechnology-driven biosensors utilizing NPs, quantum dots, nanowires, nanotubes, and nanocomposites represent a significant advancement over traditional methods, enabling sensitive and specific detection across various analytical challenges.

15.9 Applications of Nano Biosensors for Abiotic Stress Detection in Crops

Nanobiosensors have emerged as powerful tools for detecting abiotic stress in crops, offering unprecedented sensitivity, specificity, and real-time monitoring capabilities. These nanotechnology-enabled devices integrate biological recognition elements with NMs-based transducers to detect stress-induced biomarkers efficiently. Here, we explore the diverse applications of nanobiosensors in agricultural settings, particularly for the detection and management of abiotic stressors such as drought, salinity, temperature extremes, and heavy metals (Table 15.1). Nano biosensors are designed to detect stress-responsive biomolecules, such as reactive oxygen species (ROS), stress-related proteins, and metabolites, which serve as indicators of plant stress (Mishra et al. 2020). These sensors leverage the unique properties of NMs

Table 15.1 Previous studies on nanotechnology-based biosensors for early detection of abiotic stress in crop plants

Abiotic stress	Biosensor technology	NMs	Key findings	Source
Drought	Electrochemical biosensors	Au and Ag NPs	Enhanced sensitivity in detecting drought-induced biomarkers like ABA in plant tissues.	Mishra et al. (2020)
Salinity	Optical biosensors	Quantum dots	Detection of ion fluxes (e.g., Na^+, K^+) in response to salt stress in crop plants.	Pandey et al. (2020)
Temperature extremes	Thermal biosensors	Nanotubes, nanowires	Monitoring of heat stress responses through precise measurement of thermal gradients in plant leaves.	Zhang et al. (2018)
Heavy metals	Electrochemical biosensors	Nanocomposites	Trace detection of heavy metal contaminants in soil and plant tissues.	Ali et al. (2023)
Multi-stress detection	Nanoelectronic biosensors	Nanowires, quantum dots	Simultaneous detection of multiple stressors (drought, salinity) using integrated sensor arrays.	Sehgal et al. (2021)

such as NPs, nanowires, and nanotubes to enhance sensitivity and selectivity. For instance, NMs provide a large surface area for biomolecule immobilization, improving the efficiency of analyte capture and detection (Pandey et al. 2020).

When the plant is under drought stress theses nano biosensors can detect various changes in leaf water retention or stomatal conductance, and early warning signs before visible changes (Sehgal et al. 2021). Similarly, for salinity stress, sensors can monitor ion fluxes or osmolyte concentrations in plant tissues, providing insights into stress tolerance mechanisms (Zhang et al. 2018). Temperature stress can be monitored using nano biosensors equipped with thermosensitive NMs that respond to slight changes in temperature, indicating stress-induced thermal gradients within plant tissues (Li et al. 2019). Heavy metal stress, which adversely affects crop productivity and food safety, is another area where nano biosensors excel. These sensors can detect trace levels of heavy metals in soil or plant tissues, offering rapid and on-site monitoring capabilities crucial for mitigating environmental contamination. Furthermore, nanobiosensors enable remote sensing applications through integration with wireless communication technologies and unmanned aerial vehicles (UAVs). This allows for real-time monitoring of crop stress over large agricultural landscapes, facilitating timely interventions and precision agriculture practices (Mansoor et al. 2025) (Table 15.1).

15.10 Challenges and Future Directions

In the realm of nanotechnology-based biosensors for early detection of abiotic stress in crops, significant challenges persist alongside promising future directions. Key hurdles include adapting sensors for robust performance in diverse field conditions, standardizing fabrication and validation processes, and ensuring cost-effectiveness for widespread adoption. Long-term stability and durability under environmental stresses, along with addressing ethical and regulatory concerns, remain critical. Future advancements should focus on enhancing multi-analyte detection capabilities, integrating smart sensor networks with IoT technologies, and exploring innovative NMs and biological signal amplification strategies. Additionally, leveraging data analytics and AI for real-time decision-making and fostering community engagement are pivotal for advancing these biosensors in agriculture, thereby bolstering sustainable crop management and global food security.

15.11 Conclusions

Nanotechnology based biosensors represent a transformative approach in agriculture, offering unprecedented capabilities for early detection and management of abiotic stress in crops. By harnessing the unique properties of NMs and integrating them with sensitive biological recognition elements, these biosensors enable rapid, accurate, and real-time monitoring of stress factors such as drought, salinity, temperature extremes, and heavy metal contamination. Despite current challenges in field deployment, including standardization, cost-effectiveness, and regulatory considerations, ongoing advancements in nanotechnology hold immense promise. Future research should focus on enhancing sensor reliability, developing multi-stressor detection capabilities, and integrating these technologies into smart agriculture frameworks. Ultimately, the widespread adoption of nanotechnology-based biosensors has the potential to revolutionize crop management practices, improve agricultural productivity, and contribute significantly to global food security in a changing climate landscape.

References

Alharbi KK, Farrukh MA, Altalhi T (2018) Nanomaterials-based biosensors for monitoring environmental and biological systems. Environ Technol Innov 11:122–139

Ali Q, Zia MA, Kamran M, Shabaan M, Zulfiqar U, Ahmad M, Iqbal R, Maqsood MF (2023) Nanoremediation for heavy metal contamination: A review. Hybrid Advances 4:100091. https://doi.org/10.1016/j.hybadv.2023.100091

Bailey-Serres J, Parker JE, Ainsworth EA, Oldroyd GED, Schroeder JI (2019) Genetic strategies for improving crop yields. Nature 575(7781):109–118

Baker NR (2008) Chlorophyll fluorescence: a probe of photosynthesis in vivo. Annu Rev Plant Biol 59:89–113

Bita CE, Gerats T (2013) Plant tolerance to high temperature in a changing environment: scientific fundamentals and production of heat stress-tolerant crops. Front Plant Sci 4:273

Cao YC, Jin R, Mirkin CA (2002) Nanoparticles with Raman spectroscopic fingerprints for DNA and RNA detection. Science 297(5586):1536–1540

Chaves MM, Maroco JP, Pereira JS (2003) Understanding plant responses to drought — from genes to the whole plant. Funct Plant Biol 30(3):239–264

Choudhary R, Rajput VD, Ghodake G, Ahmad F, Meena M, Rehman RU, Prasad R, Sharma RK, Singh R, Seth CS (2024) Comprehensive journey from past to present to future about seed priming with hydrogen peroxide and hydrogen sulfide concerning drought, temperature, UV and ozone stresses- a review. Plant Soil. https://doi.org/10.1007/s11104-024-06499-9

Choudhary R, Ahmad F, Kaya C, Upadhyay SK, Muneer S, Kumar V, Meena M, Liu H, Upadhyay H, Seth CS (2025) Decrypting proteomics, transcriptomics, genomics, and integrated omics for augmenting the abiotic, biotic, and climate change stress resilience in plants. J Plant Physiol. https://doi.org/10.1016/j.jplph.2025.154430

Cinteza LO, Olaru A (2017) Recent advances in nanomaterials for sensing applications. Curr Opin Green Sustain Chem 5:75–81

Clemens S (2006) Toxic metal accumulation, responses to exposure and mechanisms of tolerance in plants. Biochimie 88(11):1707–1719

Dimkpa CO, McLean JE, Martineau N, Britt DW, Haverkamp R, Anderson AJ (2012) Silver nanoparticles disrupt wheat (Triticum aestivum L.) growth in a sand matrix. Environ Sci Technol 46(1):752–760

Fahad S, Bajwa AA, Nazir U, Anjum SA, Farooq A, Zohaib A, Sadia S (2015) Crop production under drought and heat stress: plant responses and management options. Front Plant Sci 6:1141

Fleury D, Jefferies S, Kuchel H, Langridge P (2010) Genetic and genomic tools to improve drought tolerance in wheat. J Exp Bot 61(12):3211–3222

Flexas J, Bota J, Galmés J, Medrano H, Ribas-Carbó M (2006) Keeping a positive carbon balance under adverse conditions: responses of photosynthesis and respiration to water stress. Physiol Plant 127(3):343–352

Gao W, Emaminejad S, Nyein HYY, Challa S, Chen K, Peck A et al (2016) Fully integrated wearable sensor arrays for multiplexed in situ perspiration analysis. Nature 529(7587):509–514

Heller I, Janssens AM, Männik J, Minot ED, Lemay SG, Dekker C (2009) Identifying the mechanism of biosensing with carbon nanotube transistors. Nano Lett 9(2):655–660

Jones HG, Serraj R, Loveys BR, Xiong L, Wheaton A, Price AH (2010) Thermal infrared imaging of crop canopies for the remote diagnosis and quantification of plant responses to water stress in the field. Funct Plant Biol 36(11):978–989

Khodakovskaya M, Dervishi E, Mahmood M, Xu Y, Li Z, Watanabe F (2009) Carbon nanotubes are able to penetrate plant seed coat and dramatically affect seed germination and plant growth. ACS Nano 3(10):3221–3227

Khodakovskaya MV, de Silva K, Biris AS, Dervishi E, Villagarcia H (2012) Carbon nanotubes induce growth enhancement of tobacco cells. ACS Nano 6(3):2128–2135

Khot LR, Sankaran S, Maja JM, Ehsani R, Schuster EW (2012) Applications of nanomaterials in agricultural production and crop protection: a review. Crop Prot 35:64–70

Kumar S, Nehra M, Dilbaghi N (2012) Nanotechnology-based approaches for sensing and destroying pesticides. J Nanosci Nanotechnol 12(6):4133–4148

Kumar D, Mariyam S, Gupta KJ, Thiruvengadam M, Ghodake GS, Xing B, Seth CS (2024a) Comparative investigation on chemical and green synthesized titanium dioxide nanoparticles against chromium (VI) stress eliciting differential physiological, biochemical, and cellular attributes in Helianthus annuus L. Sci Total Environ 930:172413. https://doi.org/10.1016/j.scitotenv.2024.172413

Kumar D, Singh R, Upadhyay SK, Verma KK, Tripathi RM, Liu H, Dhankher OP, Tripathi RD, Sahi SV, Seth CS (2024b) Review on interactions between nanomaterials and phytohormones:

novel perspectives and opportunities for mitigating environmental challenges. Plant Sci 340:111964. https://doi.org/10.1016/j.plantsci.2023.111964

Kumar D, Biswas JK, Mulla SI, Singh R, Shukla R, Ahanger MA, Shekhawat GS, Verma KK, Siddiqui MW, Seth CS (2024c) Micro and nanoplastics pollution: sources, distribution, uptake in plants, toxicological effects, and innovative remediation strategies for environmental sustainability. Plant Physiol Biochem 213:108795. https://doi.org/10.1016/j.plaphy.2024.108795

Kumari A, Lakshmi GA, Krishna GK, Patni B, Prakash S, Bhattacharyya M, Singh SK, Verma KK (2022) Climate change and its impact on crops: a comprehensive investigation for sustainable agriculture. Agronomy 12:3008. https://doi.org/10.3390/agronomy12123008

Lesk C, Rowhani P, Ramankutty N (2016) Influence of extreme weather disasters on global crop production. Nature 529(7584):84–87

Li Y, Tan W, Zhang H (2019) Advances in nanobiosensors for environmental monitoring. Nano Today 27:6–12

Liu Q, Zhang H, Li J, Yang J (2019) Nanotechnology in agriculture: opportunities, toxicological implications, and occupational risks. NanoImpact 14:100161

Lobell DB, Schlenker W, Costa-Roberts J (2011) Climate trends and global crop production since 1980. Science 333(6042):616–620

Mahajan S, Tuteja N (2005) Cold, salinity, and drought stresses: an overview. Arch Biochem Biophys 444(2):139–158

Mahlein AK (2016) Plant disease detection by imaging sensors - parallels and specific demands for precision agriculture and plant phenotyping. Plant Dis 100(2):241–251

Mansoor S, Iqbal S, Popescu SM, Kim SL, Chung YS and Baek J-H (2025) Integration of smart sensors and IOT in precision agriculture: trends, challenges and future prospectives. Front. Plant Sci. 16:1587869. https://doi.org/10.3389/fpls.2025.1587869

Mariyam S, Upadhyay SK, Chakraborty K, Verma KK, Duhan JS, Muneer S, Meena M, Sharma RK, Ghodake G, Seth CS (2024) Nanotechnology, a frontier in agricultural science, a novel approach in abiotic stress management and convergence with new age medicine-a review. Sci Total Environ 912:169097. https://doi.org/10.1016/j.scitotenv.2023.169097

Medintz IL, Uyeda HT, Goldman ER, Mattoussi H (2005) Quantum dot bioconjugates for imaging, labelling and sensing. Nat Mater 4(6):435–446

Miralles P, Johnson E, Church TL, Harris AT (2012) Multiwalled carbon nanotubes in alfalfa and wheat: toxicology and uptake. J R Soc Interface 9(77):3514–3527

Mishra S, Srivastava S, Tripathi RD, Trivedi PK (2014) Thiol metabolism and abiotic stress tolerance in crop plants. Front Plant Sci 5:562

Mishra A, Pathak N, Sahoo S, Singh S, Gupta A (2020) Nanobiosensors: concepts and applications in agricultural biotechnology. Curr Opin Biotechnol 63:17–24

Mittler R (2006) Abiotic stress, the field environment and stress combination. Trends Plant Sci 11(1):15–19

Munns R, Tester M (2008) Mechanisms of salinity tolerance. Annu Rev Plant Biol 59:651–681

Pandey PC, Singh K, Singh J, Singh SP (2020) Recent advances in nanotechnology-based biosensors for agriculture and food industry. Biosens Bioelectron 159:112214

Pareek A, Singla-Pareek SL, Sopory SK (2010) Analyzing salt stress-responsive proteomes and determining gene functions using transgenic plants for improving plant tolerance to salinity. Methods Mol Biol 639:247–269

Patolsky F, Zheng G, Lieber CM (2004) Nanowire-based nanoelectronic devices in the life sciences. MRS Bull 29(05):330–335

Rajput VD, Minkina T, Kumari A, Harish, Singh VK, Verma KK, Mandzhieva S, Sushkova S, Srivastava S, Keswani C (2021) Coping with the challenges of abiotic stress in plants: new dimensions in the field application of nanoparticles. Plan Theory 10:1221. https://doi.org/10.3390/plants10061221

Ray PP, Mukherjee M, Shu L (2012) Internet of Things for agriculture technologies. IEEE Internet Things J 1(1):3–13

Sehgal A, Sita K, Bhandari K (2021) Application of nanobiosensors for abiotic stress detection in crops. Environ Nanotechnol Monit Manag 15:100406

Sperling RA, Rivera Gil P, Zhang F, Zanella M, Parak WJ (2008) Biological applications of gold nanoparticles. Chem Soc Rev 37(9):1896–1908

Tuteja N, Chandra M, Tuteja R (2013) Nanotechnology: prospective tool for enhancing sustainability in agriculture. Environ Sci Technol 47(18):9393–9405

Verma KK, Singh P, Song X-P, Malviya MK, Singh RK, Chen G-L, Solomon S, Li YR (2020) Mitigating climate change for sugarcane improvement: role of silicon in alleviating abiotic stresses. Sugar Tech 22(5):741–749. https://doi.org/10.1007/s12355-020-00831-0

Verma KK, Song XP, Joshi A, Rajput VD, Singh M, Sharma A, Singh RK, Li DM, Arora J, Minkina T, Li YR (2022) Nanofertilizer possibilities for healthy soil, water and food in future: an overview. Front Plant Sci 13:865048. https://doi.org/10.3389/fpls.2022.865048

Verma KK, Song XP, Kumari A, Jagadesh M, Singh SK, Bhatt R, Singh M, Seth CS, Li YR (2024a) Climate change adaptation: challenges for agricultural sustainability. Plant Cell Environ:1–12. https://doi.org/10.1111/pce.15078

Verma KK, Song X-P, Singh S, Kumari A, Arora J, Singh SK, Solanki MK, Seth CS, Li Y-R (2024b) Synergistic interactions of nanoparticles and plant growth promoting rhizobacteria enhancing soil-plant systems: a multigenerational perspective. Front Plant Sci 15:1376214. https://doi.org/10.3389/fpls.2024.1376214

Wang P, Lombi E, Zhao FJ, Kopittke PM (2016) Nanotechnology: a new opportunity in plant sciences. Trends Plant Sci 21(8):699–712

Wei F, Lillehoj PB, Ho CM (2014) DNA diagnostics: nanotechnology-enhanced electrochemical detection of nucleic acids. Pediatr Res 75(1–2):264–270

Zandalinas SI, Mittler R, Balfagón D, Arbona V, Gómez-Cadenas A (2018) Plant adaptations to the combination of drought and high temperatures. Physiol Plant 162(1):2–12

Zhang Y, Wang M, Su X (2018) Recent advances in nanotechnology-based biosensors for precision agriculture and food safety. Trends Food Sci Technol 76:36–46

Zhou L, Du W, Xi C, Wang S, Wu H (2017) Electrochemical sensors and biosensors based on nanomaterials and nanostructures. Anal Chem 89(1):123–138

Zhu Z, Garcia-Gancedo L, Flewitt AJ, Xie H, Moussy F, Milne WI (2015) Development of polypyrrole/carbon nanotube chemiresistive biosensors for detection of nitric oxide released from endothelial cells. Biosens Bioelectron 67:694–699

Chapter 16
Environmental and Safety Considerations of Nano-formulations for Sustainable Agriculture

Bharat M. Bhalerao, Krishan K. Verma, and Vishnu D. Rajput

16.1 Introduction

In emerging nations like India, where agriculture supports a significant portion of the population, the demand for agricultural products is high, leading to pressure on developing new agricultural practices. Chemical fertilizers, including high-yielding synthetics, bio-manures, and farmyard manures, have become integral to meeting this demand, providing essential macro- and micronutrients lacking in soil (Sharma et al. 2020). However, prolonged agrochemical use has led to adverse effects on human and soil health, contributing to pollution, soil erosion, nutrient imbalances, and reduced biodiversity (Singh et al. 2010; Verma et al. 2024a). Nanotechnology presents a potential solution to address these challenges more effectively. Nanomaterials, due to their small size and increased surface area, offer versatile applications in precision agriculture, including slow-release fertilizers, nano-herbicides, nano-fungicides, and nano-sensors for analysis (Choudhury et al. 2017; Kumar et al. 2024b). Many nations recognize the role of nanotechnology in meeting growing population demands and preserving natural resources (Raliya and Saharan 2016).

The development of environmentally safe and sustainable nano-materials holds promise for improving fertilization, plant growth regulation, and pest control,

B. M. Bhalerao (✉)
Department of Biochemistry, Post Graduate Institute, Mahatma Phule Krishi Vidyapeeth, Rahuri, Maharashtra, India

K. K. Verma
Sugarcane Research Institute, Guangxi Academy of Agricultural Sciences, Nanning, Guangxi, China

V. D. Rajput
Academy of Biology and Biotechnology, Southern Federal University, Rostov-on-Don, Russia

C. S. Seth et al. (eds.), *Nano-formulations for Sustainable Agriculture*,
Nanotechnology in the Life Sciences, https://doi.org/10.1007/978-3-031-90617-6_16

thereby reducing environmental pollution (Verma et al. 2022a, b). Nanoparticles'vast surface area makes them attractive for addressing issues conventional methods cannot, offering new agrochemical agents and delivery systems while decreasing pesticide usage (Kumar et al. 2021). Applications of nanotechnology in agriculture span production, processing, storage, packing, and transportation, with precision farming techniques offering opportunities to increase yields while minimizing environmental harm (Gogos et al. 2012). Despite its potential, concerns regarding risk assessment and environmental impact persist and must be addressed to ensure the responsible implementation of nanotechnology in agriculture. Nano-formulations represent a revolutionary advancement in agricultural technology, offering innovative solutions to enhance crop productivity and sustainability. These formulations employ nanoparticles, which are particles with dimensions measured in nanometers, to improve the delivery and effectiveness of agricultural inputs such as fertilizers, pesticides, and growth stimulants (Verma et al. 2024b). One of the primary advantages of nano-formulations is their ability to enhance the efficiency of agricultural inputs. For instance, nano-fertilizers can release nutrients in a controlled manner, ensuring a steady supply to plants and minimizing nutrient losses due to leaching and volatilization (Sekhon 2014; Rajput et al. 2021a).

This leads to improved nutrient uptake and higher crop yields. Similarly, nanopesticides offer targeted delivery systems that increase pest control efficacy while reducing the amount of active ingredient needed, thereby lowering the environmental impact (Kah et al. 2018). Moreover, the small size and large surface area of nanoparticles enable them to interact more effectively with plant surfaces and soil particles. This improves the solubility and bioavailability of agricultural inputs, ensuring that they are more readily absorbed by plants (Servin et al. 2015; Rajput et al. 2021b). Despite these benefits, the use of nano-formulations in agriculture raises significant environmental and safety concerns. The introduction of nanoparticles into the environment can lead to unforeseen ecological effects, particularly on non-target organisms such as soil microorganisms, insects, and aquatic life (Gogos et al. 2012; Choudhary et al. 2024).

The persistence and bioaccumulation of nanoparticles in the environment are also critical issues that require careful assessment (Maurer-Jones et al. 2013). Human health risks associated with exposure to nanoparticles used in agriculture must be addressed through rigorous safety evaluations and regulatory measures. Potential risks include respiratory and dermal exposure for agricultural workers, as well as residual nanoparticle consumption through treated crops (Grillo et al. 2015). To mitigate these risks, ongoing research focuses on the development of environmentally benign and biodegradable nanoparticles. Innovations in this field aim to design nanoparticles that can degrade into non-toxic byproducts, thereby reducing their ecological footprint (Parisi et al. 2015). Additionally, interdisciplinary research and international collaboration are essential for establishing comprehensive guidelines and regulations to ensure the safe and sustainable use of nano-formulations in agriculture. Nano-formulations hold great promise for enhancing agricultural productivity and sustainability by improving the efficiency and effectiveness of agricultural inputs. However, careful consideration of environmental and safety impacts is crucial to realize their full potential responsibly. By balancing innovation with precaution, the agricultural sector can harness the benefits of nanotechnology while safeguarding environmental and human health (Fig. 16.1).

16.2 Overview of Nano-formulations in Agriculture

16.2.1 Types of Nanoformulations

Nano-formulations in agriculture employ nanotechnology to improve the efficiency and effectiveness of fertilizers, pesticides, and growth stimulants by incorporating particles measured in nanometers. Nano pesticides enhance pesticide delivery, stability, and efficacy, allowing for lower doses and reducing environmental impact. They release active ingredients slowly and precisely at the target site, minimizing

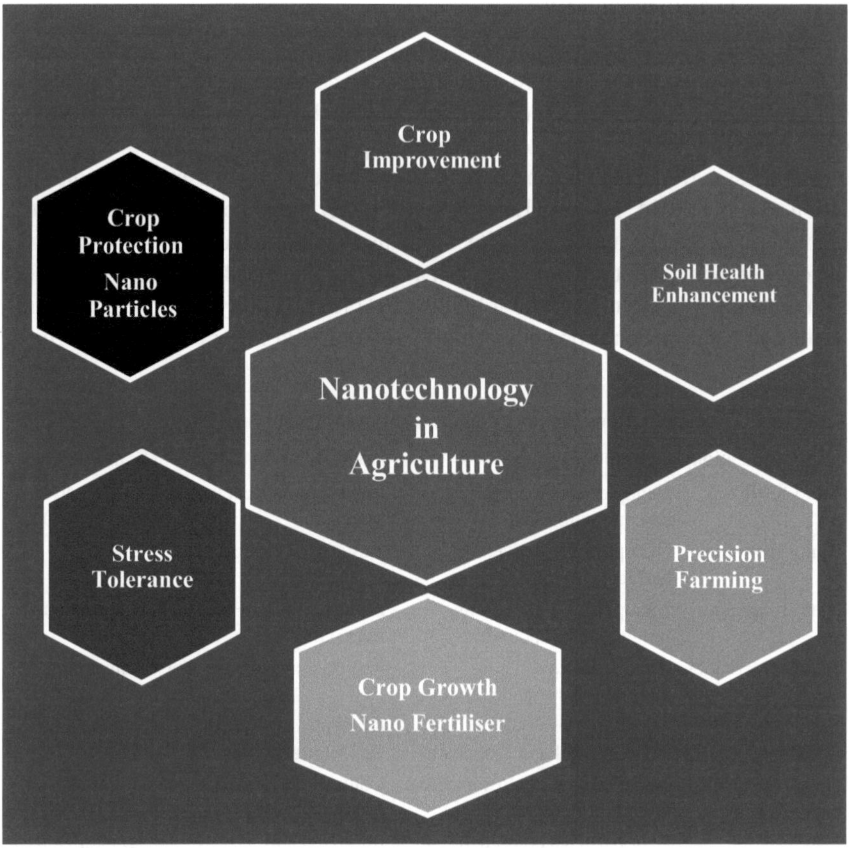

Fig. 16.1 Various applications of nanoformulations in agriculture

environmental contamination (Kah et al. 2018; Ghormade et al. 2011; Mariyam et al. 2024 Nanofertilizers (NFs) offer controlled and sustained nutrient release, improving nutrient availability and reducing wastage. This leads to increased crop yields and a reduced environmental footprint (Sekhon 2014; Liu and Lal 2015). Nano sensors monitor soil health, plant growth, and environmental conditions, enabling precise interventions and optimized input use (Rai et al. 2012). Nano encapsulated delivery systems protect active ingredients from environmental degradation and enhance plant uptake, increasing efficiency and reducing application frequency (Chhipa 2017; Kumar et al. 2024a). These nano-formulations promise to transform agriculture by improving input precision and efficiency, boosting crop yields, and reducing environmental impact. Ongoing research is essential to understand their long-term effects and establish appropriate regulatory frameworks (Parisi et al. 2015) (Table 16.1).

16.2.2 Advantages of Nanoformulations Over Traditional Methods

Nano-formulations in agriculture offer significant advantages over traditional methods, enhancing efficiency, precision, and sustainability. Nano-fertilizers provide controlled and sustained nutrient release, improving plant uptake and reducing nutrient losses from leaching and volatilization, which mitigates environmental pollution (Sekhon 2014; Liu and Lal 2015). Similarly, nano pesticides deliver active ingredients precisely at the target site, reducing the required amounts and minimizing environmental contamination (Kah et al. 2018). The small size and large surface area of nanoparticles enhance the solubility and bioavailability of active ingredients, leading to better penetration and absorption by plants at lower dosages (Ghormade

Table 16.1 Outlining various nanoformulations used in sustainable agriculture

Nanoformulation	Description	Application	Source
Nano-fertilizers	Nanoparticles loaded with essential nutrients	Enhanced nutrient uptake by plants	Raliya and Saharan (2016)
Nano-herbicides	Nanoparticles delivering herbicidal agents	Targeted weed control	Choudhury et al. (2017)
Nano-fungicides	Nanoparticles containing fungicidal compounds	Effective control of fungal diseases	Nizamani et al. (2024)
Nano-insecticides	Nanoparticles carrying insecticidal substances	Precise pest management	Kumar et al. (2021)
Nano-sensors	Nanoparticles used for detecting soil and crop status	Real-time monitoring of agricultural conditions	Sharma et al. (2020)
Nano-encapsulants	Nanoparticles enclosing active agricultural agents	Controlled release for prolonged efficacy	Gogos et al. (2012)
Nano-biofertilizers	Nanoparticles combined with beneficial microorganisms	Improved soil fertility and plant growth	Tripathi et al. (2017)

et al. 2011). Traditional formulations lack this efficiency, often requiring higher concentrations. Nanoencapsulation protects active ingredients from environmental degradation, extending their lifespan and reducing application frequency. This results in lower labor costs and decreased exposure to harmful chemicals (Chhipa 2017). Nano sensors provide high-resolution monitoring of soil health, nutrient levels, and pest presence, allowing for precise and timely interventions, a level of precision hard to achieve with traditional methods (Rai et al. 2012). Overall, nano-formulations reduce the environmental footprint of agriculture by lowering the quantities of chemicals needed, thereby promoting ecological balance (Parisi et al. 2015). These advancements in nanotechnology make agricultural practices more efficient, precise, and sustainable.

16.3 Environmental Considerations

16.3.1 Environmental Fate and Behavior of Nano-formulations

The environmental fate and behavior of nano-formulations play a crucial role in determining their impact on ecosystems and human health. Understanding how these formulations interact with the environment is essential for assessing their long-term implications and ensuring their safe and sustainable use. Nano-formulations, particularly those used in agriculture, can undergo various processes in the environment, including dispersion, transformation, and transport. These processes are influenced by factors such as nanoparticle size, surface properties, and environmental conditions. Dispersion is the initial step in the environmental fate of nano-formulations, where nanoparticles are released into the environment through application or waste disposal. Once dispersed, nanoparticles can undergo transformation processes such as aggregation, dissolution, or chemical reactions, altering their physicochemical properties and behavior (Nowack and Bucheli 2007).

Transport mechanisms, such as advection, diffusion, and sedimentation, determine the movement of nanoparticles in the environment. Nanoparticles can travel through soil, water, and air, potentially spreading over large distances and interacting with various environmental compartments (Gottschalk et al. 2009). The fate of nano-formulations in the environment is also influenced by their interactions with biota. Nanoparticles can be taken up by plants and animals through processes such as absorption, ingestion, or inhalation, leading to potential bioaccumulation and biomagnification in food chains (Nel et al. 2006; Kumar et al. 2024a). The behavior of nano-formulations in aquatic environments is of particular concern due to their potential to accumulate in sediments and affect aquatic organisms. Nanoparticles can adsorb to sediment particles, altering sediment properties and impacting benthic ecosystems (Unrine et al. 2008). Furthermore, the fate and behavior of nano-formulations can be influenced by environmental factors such as pH, temperature, and organic matter content. Changes in these factors can affect nanoparticle stability, mobility, and toxicity, highlighting the importance of considering environmental variability in risk assessments (Navarro et al. 2008).

16.3.2 Potential Environmental Impacts

Nano-formulations in agriculture have the potential to significantly impact the environment, including soil, water, and air quality. Understanding these potential impacts is essential for assessing the risks associated with their use and implementing appropriate risk management strategies. Nano-formulations in agriculture have the potential to impact soil, water, and air quality, with implications for ecosystem health, agricultural productivity, and human well-being. Understanding these potential impacts is critical for ensuring the safe and sustainable use of nano-formulations in agriculture.

16.3.2.1 Impact of Soil, Water and Air

Nano-formulations applied to soil can affect soil properties, microbial communities, and plant growth. Nanoparticles may alter soil structure and nutrient cycling, potentially leading to changes in soil fertility and productivity. Additionally, nanoparticles may interact with soil microorganisms, affecting soil microbial diversity and activity (Zhang et al. 2015). These changes could have implications for soil health, ecosystem functioning, and agricultural productivity.

Nano-formulations can enter water bodies through runoff, leaching, or direct application to aquatic environments. Once in water, nanoparticles may undergo dispersion, aggregation, and sedimentation processes, leading to potential accumulation in sediments and aquatic organisms (Wang et al. 2019). Nanoparticles can also impact water quality by affecting nutrient cycling, algae growth, and aquatic ecosystem dynamics (Navarro et al. 2008; Kumar et al. 2024c). Furthermore, nanoparticles may pose risks to aquatic organisms, including fish, invertebrates, and algae, through direct toxicity or indirect effects on food webs (Navarro et al. 2008).

During application and handling, nano-formulations may generate airborne particles that can disperse into the atmosphere. These airborne nanoparticles may pose risks to human health and environmental quality, particularly in agricultural settings where workers may be exposed to high concentrations of nanoparticles (Powers et al. 2015). Additionally, nanoparticles in the atmosphere may undergo long-range transport and deposition, potentially affecting ecosystems far from the source of emission (Wiesner et al. 2006).

16.3.3 Safety Considerations

Nano-formulations in agriculture pose risks to human health through inhalation, dermal contact, and oral ingestion. Workers handling these formulations may inhale airborne nanoparticles, leading to respiratory irritation and potential long-term respiratory diseases (Maynard et al. 2006; Lam et al. 2004). Skin contact can cause irritation and allergic reactions, while accidental ingestion may lead to gastrointestinal effects and systemic distribution of nanoparticles in the body (Cross et al. 2007; Bouwmeester et al. 2011). Once inside the body, nanoparticles can accumulate in organs, potentially causing cytotoxicity, genotoxicity, and immunotoxicity (Nel et al. 2006; Oberdörster et al. 2005). These risks highlight the importance of safe handling, application, and disposal of nano-formulations to protect human health.

16.4 Sustainable Practices and Future Directions

The integration of nano-formulations into sustainable agriculture practices offers innovative solutions to address global agricultural challenges. NFs enable targeted nutrient delivery, enhancing nutrient use efficiency and reducing environmental pollution (Raliya et al. 2015). Nanopesticides provide precise pest management while minimizing pesticide usage and preserving beneficial organisms (Khot et al. 2012). Nano-formulations also contribute to water conservation efforts by improving soil moisture retention (Hussain et al. 2016). Additionally, they enhance crop resilience to climate change through improved stress tolerance (Tripathi et al. 2017; Choudhary et al. 2025). Nanotechnology enables real-time monitoring of soil and crop health, facilitating precision agriculture practices (Zhang et al. 2019). Sustainability metrics, such as life cycle assessment, guide decision-making for sustainable agricultural development (Sekhon 2014).

Overall, nano-formulations contribute to more efficient, resilient, and environmentally friendly agricultural systems. Innovations in nanoformulations for agriculture are continually emerging, aiming to minimize environmental impact while enhancing agricultural productivity. Nano-fertilizers offer precise nutrient delivery, reducing fertilizer runoff and environmental pollution while improving crop yields (Raliya et al. 2015). Nano pesticides provide targeted pest management, minimizing non-target effects and pesticide resistance while reducing overall pesticide use (Khot et al. 2012). Nano-formulations also contribute to water conservation by enhancing soil moisture retention, particularly in arid regions, thereby promoting sustainable water management (Hussain et al. 2016). Advanced nanoformulations focus on improving soil health and fertility, enhancing soil structure and nutrient availability, and promoting microbial diversity (Kah et al. 2018). Additionally, nano sensors and nanoprobes enable real-time monitoring of environmental parameters, facilitating timely interventions and sustainable management practices (Zhang et al. 2019; Mariyam et al. 2024). These innovations leverage nanotechnology to minimize environmental impact while maximizing agricultural sustainability.

16.5 Future Trends in Nano-formulation Development

Future trends in nano-formulation development are anticipated to revolutionize agriculture by addressing key challenges and enhancing sustainability. These trends will leverage cutting-edge nanotechnology to develop innovative formulations that improve crop productivity, reduce environmental impact, and ensure food security. Advanced delivery systems will enable precise targeting and controlled release of nutrients, pesticides, and growth regulators (Khot et al. 2012). Biocompatible nanomaterials derived from natural polymers will enhance environmental safety and minimize ecological impact (Raliya et al. 2015). Integration with precision agriculture technologies will enable real-time monitoring and management of nano-formulations, facilitating precise application and optimization (Zhang et al. 2019). Future nano-formulations will feature multifunctional nanostructures that offer comprehensive solutions for sustainable crop management (Tripathi et al. 2017). Understanding the biological interactions of nano-formulations with plants, microbes, and the environment will ensure their safe and responsible use in agriculture (Sekhon 2014). By embracing these future trends, agriculture can harness the full potential of nanotechnology to build a more sustainable and resilient food system.

16.6 Challenges

Challenges in nanoformulations for sustainable agriculture encompass various aspects, including environmental concerns, safety issues, regulatory hurdles, and technological limitations. Addressing these challenges is crucial to realizing the potential benefits of nano-enabled agricultural practices while mitigating potential risks. One significant challenge lies in understanding the environmental fate and behavior of nanoparticles used in nano formulations. Research indicates that nanoparticles may exhibit unique properties that could lead to unintended environmental impacts, such as accumulation in soil and water bodies, potentially affecting ecosystems. Additionally, concerns about nanoparticle toxicity and ecotoxicity raise questions about their long-term effects on environmental health (González-Melendi et al. 2008). Ensuring the safety of nanoformulations for human health and agricultural workers is paramount. There are concerns regarding potential adverse health effects associated with nanoparticle exposure, including respiratory issues, skin irritation, and systemic toxicity (Li et al. 2013).

Minimizing human exposure and understanding the risks associated with nano-enabled products are critical challenges. The regulatory landscape for nano-enabled agricultural products is complex and evolving. Regulatory agencies face challenges in developing appropriate standards and guidelines for assessing the safety and efficacy of nano formulations (Kuzma and Kokotovich 2018). Harmonizing regulations across different regions and ensuring compliance with existing frameworks present additional hurdles. Technological limitations, such as scalability, reproducibility, and cost-effectiveness, pose challenges to the widespread adoption of nanoformulations in agriculture. Developing scalable production methods for nanoparticles and ensuring consistent performance under varying environmental conditions are key technological challenges (Xu et al. 2016). Addressing these challenges requires multidisciplinary collaboration, robust research efforts, and proactive regulatory frameworks. By overcoming these hurdles, nanoformulations have the potential to contribute significantly to sustainable agriculture by enhancing crop productivity, minimizing environmental impact, and promoting resource efficiency.

16.7 Conclusion

The integration of nano-formulations into sustainable agriculture holds immense promise for addressing global agricultural challenges while ensuring environmental and human safety. Nano-formulations offer innovative solutions for improving nutrient management, pest control, water conservation, and soil health, ultimately contributing to sustainable agricultural practices. However, the widespread adoption of nano-formulations necessitates careful consideration of environmental and safety considerations. Environmental fate and behavior, potential impacts on soil, water, and air quality, as well as human health risks associated with nano-formulations, must be thoroughly evaluated and mitigated through robust regulatory frameworks and guidelines. Safety protocols, risk assessment, and regulatory compliance are essential to ensure the responsible use of nano-formulations in agriculture while minimizing potential risks to human health and the environment. Future trends in nano-formulation development, such as smart delivery systems, biocompatible materials, precision agriculture technologies, multifunctional nanostructures, and understanding biological interactions, offer promising avenues for further enhancing the sustainability and effectiveness of nano-enabled agricultural practices. By integrating nano-formulations into sustainable agriculture practices and addressing environmental and safety considerations proactively, we can harness the full potential of nanotechnology to build a more resilient, efficient, and environmentally friendly food system that meets the needs of present and future generations.

References

Bouwmeester H, Dekkers S, Noordam MY, Hagens WI, Bulder AS, de Heer C et al (2011) Review of health safety aspects of nanotechnologies in food production. Regul Toxicol Pharmacol 53(1):52–62

Chhipa H (2017) Nanofertilizers and nanopesticides for agriculture. Environ Chem Lett 15:15–22

Choudhary R, Rajput VD, Ghodake G, Ahmad F, Meena M, Rehman RU, Prasad R, Sharma RK, Singh R, Seth CS (2024) Comprehensive journey from past to present to future about seed priming with hydrogen peroxide and hydrogen sulfide concerning drought, tempera-

ture, UV and ozone stresses- a review. Plant Soil 500(1–2):351–373. https://doi.org/10.1007/s11104-024-06499-9
Choudhary R, Ahmad F, Kaya C, Upadhyay SK, Muneer S, Kumar V, Meena M, Liu H, Upadhyay H, Seth CS (2025) Decrypting proteomics, transcriptomics, genomics, and integrated omics for augmenting the abiotic, biotic, and climate change stress resilience in plants. J Plant Physiol. https://doi.org/10.1016/j.jplph.2025.154430
Choudhury SR, Goswami A, Barua S (2017) Nanotechnology in agriculture. Int J Agric Environ Biotechnol 10(1):139–152
Cross SE, Innes B, Roberts MS, Tsuzuki T (2007) Human skin penetration of sunscreen nanoparticles: in-vitro assessment of a novel micronized zinc oxide formulation. Skin Pharmacol Physiol 20(3):148–154
Ghormade V, Deshpande MV, Paknikar KM (2011) Perspectives for nano-biotechnology enabled protection and nutrition of plants. Biotechnol Adv 29(6):792–803
Gogos A, Knauer K, Bucheli TD (2012) Nanomaterials in plant protection and fertilization: current state, foreseen applications, and research priorities. J Agric Food Chem 60(39):9781–9792
González-Melendi P, Fernández-Pacheco R, Coronado MJ, Corredor E, Testillano PS, Risueño MC et al (2008) Nanoparticles as smart treatment-delivery systems in plants: assessment of different techniques of microscopy for their visualization in plant tissues. Ann Bot 101(1):187–195
Gottschalk F, Sonderer T, Scholz RW, Nowack B (2009) Modeled environmental concentrations of engineered nanomaterials (TiO2, ZnO, Ag, CNT, Fullerenes) for different regions. Environ Sci Technol 43(24):9216–9222
Grillo R, Pereira AES, Nishisaka CS, de Lima R, Oehlke K, Greiner R, Fraceto LF (2015) Chitosan/tripolyphosphate nanoparticles loaded with paraquat herbicide: an environmentally safer alternative for weed control. J Hazard Mater 278:163–171
Hussain A, Hussain I, Singh NB (2016) Hydrogels: smart materials for water remediation. Environ Chem Lett 14(1):1–14
Kah M, Kookana RS, Gogos A, Bucheli TD (2018) A critical evaluation of nanopesticides and nanofertilizers against their conventional analogues. Nat Nanotechnol 13(8):677–684
Khot LR, Sankaran S, Maja JM, Ehsani R, Schuster EW (2012) Applications of nanomaterials in agricultural production and crop protection: a review. Crop Prot 35:64–70
Kumar V, Sharma KL, Misra S (2021) Nanotechnology: a promising solution for agriculture. J Pharmacogn Phytochem 10(3):3087–3094
Kumar D, Biswas JK, Mulla SI, Singh R, Shukla R, Ahanger MA, Shekhawat GS, Verma KK, Siddiqui MW, Seth CS (2024a) Micro and nanoplastics pollution: sources, distribution, uptake in plants, toxicological effects, and innovative remediation strategies for environmental sustainability. Plant Physiol Biochem 213:108795. https://doi.org/10.1016/j.plaphy.2024.108795
Kumar D, Mariyam S, Gupta KJ, Thiruvengadam M, Ghodake GS, Xing B, Seth CS (2024b) Comparative investigation on chemical and green synthesized titanium dioxide nanoparticles against chromium (VI) stress eliciting differential physiological, biochemical, and cellular attributes in Helianthus annuus L. Sci Total Environ 930:172413. https://doi.org/10.1016/j.scitotenv.2024.172413
Kumar D, Singh R, Upadhyay SK, Verma KK, Tripathi RM, Liu H, Dhankher OP, Tripathi RD, Sahi SV, Seth CS (2024c) Review on interactions between nanomaterials and phytohormones: novel perspectives and opportunities for mitigating environmental challenges. Plant Sci 340:111964. https://doi.org/10.1016/j.plantsci.2023.111964
Kuzma J, Kokotovich A (2018) Chemical oversight and nanotechnology: navigating policy options for innovation. J Nanopart Res 20(2):33
Lam CW, James JT, McCluskey R, Hunter RL (2004) Pulmonary toxicity of single-wall carbon nanotubes in mice 7 and 90 days after intratracheal instillation. Toxicol Sci 77(1):126–134
Li J, Chang PR, Huang J, Wang Y (2013) Advances in sustainable polymer composites: nanotechnology and beyond. In: Natural polymers. Springer, Berlin, Heidelberg, pp 569–586
Liu R, Lal R (2015) Potentials of engineered nanoparticles as fertilizers for increasing agronomic productions. Sci Total Environ 514:131–139

Mariyam S, Upadhyay SK, Chakraborty K, Verma KK, Duhan JS, Muneer S, Meena M, Sharma RK, Ghodake G, Seth CS (2024) Nanotechnology, a frontier in agricultural science, a novel approach in abiotic stress management and convergence with new age medicine-a review. Sci Total Environ 912:169097. https://doi.org/10.1016/j.scitotenv.2023.169097

Maurer-Jones MA, Gunsolus IL, Murphy CJ, Haynes CL (2013) Toxicity of engineered nanoparticles in the environment. Anal Chem 85(6):3036–3049

Maynard AD, Warheit DB, Philbert MA (2006) The new toxicology of sophisticated materials: nanotoxicology and beyond. Toxicol Sci 105(2):163–172

Navarro E, Baun A, Behra R, Hartmann NB, Filser J, Miao AJ et al (2008) Environmental behavior and ecotoxicity of engineered nanoparticles to algae, plants, and fungi. Ecotoxicology 17(5):372–386

Nel A, Xia T, Madler L, Li N (2006) Toxic potential of materials at the nanolevel. Science 311(5761):622–627

Nizamani MM, Hughes AC, Zhang HL, Wang Y (2024) Revolutionizing agriculture with nanotechnology: Innovative approaches in fungal disease management and plant health monitoring. 928:172473. https://doi.org/10.1016/j.scitotenv.2024.172473

Nowack B, Bucheli TD (2007) Occurrence, behavior and effects of nanoparticles in the environment. Environ Pollut 150(1):5–22

Oberdörster G, Oberdörster E, Oberdörster J (2005) Nanotoxicology: an emerging discipline evolving from studies of ultrafine particles. Environ Health Perspect 113(7):823–839

Parisi C, Vigani M, Rodríguez-Cerezo E (2015) Agricultural nanotechnologies: what are the current possibilities? Nano Today 10(2):124–127

Powers KW, Palazuelos M, Moudgil BM, Roberts SM (2015) Characterization of the size, shape, and state of dispersion of nanoparticles for toxicological studies. Nanotoxicology 4(1):10–18

Rai M, Ingle AP, Gaikwad S (2012) Nanotechnology for plant disease management. J Plant Pathol 94(2):223–236

Rajput VD, Singh A, Minkina TM, Shende SS, Kumar P, Verma KK, Bauer T, Gorobtsova O, Deneva S, Sindireva A (2021a) Potential applications of nanobiotechnology in plant nutrition and protection for sustainable agriculture. In: Ingle AP (ed) Nanotechnology in plant growth promotion and protection: recent advances and impacts, Ist edn. John Wiley & Sons, Hoboken, Chichester

Rajput VD, Singh A, Minkina T, Rawat S, Mandzhieva S, Sushkova S, Shuvaeva V, Nazarenko O, Rajput P, Komariah, Verma KK, Singh AK, Rao M, Upadhyay SK (2021b) Nano-enabled products: challenges and opportunities for sustainable agriculture. Plan Theory 10:2727. https://doi.org/10.3390/plants10122727

Raliya R, Saharan V (2016) Nanofertilizers and nanopesticides: emerging contaminants or opportunities for agricultural sustainability? Front Plant Sci 7:19

Raliya R, Nair R, Chavalmane S (2015) Engineered nanoparticle mediated intracellular delivery of micronutrients in plants. J Nanopart Res 17(1):17

Sekhon BS (2014) Nanotechnology in agri-food production: an overview. Nanotechnol Sci Appl 7:31–53

Servin AD, Elmer W, Mukherjee A, De La Torre-Roche R, Hamdi H, White JC, Bindraban PS (2015) A review of the use of engineered nanomaterials to suppress plant disease and enhance crop yield. J Nanopart Res 17(2):92

Sharma S, Sharma P, Sharma R, Singh N, Kumar A, Bhardwaj R (2020) Nanotechnology in agriculture and food industry: status and application. J Pharmacogn Phytochem 9(3):1112–1116

Singh DK, Sharma SK, Prasad SM (2010) Use of agro-chemicals and their impact on soil and microbial diversity of North-East India. Indian J Sci Technol 3(6):631–633

Tripathi DK, Singh S, Singh VP, Prasad SM, Chauhan DK (2017) In: Dubey NK, Tripathi DK (eds) Role of nanomaterials in plants under challenging environments. Springer

Unrine JM, Colman BP, Bone AJ, Gondikas AP, Matson CW (2008) Biotic and abiotic interactions in aquatic microcosms determine fate and toxicity of Ag nanoparticles. Part 1. Aggregation and dissolution. Environ Sci Technol 47(1):692–700

Verma KK, Song XP, Joshi A, Rajput VD, Singh M, Sharma A, Singh RK, Li DM, Arora J, Minkina T, Li YR (2022a) Nanofertilizer possibilities for healthy soil, water and food in future: an overview. Front Plant Sci 13:865048. https://doi.org/10.3389/fpls.2022.865048

Verma KK, Song XP, Joshi A, Tian DD, Rajput VD, Singh M, Arora J, Minkina TM, Li YR (2022b) Recent trends in nano-fertilizer for advancing sustainable agriculture under the era of climate change to ensure food security for future generations. Nano 12:173. https://doi.org/10.3390/nano12010173

Verma KK, Song XP, Kumari A, Jagadesh M, Singh SK, Bhatt R, Singh M, Seth CS, Li YR (2024a) Climate change adaptation: challenges for agricultural sustainability. Plant Cell Environ:1–12. https://doi.org/10.1111/pce.15078

Verma KK, Song X-P, Singh S, Kumari A, Arora J, Singh SK, Solanki MK, Seth CS, Li Y-R (2024b) Synergistic interactions of nanoparticles and plant growth promoting rhizobacteria enhancing soil-plant systems: a multigenerational perspective. Front Plant Sci 15:1376214. https://doi.org/10.3389/fpls.2024.1376214

Wang P, Menzies NW, Lombi E (2019) Fate of nanoparticles in soil environments. Wiley Interdiscip Rev Nanomed Nanobiotechnol 11(3):e1532

Wiesner MR, Lowry GV, Alvarez P, Dionysiou D, Biswas P (2006) Assessing the risks of manufactured nanomaterials. Environ Sci Technol 40(14):4336–4345

Xu P, Zeng GM, Huang DL, Feng CL, Hu S, Zhao MH et al (2016) Use of iron oxide nanomaterials in wastewater treatment: a review. Sci Total Environ 424:1–10

Zhang W, Zhang Z, Zhang Y, Zhang H (2015) Insights into the eco-toxicity of graphene oxide to plants. Environ Sci Nano 2(6):615–620

Zhang M, Duan X, Li Q, Zeng G (2019) Application of nanomaterials in precision agriculture: review and future prospects. Int J Agric Biol Eng 12(2):1–10

Chapter 17
Emerging Nanoformulated Approaches for Sustainable Agriculture: Current Understanding and Future Directions

Haitam Lahmamsi, Samir Ananou, Mohammed Taoussi, Ikram Legrifi, Mohammed Radi, Abdessalem Tahiri, and Rachid Lahlali

17.1 Introduction

Due to the intensity of environmental pressure and growing population, addressing food insecurity and agriculture sustainability is becoming a global challenge. Since over 800 million people suffer from severe hunger while the projection of the human population in 2050 exceeds 9.7 billion, an increase by 50–70% in the productivity must be achieved to meet the expected global demand (Sharma et al. 2023; Verma et al. 2024a). From the green revolution to nowadays the world has witnessed an unprecedented growth in farmland areas due to the use of agrochemicals, which tremendously increased the yield and productivity, however a serious reconsideration of their use was made due to the adverse effects of excessive and irrational use of agrochemicals not only on the human health but also the environment including ground water pollution and soil microbiota imbalance, etc. (Verma et al. 2022a, b).

H. Lahmamsi
Phytopathology Unit, Department of Plant Protection, Ecole Nationale d'Agriculture de Meknès, Meknès, Morocco

Laboratory of Microbial Biotechnology and Bioactive Molecules, Faculty of Sciences and Techniques, University of Sidi Mohamed Ben Abdellah, Fez, Morocco

S. Ananou
Laboratory of Microbial Biotechnology and Bioactive Molecules, Faculty of Sciences and Techniques, University of Sidi Mohamed Ben Abdellah, Fez, Morocco

M. Taoussi · M. Radi
Phytopathology Unit, Department of Plant Protection, Ecole Nationale d'Agriculture de Meknès, Meknès, Morocco

Laboratory of Environment and Valorization of Microbial and Plant Ressources, Faculty of Sciences, Moulay Ismail University, Zitoune, Meknes, Morocco

C. S. Seth et al. (eds.), *Nano-formulations for Sustainable Agriculture*, Nanotechnology in the Life Sciences, https://doi.org/10.1007/978-3-031-90617-6_17

Moreover, the climate change, water shortage and the decline of agriculture economy put more pressure and prompt an urgent technological intervention to search for ecofriendly alternatives with competitive yields and crop protection (Kumari et al. 2022). This led a growing tendency toward Sustainable agriculture that involves the balanced use of natural resources while preserving the ecosystem and the environment (Raliya et al. 2017). Starting from the principle that matter properties such as: heat resistance, bioavailability, magnetic properties, surface area and many other, multiplies in the nano scale, nanotechnology stands for the use of a newly introduced scientific concept that manipulate the physico-chemical parameters of matter at molecular level, it could be considered as the art of dealing with matter manipulation of nanomaterial at nanoscale in the range of 1–100 nm (DeRosa et al. 2010; Chen and Yada 2011; Abobatta 2018; de Francisco and García-Estepa 2018).

Its application in agriculture is represented by formulating fertilizers, pesticides and more interestingly biocontrol agents or plant growth promoters with organic or inorganic nanoformulation or nanocapsules that have multiple advantages such as the deferred delivery, enhanced solubility, controlled absorption in addition to its effectiveness and ecofriendly nature (Sharon et al. 2010; Verma et al. 2024b). The integration of nanotechnology into agriculture aligns with the global goals of sustainability that promotes practices enhancing production yields while safeguarding environment. As the world seeks solutions for sustainable food production, nanotechnology stands out as a beacon of hope, offering a path towards a more resilient and prosperous future in agriculture. This chapter emphasize the current understanding of nanoformulated approaches and its different applications in agriculture, bottlenecks and future perspectives are also mentioned.

17.2 Exploring Nanofertilizers Engineering to Enhance Plant Productivity

Plants cannot use a significant amount of inorganic fertilizers that are incorporated into the soil due to loss. For instance, soils lose and/or fix 40–70%, 80–90%, and 50–90% of nitrogen (N), phosphorous (P), and potassium (K) fertilizers,

I. Legrifi
Phytopathology Unit, Department of Plant Protection, Ecole Nationale d'Agriculture de Meknès, Meknès, Morocco

Laboratory of Functional Ecology and Environmental Engineering, Sidi Mohamed Ben Abdellah University, Fez, Morocco

A. Tahiri · R. Lahlali (✉)
Phytopathology Unit, Department of Plant Protection, Ecole Nationale d'Agriculture de Meknès, Meknès, Morocco
e-mail: rlahlali@enameknes.ac.ma

respectively, causing financial losses, as a result, soil nutrient balance may be adversely affected as additional fertilizers will be added to the soil to make up for those that are lost (El-Saadony et al. 2021). The improper use of inorganic fertilizers is mitigated by slow-released nano-fertilizers. The slow rate at which supplements disperse throughout crop production makes these slowly released nano-fertilizers a promising substitute for soluble inorganic fertilizers and it can help minimize nutrient loss, reduce soil degradation, and enhance soil fertility and crop productivity in the long run (Verma et al. 2023; Garg et al. 2023).

As a result, plants may absorb most of the nutrients they need without experiencing any losses (Guo et al. 2018). That's why engineering nanoparticles-based nanofertilizers, whether biological or chemical, offer significant benefits in agriculture, as it enhances crop nutrition management by improving abiotic stress tolerance and increasing agricultural productivity (Rajput et al. 2021; Kumar et al. 2024c). Nano fertilizers are composed of nanoparticles containing essential nutrients like N, P, K, Fe, and Mn, which are delivered to plants efficiently (Nongbet et al. 2022). In addition, fertilizer particles with nanomaterials on their surface are more effective in controlling nutrient release because they have a greater surface tension than conventional fertilizer particles (El-Saadony et al. 2021).

The large surface area to volume ratio and penetration capabilities of these novel fertilizers provide them an edge over traditional chemical fertilizers (Garg et al. 2023) and the use of Nano fertilizers can stimulate nitrogen metabolism, seed germination, photosynthesis, protein synthesis, and stress tolerance in plants (Verma et al. 2023). Zeolites, aluminium (Al), carbon (C), copper (Cu), zinc (Zn), nitrogen (N), silica (Si), iron (Fe), silver (Ag), magnesium (Mg), sodium (Na), potassium (K), calcium (Ca), and manganese (Mn) are common basic materials that may be nanostructured and used as NFs. Banana peels (Hussain et al. 2022) and other elements derived from plants are also utilized in the production of NFs. NFs have recently been created from natural zeolite, which is made up of more than 50 minerals. This is mostly because it is inexpensive and widely accessible (Eroglu et al. 2017).

Many micronutrients, such as iron, zinc, copper, and silica, have been produced at the nanoscale and are employed in the regulation of plant development (El-Saadony et al. 2021). In 2018, Heidari et al. looked into the effects of plant growth-promoting rhizobacteria (PGPR) on nano Fe-chelated corn on grain yield, growth, and physiological reactions. They found that applying nano and *Azospirillum brasilens* topically enhanced the growth and yield of maize plants (Heidari et al. 2018). A Second research by Mokarram-Kashtiban et al. 2019 found that PGPR, or arbuscular mycorrhizal, when added to plants together with a little number of Fe-NPs significantly increased the phytoremediation of heavy metals, improving the root zone and leaf area of young plants (El-Ramady et al. 2022). Additionally, proline, soluble sugars, and antioxidant enzymes known as bioagents as well as nano Zn-Fe oxide greatly enhanced seed growth, photosynthesis, and osmolyte content in salt-stressed wheat plants (Mokarram-Kashtiban et al. 2019).

17.3 Biosynthesized Nanoparticles

Utilizing a combination of one or more microorganisms that fix atmospheric nitrogen, produce compounds that promote growth, and solubilize phosphorus, biofertilizers are formulations that boost soil productivity (Kumari and Singh 2020). On the other hand, biosynthesized nanoparticles (NPs) are produced from biomolecules, algae, microbes, and plants, are gaining attention from researchers they provide an environmentally friendly and long-lasting address to farming challenges (Kumar et al. 2024b). Microorganisms, including fungus and bacteria, can create NPs during incubation in the medium. By using various metallic ion reductases to convert metallic ions into metallic NPs, these organisms may either employ certain metallic ions as growth nutrients or reduce the damaging effects of noble metallic ions (Hulkoti and Taranath 2014) and microorganisms, biomolecular activity, and physiological activities all have an effect on the effectiveness of biosynthetic processes such as reductive activity and metal ion bonding (Xu et al. 2021) (Fig. 17.1).

However, because it can be done quickly and doesn't require incubating cells, plant-based synthesis of biosynthesized NPs is being researched more than microorganism-based synthesis. Extracts from plants, including flowers, leaves, fruits, and roots, contain alkaloids, vitamins, terpenoids, polyphenols, phenolic acids, and flavonoids. These chemicals stabilize electrostatic force, control crystal formation, and size particles. Synthetic characteristics like solvent type, pH, temperature, reaction duration, and plant ratio extracts and metalic ions affect NP shape and size. Researchers produced Ag-NPs from cucumber leaf extract, identifying 245 metabolites using GC-MS (Bergal et al. 2022; Choudhary et al. 2024). A study on producing ZnO-NPs from *Malva parviflora* leaves found that increasing the leaf concentration improves the synthesis process (Iqbal et al. 2022). The extracted

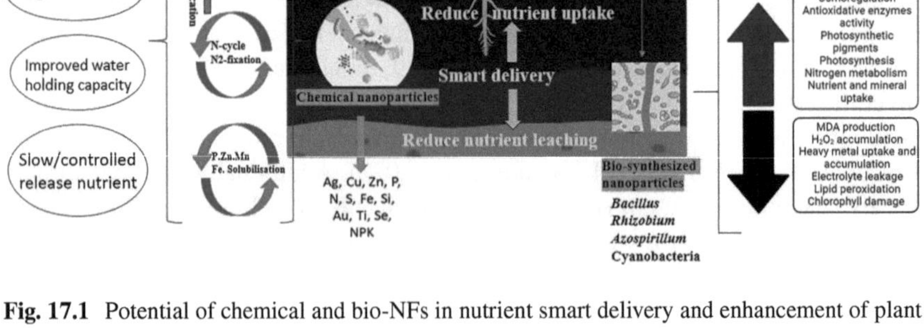

Fig. 17.1 Potential of chemical and bio-NFs in nutrient smart delivery and enhancement of plant growth and productivity

concentration directly affects the decrease rate of Au-NPs produced from *Combretum erythrophyllum* leaf extract (Fanoro et al. 2021).

While there are three categories of microorganisms—nitrogen-fixing rhizobacteria, plant growth-promoting rhizobacteria (PGPR), and arbuscular mycorrhizal fungi—that can all be utilized in nano-biofertilizers (NBFs), PGPR have been predominantly employed as biofertilizers. The key aspects of NBFs revolve around their interactions with microorganisms, longevity, and transportation mechanisms (Malusa and Vassilev 2014). It contains nutrients as well as plant growth microorganisms (*Azolla,* Cyanobacteria, *Pseudomonas Azospirihum, Bacillus, Azotobacetera, Azotobacter, Beijerinckia, Rhizobium*, etc.) that help in phosphate solubilizing, restoring soil nutrient richness, fixing atmospheric nitrogen and solubilizing insoluble complex organic matter into simple compounds (Dineshkumar et al. 2018; Itelima et al. 2018). The following interactions may occur between plants, NPs (heavy metals), and microbes, such as *Pseudomonas, Bacillus, Azotobacter, Rhizobium,* etc. The advantages of biologically produced fertilizers utilizing microorganisms include their ability to reduce NP toxicity, influence plant reactions to NFs, and establish a close connection between plants and NBFs. Consequently, NPs can penetrate plant cells through the stomatal or vascular pathways, potentially enhancing plant cell metabolism and contributing to higher crop yields. The stomatal route may be particularly efficient due to its rapid transport speed and large size exclusion threshold (Boddupalli et al. 2017).

Though NPs-based fertilizers have considerable relevance in agriculture, various drawbacks were discovered when they were employed directly into the soil, such as low stability, poor shelf-life, environmental effect and performance characteristics, like pH, temperature, and radiation (Itelima et al. 2018). An enhancement of safflower grain yields was observed when integrated treatments via organic fertilizer, bio-fertilizer, and bio-phosphate fertilizer, combined with 50% chemical fertilizer (Torabian et al. 2016) in another study, the application of silver (Ag) and gold (Au) NPs coated bio-fertilizer employing *Pseudomonas fluorescens, Bacillus subtilis,* and *Paenibacillus elgii* microorganisms has demonstrated better potential in plant development in various agricultural plants (Rahman and Zhang 2018). The biogenic production of ZnO NPs with the presence of microorganisms of *Pseudomonas aeruginosa* had broad-spectrum antibacterial characteristics, which can be further utilized for crop resistance (Rahman and Zhang 2018). With the treatment of leguminous crops (*Cajanus cajan*), (Jyoti et al. 2017) created copper NPs coated bio-fertilizer (*Piriformospora indica*) and showed that vitality and plant development could be boosted effectively and efficiently (Jyoti et al. 2017). The non-toxic titanium dioxide (TiO_2 NPs) nano-bio fertilizer has a significant impact on the biological processes of plants, controlling the effects of Cd on biomass production and chlorophyll concentration because cadmium (Cd) can causes oxidative stress, which can be hazardous to reactive oxygen species, it has a detrimental effect on the amount of chlorophyll (Al-Mamun et al. 2021; Kumar et al. 2024a) (Fig. 17.1).

17.4 Chemical Nanofertilizers

17.4.1 *Different Engineering of Organic Nanofertilizers*

17.4.1.1 Macronutrient Nanofertilizers

The development of macronutrient NFs is crucial in fertilizer as it provides the right nutrients to plants, reduces transportation costs, and increases food production. It significantly improves crop development and yields, reduces nutrient losses, and mitigates environmental impacts. These NFs consist of specific NFs (Zulfiqar et al. 2019). It has been found that the increased surface to volume ratio of NPs decreases the quantity and improves the efficacy of macronutrient NFs when compared to traditional fertilizers (Maghsoodi et al. 2020). A noticed uptake process includes NPs entering the intercellular space via the apoplastic pathway. NPs can enter epidermal and cortical cells and accumulate uniformly or in clusters in the endodermis (Chhipa 2017). Commercially available phosphorous (P)-based Nano fertilizers include MAP, DAP, SSP, and TSP, which are water-soluble phosphate salts ideal for plant application (Kumar et al. 2024b).

According to Maghsoodi et al. (2020), phosphorous NPs are ecologically acceptable and may be delivered to the root zone by irrigation or spraying (Maghsoodi et al. 2020). Regarding Nitrogen fertilizers, they sure do have impact plant growth and development and contribute to maintaining an appropriate supply of healthy food (Erisman et al. 2008). Due to quick volatilization and leaching, these commercially available NFs (urea) evaporate after roughly 75% application. The downsides of nitrogen fertilizers include their poor utilization and the loss of N to the environment, which contributes to eutrophication and high levels of greenhouse gas emissions (Kahrl et al. 2010).

Recently, an improved N NFs was created employing urea-coated hydroxyapatite NPs (rod-shaped particles with an average aspect ratio of 10) for targeted distribution and gradual release (Kottegoda et al. 2017). Several biological techniques have been investigated to promote plant P absorption, including bacterial species, P-solubilizing bacteria, and mycorrhiza. However, P-mobilising bacteria are not useful because of low soil organic carbon, restricted energy supply, and excessive water evaporation, which all have an impact on their survival, Nanotechnology researchers are studying approaches that will increase the delayed release and absorption efficiency of P fertilizers, retain P in its useful state in the long term, and allow for the mobilization of native P to plants (Maghsoodi et al. 2020).

Tarafdar et al. (2012) employed tricalcium phosphate as a precursor salt to synthesize fungal-mediated P NPs. Liu and Lal (2015) then evaluated the effect of carboxymethyl cellulose-stabilized hydroxyapatite NPs on soybeans. P NFs demonstrated encouraging results for seed treatment when natural raw phosphorite was treated by ultrasonic dispersion. Although P is frequently found in sufficient proportions in the soil, farmers have a high need for P fertilizers since the amount of P is not consistent for the plants. Zinc NPs were utilized to improve the efficiency of converting native P into plant-available P. Enzymes like phosphatase and phytase require zinc as a cofactor. Adding ZnO NPs to legumes and grains enhanced enzyme activity and P absorption by 11% (Raliya and Tarafdar 2013) which also improved the yield, the biomass, plant growth and the nutritional significance of cereals and legumes (Raliya et al. 2017). Cotton plants have also been shown to get P supplements in addition to ZnO NP exposure (Venkatachalam et al. 2017). To increase P absorption, either micronutrient-based NPs (such as Fe and Zn) or the production of P-based NPs are used to encourage native P mobilization. This not only improves plant development but also helps to address other challenges, such as the ecological consequences of P-induced eutrophication and the restricted P availability (Nongbet et al. 2022; Mariyam et al. 2024).

17.4.2 Nano-hydroxyapatite

Many researchers are now looking into nano-hydroxyapatite based Nano fertilizers as an alternative to typical water-soluble phosphorus (P) fertilizers (Maghsoodi et al. 2020). It was a significant component in human bones, teeth, and hard tissues and has the potential to increase the utilization of P fertilizer (Gómez-Morales et al. 2013). Commercially available P fertilizers are easily soluble in soil solution, and mobile soluble phosphates can be combined with surface water bodies via runoff, producing eutrophication (Maghsoodi et al. 2020). Because of its poor solubility, nHA may be able to provide nutritional P in the same way as traditional P fertilizer does, while also reducing the potential of contamination. It has the ability to promote delayed urea or nitrogen release by building a strong binding with urea (Kottegoda et al. 2011). Researchers have used several ways to manufacture synthetic nHA, including wet chemical precipitation (Gypser and Freese 2020), mechanochemical (Vahdat et al. 2020), hydrothermal (Ali et al. 2021), hydrolysis (Rop et al. 2020) and sol–gel method (Maghsoodi et al. 2020). The produced nHA from carboxymethyl cellulose (CMC) solution was applied to the soybean plant to assess its fertilizing efficacy (Liu and Lal 2014) and the growth rate of nHA-treated soybean plants was found to be 32.6% greater than that of traditional P fertilizer-treated soybeans.

Kottegoda et al. (2017) developed another nitrogen-based SRF using a urea-modified hydroxyapatite nanohybrid with a weight ratio of urea to hydroxyapatite (6:1) (Kottegoda et al. 2017). Nanohybrids were shown to cut N consumption by 50% due to their potential slow-release qualities and boost plant growth when tested on rice fields. Rop et al. (2018) created another SRF based on a combination of nHA and water-soluble fertilizer (urea, $(NH_4)2HPO_4$, and K_2SO_4) in water hyacinth cellulose-graft-poly(acrylamide) polymer hydrogel for NPK (Rop et al. 2018). The chemical interaction of nHA, urea, and monomer was investigated using FTIR spectra, and the results revealed that HA and urea formed a complex via hydrogen (H) binding, contributing to a slower nitrogen release rate. The scientists reported that SRF could release N and P at a slower pace, while K release was much faster.

17.4.3 Nano-zeolite

In general, zeolite promotes crop cultivation by improving soil conditions by enhancing water use efficiency with nutrients, fertility, biological activity, and lowering the volatilization of ammonia and soil salinity (Sangeetha and Baskar 2016; Choudhary et al. 2024). It can also improve nitrogen concentration in soil by increasing nutrient retention capacity. Because of the slow breakdown and disintegration rate in soil, plants may absorb nutrients for a longer length of time as their availability rises (Rai et al. 2012). After certain modifications to the structure, zeolite was gradually fed to the plants as a fertilizer, taking into account all of its benefits (Al-Mamun et al. 2021). Manikandan and Subramanian (2016) investigated zeolite-based Nano fertilizers and applied them to maize (*Zea mays*) plants grown in various soil types. The nano-zeolite has a large surface area, and its covering regulates the release of nitrogen from the fertilizer. The synthetic fertilizer was demonstrated to have an overly delayed release in soil without damaging the environment (Manikandan and Subramanian 2016). It was discovered that there was a significant rise in water absorbency, water retention capacity, swelling ratio, and equilibrium water content of ZNC compared to NZ, therefore ZNC was more favorable than NZ, and the use of these produced nanocomposite materials is ecologically friendly and safe (Lateef et al. 2016).

The influence of zinc oxide NPs and zeolite on soil nutrient availability was investigated in terms of nitrogen absorption by herbage from biogas slurry in the work of Yuvaraj and Subramanian 2018, in this study, three distinct ZnO NP concentrations (1.4–3.6 mg kg^{-1} soil) were mixed with zeolite (141–423 mg kg^{-1} soil) in the presence of AS and applied to herbage (Yuvaraj and Subramanian 2018). Zeolite and ZnO NPs were shown to considerably improve mineral nitrogen (N) content in soil compared to AS alone. This was discovered owing to increased nutrient mobilization in the soli through the influence of an extracellular enzyme (phosphatase, urease) and the soil microbial community. Furthermore, soil pH and electrical conductivity had no effect on zeolite or ZnO NPs due to their larger surface area or availability of cation exchange sites. This can be due to the adsorption of all cations on the sites, which reduces the number of cations in the soil solution.

The most plausible explanation for this outcome is that the organic matter/carbon content of the AS utilized in this investigation was significantly lower (Aziz et al. 2019). Zeolite-based urea has recently been widely employed in the development of controlled-release fertilizers (CRFs) due to its intrinsic cation exchange capability and inexpensive cost, which efficiently limits nutrient release rate. SEM, FT-IR, PSD, crushing strength, structural stability, and elemental analysis were all used to characterize the produced CRFs. Zeolite coated urea with CRF was made using a variety of acceptable binders, including potato starch, maize starch, acrylic polymer, white cement, and bentonite clay. In the presence of acrylic polymer to function as a binder (UP-AP), zeolite-coated urea outperformed the others (Dubey and Mailapalli 2019).

17.4.4 Micronutrient

In this category, the nutrient is required in minute quantities (≤100 ppm) yet important for maintaining the viability of several metabolic processes in plants. The micronutrients include iron (Fe), manganese (Mn), zinc (Zn), copper (Cu), molybdenum (Mo), and titanium dioxide (TiO_2). Micronutrients were frequently added to NPK fertilizers as soluble salt at low rates to promote crop absorption. It was also involved in the synthesis of protein and carbs, control of auxins, and protecting plants from dangerous infections (Zulfiqar et al. 2019).

Iron (Fe) serves as a vital micronutrient in Nano fertilizer essential for optimal plant growth. As per Hoagland et al. (1950), most plants thrive with 1–5 mg L^{-1} of Fe in the soil solution. Ghafariyan et al. (2013) discovered that super-paramagnetic FeNPs, when applied at low concentrations, effectively addressed chlorosis due to Fe deficiency in greenhouse-grown soybeans irrigated with water. The use of FeNPs resulted in comparable enhancements in chlorophyll levels in soybean sub-apical leaves to those achieved by commercial Fe-EDTA (below 45 mg L^{-1}) (Ghafariyan et al. 2013). Delfani et al. (2014) found in another study that applying 0.5 g L^{-1} Fe NPs to black-eyed peas produced exceptional results. They reported a 34% increase in the Fe content of the leaves, a 47% increase in the number of pods per plant, a 7% increase in grain weight of 1000-seed, and a 10% enrichment in chlorophyll content over-controlled (Delfani et al. 2014). The physiochemical impact of Fe NPs on black-eyed peas was evaluated, showing significant improvements in yield, leaf Fe content, stem Mg content, plasma membrane integrity, and chlorophyll content after 85 days (Delfani et al. 2014). Additionally, Patel et al. (2017) found that calcium alginate functionalized with Fe nanocarriers demonstrated controlled release of Fe, dependent on specific parameters such as pH and temperature of the medium (Patel et al. 2017).

Zinc (Zn) is an essential mineral element crucial for the proper development of all plants. It plays a vital role in activating enzymes like RNA polymerase and superoxide dismutase, regulating cell proliferation and differentiation, and enhancing chloroplast growth. However, while zinc has numerous beneficial effects on plants, including promoting growth and defense mechanisms, excessive levels can be harmful. Therefore, careful tuning of zinc levels is necessary to ensure optimal plant performance (Sturikova et al. 2018). While various types of ZnNPs, including ZnS, ZnSe, and quantum dots like CdSe/ZnS, are available, ZnONPs have emerged as a particularly potent zinc supplement for plants that has been extensively investigated in recent years (Sturikova et al. 2018).

Regarding copper in agriculture, plants need small amounts of copper for optimal growth, but excessive concentrations can lead to toxicity. CuNP-mediated fertilizers ensure rapid nutrient release, making them easily accessible to plants. The use of CuNPs-based fertilizers is on the rise in agriculture due to their dual function as insecticides and fertilizers (Wang et al. 2020). Wang et al. (2020) utilized three different sources of copper, namely nano CuO (nCuO), bulk CuO (bCuO), and $CuSO_4$, in the cultivation of Chinese onions—scallions. Their study revealed that both nCuO and $CuSO_4$ had beneficial effects, increasing leaf allicin (an antioxidant) levels by 56–187% and 42–90%, respectively (Wang et al. 2020). Molybdenum (Mo) plays a vital role in plant growth, with levels as low as 0.01 mg/L in soil solutions being essential for proper plant metabolism. Colloidal MoNPs and microbe functionalized colloidal Mo NPs for chickpea cultivation, demonstrating their effectiveness in improving the performance, yield, and disease resistance of legumes, especially in challenging environmental conditions (Taran et al. 2014).

Photocatalytic materials, particularly titanium dioxide (TiO_2), offer a sustainable solution to environmental challenges in agriculture. TiO_2 serves as a model photocatalytic material due to its photoactivity, chemical stability, adjustable hydrophilicity, and biocompatibility, promoting sustainability in agriculture. TiO_2 based photocatalytic nanoparticles (TiO_2 NPs) have advantages over metallic and carbon-based NMs. Research by Rodríguez-González et al. (2019) revealed that TiO_2 NPs did not affect the germination of various crops, like rice, lettuce, radish, cucumber, tomato, or pea, but did enhance root elongation at a lower dose. Additionally, TiO_2 NPs improved barley yield and increased the content of amino acids, Ca, Mg, and Zn in kernels when applied at lower doses (Rodríguez-González et al. 2019; Al-Mamun et al. 2021).

17.5 Application of Nanoformulation-Based Pesticide or Biocontrol Agents

Nano/microencapsulation of biocontrol agents (biological and chemical) represents a cutting-edge approach in agricultural practices. By encapsulating these agents at the nano or micro scale, their efficacy and stability can be significantly enhanced. Encapsulation has been described as simply the process of embedding an active core by physical action and then stabilizing it through chemical methods (Mariello et al. 2022). Encapsulation is a sophisticated process protecting BCAs against biotic and abiotic stresses, and preserving sustained metabolic action of BCAs through storage and after application. A further important characteristic is the release of the formulated microbes or biomolecules in a controlled manner (Riseh et al. 2021). A variety of biopolymers have been used to encapsulate BCAs, generating considerable research attention as potential and emerging technologies (Baranwal et al. 2022). In this connection, various biopolymers are currently employed for encapsulating BCAs, notably chitosan, alginate, gelatin, gellan gum, xanthan gum, etc. (Klein and Poverenov 2020).

The survival of encapsulated cells varies with the concentration and coating type, the dimension of the capsule and the number of cells initially involved. The encapsulated cells may comprise one or a combination of substances. Particularly formulated to prevent the release into the environment of nuclear material until the ideal time (Riseh et al. 2021). Encapsulation represents a major technique in nanotechnology. Because of the benefits provided by nanostructures, in recent years the application of nano-encapsulation has increased for releasing bioactive compounds (Rehman et al. 2019). In this regard, nanoencapsulation in particular is best described as the embedding or entrapment of bioactive substances supports of nanometric diameter (Bayraktar et al. 2017).

Encapsulation of a material is achieved using a diverse range of technologies, like emulsion, freeze-drying, coating by extrusion, cooling by spray, coating on fluidized bed, coacervation, spray-drying and heat gelling. A number of considerations, such as the physiochemical parameters of the encapsulation or base materials and their applicability, are important in selecting the appropriate encapsulation process. In particular, using emulsion, spraying-drying and extrusion processes have proved highly successful in microencapsulating plant growth-promoting bacteria (Riseh et al. 2021).

Moreover, in agriculture, nanotechnologies can be used to reduce biotic stress factors by means of nanomaterial-based pesticides and the use of agrochemical encapsulation to make pesticides perform better, to enhance plant growth and crop performance by slowly releasing fertilizers or water into plants, and for use in biosensors (Hassanisaadi et al. 2022). Phytopathogens including bacteria, fungi, viruses, nematodes and other pests are important limiting factors affecting agricultural productivity. Nanotechnologies can be used to treat different diseases brought about by bacteria, fungi, viruses and nematodes (Mishra et al. 2022).

However, nano-biotechnology represents a viable alternative for sustainable agriculture. In recent years, the use of chemical pesticides has provided very effective and convenient control over many plant diseases, but their growth has resulted in environmental disturbances with non-specific impacts posing a serious risk over beneficial microbes and human health (Deveci et al. 2021; Tang et al. 2021). The term nanobiotechnology was created by Lynn W. Jelinski, director of the biotechnology program at Cornell, USA. More recently, it has gained prominence by offering technology to mitigate increasing environmental pollution from chemical pesticides, which are damaging human health and biodiversity (AL-Ahmadi 2019; Pramanik et al. 2020).

In nano-biotechnology, biological agents like microbes, plant extracts and enzymes have been combined into NPs (1–100 nm in size) (Auría-Soro et al. 2019; Hyman and Denyes 2021). Nanoparticles have attractive potential in the fields of ultra-sensitive biomolecular detection, disease diagnosis, and the development of anti-microbial and therapeutic products. Integrating nanotechnology with biological science represents a promising new science of the present era (Dutta et al. 2022). Nanoparticles exercising their suppressive activity against plant pathogens have included non-metals, metalloids, oxide metals and nano-carbon materials (Elmer et al. 2018). Silver, titanium oxide, copper, gold, zinc oxide, iron and particularly silver have been the most extensively tested and widely used NPs (Iravani et al. 2014; Kandi and Kandi 2015).

NPs showed some efficacy on numerous pathogenic microorganisms, notably *Rhizoctonia solani*, *Sclerotinia sclerotiorum*, *Fusarium oxysporum*, *Sclerotium rolfsii* (Kaman and Dutta 2019; Abdelrhim et al. 2021), *Colletotrichum sp.* (Goswami et al. 2020), *Sarocladium oryzae* (Das and Dutta 2021), *Xanthomonas oryzae* pv.

Oryzae (Abdallah et al. 2019; Ogunyemi et al. 2020), and viruses such as bean yellow mosaic virus (BYMV) (Elbeshehy et al. 2015), potato virus Y (PVY) (Noha et al. 2018), cucumber mosaic virus (CMV) (El-Sawy et al. 2017), etc.

Nanomaterials mechanism of action vis-a-vis organisms comprises cell membrane perturbation, aggregation, suppression of H^+-ATPase activity, inhibition of messenger RNA and protein synthesis, suppression of the production of toxins, blockage of the growth of microbes and inhibition of nutritional flux (Dakal et al. 2016; Malerba and Cerana 2016). NFs also represent a further major aspect of plant protection products, offering the potential to boost plant resistance to various diseases. As such, nano-fertilizers have been investigated in a number of plant-disease systems, notably fusarium wilt in chrysanthemum, red root rot in tea, verticillium wilt in brinjal, crown and root rot in asparagus, fusarium wilt in tomato, etc. (Elmer et al. 2018).

Nanoparticle synthesis performed by various techniques using biological, chemical and physical methods (Gopinath and Velusamy 2013). The most favored methods for synthesizing NPs are likely to provide higher properties to the encapsulated material and to be cost-efficient with no risk of toxicity (Ghiuță et al. 2018). Nanoparticles based on conventional chemical processes have been largely ignored in the clinical and food sectors, since they generate toxicity because of the chemicals used in formulation. Nanoformulations basing on biogenically generated approaches (low-cost and eco-friendly) have gained in importance with the removal of chemical toxicity (Wei et al. 2012).

Pesticide nanoformulations require a wide range of properties like increased potency and durability, enhanced wettability and dispersibility, biodegradability with minimal active ingredients, and non-toxicity with convenient pesticidal performance. Given the fast progress in nanopesticides technology, researchers remain highly interested in developing less toxic, highly target-specific nanopesticides while maintaining high efficacy. As a result, nano-pesticides specific to a target are expected to contribute to reduced damage to non-target crops and less release of the pesticide into the atmosphere (Kumar et al. 2019).

The ability of bacteria actively to attenuate metal ions has been a major factor in their use to produce antagonistic NPs for controlling insect pests. A number of bacterial species (*Bacillus* sp., *Pseudomonas* sp., *Corynebacterium* sp., *Shewanella* sp.) have been employed to synthesize various NPs with metal inorganics, i.e., Ag, Au, Al, ZnO, MnO, TiO_2 over the past two decades. Utilization of such bacterially generated NPs has enhanced agriculture productivity while minimizing losses during crop processing, transport and storage (Kumari et al. 2019).

One Bacillus species, *strain GP23* isolated from marine soil, was employed to produce silver nanoparticles (Ag NPs) using the extracellular process, suppressing *Fusarium oxysporum*, causing wilt disease in bananas, legumes, tobacco, sweet potatoes, cucurbits, and tomatoes (Gopinath and Velusamy 2013). Ag NPs produced extracellularly by *Bacillus strain SBT8* exhibit antimicrobial activity and function as a bio-catalyst (Yurtluk et al. 2018). The NPs made whit silver nitrate constituting compounds using *B. amyloliquefaciens* and *B. subtilis* demonstrated antibacterial activity on gram-positive and gram-negative bacteria, along as antifungal activity on *Candida albicans* (Ghiuță et al. 2018).

Bt nanoscale chitinases have shown the ability to deliver nematicidal preparations to manage *Caenorhabditis elegans*. Nanoscale chitinases from *B. thuringiensis* could therefore serve as a biopesticide dispensing system to boost agricultural applications (Qin et al. 2020). Using the nanotechnology technique, entomopathogenic (Bt) bacteria can be decrease in size to formulate Bt nanoparticles, which have subsequently been used as biocontrol agents (Swamy and Asokan 2013). Alternatives to NPs, Bt nanocomposites with Bt bioactives in various nano-formulations such as nano-capsules, nano-emulsion and nano-suspension were employed as nano-pesticides (Damalas and Koutroubas 2018).

The Cry protein obtained by Bt is the principal element of a large group of proteins with different toxic activities and various characteristics. The majority of these proteins are toxic to *Diptera*, *Coleoptera* and *Lepidoptera*, and some of them also act on nematodes and *Hymenoptera*. The loading of Cry proteins in NMs to produce nanocomposites potentially proved easy, and Cry protein bioactivity was sustained through NMs that boosted stability too (Devi et al. 2019). Nano-crystallization of the cyclic lipopeptides taken from *B. subtilis* increases storage condition stability by inhibiting both oxidation and anti-fungal properties via controlled release of the cyclic lipopeptides. These cyclic lipopeptide nanocrystals (solid lipid NPs) inhibit the development of *Aspergillus carbonarius* and *A. niger* spores and hyphae (Lin et al. 2020).

17.6 Usage of Nanoformulations in Bioremediation and Abiotic Stress Mitigation

17.6.1 Nanobioremediation for Agricultural Contaminants

Agricultural activities often lead to soil and water contamination, posing significant risks to ecosystem health and human well-being (Brusseau et al. 2019). Nanobioremediation, employing engineered NMs for environmental clean-up, has emerged as a promising alternative due to its high efficiency, specificity, and eco-friendliness (Benjamin et al. 2019).

17.6.1.1 Heavy Metals

The accumulation of heavy metals in the soil and environment arises from anthropogenic activities and perturbed biogeochemical cycles. Although heavy metals including As, Cd, and Pb, have no defined biological functions, their presence exerts toxic effects on the biotic components of ecosystems, even at low concentrations (Asati et al. 2016). In particular, in acid soils characterized by low nutrient levels, heavy metal toxicity is enhanced (Lwin et al. 2018). The production of reactive oxygen species by heavy metals causes damage to essential macromolecules, including proteins and nucleic acids, in plants, and microorganisms (Berni et al. 2019). A common strategy to remediate heavy metal contamination is to immobilize heavy metal molecules (Derakhshan Nejad et al. 2018). Bio-organic NPs synthesized by biological organisms, have become essential agents for the removal of heavy metals from soil matrices.

In the last few years, carbon nanotubes (CNTs) have emerged as a promising solution for the remediation of heavy metal. CNTs exist in two primary forms: single-walled and multi-walled variants (Martel et al. 1998). Their efficacy in treating heavy metal surpasses that of conventional methods, owing to their rapid adsorption kinetics facilitated by a substantial specific surface area and heightened adsorption capacity, particularly notable in the case of heavy metal ions like Cr^{7+}, Cu^{2+}, Ti^{+}, Pb^{2+}, and Mn^{7+} (Stafiej and Pyrzynska 2007). Furthermore, the adsorption properties of CNTs towards heavy metals can be augmented through various surface modifications, including endohedral filling, chemical alteration, or heat treatment, thereby introducing functional groups like OH, NH, and COOH (Kumar et al. 2021).

A research investigation was undertaken to explore the efficacy of a graphene oxide (rGO) modified electrode incorporating Au nanoparticles (rGO-Au NPs) for simultaneous detection of toxic metal ions, notably Cd^{2+}, Pb^{2+}, Cu^{2+}, and Hg^{2+}. The synthesis of nanoparticles employed a vegetable extract derived from *Abemoschus esculentus* as a reducing agent. In parallel, various bacterial strains, such as *B. subtilis, Staphylococcus aureus, Rizobium gallium,* and *Pseudomonas aeruginosa*, were introduced due to their demonstrated high adsorptive potential for bioremediation of toxic metal ions, effectively scavenging heavy metal ions (Gnanaprakasam et al. 2016). In the same context, magnetic nanoparticles composed of Fe_3O_4, coated with phthalic acid and subjected to treatment with *S. aureus*, were deployed for the remediation of Ni, Pb, and Cu contaminants. Remarkably, these particles showed substantial efficiency, achieving removal efficiencies ranging from 92.6% to 97.5% for Ni^{2+} from 99.4% to 100% for Pb^{2+}, and from 83% to 89% for Cu^{2+}. Particularly, it was revealed that the functional groups present both at the core of the NPs and on the microbial surface played a key role in facilitating the elimination of heavy metals (Mahmoud et al. 2016).

Furthermore, selenium NPs, combined with the probiotic bacterium *Lactobacillus casei*, have been applied for the remediation of cadmium-contaminated water and land. Remarkable results revealed a cadmium absorption efficiency of 65% when using *L. casei* in combination with selenium NPs, a figure significantly higher than that achieved by *L. casei* alone, which ranged from 43% to 78% (Dong et al. 2013). Cao et al. (2020) reported the utility of polyvinylpyrrolidone-coated iron oxide NPs for metal elimination, focusing particularly on Pb^{2+} and Cd^{2+}, via the interaction of bacteria with *Halomonas* sp. Various treatment modalities, involving the administration of NPs and bacteria individually as well as in combination, were evaluated. Following a 24 h treatment duration, 100% removal efficiency for Cd^{2+} was achieved in the combined treatment, a percentage that remained consistent after 48 h. In contrast, treatment with NPs alone resulted in a 60% removal of Cd^{2+} and 80% removal of Pb^{2+}. In fact, the introduction of *Halomonas* sp. enhanced Cd^{2+} removal to 80%, while the removal efficiency for Pb^{2+} remained at 80%, reflecting the outcomes observed with NPs treatment alone (Cao et al. 2020).

The use of biogenic manganese oxide (BMO) derived from *Pseudomonas putida* MnB1 has emerged as a promising avenue for heavy metal remediation in environmental settings. BMO exhibits remarkable enhanced performance in heavy metal adsorption in comparison to chemically synthesized manganese oxide (birnessite). This superiority is attributed to BMO's small particle size, amorphous nature, and expansive surface area, collectively rendering it an exceptional adsorbent. Specifically, BMO demonstrates a markedly accelerated adsorption rate, approximately 7–8 times faster than birnessite, particularly evident in the adsorption of cadmium and zinc. Moreover, variations in temperature and pH further augment BMO's efficacy in heavy metal adsorption (Zhou et al. 2015).

17.6.1.2 Organic Contaminants (Including Pesticides and Herbicides)

Organic pollutants, particularly persistent organic pollutants, pose important threats to both human health and environmental integrity, necessitating their remediation from soil matrices. Various strategies employing NPs in conjunction with microbial agents have been explored for the efficient removal of these contaminants. For instance, magnetic nanoparticles, when combined with *Rhodococcus engthropolis*, facilitated dibenzothiophene (DBT) desulfurization (Ansari et al. 2009). *Alcaligenes faecelis*, coupled with Fe_2O_3 NPs, exhibited enhanced degradation of hydrocarbon compounds associated with crude oil contamination (Ningombam et al. 2024). Additionally, Bimetallic (Pd/nFe) nanoparticles, employed alongside *Sphingomonas wittichii*, demonstrated efficacy in the bioremediation of NBR 2, 3, 7, 8-tetrachlorodibenzo-p-dioxin hydrocarbons (Bokare et al. 2012). Furthermore, *Sphingomonas strain* NM05, utilized for hexachlorocyclohexane (HCH) degradation, demonstrated a significant enhancement in degradation efficacy upon treatment with Pd/FeO bimetallic NPs (Singh et al. 2013). Moreover, NPs not only augment the remediation capabilities of microbes but also contribute to soil health improvement. Silicon NPs, for instance, have been shown to enhance soil microflora and biomass, thereby promoting the growth of rhizospheric microbes (Rajput et al. 2022).

The integration of NPs with microbial entities holds promise for advancing sustainable pesticide mitigation strategies in soil remediation efforts. For instance, Zhuang et al. (2015) investigated the use of *Streptomyces* sp. N01 immobilized on bamboo-carbon supported Fe_3O_4 nanoparticles (Fe_3O_4/BC) for the degradation of the heterocyclic aromatic organic compound quinoline. Their study demonstrated the efficacy of this immobilized system across varying quinoline concentrations (100–400 mg/L), temperatures (20–45 °C), and pH levels (5–10). Moreover, the Fe_3O_4/BC immobilized cells exhibited enhanced tolerance to pesticide concentrations when subjected to changes in pH and temperature. Notably, in the ninth cycle, a pesticide removal rate of 85.3% was achieved.

Similarly, recombinant *Escherichia coli* cells expressing the atrazine-dechlorinating enzyme were utilized for pesticide removal. These cells were encapsulated within a porous gel matrix formed using siliconized silica nanoparticles (Si NPs). The Si NPs were functionalized through silanization with methyltrimethoxysilane (MTMS) and polyethylene glycol, resulting in the formation of a porous gel network acting as a cross-linker. Stability assessments conducted at temperatures of 23 °C and 45 °C revealed the robust structural integrity of the gel, particularly at higher temperatures. Importantly, the degradation efficiency of the pesticide by the encapsulated cells was found to be three times greater than that achieved by free cells (Reátegui et al. 2012).

17.7 Nanobiotechnology in Crop Abiotic Stress Management

Abiotic stresses exert a serious negative effect on plant growth, resulting in an important reduction in crop yield, often by as much as 50% (Farooq et al. 2009; Zahra et al. 2021). These stresses perturb the metabolism of plants, leading to diminished development, production and growth (Nadeem et al. 2022). Among the various impact of abiotic stresses, the excessive production of reactive oxygen species (ROS) stands out as a major contributor to crop losses (Mansoor et al. 2022; Choudhary et al. 2025). Efforts to mitigate the detrimental effects of abiotic stresses have involved the use of Osmo-protectants, nutrients, and stress signaling compounds, which have shown efficacy in scavenging ROS and enhancing tolerance against abiotic stresses (Arshad et al. 2017). However, recent studies suggest that delivering these substances in nano-sized forms may offer even greater effectiveness. Further evidence suggests that administration of the above substances in nanomaterial form can enhance plant tolerance to a range of abiotic stresses, including salinity (Oliveira et al. 2016), drought (Zhao et al. 2020), and heat stresses (Wu et al. 2017).

17.7.1 Salt Stress

Salt stress affects more than 20% of cultivated areas, primarily through specific ion toxicity, mineral imbalance, and salinity-induced osmotic stress, consequently impeding plant growth (Ondrasek et al. 2022). The utilization of NPs has demonstrated significant effects on plant growth under salt stress conditions by modulating protein and carbohydrate synthesis, and enhancing the antioxidant enzymes activities such as catalase (CAT) guaiacol peroxidase (GPX), and ascorbate peroxidase (APX) (Ghosh et al. 2016; Zia-ur-Rehman et al. 2023). This action aids in decreasing the effect of salt stress by detoxifying ROS and regulating hormonal responses (Fig. 17.2).

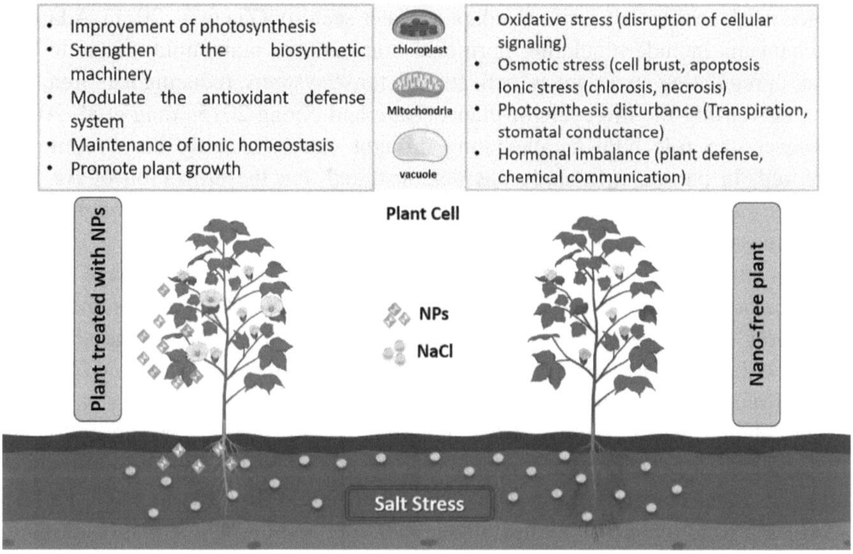

Fig. 17.2 Mechanisms in plant cells involved in the treatment of salt stress with NPs

Studies have shown that under salt stress, the uptake of Cl^- and Na^+ increases, leading to oxidative damage and restricting the uptake of essential ions like Ca^{2+} and K^+ (Gul et al. 2019). NPs application has been found to mitigate Cl^- and Na^+ accumulation in plants under salinity, with ZnO-NPs, for instance, enhancing the uptake of Ca^{2+} and K^+ while decreasing Cl^- and Na^+ accumulation (Ahmad and Akhtar 2019). Furthermore, NPs of various elements, including manganese (Mn), copper (Cu), iron (Fe), cesium (Cs), and cobalt (Co), have been shown to support antioxidant enzyme activity, including peroxidase (POD) catalase (CAT), thereby alleviating oxidative stress under salt stress conditions (Rico et al. 2015).

Moreover, the low concentration application of silver nanoparticles enhanced the antioxidant enzymatic function and improved the growth of plants subjected to salt stress (Sami et al. 2020). Similarly, TiO_2 NPs have been shown to ameliorate the tolerance to salt in *Dracocephalum moldavica* by activating the antioxidant defense system, which deceases salinity-induced oxidative harm and promotes the growth of plants (Gohari et al. 2020). Overall, the utilization of NPs holds promise for improving plant salt tolerance via the maintenance of tissue water status, ion homeostasis, the antioxidant system activation, and other mechanisms.

17.7.2 *Drought Stress*

Drought stands as a significant abiotic stressor, with profound repercussions on the growth and productivity of plants (Tahi et al. 2007). This stressor induces a myriad of responses in crop plants at biochemical, physiological, morphological, and molecular levels as they strive to adapt to water scarcity (Zia et al. 2021). Adaptive mechanisms include regulating hormonal homeostasis, maintaining osmotic balance, upregulating stress genes, activating defense systems, reducing leaf area, and even decreasing the life cycle of plants (Fang and Xiong 2015; Yang et al. 2021). However, the role NPs in alleviating drought stress has only been minimally explored. In particular, a study has demonstrated that the utilization of Ag NPs enhances antimicrobial activity against *S. aureus* and *E. coli* under drought stress conditions in *Tephrosia apollinea* (Ali et al. 2019).

This improvement coincides with a reduction in membrane damage and an enhancement in hydrogen peroxide (H_2O_2) contents in *T. apollinea* roots under progressively severe drought conditions. ZnO NPs have demonstrated the ability to improve the seedlings growth and improve nutrient utilization efficiency of plants under drought conditions (Sedghi et al. 2013). Additionally, Dimkpa et al. (2019) observed improved nutrient acquisition during drought stress in sorghum following the application of ZnO-NPs. Furthermore, the use of Si NPs has been shown to alleviate water-deficit stress in wheat (Baybordi 2005). In a recent study, Fe-NPs were used as a treatment to wheat under cadmium and drought stress, resulting in improved photosynthesis rates and reduced oxidative stress (Adrees et al. 2020). Thus, NPs offer potential in mitigating the adverse drawbacks of drought stress. However, more researches into the associated molecular, biochemical, and physiological mechanisms are warranted.

17.7.3 Thermal Stress

The utilization of NMs has proven highly efficient in enhancing the tolerance of plant to heat stress (Hassan et al. 2018). For instance, Iqbal et al. (2019) demonstrated a significant enhancement in leaf area and the biomass of wheat under heat stress through the application of Ag NPs produced with moringa extract. These Ag NPs were found to balance the water content in the tissues and enhance the content of chlorophyll compared to the control, thereby mitigating the adverse effects of heat stress on wheat crops (Iqbal et al. 2019). Selenium (Se) is recognized for its essential effect in the antioxidant defense system of plants. Utilization of Se NPs at low concentrations has been demonstrated to stimulate the antioxidant defense system by improving the antioxidant enzymes activities and enhancing photosynthetic pigments in grain sorghum (Djanaguiraman et al. 2018) and tomato (Haghighi et al. 2014). Moreover, the application of Se NPs synthesized with *L. casei* has been shown to increase the tolerance to heat stress in chrysanthemum by activating the antioxidant defense system (Seliem et al. 2020). Additionally, the treatment with NPs of titanium dioxide (TiO_2) has been found to contribute to tomato heat tolerance by regulating stomatal oscillation (Qi et al. 2013).

Nanomaterials utilization has emerged as an effective strategy for inducing cold tolerance in various plant species. For instance, the application of Se NPs has been shown to enhance cold tolerance in tomato seedling by activating antioxidant defense system, improving photosynthetic pigments, and promoting plant growth under cold stress conditions (Sayed et al. 2024). Similarly, the use of TiO_2 NPs has demonstrated promising results in enhancing cold tolerance in chickpea (Mohammadi et al. 2014). This enhancement was attributed to an important enhancement in the chlorophyll contents, antioxidant enzymes activities, and a decrease in ROS and oxidative damage in comparison with control plants under cold conditions.

The use of chitosan NPs has demonstrated effectiveness in decreasing ROS and promoting the osmoprotectants accumulation in banana plants subjected to cold stress (Wang et al. 2021). Additionally, foliar application NPs of zinc oxide (ZnO) has been observed to mitigate chilling stress in rice plants by modulating the anti-oxidative system and transcription factors associated with the chilling response (Song et al. 2021). Similarly, the utilization of silicon nanoparticles has shown promise in enhancing the photosynthetic capacity of sugarcane plants subjected to cold stress conditions (Elsheery et al. 2020).

17.8 Challenges and Future Directions

As mentioned earlier many challenges face the introduction of nanotechnology in sustainable agriculture, limiting its use on a large scale. For instance, the main challenge in its commercialisation and globalization is the toxicological aspect of nano formulations since researches on this aspect in a lot of new technologies are very few, in addition it is very critical consideration and need a thorough management and evaluation, mainly the metal-based nano formulations that could be translocated and accumulated in plants and possibly poses a threat to human health through the food chain, also their interactions with soil could lead to toxic ions liberation and thus soil pollution. Also, the young age of this technology in agriculture makes it essential to do more and more researches to understand the fate, toxicity and transport of nanoformulations to assess environmental risks associated with their use in agriculture. Thus, overcoming the challenges related to the toxicology of nanomaterials, their impact on the environment and the effect on human health is an essential step ensuring the safe manipulation and use of NMs in agricultural practices by conducting thorough risk assessments, developing robust safety protocols and implementing precise application methods.

17.9 Conclusion

In conclusion, the integration of nanotechnology in sustainable agriculture as an ecofriendly solution presents immense potential to enhance productivity and environmental stewardship, either by its use as NFs, NPs or nanocarriers of bioactive ingredients such as plant extracts and essential oils or microbial biocontrol agents and plant growth promoters. In addition to their potential use as abiotic stress mitigators. However, to realize these benefits effectively, it is essential to address the challenges facing its common use such as environmental impact, toxicity, and safe application of nano-enabled products. By prioritizing research, implementing stringent safety measures, and fostering responsible practices, the agricultural sector can leverage nanotechnology to achieve sustainable and resilient farming systems that balance productivity with environmental and human health considerations. Embracing innovation while ensuring safety and sustainability will be key to unlocking the full potential of nanotechnology in sustainable agriculture in the era of climate change.

Acknowledgements This research received financial support from the Phytopathology Unit within the Department of Plant Protection and Environment at The Ecole Nationale d'Agriculture de Meknès-Morocco.

References

Abdallah Y, Ogunyemi SO, Abdelazez A, Zhang M, Hong X, Ibrahim E, Hossain A, Fouad H, Li B, Chen J (2019) The green synthesis of MgO nano-flowers using Rosmarinus officinalis L. (Rosemary) and the antibacterial activities against Xanthomonas oryzae pv. oryzae. Biomed Res Int 2019:5620989

Abdelrhim AS, Mazrou YSA, Nehela Y, Atallah OO, El-Ashmony RM, Dawood MFA (2021) Silicon dioxide nanoparticles induce innate immune responses and activate antioxidant machinery in wheat against Rhizoctonia solani. Plan Theory 10:2758

Abobatta WF (2018) Nanotechnology application in agriculture. Acta Sci Agric 2

Adrees M, Khan ZS, Ali S, Hafeez M, Khalid S, ur Rehman MZ, Hussain A, Hussain K, Chatha SAS, Rizwan M (2020) Simultaneous mitigation of cadmium and drought stress in wheat by soil application of iron nanoparticles. Chemosphere 238:124681

Ahmad, I., Akhtar, M.S. (2019). Use of Nanoparticles in Alleviating Salt Stress. In: Akhtar, M. (eds) Salt Stress, Microbes, andPlant Interactions: Causes and Solution. Springer, Singapore. https://doi.org/10.1007/978-981-13-8801-9_9

Saud AL-Ahmadi M (2019) Pesticides, Anthropogenic Activities, and the Health of Our Environment Safety. Pesticides - Use and Misuse and Their Impact in the Environment. Rijeka. IntechOpen. Available at: https://doi.org/10.5772/intechopen.84161

Ali MA, Mosa KA, El-Keblawy A, Alawadhi H (2019) Exogenous production of silver nanoparticles by Tephrosia apollinea living plants under drought stress and their antimicrobial activities. Nano 9:1716

Ali AF, Alrowaili ZA, El-Giar EM, Ahmed MM, El-Kady AM (2021) Novel green synthesis of hydroxyapatite uniform nanorods via microwave-hydrothermal route using licorice root extract as template. Ceram Int 47:3928–3937

Al-Mamun MR, Hasan MR, Ahommed MS, Bacchu MS, Ali MR, Khan MZH (2021) Nanofertilizers towards sustainable agriculture and environment. Environ Technol Innov 23:101658. https://doi.org/10.1016/j.eti.2021.101658

Ansari F, Grigoriev P, Libor S, Tothill IE, Ramsden JJ (2009) DBT degradation enhancement by decorating Rhodococcus erythropolis IGST8 with magnetic Fe_3O_4 nanoparticles. Biotechnol Bioeng 102:1505–1512

Arshad MS, Farooq M, Asch F, Krishna JSV, Prasad PVV, Siddique KHM (2017) Thermal stress impacts reproductive development and grain yield in rice. Plant Physiol Biochem 115:57–72

Asati A, Pichhode M, Nikhil K (2016) Effect of heavy metals on plants: an overview. Int J Appl Innov Eng Manag 5:56–66

Auría-Soro C, Nesma T, Juanes-Velasco P, Landeira-Viñuela A, Fidalgo-Gomez H, Acebes-Fernandez V, Gongora R, Almendral Parra MJ, Manzano-Roman R, Fuentes M (2019) Interactions of nanoparticles and biosystems: microenvironment of nanoparticles and biomolecules in nanomedicine. Nano 9:1365

Aziz Y, Shah GA, Rashid MI (2019) ZnO nanoparticles and zeolite influence soil nutrient availability but do not affect herbage nitrogen uptake from biogas slurry. Chemosphere 216:564–575

Baranwal J, Barse B, Fais A, Delogu GL, Kumar A (2022) Biopolymer: a sustainable material for food and medical applications. Polymers 14:983

Baybordi A (2005) Effect of zinc, iron, manganese and copper on wheat quality under salt stress conditions. J Water Soil 140:150–170

Bayraktar O, Erdoğan İ, Köse MD, Kalmaz G (2017) Nanocarriers for plant-derived natural compounds. In: Nanostructures for antimicrobial therapy. Amsterdam. Elsevier, pp 395–412

Benjamin SR, De Lima F, Florean EOPT, Guedes MIF (2019) Current trends in nanotechnology for bioremediation. Int J Environ Pollut 66:19–40

Bergal A, Matar GH, Andaç M (2022) Olive and green tea leaf extracts mediated green synthesis of silver nanoparticles (AgNPs): comparison investigation on characterizations and antibacterial activity. BioNanoScience 12:307–321

Berni R, Luyckx M, Xu X, Legay S, Sergeant K, Hausman J-F, Lutts S, Cai G, Guerriero G (2019) Reactive oxygen species and heavy metal stress in plants: impact on the cell wall and secondary metabolism. Environ Exp Bot 161:98–106

Boddupalli A, Tiwari R, Sharma A, Singh S, Prasanna R, Nain L (2017) Elucidating the interactions and phytotoxicity of zinc oxide nanoparticles with agriculturally beneficial bacteria and selected crop plants. Folia Microbiol 62:253–262

Bokare V, Murugesan K, Kim J-H, Kim E-J, Chang Y-S (2012) Integrated hybrid treatment for the remediation of 2, 3, 7, 8-tetrachlorodibenzo-p-dioxin. Sci Total Environ 435:563–566

Brusseau ML, Ramirez-Andreotta M, Pepper IL, Maximillian J (2019) Environmental impacts on human health and well-being. In: Environmental and pollution science. Elsevier, pp 477–499. https://doi.org/10.1016/B978-0-12-814719-1.00026-4

Cao X, Alabresm A, Chen YP, Decho AW, Lead J (2020) Improved metal remediation using a combined bacterial and nanoscience approach. Sci Total Environ 704:135378

Chen H, Yada R (2011) Nanotechnologies in agriculture: new tools for sustainable development. Trends Food Sci Technol 22:585–594

Chhipa H (2017) Nanofertilizers and nanopesticides for agriculture. Environ Chem Lett 15:15–22

Choudhary R, Rajput VD, Ghodake G, Ahmad F, Meena M, Rehman RU, Prasad R, Sharma RK, Singh R, Seth CS (2024) Comprehensive journey from past to present to future about seed priming with hydrogen peroxide and hydrogen sulfide concerning drought, temperature, UV and ozone stresses- a review. Plant Soil 500(1–2):351–373. https://doi.org/10.1007/s11104-024-06499-9

Choudhary R, Ahmad F, Kaya C, Upadhyay SK, Muneer S, Kumar V, Meena M, Liu H, Upadhyay H, Seth CS (2025) Decrypting proteomics, transcriptomics, genomics, and integrated omics for augmenting the abiotic, biotic, and climate change stress resilience in plants. J Plant Physiol. https://doi.org/10.1016/j.jplph.2025.154430

Dakal TC, Kumar A, Majumdar RS, Yadav V (2016) Mechanistic basis of antimicrobial actions of silver nanoparticles. Front Microbiol 7:231711

Damalas CA, Koutroubas SD (2018) Current status and recent developments in biopesticide use. Agriculture 8:13

Das A, Dutta P (2021) Antifungal activity of biogenically synthesized silver and gold nanoparticles against sheath blight of rice. J Nanosci Nanotechnol 21:3547–3555

de Francisco EV, García-Estepa RM (2018) Nanotechnology in the agrofood industry. J Food Eng 238:1–11

Delfani M, Baradarn Firouzabadi M, Farrokhi N, Makarian H (2014) Some physiological responses of black-eyed pea to iron and magnesium nanofertilizers. Commun Soil Sci Plant Anal 45:530–540

Derakhshan Nejad Z, Jung MC, Kim K-H (2018) Remediation of soils contaminated with heavy metals with an emphasis on immobilization technology. Environ Geochem Health 40:927–953

DeRosa MC, Monreal C, Schnitzer M, Walsh R, Sultan Y (2010) Nanotechnology in fertilizers. Nat Nanotechnol 5:91

Deveci HA, Nur G, Deveci A, Kaya I, Kaya MM, Kükürt A, Gelen V, Başer ÖF, Karapehlivan M (2021) An overview of the biochemical and histopathological effects of insecticides. Rijeka. IntechOpen

Devi PSV, Duraimurugan P, Chandrika K (2019) Bacillus thuringiensis-based nanopesticides for crop protection. In: Nano-biopesticides today and future perspectives. Elsevier, pp 249–260

Dimkpa CO, Singh U, Bindraban PS, Elmer WH, Gardea-Torresdey JL, White JC (2019) Zinc oxide nanoparticles alleviate drought-induced alterations in sorghum performance, nutrient acquisition, and grain fortification. Sci Total Environ 688:926–934

Dineshkumar R, Kumaravel R, Gopalsamy J, Sikder MNA, Sampathkumar P (2018) Microalgae as bio-fertilizers for rice growth and seed yield productivity. Waste Biomass Valor 9:793–800

Djanaguiraman M, Belliraj N, Bossmann SH, Prasad PVV (2018) High-temperature stress alleviation by selenium nanoparticle treatment in grain sorghum. ACS Omega 3:2479–2491

Dong H, Rowland I, Thomas LV, Yaqoob P (2013) Immunomodulatory effects of a probiotic drink containing Lactobacillus casei Shirota in healthy older volunteers. Eur J Nutr 52:1853–1863

Dubey A, Mailapalli DR (2019) Zeolite coated urea fertilizer using different binders: fabrication, material properties and nitrogen release studies. Environ Technol Innov 16:100452

Dutta P, Kumari A, Mahanta M, Biswas KK, Dudkiewicz A, Thakuria D, Abdelrhim AS, Singh SB, Muthukrishnan G, Sabarinathan KG (2022) Advances in nanotechnology as a potential alternative for plant viral disease management. Front Microbiol 13:935193

Elbeshehy EKF, Elazzazy AM, Aggelis G (2015) Silver nanoparticles synthesis mediated by new isolates of Bacillus spp., nanoparticle characterization and their activity against Bean Yellow Mosaic Virus and human pathogens. Front Microbiol 6:141370

Elmer W, Ma C, White J (2018) Nanoparticles for plant disease management. Curr Opin Environ Sci Health 6:66–70

El-Ramady HR, Verma KK, Rajput VD, Minkina T, Elbehiry F, Elbasiouny HY, Elsakhawy T, Omara AED, Amer MM (2022) Sources of silicon and nano-silicon in soil and plants. In: Silicon and nano-silicon in environmental stress management and crop quality improvement: progress and prospects. Academic, Elsevier. https://doi.org/10.1016/B978-0-323-91225-9.00003-0

El-Saadony MT, ALmoshadak AS, Shafi ME, Albaqami NM, Saad AM, El-Tahan AM, Desoky ESM, Elnahal ASM, Almakas A, Abd El-Mageed TA, Taha AE, Elrys AS, Helmy AM (2021) Vital roles of sustainable nano-fertilizers in improving plant quality and quantity-an updated review. Saudi J Biol Sci 28:7349–7359. https://doi.org/10.1016/j.sjbs.2021.08.032

El-Sawy, M. M., Elsharkawy, M. M., Abass, J. M., & Kasem, M. H. (2017). Antiviral activity of 2–nitromethyl phenol, zinc nanoparticles and seaweed extract against cucumber mosaic virus (CMV) in Eggplant. J Virol Antivir Res, 6(2)

Elsheery NI, Sunoj VSJ, Wen Y, Zhu JJ, Muralidharan G, Cao KF (2020) Foliar application of nanoparticles mitigates the chilling effect on photosynthesis and photoprotection in sugarcane. Plant Physiol Biochem 149:50–60

Erisman JW, Sutton MA, Galloway J, Klimont Z, Winiwarter W (2008) How a century of ammonia synthesis changed the world. Nat Geosci 1:636–639

Eroglu N, Emekci M, Athanassiou CG (2017) Applications of natural zeolites on agriculture and food production. J Sci Food Agric 97:3487–3499

Fang Y, Xiong L (2015) General mechanisms of drought response and their application in drought resistance improvement in plants. Cell Mol Life Sci 72:673–689

Fanoro OT, Parani S, Maluleke R, Lebepe TC, Varghese JR, Mavumengwana V, Oluwafemi OS (2021) Facile green, room-temperature synthesis of gold nanoparticles using Combretum erythrophyllum leaf extract: antibacterial and cell viability studies against normal and cancerous cells. Antibiotics (Basel) 10:893

Farooq, M., Wahid, A., Kobayashi, N. S. M. A., Fujita, D. B. S. M. A., & Basra, S. M. (2009). Plant drought stress: effects, mechanisms and management. In Sustainable agriculture (pp. 153-188). Dordrecht: Springer Netherlands.

Garg D, Sridhar K, Stephen Inbaraj B, Chawla P, Tripathi M, Sharma M (2023) Nano-biofertilizer formulations for agriculture: a systematic review on recent advances and prospective applications. Bioengineering 10:1–33. https://doi.org/10.3390/bioengineering10091010

Ghafariyan MH, Malakouti MJ, Dadpour MR, Stroeve P, Mahmoudi M (2013) Effects of magnetite nanoparticles on soybean chlorophyll. Environ Sci Technol 47:10645–10652

Ghiuță I, Cristea D, Croitoru C, Kost J, Wenkert R, Vyrides I, Anayiotos A, Munteanu D (2018) Characterization and antimicrobial activity of silver nanoparticles, biosynthesized using Bacillus species. Appl Surf Sci 438:66–73

Ghosh M, Jana A, Sinha S, Jothiramajayam M, Nag A, Chakraborty A, Mukherjee A, Mukherjee A (2016) Effects of ZnO nanoparticles in plants: cytotoxicity, genotoxicity, deregulation of antioxidant defenses, and cell-cycle arrest. Mutat Res Genet Toxicol Environ Mutagen 807:25–32

Gnanaprakasam P, Jeena SE, Premnath D, Selvaraju T (2016) Simple and robust green synthesis of Au NPs on reduced graphene oxide for the simultaneous detection of toxic heavy metal ions and bioremediation using bacterium as the scavenger. Electroanalysis 28:1885–1893

Gohari G, Mohammadi A, Akbari A, Panahirad S, Dadpour MR, Fotopoulos V, Kimura S (2020) Titanium dioxide nanoparticles (TiO2 NPs) promote growth and ameliorate salinity stress effects on essential oil profile and biochemical attributes of Dracocephalum moldavica. Sci Rep 10:912

Gómez-Morales J, Iafisco M, Delgado-López JM, Sarda S, Drouet C (2013) Progress on the preparation of nanocrystalline apatites and surface characterization: overview of fundamental and applied aspects. Prog Cryst Growth Charact Mater 59:1–46

Gopinath V, Velusamy P (2013) Extracellular biosynthesis of silver nanoparticles using Bacillus sp. GP-23 and evaluation of their antifungal activity towards Fusarium oxysporum. Spectrochim Acta A Mol Biomol Spectrosc 106:170–174

Goswami, R., Bhattacharyya, A., & Dutta, P. (2020). Nanotechnological approach for management of anthracnose and crown rot diseases of banana. 50(4). J Mycol Pl Pathol, 335–346

Gul M, Wakeel A, Steffens D, Lindberg S (2019) Potassium-induced decrease in cytosolic Na+ alleviates deleterious effects of salt stress on wheat (Triticum aestivum L.). Plant Biol 21:825–831

Guo H, White JC, Wang Z, Xing B (2018) Nano-enabled fertilizers to control the release and use efficiency of nutrients. Curr Opin Environ Sci Health 6:77–83

Gypser S, Freese D (2020) Phosphorus release from vivianite and hydroxyapatite by organic and inorganic compounds. Pedosphere 30:190–200

Haghighi M, Abolghasemi R, da Silva JAT (2014) Low and high temperature stress affect the growth characteristics of tomato in hydroponic culture with Se and nano-Se amendment. Sci Hortic 178:231–240

Hassan NS, Salah El Din TA, Hendawey MH, Borai IH, Mahdi AA (2018) Magnetite and zinc oxide nanoparticles alleviated heat stress in wheat plants. Curr Nanomater 3:32–43

Hassanisaadi M, Barani M, Rahdar A, Heidary M, Thysiadou A, Kyzas GZ (2022) Role of agrochemical-based nanomaterials in plants: biotic and abiotic stress with germination improvement of seeds. Plant Growth Regul 97:375–418

Heidari M, Salmanpour I, Ghorbani H, Asghari HR (2018) Iron chelate and Rhizobactria changed growth, grain yield, and physiological characteristics in maize. Sci Agric Bohem 49:245–254. https://doi.org/10.2478/sab-2018-0031

Hoagland H (1950) Stress and the adrenal cortex with special reference to potassium metabolism. Psychosomatic Medicine 12(3):142–148

Hulkoti NI, Taranath TC (2014) Biosynthesis of nanoparticles using microbes—a review. Colloids Surf B: Biointerfaces 121:474–483

Hussain N, Bilal M, Iqbal HMN (2022) Carbon-based nanomaterials with multipurpose attributes for water treatment: greening the 21st-century nanostructure materials deployment. Biomater Polymers Horizon 1:48–58

Hyman, P., Denyes, J. (2021). Bacteriophages in Nanotechnology: History and Future. In: Harper, D.R., Abedon, S.T., Burrowes, B.H., McConville, M.L. (eds) Bacteriophages. Springer, Cham. https://doi.org/10.1007/978-3-319-41986-2_22

Iqbal M, Raja NI, Mashwani Z-U-R, Hussain M, Ejaz M, Yasmeen F (2019) Effect of silver nanoparticles on growth of wheat under heat stress. Iran J Sci Technol Trans A: Sci 43:387–395

Iqbal T, Raza A, Zafar M, Afsheen S, Kebaili I, Alrobei H (2022) Plant-mediated green synthesis of zinc oxide nanoparticles for novel application to enhance the shelf life of tomatoes. Appl Nanosci 12:179–191

Iravani S, Korbekandi H, Mirmohammadi SV, Zolfaghari B (2014) Synthesis of silver nanoparticles: chemical, physical and biological methods. Res Pharm Sci 9:385–406

Itelima, J. U., Bang, W. J., Onyimba, I. A., & Oj, E. (2018). A review: biofertilizer; a key player in enhancing soil fertility and crop productivity. J Microbiol Biotechnol Rep 2(1):22–28

Jyoti, Rajak Manisha, Bawaskar Dnyaneshwar, Rathod Gauravi, Agarkar Dipali, Nagaonkar Aniket, Gade Mahendra, Rai (2017) Interaction of copper nanoparticles and an endophytic growth promoter Piriformospora indica with Cajanus cajan Abstract Journal of the Science of Food and Agriculture 97(13):4562–4570. https://doi.org/10.1002/jsfa.2017.97.issue-1310.1002/jsfa.8324

Kahrl F, Li Y, Su Y, Tennigkeit T, Wilkes A, Xu J (2010) Greenhouse gas emissions from nitrogen fertilizer use in China. Environ Sci Pol 13:688–694

Kaman PK, Dutta P (2019) Synthesis, characterization and antifungal activity of biosynthesized silver nanoparticle. Indian Phytopathol 72:79–88

Kandi V, Kandi S (2015) Antimicrobial properties of nanomolecules: potential candidates as antibiotics in the era of multi-drug resistance. Epidemiol Health 37:e2015020

Klein M, Poverenov E (2020) Natural biopolymer-based hydrogels for use in food and agriculture. J Sci Food Agric 100:2337–2347

Kottegoda N, Munaweera I, Madusanka N, Karunaratne V (2011) A green slow-release fertilizer composition based on urea-modified hydroxyapatite nanoparticles encapsulated wood. Curr Sci 101(1):73–78

Kottegoda N, Sandaruwan C, Priyadarshana G, Siriwardhana A, Rathnayake UA, Berugoda Arachchige DM, Kumarasinghe AR, Dahanayake D, Karunaratne V, Amaratunga GAJ (2017) Urea-hydroxyapatite nanohybrids for slow release of nitrogen. ACS Nano 11:1214–1221

Kumar S, Nehra M, Dilbaghi N, Marrazza G, Hassan AA, Kim K-H (2019) Nano-based smart pesticide formulations: emerging opportunities for agriculture. J Control Release 294:131–153

Kumar, P., Kumar, A., Kumar, R. (2021). Phytoremediation and Nanoremediation. In: Kumar, R., Kumar, R., Kaur, G. (eds) New Frontiers of Nanomaterials in Environmental Science. Springer, Singapore. https://doi.org/10.1007/978-981-15-9239-3_13

Kumar D, Mariyam S, Gupta KJ, Thiruvengadam M, Ghodake GS, Xing B, Seth CS (2024a) Comparative investigation on chemical and green synthesized titanium dioxide nanoparticles against chromium (VI) stress eliciting differential physiological, biochemical, and cellular attributes in Helianthus annuus L. Sci Total Environ 930:172413. https://doi.org/10.1016/j.scitotenv.2024.172413

Kumar D, Singh R, Upadhyay SK, Verma KK, Tripathi RM, Liu H, Dhankher OP, Tripathi RD, Sahi SV, Seth CS (2024b) Review on interactions between nanomaterials and phytohormones: novel perspectives and opportunities for mitigating environmental challenges. Plant Sci 340:111964. https://doi.org/10.1016/j.plantsci.2023.111964

Kumar D, Biswas JK, Mulla SI, Singh R, Shukla R, Ahanger MA, Shekhawat GS, Verma KK, Siddiqui MW, Seth CS (2024c) Micro and nanoplastics pollution: sources, distribution, uptake

in plants, toxicological effects, and innovative remediation strategies for environmental sustainability. Plant Physiol Biochem 213:108795. https://doi.org/10.1016/j.plaphy.2024.108795
Kumari R, Singh DP (2020) Nano-biofertilizer: an emerging eco-friendly approach for sustainable agriculture. Proc Natl Acad Sci India Sect B Biol Sci 90:733–741
Kumari B, Mallick MA, Solanki MK, Hora A, Mani M (2019) Applying nanotechnology to bacteria: an emerging technology for sustainable agriculture. In: Role of plant growth promoting microorganisms in sustainable agriculture and nanotechnology. Elsevier, pp 121–143
Kumari A, Lakshmi GA, Krishna GK, Patni B, Prakash S, Bhattacharyya M, Singh SK, Verma KK (2022) Climate change and its impact on crops: a comprehensive investigation for sustainable agriculture. Agronomy 12:3008. https://doi.org/10.3390/agronomy12123008
Lateef A, Nazir R, Jamil N, Alam S, Shah R, Khan MN, Saleem M (2016) Synthesis and characterization of zeolite based nano–composite: an environment friendly slow release fertilizer. Microporous Mesoporous Mater 232:174–183
Lin N, Wang C, Ding J, Su L, Xu L, Zhang B, Zhang Y, Fan J (2020) Efficacy of nanoparticle encapsulation on suppressing oxidation and enhancing antifungal activity of cyclic lipopeptides produced by Bacillus subtilis. Colloids Surf B: Biointerfaces 193:111143
Liu R, Lal R (2014) Synthetic apatite nanoparticles as a phosphorus fertilizer for soybean (Glycine max). Sci Rep 4:5686
Liu R, Lal R (2015) Potentials of engineered nanoparticles as fertilizers for increasing agronomic productions. Sci Total Environ 514:131–139
Lwin CS, Seo B-H, Kim H-U, Owens G, Kim K-R (2018) Application of soil amendments to contaminated soils for heavy metal immobilization and improved soil quality—a critical review. Soil Sci Plant Nutr 64:156–167
Maghsoodi MR, Ghodszad L, Lajayer BA (2020) Dilemma of hydroxyapatite nanoparticles as phosphorus fertilizer: potentials, challenges and effects on plants. Environ Technol Innov 19:100869
Mahmoud ME, Abdou AEH, Mohamed SMS, Osman MM (2016) Engineered staphylococcus aureus via immobilization on magnetic Fe_3O_4-phthalate nanoparticles for biosorption of divalent ions from aqueous solutions. J Environ Chem Eng 4:3810–3824
Malerba M, Cerana R (2016) Chitosan effects on plant systems. Int J Mol Sci 17:996
Malusa E, Vassilev N (2014) A contribution to set a legal framework for biofertilisers. Appl Microbiol Biotechnol 98:6599–6607
Manikandan A, Subramanian KS (2016) Evaluation of zeolite based nitrogen nano-fertilizers on maize growth, yield and quality on inceptisols and alfisols. Int J Plant Soil Sci 9:1–9
Mansoor S, Ali Wani O, Lone JK, Manhas S, Kour N, Alam P, Ahmad A, Ahmad P (2022) Reactive oxygen species in plants: from source to sink. Antioxidants (Basel) 11:225
Mariello M, Kim K, Wu K, Lacour SP, Leterrier Y (2022) Recent advances in encapsulation of flexible bioelectronic implants: materials, technologies, and characterization methods. Adv Mater 34:2201129
Mariyam S, Upadhyay SK, Chakraborty K, Verma KK, Duhan JS, Muneer S, Meena M, Sharma RK, Ghodake G, Seth CS (2024) Nanotechnology, a frontier in agricultural science, a novel approach in abiotic stress management and convergence with new age medicine-a review. Sci Total Environ 912:169097. https://doi.org/10.1016/j.scitotenv.2023.169097
Martel R, Schmidt T, Shea HR, Hertel T, Avouris P (1998) Single-and multi-wall carbon nanotube field-effect transistors. Appl Phys Lett 73:2447–2449
Mishra P, Kumari S, Saini P, Meena RK (2022) Role of nanotechnology in management of plant viral diseases. Mater Today Proc 69:1–10
Mohammadi R, Maali-Amiri R, Mantri NL (2014) Effect of TiO_2 nanoparticles on oxidative damage and antioxidant defense systems in chickpea seedlings during cold stress. Russ J Plant Physiol 61:768–775
Mokarram-Kashtiban S, Hosseini SM, Tabari Kouchaksaraei M, Younesi H (2019) The impact of nanoparticles zero-valent iron (nZVI) and rhizosphere microorganisms on the phytoremediation ability of white willow and its response. Environ Sci Pollut Res 26:10776–10789

Nadeem M, Anwar-ul-Haq M, Saqib M, Maqsood M, He Z (2022) Ameliorative effect of silicic acid and silicates on oxidative, osmotic stress, and specific ion toxicity in spring wheat (Triticum aestivum L.) genotypes. J Soil Sci Plant Nutr 22:2334–2345

Ningombam L, Mana T, Apum G, Ningthoujam R, Singh YD (2024) Nano-bioremediation: a prospective approach for environmental decontamination in focus to soil, water and heavy metals. Environ Nanotechnol Monit Manag 21(1):100931

Noha K, Bondok AM, El-Dougdoug KA (2018) Evaluation of silver nanoparticles as antiviral agent against ToMV and PVY in tomato plants. Sciences 8:100–111

Nongbet A, Mishra AK, Mohanta YK, Mahanta S, Ray MK, Khan M, Baek KH, Chakrabartty I (2022) Nanofertilizers: a smart and sustainable attribute to modern agriculture. Plan Theory 11:1–20. https://doi.org/10.3390/plants11192587

Ogunyemi SO, Zhang M, Abdallah Y, Ahmed T, Qiu W, Ali MA, Yan C, Li B (2020) The biosynthesis of three metal oxide nanoparticles (ZnO, MnO_2, and MgO) and their antibacterial activity against the bacterial leaf blight pathogen. Front Microbiol 11:588326

Oliveira HC, Gomes BCR, Pelegrino MT, Seabra AB (2016) Nitric oxide-releasing chitosan nanoparticles alleviate the effects of salt stress in maize plants. Nitric Oxide 61:10–19

Ondrasek G, Rathod S, Manohara KK, Gireesh C, Anantha MS, Sakhare AS, Parmar B, Yadav BK, Bandumula N, Raihan F (2022) Salt stress in plants and mitigation approaches. Plan Theory 11:717

Patel S, Bajpai AK, Bajpai J, Saini RK, Acharya S (2017) Facile preparation of iron loaded calcium alginate nanocarriers and study of controlled release of iron. J Environ Chem Eng 5:5337–5346

Pramanik P, Krishnan P, Maity A, Mridha N, Mukherjee A, Rai V (2020) Application of nanotechnology in agriculture. Environ Nanotechnol 4:317–348

Qi M, Liu Y, Li T (2013) Nano-$TiO_{(2)}$ improve the photosynthesis of tomato leaves under mild heat stress. Biol Trace Elem Res 156:323–328

Qin J, Tong Z, Zhan Y, Buisson C, Song F, He K, Nielsen-LeRoux C, Guo S (2020) A Bacillus thuringiensis chitin-binding protein is involved in insect peritrophic matrix adhesion and takes part in the infection process. Toxins (Basel) 12:252

Rahman KMA, Zhang D (2018) Effects of fertilizer broadcasting on the excessive use of inorganic fertilizers and environmental sustainability. Sustain For 10:759

Rai, V. , Acharya, S. and Dey, N. (2012) Implications of Nanobiosensors in Agriculture. Journal of Biomaterials and Nanobiotechnology, 3:315–324. https://doi.org/10.4236/jbnb.2012.322039

Rajput VD, Singh A, Minkina TM, Shende SS, Kumar P, Verma KK, Bauer T, Gorobtsova O, Deneva S, Sindireva A (2021) Potential applications of nanobiotechnology in plant nutrition and protection for sustainable agriculture. In: Ingle AP (ed) Nanotechnology in plant growth promotion and protection: recent advances and impacts, Ist edn. Wiley

Rajput, V. D., Minkina, T., Kumari, A., Shende, S. S., Ranjan, A., Faizan, M., ... & Kızılkaya, R. (2022). A review on nanobioremediation approaches for restoration of contaminated soil. Eurasian Journal of Soil Science, 11(1):43–60

Raliya R, Tarafdar JC (2013) ZnO nanoparticle biosynthesis and its effect on phosphorous-mobilizing enzyme secretion and gum contents in Clusterbean (Cyamopsis tetragonoloba L.). Agric Res 2:48–57

Raliya R, Saharan V, Dimkpa C, Biswas P (2017) Nanofertilizer for precision and sustainable agriculture: current state and future perspectives. J Agric Food Chem 66:6487–6503

Reátegui E, Reynolds E, Kasinkas L, Aggarwal A, Sadowsky MJ, Aksan A, Wackett LP (2012) Silica gel-encapsulated AtzA biocatalyst for atrazine biodegradation. Appl Microbiol Biotechnol 96:231–240

Rehman A, Ahmad T, Aadil RM, Spotti MJ, Bakry AM, Khan IM, Zhao L, Riaz T, Tong Q (2019) Pectin polymers as wall materials for the nano-encapsulation of bioactive compounds. Trends Food Sci Technol 90:35–46

Rico, C.M., Peralta-Videa, J.R., Gardea-Torresdey, J.L. (2015). Chemistry, Biochemistry of Nanoparticles, and Their Role in Antioxidant Defense System in Plants. In: Siddiqui, M.,

Al-Whaibi, M., Mohammad, F. (eds) Nanotechnology and Plant Sciences. Springer, Cham. https://doi.org/10.1007/978-3-319-14502-0_1
Riseh RS, Skorik YA, Thakur VK, Pour MM, Tamanadar E, Noghabi SS (2021) Encapsulation of plant biocontrol bacteria with alginate as a main polymer material. Int J Mol Sci 22. https://doi.org/10.3390/ijms222011165
Rodríguez-González V, Terashima C, Fujishima A (2019) Applications of photocatalytic titanium dioxide-based nanomaterials in sustainable agriculture. J Photochem Photobiol C: Photochem Rev 40:49–67
Rop K, Karuku GN, Mbui D, Michira I, Njomo N (2018) Formulation of slow release NPK fertilizer (cellulose-graft-poly (acrylamide)/nano-hydroxyapatite/soluble fertilizer) composite and evaluating its N mineralization potential. Ann Agric Sci 63:163–172
Rop K, Mbui D, Karuku GN, Michira I, Njomo N (2020) Characterization of water hyacinth cellulose-g-poly (ammonium acrylate-co-acrylic acid)/nano-hydroxyapatite polymer hydrogel composite for potential agricultural application. Results Chem 2:100020
Sami, F., Siddiqui, H., Hayat, S. (2020). Impact of Silver Nanoparticles on Plant Physiology: A Critical Review. In: Hayat, S., Pichtel, J., Faizan, M., Fariduddin, Q. (eds) Sustainable Agriculture Reviews 41. Sustainable Agriculture Reviews, vol 41. Springer, Cham. https://doi.org/10.1007/978-3-030-33996-8_6
Sangeetha C, Baskar P (2016) Zeolite and its potential uses in agriculture: a critical review. Agric Rev 37:101–108
Sayed EG, Desoukey SF, Desouky AF, Farag MF, El-kholy RI, Azoz SN (2024) Synergistic influence of arbuscular mycorrhizal fungi inoculation with nanoparticle foliar application enhances chili (Capsicum annuum L.) antioxidant enzymes, anatomical characteristics, and productivity under cold-stress conditions. Plan Theory 13:517
Sedghi M, Hadi M, Toluie SG (2013) Effect of nano zinc oxide on the germination parameters of soybean seeds under drought stress. Ann West Univ Timisoara Ser Biol 16:73
Seliem MK, Hafez Y, El-Ramady H (2020) Using nano-selenium in reducing the negative effects of high temperature stress on Chrysanthemum morifolium Ramat. J Sustain Agric Sci 46:47–60
Sharma B, Tiwari S, Kumawat KC, Cardinale M (2023) Nano-biofertilizers as bio-emerging strategies for sustainable agriculture development: potentiality and their limitations. Sci Total Environ 860:160476. https://doi.org/10.1016/j.scitotenv.2022.160476
Sharon, M., Choudhary, A. K., & Kumar, R. (2010). Nanotechnology in agricultural diseases and food safety. Journal of Phytology, 2(4): 83–92
Singh R, Manickam N, Mudiam MKR, Murthy RC, Misra V (2013) An integrated (nano-bio) technique for degradation of γ-HCH contaminated soil. J Hazard Mater 258:35–41
Song Y, Jiang M, Zhang H, Li R (2021) Zinc oxide nanoparticles alleviate chilling stress in rice (Oryza Sativa L.) by regulating antioxidative system and chilling response transcription factors. Molecules 26:2196
Stafiej A, Pyrzynska K (2007) Adsorption of heavy metal ions with carbon nanotubes. Sep Purif Technol 58:49–52
Sturikova H, Krystofova O, Huska D, Adam V (2018) Zinc, zinc nanoparticles and plants. J Hazard Mater 349:101–110
Swamy HM, Asokan R (2013) Bacillus thuringiensis as 'nanoparticles'-a perspective for crop protection. Nanosci Nanotechnol-Asia 3:102–105
Tahi H, Wahbi S, Wakrim R, Aganchich B, Serraj R, Centritto M (2007) Water relations, photosynthesis, growth and water-use efficiency in tomato plants subjected to partial rootzone drying and regulated deficit irrigation. Plant Biosyst 141:265–274
Tang G, Tian Y, Niu J, Tang J, Yang J, Gao Y, Chen X, Li X, Wang H, Cao Y (2021) Development of carrier-free self-assembled nanoparticles based on fenhexamid and polyhexamethylene biguanide for sustainable plant disease management. Green Chem 23:2531–2540
Tarafdar JC, Raliya R, Rathore I (2012) Microbial synthesis of phosphorous nanoparticle from tricalcium phosphate using Aspergillus tubingensis TFR-5. J Bionanosci 6:84–89

Taran NY, Gonchar OM, Lopatko KG, Batsmanova LM, Patyka MV, Volkogon MV (2014) The effect of colloidal solution of molybdenum nanoparticles on the microbial composition in rhizosphere of Cicer arietinum L. Nanoscale Res Lett 9:1–8

Torabian S, Zahedi M, Khoshgoftar AH (2016) Effects of foliar spray of two kinds of zinc oxide on the growth and ion concentration of sunflower cultivars under salt stress. J Plant Nutr 39:172–180

Vahdat A, Ghasemi B, Yousefpour M (2020) Mechanical properties of the hydroxyapatite and magnetic nanocomposite of hydroxyapatite adsorbents. S Afr J Chem Eng 33:90–94

Venkatachalam P, Priyanka N, Manikandan K, Ganeshbabu I, Indiraarulselvi P, Geetha N, Muralikrishna K, Bhattacharya RC, Tiwari M, Sharma N (2017) Enhanced plant growth promoting role of phycomolecules coated zinc oxide nanoparticles with P supplementation in cotton (Gossypium hirsutum L.). Plant Physiol Biochem 110:118–127

Verma KK, Song XP, Joshi A, Rajput VD, Singh M, Sharma A, Singh RK, Li DM, Arora J, Minkina T, Li YR (2022a) Nanofertilizer possibilities for healthy soil, water and food in future: an overview. Front Plant Sci 13:865048. https://doi.org/10.3389/fpls.2022.865048

Verma KK, Song XP, Joshi A, Tian DD, Rajput VD, Singh M, Arora J, Minkina TM, Li YR (2022b) Recent trends in nano-fertilizer for advancing sustainable agriculture under the era of climate change to ensure food security for future generations. Nano 12:173. https://doi.org/10.3390/nano12010173

Verma KK, Song X, Degu HD, Guo D, Joshi A, Huang H, Xu L, Singh M, Huang D, Rajput VD, Li Y (2023) Recent advances in nitrogen and nano-nitrogen fertilizers for sustainable crop production: a mini-review. Chem Biol Technol Agric 10:1–14. https://doi.org/10.1186/s40538-023-00488-3

Verma KK, Song XP, Kumari A, Jagadesh M, Singh SK, Bhatt R, Singh M, Seth CS, Li YR (2024a) Climate change adaptation: challenges for agricultural sustainability. Plant Cell Environ:1–12. https://doi.org/10.1111/pce.15078

Verma KK, Song X-P, Singh S, Kumari A, Arora J, Singh SK, Solanki MK, Seth CS, Li Y-R (2024b) Synergistic interactions of nanoparticles and plant growth promoting rhizobacteria enhancing soil-plant systems: a multigenerational perspective. Front Plant Sci 15:1376214. https://doi.org/10.3389/fpls.2024.1376214

Wang Y, Deng C, Cota-Ruiz K, Peralta-Videa JR, Sun Y, Rawat S, Tan W, Reyes A, Hernandez-Viezcas JA, Niu G (2020) Improvement of nutrient elements and allicin content in green onion (Allium fistulosum) plants exposed to CuO nanoparticles. Sci Total Environ 725:138387

Wang A, Li J, Al-Huqail AA, Al-Harbi MS, Ali EF, Wang J, Ding Z, Rekaby SA, Ghoneim AM, Eissa MA (2021) Mechanisms of chitosan nanoparticles in the regulation of cold stress resistance in banana plants. Nano 11:2670

Wei X, Luo M, Li W, Yang L, Liang X, Xu L, Kong P, Liu H (2012) Synthesis of silver nanoparticles by solar irradiation of cell-free bacillus amyloliquefaciens extracts and $AgNO_3$. Bioresour Technol 103:273–278

Wu H, Tito N, Giraldo JP (2017) Anionic cerium oxide nanoparticles protect plant photosynthesis from abiotic stress by scavenging reactive oxygen species. ACS Nano 11:11283–11297

Xu L, Zhu Z, Sun D-W (2021) Bioinspired nanomodification strategies: moving from chemical-based agrosystems to sustainable agriculture. ACS Nano 15:12655–12686

Yang X, Lu M, Wang Y, Wang Y, Liu Z, Chen S (2021) Response mechanism of plants to drought stress. Horticulturae 7:50

Yurtluk T, Akçay FA, Avcı A (2018) Biosynthesis of silver nanoparticles using novel Bacillus sp. SBT8. Prep Biochem Biotechnol 48:151–159

Yuvaraj M, Subramanian KS (2018) Development of slow release Zn fertilizer using nano-zeolite as carrier. J Plant Nutr 41:311–320

Zahra N, Hafeez MB, Shaukat K, Wahid A, Hussain S, Naseer R, Raza A, Iqbal S, Farooq M (2021) Hypoxia and anoxia stress: plant responses and tolerance mechanisms. J Agron Crop Sci 207:249–284

Zhao L, Lu L, Wang A, Zhang H, Huang M, Wu H, Xing B, Wang Z, Ji R (2020) Nano-biotechnology in agriculture: use of nanomaterials to promote plant growth and stress tolerance. J Agric Food Chem 68:1935–1947

Zhou D, Kim D-G, Ko S-O (2015) Heavy metal adsorption with biogenic manganese oxides generated by pseudomonas putida strain MnB1. J Ind Eng Chem 24:132–139

Zhuang H, Han H, Xu P, Hou B, Jia S, Wang D, Li K (2015) Biodegradation of quinoline by Streptomyces sp. N01 immobilized on bamboo carbon supported Fe_3O_4 nanoparticles. Biochem Eng J 99:44–47

Zia R, Nawaz MS, Siddique MJ, Hakim S, Imran A (2021) Plant survival under drought stress: implications, adaptive responses, and integrated rhizosphere management strategy for stress mitigation. Microbiol Res 242:126626

Zia-ur-Rehman M, Anayatullah S, Irfan E, Hussain SM, Rizwan M, Sohail MI, Jafir M, Ahmad T, Usman M, Alharby HF (2023) Nanoparticles assisted regulation of oxidative stress and antioxidant enzyme system in plants under salt stress: a review. Chemosphere 314:137649

Zulfiqar F, Navarro M, Ashraf M, Akram NA, Munné-Bosch S (2019) Nanofertilizer use for sustainable agriculture: advantages and limitations. Plant Sci 289:11027012

Zhao L, [illegible]
Zhao D, Kar D-G, [illegible]
Zhang [illegible]
[illegible]
[illegible]
[illegible]

Chapter 18
Nanobioformulations: Paving the Way for Sustainable Crop Management Beyond the Fields

Saburi Abimbola Atanda, Idorenyin Ugochi Nwaehujor, Israel Oluwasanmi Lawal, Ladipo Margaret Kudirat, and Foluso Oyedotun Agunbiade

18.1 Introduction

18.1.1 Overview of Current Challenges in Pre and Post-harvest Crop Management

Currently, the challenges in pre-harvest and post-harvest crop management include reducing post-harvest losses of agricultural produce while increasing production (Hashim et al. 2022). Sustainable agricultural practices are needed to meet feeding needs, reduce the negative impact of agriculture on the environment, and successfully yield crops under stress conditions (Violante and Pérez 2022; Verma et al. 2024a). Pathogenic attacks during post-harvest storage are a prime factor for post-harvest loss, and the inappropriate use of chemical pesticides has adverse impacts on environment and health (Bezabih et al. 2022; Jagadesh et al. 2024). The world's population growth, global climate change, and limited natural resources pose serious challenges to food security and availability in the near future (Kumari et al. 2022). Specific pre-harvest and post-harvest crop management challenges faced by

S. A. Atanda (✉) · I. U. Nwaehujor · I. O. Lawal
Perishable Crops Research Deaprtment Nigerian Stored Products Research Institute, Ilorin, Nigeria
e-mail: atandasa@nspri.gov.ng

L. M. Kudirat
Department of Chemical Sciences, Yaba College of Technology, Yaba, Nigeria

F. O. Agunbiade
Perishable Crops Research Deaprtment Nigerian Stored Products Research Institute, Ilorin, Nigeria

Deartment of Chemistry, University of Lagos, Akoka-Yaba, Nigeria

C. S. Seth et al. (eds.), *Nano-formulations for Sustainable Agriculture*,
Nanotechnology in the Life Sciences, https://doi.org/10.1007/978-3-031-90617-6_18

farmers in developing countries include limited capital and technology for food storage, processing, and distribution, which in turn lead to irredeemable losses (Demoule 2022). Additionally, up to 70% of food grains produced can be lost during post-harvest management activities such as storage, transport, milling, drying, and threshing (Hashim et al. 2022). Inappropriate collection, transport, storage, and pest control systems contribute to post-harvest losses of cereal grains in Ethiopia (Yimer 2022). Farmers in Kenya, Tanzania, Nigeria and other West African countries face challenges related to the adoption of new storage technologies, concerns about food safety, and liquidity constraints affecting storage and sales decisions (Channa 2019; Abdulbaki et al. 2023).

Climate change and environmental factors have significant impacts on both pre-harvest and post-harvest crop management practices; rising temperatures, changing rainfall patterns, and extreme weather events such as droughts and floods can reduce crop yields (Verma et al. 2020; Annie and Pal 2023). Prasad et al. (2023) explained that changing climatic conditions directly and indirectly affect the quality, maturity, and nutritional value of vegetables, which are important factors in post-harvest management. Additionally, climate change can also affect pest management in the post-harvest supply chain, leading to increased damage and contamination potential, as well as changes in species distribution and composition (Mditshwa et al. 2023). To mitigate the impacts of climate change, researchers have suggested implementing microclimatic modifications, conducting sensitivity analyses of crop simulation models, and adopting sustainable agricultural practices such as cover cropping, organic and inorganic fertilizers, and improved seed varieties (Gerken and Morrison 2022; Alhassan et al. 2023; Verma et al. 2023). These approaches can help farmers make informed decisions about crop management and adaptation measures, ultimately leading to increased productivity and resilience in the face of climate change (Fig. 18.1).

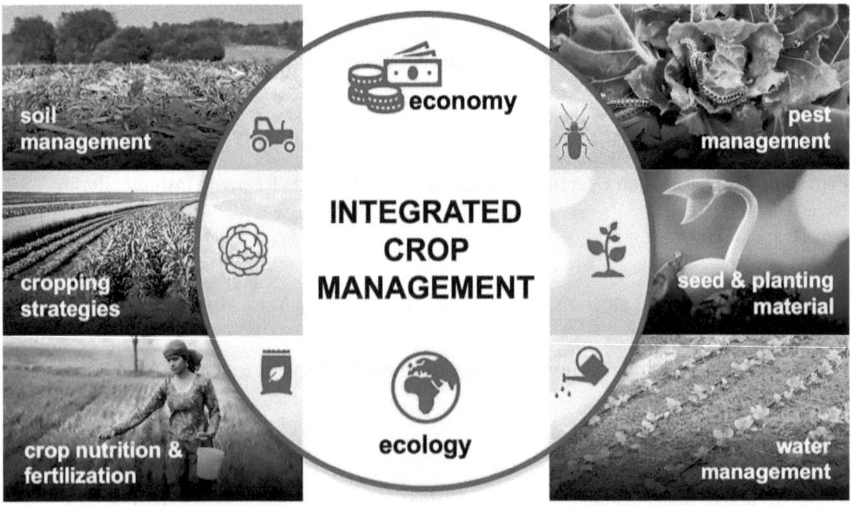

Fig. 18.1 An integrated approach considering socio-economic and environmental factors across the entire farm system (CAB International, formerly Commonwealth Agricultural Bureaux)

18.2 Nanobioformulations and Their Potential Applications in Agriculture

Nanobioformulations have emerged as a promising technology in agriculture, with potential applications in both pre-harvest and post-harvest stages of crop production. These formulations, prepared using plant growth-promoting microbes (PGPMs) and nanotechnology, offer eco-friendly and targeted solutions for sustainable agriculture. They can enhance functional activities, decipher the interaction between plants, microbes, pests, and pathogens, and improve crop yield and disease control (Verma et al. 2022a, 2024b). NBFs can also help in understanding microbial diversity, metabolic systems, and genetic potential of soil microbes, leading to the development of potent biological products like biocontrol agents, biostimulants, biopesticides, and biofertilizers (Kumari et al. 2023a, b). Nanotechnology can also be applied in seed storage, seed germination, promoting plant growth, enhancing stress tolerance, precision agriculture, and managing post-harvest fruits and vegetables (Rajput et al. 2023).

These applications of NBFs in agriculture offer a green alternative to conventional chemical fertilizers and synthetic pesticides, thereby promoting sustainable and environmental friendly practices (Bidyarani et al. 2022; Babu et al. 2022; Anguiano-Cabello et al. 2023; Bairwa et al. 2023). Nanotechnology holds immense potential in agriculture, enabling precision management through nanoscale sensors, targeted delivery of agrochemicals and enhanced pest and disease control via antimicrobial packaging, edible coating and films among others. NMs contribute to soil remediation to improve pre-harvest and post-harvest practices, enhancing crop quality and sustainability. Harnessing these advancements can revolutionize agricultural productivity, ensuring food security and environmental stewardship for future generations (El-Ramady et al. 2022) (Fig. 18.2).

18.3 Nanobioformulations: Concepts and Components

Nanobioformulations in the context of agriculture refer to the use of nanotechnology and biotechnology to develop formulations that enhance crop yield and promote sustainable agriculture. These formulations are prepared using PGPM such as *Pseudomonas*, *Trichoderma, Bacillus*, *Rhizobium*, and *Azotobacter spp*. (Bidyarani et al. 2022). Nanotechnology enables the delivery of agrochemical agents in the form of NPs, reducing the use of chemical-based pesticides and fertilizers (Mishra et al. 2022). The biosynthesis of NPs using microorganisms provides a safer and environmental friendly approach (Rajput et al. 2021; Ahmad et al. 2022). NBFs have the potential to improve agricultural productivity by utilizing limited resources and optimizing the allocation of water and agrochemicals through nanosensors and automated systems. These formulations can also be used as NFs, nano-fungicides, and nano-pesticides, offering enhanced nutrient uptake, disease management, and

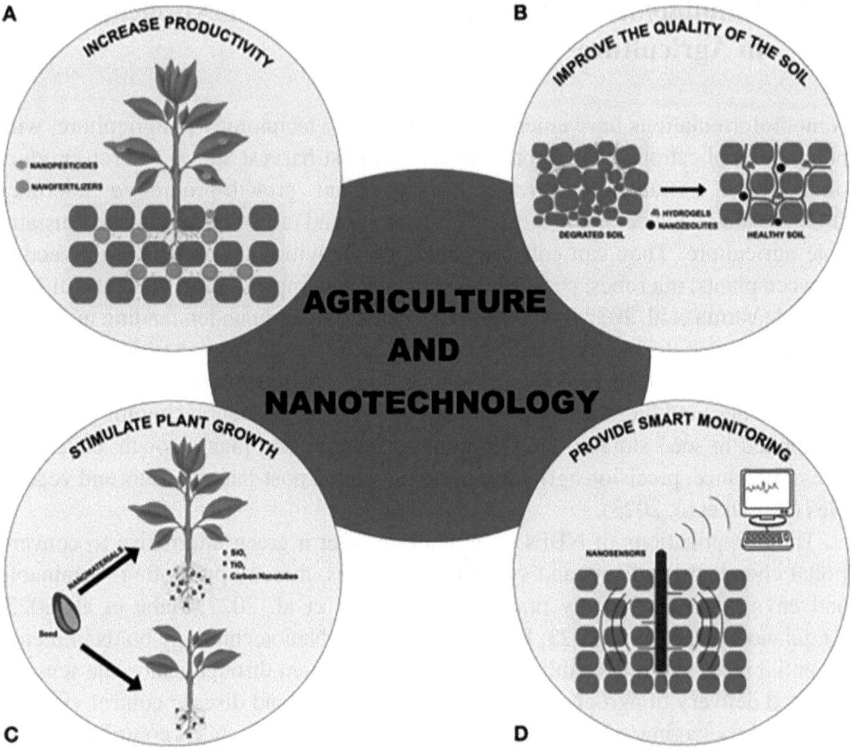

Fig. 18.2 Nanotechnology holds promise in agriculture across various applications: (**a**) Increasing productivity is possible with the use of nanopesticides and nanofertilizers. (**b**) Soil quality can be improved through the application of nanozeolites and hydrogels. (**c**) Nanomaterials, like SiO_2, TiO_2, and carbon nanotubes have the potential to stimulate plant growth. (**d**) Smart monitoring systems, facilitated by nanosensors and wireless communication devices, offer real-time insights into crop conditions (Fraceto et al. 2016; Kumar et al. 2024a)

reduced environmental impact (Nandini and Geetha 2021). Nanobiotechnology in agriculture aims to develop smart applications for crop improvement, including nano-diagnostics for rapid and precise detection of pathogens. Overall, NBFs contribute to sustainable agriculture by reducing chemical inputs, improving crop yield, and protecting the environment.

18.4 Key Components and Characteristics of Nanobioformulations

NBFs in the context of agriculture are composed of nano-sized formulations prepared from PGPM such as *Pseudomonas, Trichoderma, Bacillus*, *Rhizobium*, and *Azotobacter* spp. These formulations have the potential to improve functional activities in an eco-friendly and targeted manner, leading to sustainable agriculture. They help in deciphering the concept and mechanism of plant-microbe interactions, as well as understanding microbial diversity, metabolic systems, and genetic potential of soil microbes. NBFs offer a green alternative to conventional chemical-based formulations, promoting plant growth, enhancing crop yield, and controlling plant diseases while maintaining soil fertility (Bidyarani et al. 2022). They are part of the emerging field of nanobiotechnology, which combines the benefits of nanotechnology and green chemistry. NBFs have distinctive physical, chemical, and optical properties that make them more effective and efficient than their bulkier counterparts (Bairwa et al. 2023). These formulations are also used in various agricultural sectors, such as seed science, fertilizer industry, water management, and quality enhancement of agricultural products (Kumar et al. 2023). The use of microorganisms in the biosynthesis of NPs is a safer and eco-friendly approach, providing new dimensions of growth and production in sustainable agriculture (Mishra et al. 2023). Nanotechnology in agriculture aims to decrease the use of chemical fertilizers, improve nutrient management, and serve as a sensor to track soil quality (Ahmad et al. 2023).

18.5 Overview of Nanomaterials Used in Crop Management

Nanomaterials (NMs) used in crop management include nano-pesticides, nano-herbicides, nano-fertilizers, and different nanoforms of agrochemicals (Tripathi et al. 2023). Zinc, iron, and silicon NPs are also used in plant systems for nutrient acquisition under salinity and drought stress (Tighe-Neira et al. 2023). NMs, such as NPs, nanocomposites, carbon nanotubes, and quantum dots have been utilized in developing nanosensors for the detection of plant pathogens (Farooq et al. 2023). NMs with antiviral, antifungal, and antibacterial properties have been used for pest and disease management in agriculture (Vinay et al. 2023). Metal NPs synthesized from chemical and biological methods have been used to improve plant abiotic stress tolerance, particularly in response to high temperatures. (Kumar et al. 2024c). Overall, NMs have the potential to enhance crop productivity, nutrient acquisition, disease management, and stress tolerance in agriculture (Fig. 18.3).

18.6 Types of Nanomaterials and Their Properties

In agricultural practices, a variety of NMs are utilized for their distinct characteristics and benefits (Fig. 18.4). These include the following:

- Nano-pesticides: These are nano-sized formulations of conventional pesticides. They offer improved efficacy, targeted delivery, reduced environmental impact, and controlled release properties compared to their macro-scale counterparts (Kah et al. 2018).

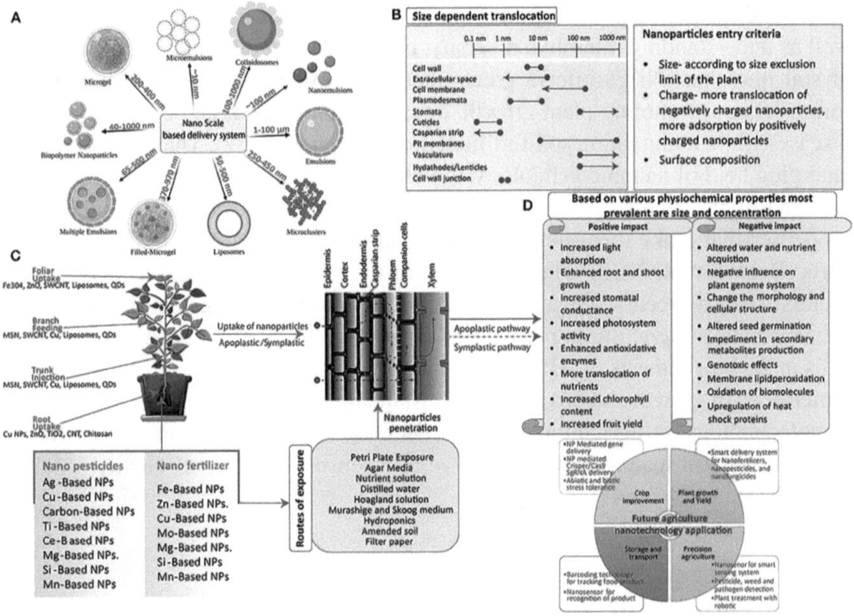

Fig. 18.3 The comprehensive overview of NMs applications in crop management

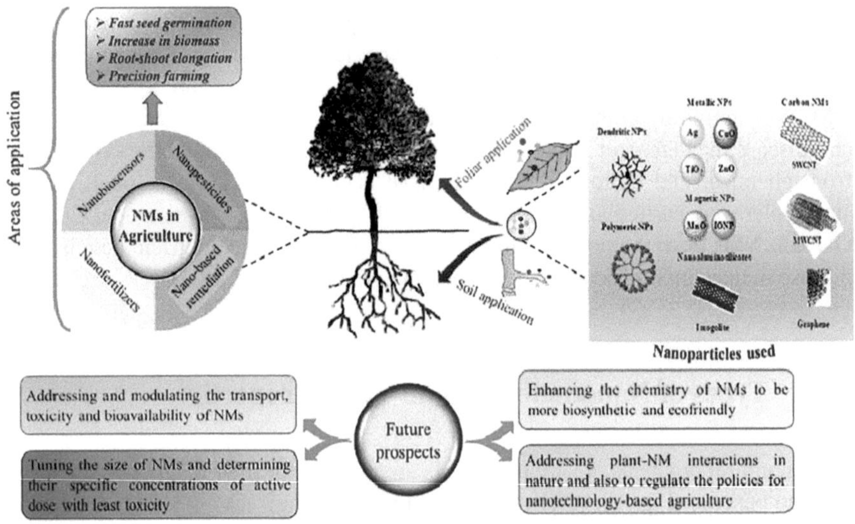

Fig. 18.4 Various NMs for agricultural nanotechnology and routes of application (Vibala et al. 2021)

- Nano-fertilizers: These are nanostructured materials used to deliver nutrients to plants more efficiently. They enhance nutrient uptake, reduce leaching and run-off, and can be tailored for slow-release, thus improving fertilizer use efficiency (Kah et al. 2018).
- Nano-zeolites and Hydrogels: These NMs are used to improve soil quality and water retention capacity. Nano-zeolites can adsorb and release nutrients, while hydrogels can retain water and nutrients in soil, reducing irrigation requirements and enhancing plant growth (Malhotra and Dolui 2019).
- Nanomaterials for Plant Growth Stimulation: Materials like silicon dioxide (SiO_2), titanium dioxide (TiO_2), and carbon nanotubes (CNTs) are used to stimulate plant growth. They can enhance photosynthesis, nutrient absorption, and stress tolerance in plants (Khodakovskaya et al. 2017).
- Nano-sensors: These are nano-sized sensors used for real-time monitoring of various parameters in soil, water, and plants. Nano-sensors provide data on nutrient levels, moisture content, pH, and pest infestations, enabling precise management practices (Malhotra and Dolui 2019).

These NMs offer several advantages in agriculture, including increased productivity, resource efficiency, and environmental sustainability.

18.7 Pre-harvest Applications of Nanobioformulations

Pre-harvest applications of NBFs in agriculture offer promising solutions to enhance crop yield, quality, and resilience. These formulations, typically composed of nano-sized materials encapsulating bioactive compounds or beneficial microorganisms, are applied before harvest to promote plant growth, mitigate stressors, and improve overall crop health (Fig. 18.5).

18.7.1 Crop Growth and Development

NMs currently being used for crop growth and development in agriculture include NFs, nano-insecticides/fungicides, nano-fungicides, nano-herbicides, carbon-based NMs, such as functionalized and nonfunctionalized, fullerene, and carbon nanotubes (CNTs). These NMs have shown various benefits, such as improved efficiency, effective control of pests and diseases, increased water absorption and retention, enhanced plant physiological and metabolic activities, and increased fruit yield (Jha and Yadav 2023. NMs also have the potential to reduce the dependence on pesticides and fertilizers, improve soil quality, and increase agricultural productivity (Shende et al. 2023; Kumar et al. 2024b). Additionally, nanobiosensors have been developed for the timely detection and remediation of phytopathogens, biotic and abiotic stressors. The use of nanotechnology in agriculture has the potential to

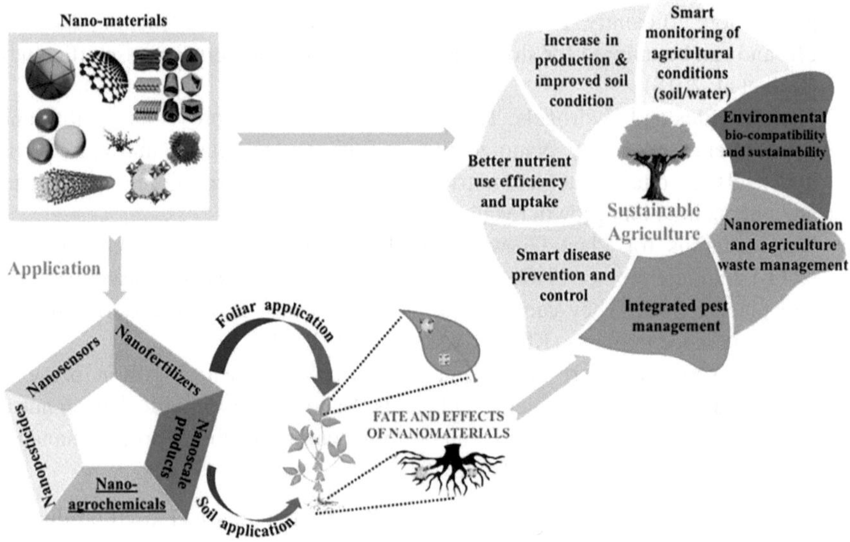

Fig. 18.5 Exploring innovative applications of nanotechnology in traditional agricultural practices (Singh et al. 2021)

revolutionize the industry by providing precise and sustainable solutions for crop production and environmental sustainability (Fig. 18.6).

18.7.2 Enhancing Seed Germination and Early Growth Stages

The application of NBFs offers a multifaceted approach to bolstering seed germination rates and nurturing the delicate early growth stages of plants. From facilitating nutrient uptake and water retention to fortifying plants against environmental stressors, the benefits are manifold (Fig. 18.7). NMs such as magnesium oxide (nMgO) and hydroxyapatite (nHAp) at optimal application rates of 100 mg L^{-1} have been found to enhance seed germination and early growth stages in water spinach and Hami melon (Wu et al. 2023; Mariyam et al. 2024). OH-functionalised multiwalled carbon nanotubes (OH-MWCNTs) at concentrations of 25–100 mg L^{-1} have shown to boost germination attributes in *Pennisetum* seedlings (Sumita and Kachhwaha 2023). Silicon dioxide NMs ($nSiO_2$) and silicon carbide NMs (nSiC) at concentrations of 10–200 mg/L have been found to increase shoot and root length of rice seedlings, as well as improve above-ground catalase activity and trace element content (Venkateswaran et al. 2023). Cobalt sulphide nanocomposites have been shown to enhance germination, shoot and root length, and germination time in *Trigonellafoenum-graecum* and *Brassica nigra* seeds (Jiang et al. 2022).

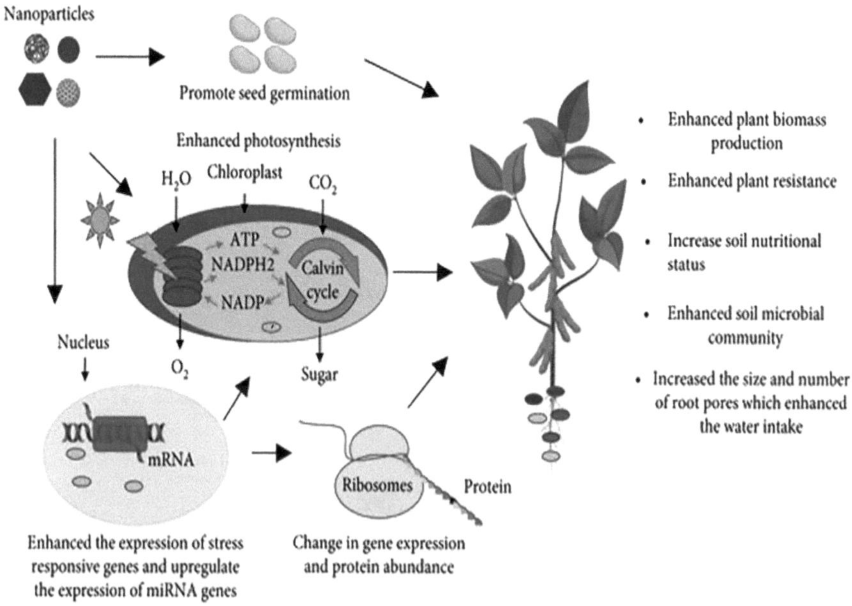

Fig. 18.6 The beneficial impact of NPs on plant growth and development: optimized concentrations enhance seed germination, photosynthesis, and stress tolerance through alterations in gene expression (Ali et al. 2021)

18.7.3 Improving Nutrient Uptake and Utilization

To achieve sustainable agricultural practices and enhance crop yields, it is crucial to prioritize the optimization of nutrient uptake and utilization. This entails guaranteeing that crops have access to a well-balanced diet of vital nutrients like nitrogen, phosphorus, and potassium, as well as micronutrients such as zinc and iron, which are essential for promoting healthy plant growth and development. Effective nutrient management can also contribute to the reduction of chemical fertilizer usage, improvement of soil health, and mitigation of environmental pollution risks. Moreover, it is imperative to take into account factors like soil type, climate, and irrigation practices when formulating a nutrient management plan to ensure that crops receive the necessary nutrients for their optimal growth (Fig. 18.8).

NMs used in improving nutrient uptake and utilization in agriculture include NFs, nano-insecticides, nano-fungicides, and nano-herbicides (Ain et al. 2023; Kumar et al. 2024a). NFs have been shown to improve nutrient use efficiency, increase crop yield, and reduce the excessive use of chemical fertilizers (Sun et al. 2022). They are designed to deliver macro and/or micronutrients to plants in a controlled release manner, allowing for targeted distribution and gradual release of nutrients. NMs can be applied to plants through seed treatment, foliar application, or soil application (Al-Antary and Ghidan 2023). Additionally, nanotechnology has the potential to enhance the capacity of plants to absorb nutrients and protect against

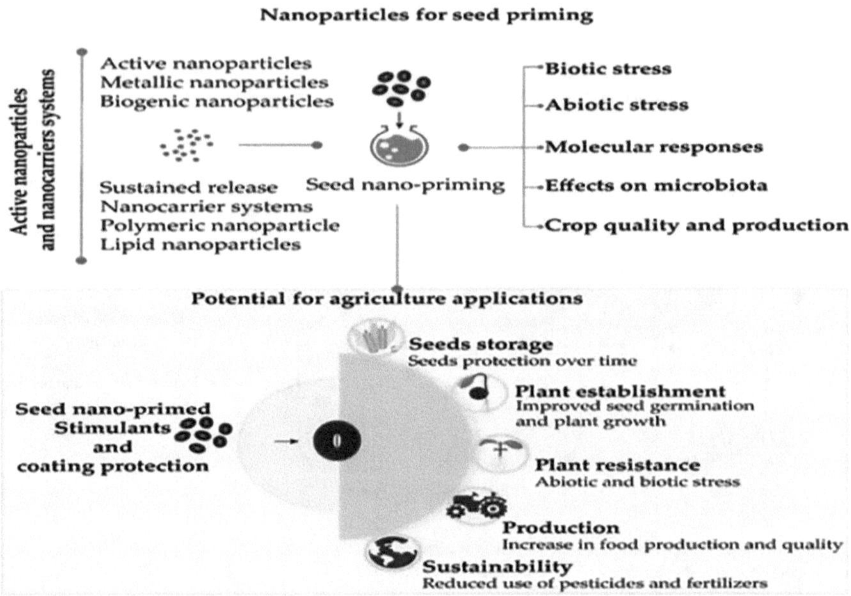

Fig. 18.7 Seed nano-priming offers numerous advantages for sustainable agriculture. Besides safeguarding seeds during storage, it enhances plant establishment in soil, reducing fertilizer requirements. Accelerated plant growth improves competition with weeds, boosting productivity and food quality. Furthermore, it enhances resilience to both abiotic and biotic stresses, potentially reducing pesticide use (Santo Pereira et al. 2021; Choudhary et al. 2024)

insect pests and microbial diseases (Mondal 2022). Nanotechnology in agriculture also includes the use of nano tools for rapid disease diagnostics and monitoring soil quality. Overall, NMs offer a promising solution for improving nutrient uptake and utilization in agriculture, leading to increased productivity and reduced environmental capacity.

18.7.4 Pest and Disease Management

Pest and disease management is a crucial aspect of agriculture that involves the use of various methods to control and prevent infestations in crops and livestock. Implementing effective pest and disease management practices is crucial for achieving healthy, productive harvests while reducing environmental harm. With the growing demand for sustainable and organic farming, it is important to implement pest and disease management strategies that are safe, efficient, and environmental—friendly. In this context, it is crucial to stay informed about the latest developments and innovations in pest and disease management, and to adopt best practices that can help farmers and growers achieve optimal results (Fig. 18.9).

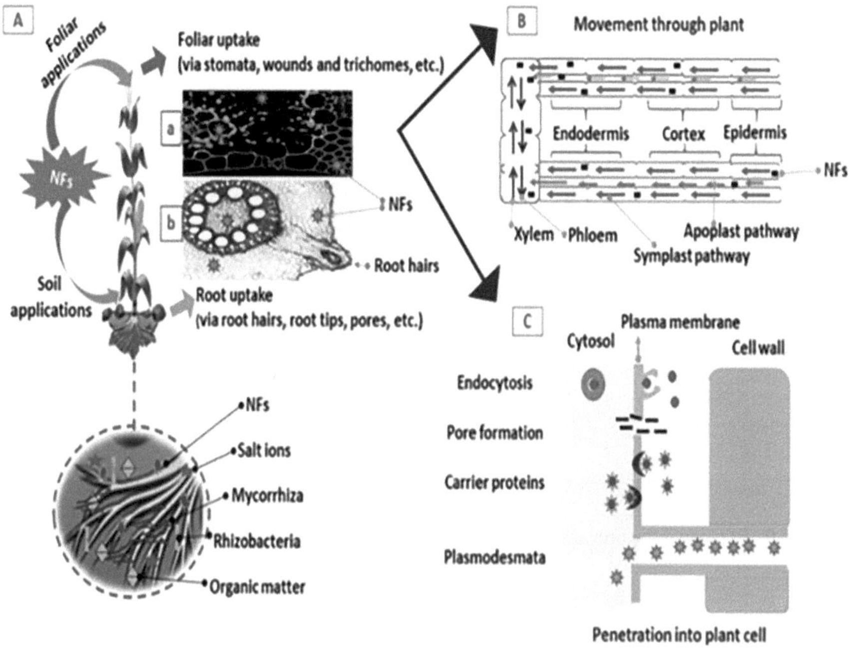

Fig. 18.8 The schematic depiction illustrates the uptake of NFs through various channels and their translocation pathways across different sections of plants. (**A**) NF characteristics influence absorption and translocation in plants, as demonstrated in transverse sections of maize leaves (a) and roots (b), sourced from publicly accessible databases. (**B**) NFs may utilize both apoplastic and symplastic pathways for upward and downward movement. (**C**) Various strategies have been proposed for the internal distribution of NFs within cells, involving endocytosis and pore formation facilitated by carrier proteins through plasmodesmata (Verma et al. 2022b)

NMs, like nanoparticles, nanocomposites, carbon nanotubes, and quantum dots have been used for the detection and management of pests and diseases in agriculture (Dutta et al. 2023). These NMs have unique bio-physiochemical and electro-optical properties that allow for the development of highly selective and sensitive nano-sensors that functions precisely to detect plant pathogens (Farooq et al. 2023). Additionally, NMs with antiviral, antifungal, and antibacterial properties have been directly used for pest and disease management in agriculture. NMs have also been employed as carriers of antipathogen molecules for crop disease management (Al-Antary and Ghidan 2023). Furthermore, NMs such as NFs, nanoinsecticides, nanofungicides, and nanoherbicides have been used for pest and disease management in agriculture (Cherian and Sheela 2023). These NMs improve efficiency and resource use efficiency in insect/pest and disease control, nutrient management, and weed control.

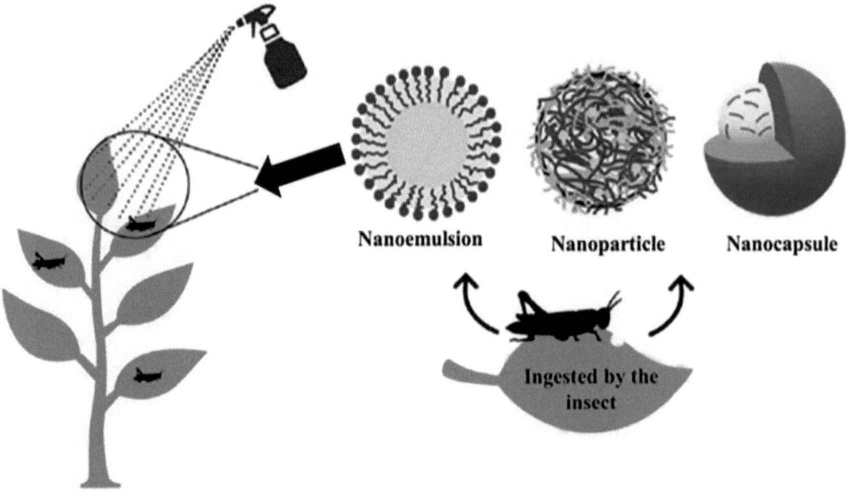

Fig. 18.9 Exploring the prevalent nanosystems in insect pest control (Pavoni et al. 2019)

18.7.5 Nanobioformulations for Eco-Friendly Pest Control

NMs used in NBFs for eco-friendly pest control in agriculture include metallic NPs, synthetic and natural polymers, nanoemulsions, lipid NPs, nanofibers, nanohydroxyapatite, nanoclays, chitosan NPs, mesoporous silica NPs, and amorphous calcium phosphate (Baliyarsingh and Pradhan 2023; Fincheira et al. 2023; Al-Khayri et al. 2023; Sabrine and Tarek 2023; Padmakumar et al. 2023). These NMs are utilized as active ingredients in the development of nanobiopesticides, which are effective alternatives to conventional chemical pesticides. Metallic NPs, particularly silver NPs, have shown success in controlling insects of stored food products. Nanocarriers, such as synthetic and natural polymers, lipid NPs, and nanofibers, are used as controlled release systems for agrochemicals, improving their efficacy and reducing environmental pollution. NFs, including nanohydroxyapatite, nanoclays, chitosan NPs, mesoporous silica NPs, and amorphous calcium phosphate, enhance nutrient release and have site-specific action. Bacterial biocontrol agents (BBCAs) are also used in the synthesis of nano-biopesticides, known as bacteria premised nanobiopesticides (B-NBPs), which are efficient alternatives to agrochemicals. These B-NBPs can be synthesized using high-pressure homogenization, jet milling, and hammer milling.

18.7.6 Disease Resistance and Prevention Using Nanotechnology

NMs used in disease resistance and prevention in agriculture include NPs of oxide, silver, and titanium oxide (Prakash et al. 2023). These NPs have been tested against *Macrophomina phaseolina*, a soil-borne infectious agent, and silver NPs showed a higher antifungal effect compared to other metallic elements and titanic oxide NPs. Nanotechnology has also been used to develop NFs, nanoinsecticides, nanofungicides, and nanoherbicides for effective control of insects, fungal diseases, and weeds. NMs can be applied to plants through seed treatment, foliar, or soil application (Dong et al. 2022). Additionally, NMs have the potential to induce disease resistance genes, deliver hormones and biomolecules, and obtain transgenic plants to resist plant diseases. Nanotools such as biosensors are being developed for the timely remediation of phytopathogens, biotic, and abiotic stressors (Choudhary et al. 2025). Overall, nanotechnology offers promising solutions for disease resistance and prevention in agriculture using various NMs.

18.8 Abiotic Stress Tolerance

Abiotic stress tolerance in agriculture refers to the ability of plants to withstand and adapt to non-living environmental factors that can adversely affect their growth, development, and productivity (Fig. 18.10). These stressors include extremes in temperature, water availability (drought or excess water), salinity, heavy metals, and various pollutants. Efforts to enhance abiotic stress tolerance in crops are crucial for ensuring food security and agricultural sustainability in the face of climate change and environmental degradation. Through breeding programs, genetic engineering, and agronomic practices, researchers and farmers aim to develop resilient crop varieties capable of thriving in diverse and challenging environments. NMs, i.e., zinc nanoparticles (ZnO NPs), iron nanoparticles (Fe_2O_3 NPs), silicon nanoparticles (SiO_2 NPs), cerium nanoparticles (CeO_2 NPs), silver nanoparticles (Ag NPs), titanium dioxide nanoparticles (TiO_2 NPs), and carbon nanoparticles (C NPs) have been used for abiotic stress tolerance in agriculture.

These NPs have been shown to improve plant physiology and metabolic pathways under environmental stresses, enhance plant growth and development, and increase antioxidant enzyme activity in plants (Sharma and Singh 2023). NPs synthesized through various approaches have been found to strengthen physical barriers, improve plant photosynthesis, activate plant defense mechanisms, and upregulate the expression of stress-related genes (Azameti and Imoro 2023). Additionally, they have the potential to alleviate nutrient inadequacy in soil, improve plant nutritional value, and boost crop productivity. The use of NPs as NFs has shown promise in managing abiotic stresses and increasing plant productivity (Yasmeen et al. 2023; Nawaz et al. 2023).

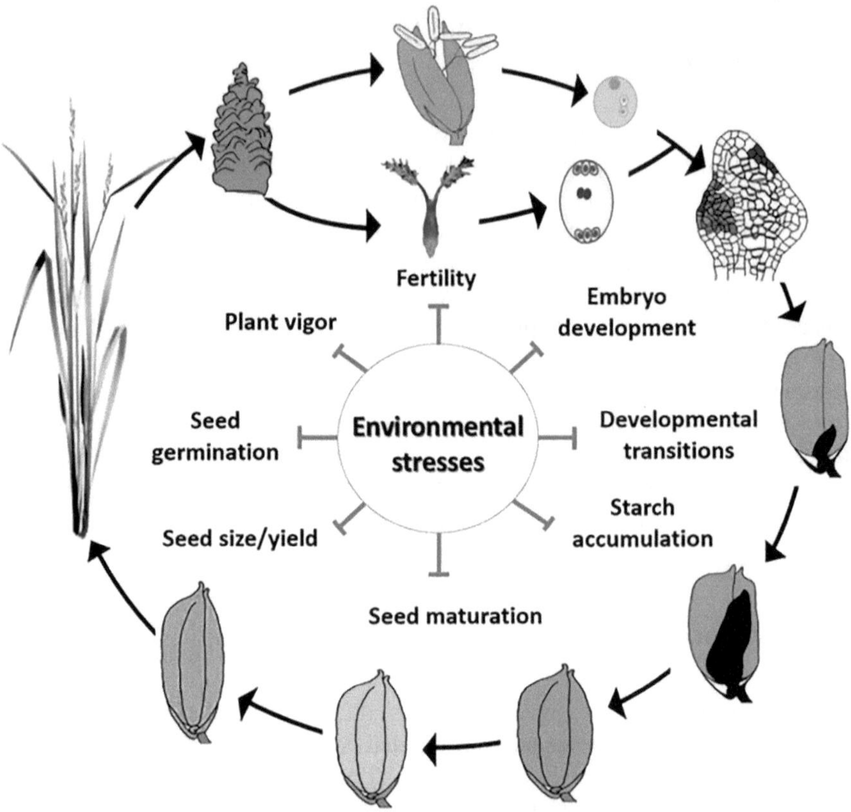

Fig. 18.10 The impact of environmental stressors on plant health, showcasing physiological, biochemical, and molecular responses to challenges such as drought, salinity, temperature extremes, and pollutants (BegcyLab)

18.8.1 Improving Resilience to Adverse Weather Conditions

NMs improve resilience to adverse weather conditions in agriculture by enhancing stress tolerance in crops, improving the efficiency of water usage and increasing plant growth and development (Weisany and Khosropour 2023). NPs stimulate the activity of enzymes, increase chlorophyll content, and enhance photosynthesis efficiency, thereby promoting stress tolerance and improving crop yield and quality (El-Saadony et al. 2022). Additionally, nanoscale agrochemicals, such as nanopesticides, nanoherbicides, and NFs, can help combat nutrient deficiencies, control plant pathogens, and reduce overall application rates, minimizing environmental pollution (Ma et al. 2023). NMs also have the potential to remediate heavy metals and organic pollutants in soil, improve crop immune responses, and strengthen plant immunity and growth (Campos et al. 2022). Furthermore, NMs offer a wider specific surface area to fertilizers and pesticides, facilitating site-targeted controlled delivery of nutrients and pesticides for increased crop protection (Qayyum et al. 2022). The integration of nanotechnology into nano-sensors has also increased their potential to sense and identify environmental conditions or impairments, supporting the development of high-tech agricultural farms.

18.9 Post-harvest Applications of Nanobioformulations

The post-harvest period presents a critical phase in agricultural production, where the quality and shelf life of harvested crops are vulnerable to deterioration due to various factors such as microbial spoilage, physiological changes, and environmental stresses. In recent years, the application of nanotechnology in agriculture has garnered significant attention, offering novel solutions to enhance post-harvest management and prolong the freshness and quality of harvested product.

18.9.1 Role of Nanobioformulations in Preserving Freshness and Quality

NBFs play a crucial role in preserving the freshness and quality of crops in agriculture. They can extend the shelf life of horticultural products by controlling the growth and development of microorganisms (Vasudevan and Arumugam 2023) NBFs prepared from PGPM can enhance crop yield and adapt to harsh climatic conditions (Halake and Haro 2022). NBFs also improve the qualities of vegetable crops by using NPs of essential nutrients as fertilizers or stimulants, making them more enriched with major nutrients and disease-free (Bidyarani et al. 2022). Additionally, NBFs can reduce the quantity of chemical fertilizers used in agriculture, minimize nutrient losses, and improve pest and nutrient management (Rani 2022). They can also serve as sensors to track soil quality and preserve the health of agricultural plants (Ahmad et al. 2023). Overall, NBFs offer a green and eco-friendly approach to enhance crop yield, control diseases, and improve the overall quality of crops in agriculture (Fig. 18.11).

18.9.2 Food Safety and Quality Assurance

Nanobioformulations have emerged as a promising frontier in the realm of food safety and quality assurance, offering innovative approaches to address challenges across the food supply chain. By integrating nanotechnology with biological components, these formulations present a versatile toolkit for enhancing food safety, extending shelf life, and preserving nutritional content (Fig. 18.12). NBFs have a

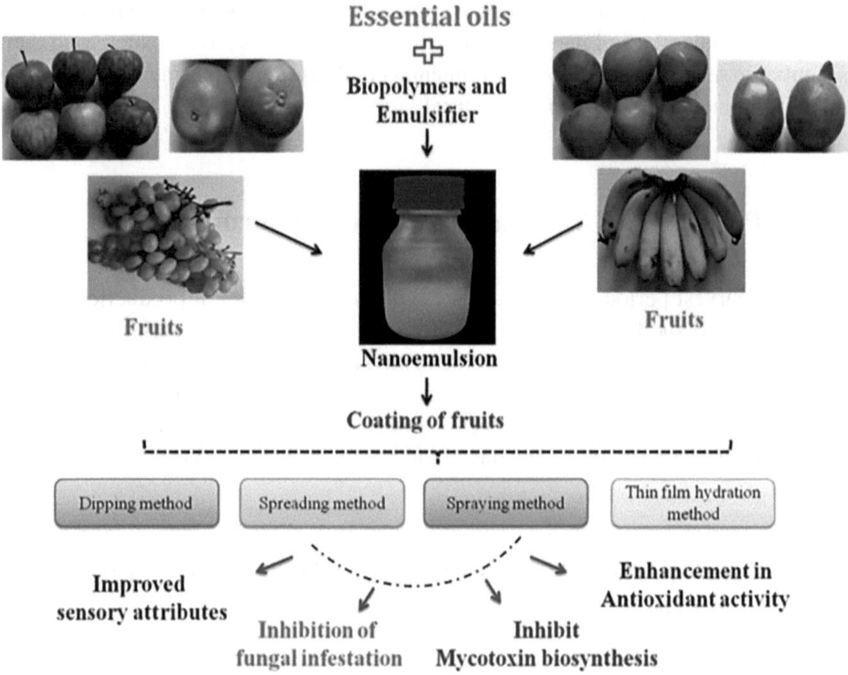

Fig. 18.11 Various methods for the application of essential oil nanoemulsion coatings on fruits and their practical implications in maintaining fruit quality. Illustrating techniques such as dipping, spraying, and vacuum impregnation, alongside the benefits of enhanced shelf life, reduced microbial growth, and preservation of nutritional attributes through nanoemulsion coatings (Das et al. 2021)

significant impact on food safety and quality assurance in agriculture. Nanotechnology offers innovative techniques such as nanosensors, NMs, nanocomposites, nanoencapsulation, and nanocoatings to ensure high-quality food products throughout the manufacturing, processing, and packaging stages (Britto et al. 2023). Nanosensors can detect carcinogens, spoilage, pathogens, and environmental changes that could lead to food degradation (Srivastava et al. 2023).

NMs, such as NPs, are used in antimicrobial packaging to maintain food stability and inhibit microbial biofilm formation (Ahmad et al. 2023). Nanoencapsulation of bioactive compounds enhances the shelf life of foods by reducing degradation (Muthukumar et al. 2023). Nanocoatings detect storage contaminants and maintain the freshness of food even after packaging is unsealed. NBFs, including NFs, nanofungicides, and nanopesticides, improve plant disease management and agricultural productivity. NPs-based sensors are used for the detection of pathogens and disease indicators. Nanotechnology has the potential to address challenges related to food quality, safety, and agricultural productivity in the face of climate change.

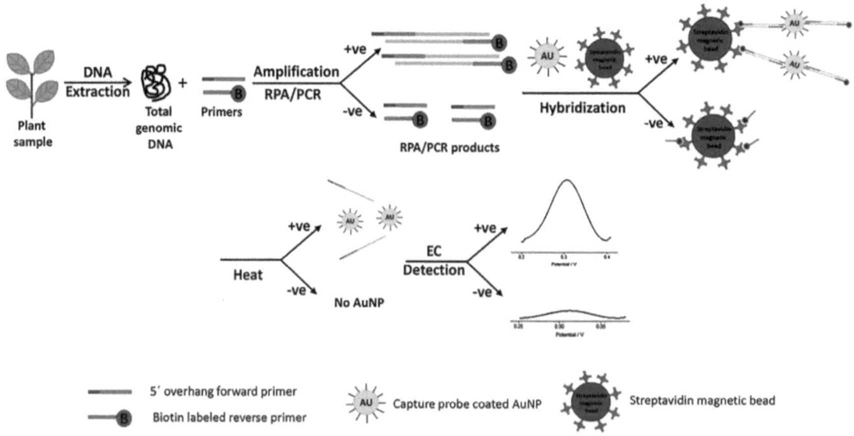

Fig. 18.12 A schematic representation of a DNA-based electrochemical bioassay designed for plant pathogen detection. Derived from and adapted with permission, the assay utilizes electrochemical detection (EC) and incorporates Au NPs for enhanced sensitivity and specificity (Lau et al. 2017)

18.10 Challenges and Future Prospects

The utilization of NBFs in agriculture presents a dual landscape of challenges and future prospects. This dynamic field holds tremendous potential for addressing key agricultural issues, yet it also confronts hurdles in implementation and development. Exploring these challenges and envisioning future possibilities is crucial for unlocking the transformative power of NBFs in enhancing crop production, sustainability, and food security.

18.10.1 Challenges in Nanobioformulations

Nanobioformulations represent a promising frontier in agricultural innovation, leveraging nanotechnology's capabilities to address pressing challenges in crop production and management. However, despite their potential, the field of NBFs in agriculture faces several current challenges. These obstacles span from regulatory concerns to practical limitations in application and efficacy. Understanding and overcoming these challenges is critical to harnessing the full potential of NBFs for sustainable agriculture. This introduction will delve into some of the key challenges currently encountered in the implementation of NBFs within agricultural systems.

18.10.2 Regulatory Concerns and Safety Issues

Nanobioformulations in crop pre and post-harvest management raise regulatory concerns and safety issues. The use of nanotechnology in agriculture, including nanobiopesticides and NFs, requires careful assessment to ensure consumer and environmental safety (Kumari et al. 2023a, b). Regulatory frameworks in various countries are being examined to determine their suitability for handling nanotechnologies in agriculture, feed, and food (Padmakumar et al. 2023). Nanobiopesticides, such as bacteria premised nanobiopesticides (B-NBPs), offer efficient alternatives to agrochemicals, but their efficacy and safety need to be addressed (Babu et al. 2022). Nanotechnology can also be used to address post-harvest losses by enhancing agricultural productivity and mitigating environmental issues (Mude et al. 2022). NMs can be used as intelligent delivery systems, sensors, and fertilizers in sustainable agriculture, but their safety and efficacy must be evaluated (Ghosh et al. 2022). The use of NMs in crop post-harvest management, such as NFs and nanopesticides, requires regulation and safety measures to ensure their safe use.

18.10.3 Ethical Considerations in the Use of nanotechnology in Agriculture

Nanotechnology has the potential to enhance crop pre and post-harvest management, but it also raises ethical concerns. The use of NMs in agriculture can improve seed storage, plant growth, and crop productivity, as well as mitigate the environmental issues caused by conventional chemical fertilizers (Srivastava 2023). However, the ethical implications of using nanotechnology in agriculture need to be considered. The development of NBFs for crop management raise questions about the safety and potential risks associated with the use of engineered nanoagroparticles (Engelmann et al. 2022). Additionally, the ethical acceptability of nanotechnology in agriculture should be evaluated in terms of sustainability, responsibility, and respect for fundamental rights (Babu et al. 2022). It is important to ensure that nanotechnology applications in agriculture are safe, ethical, and contribute to sustainable development. Further research and analysis are needed to fully understand and address the ethical implications of using nanotechnology in crop pre and post-harvest management (Ferreira and Filipe 2022).

18.11 Future Prospects and Research Directions

The integration of NBFs with sensing technologies holds promise for precision agriculture. Real-time monitoring coupled with targeted delivery of inputs could revolutionize crop management, optimizing resource use and enhancing yields sustainably. NBFs offer a pathway to more sustainable crop protection by bolstering the efficacy of biological control agents. Encapsulation and delivery systems can stabilize biopesticides, paving the way for eco-friendly pest management practices. Efforts to improve nutrient management in agriculture are buoyed by NBFs, which promise enhanced nutrient uptake and reduced environmental impact. This could lead to increased crop productivity while minimizing the ecological footprint of farming practices. In the realm of food preservation, NBFs present exciting avenues for extending shelf life and curbing post-harvest losses. From active packaging materials to novel preservation techniques, nanotechnology offers versatile solutions to enhance food security and reduce wastage.

Further research and development in crop pre and post-harvest management can focus on several areas. One area of interest is the development of new rootstocks and varieties to address the challenges associated with relying on a single variety, such as the 'Hass' avocado variety (Sciarelli 2023). Another area is the exploration of different post-harvest treatments, such as 1-methylcycloprpene, methyl jasmonate, and edible coatings, to improve the post-harvest quality and storage extension of fresh produce (Tsantili and Bai 2022). Additionally, there is a need for research on weed control programs and effective herbicide treatments to manage troublesome weeds like prickly sida in agronomic crops (Copes 2016). Lastly, the use of deep learning techniques, such as recurrent neural networks and deep learning networks, can be further explored for pre and post-harvest operations in viticulture, specifically in grape quality monitoring and image segmentation and classification of grape bunches (Patil et al. 2022).

18.12 Conclusions

The integration of NBFs in agriculture represents a pivotal advancement towards sustainable crop management and food security. By harnessing the power of nanotechnology and biotechnology, these formulations offer eco-friendly solutions to address key challenges in pre and post-harvest crop management. They optimize resource allocation, enhance crop yield, mitigate disease risks, and promote soil fertility while minimizing environmental impact. However, regulatory concerns and safety issues surrounding NMs necessitate careful evaluation and implementation of safety measures. Despite challenges, the synergy of nanotechnology with sensing technologies holds promise for precision agriculture and improved productivity. Future research endeavours should focus on developing new crop varieties, exploring innovative post-harvest treatments, and leveraging advanced techniques for crop quality monitoring. Ultimately, the adoption of NBFs in agriculture signifies a transformative step towards a more resilient, efficient, and environmentally conscious agricultural industry.

- Recapitulation of Key Findings
- Implications for Agriculture
- Call to Action for Further Research and Implementation

References

Abdulbaki M, Elemasho MK, Ajanwachuku NC, Okoroafor CH, Nwaehujor IU, Olayemi FF, Pessu PO (2023) Farmers' constraints to production and post-harvest handling of selected neglected food crops in South-east and South-south Nigeria. J Agric Food Sci 20(2):62–74. https://doi.org/10.4314/jafs.v20i2.4

Ahmad R, Fiodor A, Nain L, Pranaw K (2023) Microbially synthesized nanoparticles: implications and their applications in agriculture. In: Kumar V, Singh M (eds) Microbial nanotechnology: advances in agriculture, industry and health sectors, vol 2022. De Gruyter, Berlin, Boston, pp 25–52. https://doi.org/10.1515/9783110754476-002

Ahmad M, Hussain M, Ali S (2022) Sensitivity analysis of crop simulation models under changing climate scenarios: A review. Agricultural Systems 195:103318. https://doi.org/10.1016/j.agsy.2021.103318

Ain Q, Hussain HA, Zhang Q, Rasheed A, Imran A, Hussain S, Ahmad N, Bibi H, Ali KS (2023) Chapter thirteen - use of nano-fertilizers to improve the nutrient use efficiencies in plants. In: Sustainable plant nutrition molecular interventions and advancements for crop improvement. Academic, pp 299–321. https://doi.org/10.1016/b978-0-443-18675-2.00013-4

Al-Antary TM, Ghidan AY (2023) Chapter 17: Strengths and weaknesses of metal oxide nanoparticles in agriculture. In: Nanometal oxides in horticulture and agronomy, Nanomaterial-plant interactions. Academic, pp 353–376. https://doi.org/10.1016/b978-0-323-91809-1.00004-4

Ali S, Mehmood A, Khan N (2021) Uptake, translocation, and consequences of nanomaterials on plant growth and stress adaptation. J Nanomater 2021:6677616, 17 pages. https://doi.org/10.1155/2021/6677616

Alhassan SI, Dabaieh M, El-Shimy M (2023) Climate change adaptation at the margins: The case of Cairo, Egypt. Environmental Research 234:116569. https://doi.org/10.1016/j.envres.2023.116569

Al-Khayri JM, Banu AN, Kudesia N, Ahmed F, Arora A, Abd-Elsalam KA (2023) Entomopathogenic fungi: an eco-friendly synthesis of sustainable nanoparticles and their nanopesticide properties. Microorganisms 11(6):1617–1617. https://doi.org/10.3390/microorganisms11061617

Anguiano-Cabello JC, Laredo-Alcalá EI, Barrera-Martínez C, Hernández-Castillo FD, Ilyina A, Segura-Ceniceros P, Arredondo-Valdés R (2023) Plants and microorganism bio-compounds on agriculture applications by nanotechnology. In: Aguilar CN, Abdulhameed S, Rodriguez-Herrera R, Sugathan S (eds) Microbial biodiversity, biotechnology and ecosystem sustainability. Springer, Singapore. https://doi.org/10.1007/978-981-19-4336-2_8

Annie M, Pal RK (2023) Assessing the impact of climate change on agricultural production using crop simulation model. Int J Environ Clim Chang 13(7):538–550. https://doi.org/10.9734/ijecc/2023/v13i71906

Azameti MK, Imoro AM (2023) Nanotechnology: a promising field in enhancing abiotic stress tolerance in plants. In: Crop design. Elsevier, pp 100037–100037. https://doi.org/10.1016/j.cropd.2023.100037

Babu PJ, Longchar B, Rajasekhar A (2022) Nanobiotechnology-mediated sustainable agriculture and post-harvest management. Curr Res Biotechnol 4:326–336. https://doi.org/10.1016/j.crbiot.2022.07.004

Bairwa PK, Kumar N, Devra V, Abd-Elsalam KA (2023) Nano-biofertilizers synthesis and applications in agroecosystems. Agrochemicals 2(1):118–134. https://doi.org/10.3390/agrochemicals2010009

Baliyarsingh B, Pradhan CK (2023) Prospects of plant-derived metallic nanopesticides against storage pests - a review. J Agric Food Res 14:100687–100687. https://doi.org/10.1016/j.jafr.2023.100687

Bezabih G, Satheesh N, Fanta SW, Wale M, Atlabachew M (2022) Reducing post-harvest loss of stored grains using plant-based biopesticides: a review of past research efforts. Adv Agric 2022:1–16. https://doi.org/10.1155/2022/6946916

Bidyarani N, Vyas G, Jaiswal J, Deshmukh SK, Kumar U (2022) Recent approaches in nanobioformulation for sustainable agriculture. In: Nanotechnology in agriculture and environmental science. CRC Press, pp 166–182. https://doi.org/10.1201/9781003323945-14

Britto CGJ, John ADSA, Jayakumar A, Selvasekaran P, Chidambaram R, Zheng H (2023) Chapter 1: Nanotechnology applications for food safety: benefits and risks. In: Nanotechnology applications for food safety and quality monitoring. Academic, pp 3–30. https://doi.org/10.1016/b978-0-323-85791-8.00021-5

Campos EVR, Anderson PAES, Oliveira JLD, Villarreal GPU, Fraceto LF (2022) Nature-based nanocarrier system: an eco-friendly alternative for improving crop resilience to climate changes. Anthr Sci 1(3):396–403. https://doi.org/10.1007/s44177-022-00029-x

Channa H (2019) Behavioral responses to post-harvest challenges in East Africa: lessons from field experiments. Purdue University Graduate School. Thesis. https://doi.org/10.25394/PGS.8044382.V1

Cherian P, Sheela D (2023) Nanotechnology applications in biological engineering: plant health management. In: Nanotechnology horizons in food process engineering. Apple Academic Press, pp 363–383. https://doi.org/10.1201/9781003305385-13

Choudhary R, Rajput VD, Ghodake G, Ahmad F, Meena M, Rehman RU, Prasad R, Sharma RK, Singh R, Seth CS (2024) Comprehensive journey from past to present to future about seed priming with hydrogen peroxide and hydrogen sulfide concerning drought, temperature, UV and ozone stresses- a review. Plant Soil 500(1–2):351–373. https://doi.org/10.1007/s11104-024-06499-9

Choudhary R, Ahmad F, Kaya C, Upadhyay SK, Muneer S, Kumar V, Meena M, Liu H, Upadhyay H, Seth CS (2025) Decrypting proteomics, transcriptomics, genomics, and integrated omics for augmenting the abiotic, biotic, and climate change stress resilience in plants. J Plant Physiol. https://doi.org/10.1016/j.jplph.2025.154430

Copes JT (2016) Prickly Sida (*Sidaspinosa L.*): biology and in-crop and post-harvest management programs. In: LSU Doctoral Dissertations, p 1279. https://doi.org/10.31390/gradschool_dissertations.1279

Das S, Ghosh A, Mukherjee A (2021) Nanoencapsulation-based edible coating of essential oils as a novel green strategy against fungal spoilage, mycotoxin contamination, and quality deterioration of stored fruits: an overview. Front Microbiol 12:768414. https://doi.org/10.3389/fmicb.2021.768414

Demoule J (2022) Challenges and measures to recapitalise handling of post-harvest crops in developing countries. In: Postharvest technology. IntechOpen. https://doi.org/10.5772/intechopen.101222

Dong B, Jiang R, Chen J, Xiao Y, Lv Z, Chen W (2022) Strategic nanoparticle-mediated plant disease resistance. Crit Rev Biotechnol 43(1):22–37. https://doi.org/10.1080/07388551.2021.2007842

Dutta PK, Kumari A, Mahanta M, Upamanya GK, Heisnam P, Borua S, Kaman PK, Mishra AK, Mallik MK, Muthukrishnan G, Sabarinathan KG, Puzari KR, Vijayreddy D (2023) Nanotechnological approaches for management of soil-borne plant pathogens. Front Plant Sci 14. https://doi.org/10.3389/fpls.2023.1136233

El-Ramady HR, Verma KK, Rajput VD, Minkina T, Elbehiry F, Elbasiouny HY, Elsakhawy T, Omara AED, Amer MM (2022) Sources of silicon and nano-silicon in soil and plants. In: Silicon and nano-silicon in environmental stress management and crop quality improvement: progress and prospects. Academic Press, Elsevier. https://doi.org/10.1016/B978-0-323-91225-9.00003-0

El-Saadony MT, Saad AM, Soliman SM, Salem HM, Desoky EM, Babalghith AO, El-Tahan AM, Ibrahim OM, Ebrahim AAM, El-Mageed TAA, Elrys AS, Elbadawi A, El-Tarabily KA, AbuQamar SF (2022) Role of nanoparticles in enhancing crop tolerance to abiotic stress: a comprehensive review. Front Plant Sci 13:946717. https://doi.org/10.3389/fpls.2022.946717

Engelmann W, Hohendorff RV, Leal SWD (2022) Ethical issues in nanotechnology-II. In: Environmental, ethical, and economical issues of nanotechnology, pp 131–149. https://doi.org/10.1201/9781003261858-7

Farooq T, Hameed A, Hameed A (2023) Chapter 15: Engineered nanomaterials for plant disease diagnosis and management. In: Engineered nanomaterials for sustainable agricultural production, soil improvement and stress management, Plant biology, sustainability and climate change. Academic, pp 337–355. https://doi.org/10.1016/b978-0-323-91933-3.00021-0

Ferreira MAM, Filipe JA (2022) Ethical considerations on nanotechnology. https://doi.org/10.48550/arxiv.2202.01063

Fincheira P, Hoffmann N, Tortella G, Ruiz A, Cornejo P, Diez MC, Seabra AB, Benavides-Mendoza A, Rubilar O (2023) Eco-efficient systems based on nanocarriers for the controlled release of fertilizers and pesticides: toward smart agriculture. Nano 13(13):1978–1978. https://doi.org/10.3390/nano13131978

Fraceto L, Grillo R, Medeiros G, Scognamiglio V, Rea G, Bartolucci C (2016) Nanotechnology in agriculture: which innovation potential does it have? Front Environ Sci 4. https://doi.org/10.3389/fenvs.2016.00020

Gerken AR, Morrison WR (2022) Pest management in the post-harvest agricultural supply chain under climate change. Front Agron 4:918845. https://doi.org/10.3389/fagro.2022.918845

Ghosh S, Sarkar B, Kumar A, Thongmee S (2022) Chapter 18: Regulatory affairs, commercialization, and economic aspects of nanomaterials used for agriculture. In: Agricultural nanobiotechnology, Woodhead publishing series in food science, technology and nutrition. Woodhead Publishing, pp 479–502. https://doi.org/10.1016/b978-0-323-91908-1.00008-0

Halake NH, Haro JM (2022) Role of nanobiotechnology towards agri-food system. J Nanotechnol 2022:1–7. https://doi.org/10.1155/2022/6108610

Hashim N, Onwude DI, Maringgal B (2022) Chapter 15: Technological advances in post-harvest management of food grains. In: Research and technological advances in food science. Academic, pp 371–406. https://doi.org/10.1016/b978-0-12-824369-5.00016-6

Jagadesh M, Dash M, Kumari A, Singh SK, Verma KK, Kumar P, Bhatt R, Sharma SK (2024) Revealing the hidden world of soil microbes: metagenomic insights into plant, bacteria, and fungi interactions for sustainable agriculture and ecosystem restoration. Microbiol Res 285:127764. https://doi.org/10.1016/j.micres.2024.127764

Jha S, Yadav A (2023) Chapter 6: Assessment of carbon and fullerene nanomaterials for sustainable crop plants growth and production. In: Engineered nanomaterials for sustainable agricultural production, soil improvement and stress management, Plant biology, sustainability and climate change. Academic, pp 145–160. https://doi.org/10.1016/b978-0-323-91933-3.00015-5

Jiang Y, Yang J, Li M, Li Y, Zhou P, Wang Q, Sun Y, Zhu G, Wang Q, Zhang F, Rui Y, Lynch I (2022) Effect of silica-based nanomaterials on seed germination and seedling growth of rice (*Oryzasativa* L.). Nano 12(23):4160–4160. https://doi.org/10.3390/nano12234160

Kah M, Beulke S, Tiede K, Hofmann T (2018) Nanopesticides: State of knowledge, environmental fate, and exposure modeling. Critical Reviews in Environmental Science and Technology 48(6):542–592. https://doi.org/10.1080/10643389.2017.1398297

Khodakovskaya MV, SilvaK D, Nedosekin DA, Dervishi E (2017) Complex genetic, photothermal, and photoacoustic analysis of nanoparticle-plant interactions. Proc Natl Acad Sci USA 108(3):1028–1033

Kumar P, Dhiman K, Shaunak I, Gambhir G, Kumar A, Srivastava DK (2023) Nanotechnology applications in agriculture. In: Nanotechnology horizons in food process engineering. Apple Academic Press, pp 75–97. https://doi.org/10.1201/9781003305408-5

Kumar D, Biswas JK, Mulla SI, Singh R, Shukla R, Ahanger MA, Shekhawat GS, Verma KK, Siddiqui MW, Seth CS (2024a) Micro and nanoplastics pollution: sources, distribution, uptake in plants, toxicological effects, and innovative remediation strategies for environmental sustainability. Plant Physiol Biochem 213:108795. https://doi.org/10.1016/j.plaphy.2024.108795

Kumar D, Mariyam S, Gupta KJ, Thiruvengadam M, Ghodake GS, Xing B, Seth CS (2024b) Comparative investigation on chemical and green synthesized titanium dioxide nanoparticles against chromium (VI) stress eliciting differential physiological, biochemical, and cellular attributes in Helianthus annuus L. Sci Total Environ 930:172413. https://doi.org/10.1016/j.scitotenv.2024.172413

Kumar D, Singh R, Upadhyay SK, Verma KK, Tripathi RM, Liu H, Dhankher OP, Tripathi RD, Sahi SV, Seth CS (2024c) Review on interactions between nanomaterials and phytohormones: novel perspectives and opportunities for mitigating environmental challenges. Plant Sci 340:111964. https://doi.org/10.1016/j.plantsci.2023.111964

Kumari A, Lakshmi GA, Krishna GK, Patni B, Prakash S, Bhattacharyya M, Singh SK, Verma KK (2022) Climate change and its impact on crops: a comprehensive investigation for sustainable agriculture. Agronomy 12:3008. https://doi.org/10.3390/agronomy12123008

Kumari R, Suman K, Karmakar S, Mishra VK, Lakra SG, Saurav GK, Mahto BK (2023a) Regulation and safety measures for nanotechnology-based agri-products. Front Genome Ed 5:1200987. https://doi.org/10.3389/fgeed.2023.1200987

Kumari A, Dash M, Singh SK, Jagadesh M, Mathpal B, Mishra PK, Pandey SK, Verma KK (2023b) Soil microbes: a natural solution for mitigating the impact of climate change. Environ Monit Assess 195:1436. https://doi.org/10.1007/s10661-023-11988-y

Lau HY, Wu H, Wee EJH, Trau M, Wang Y, Botella JR (2017) Specific and sensitive isothermal electrochemical biosensor for plant pathogen DNA detection with colloidal gold nanoparticles as probes. Sci Rep 7:38896. https://doi.org/10.1038/srep38896

Ma C, Han L, Shang H, Hao Y, Xu X, White JC, Wang Z, Xing B (2023) Nanomaterials in agricultural soils: ecotoxicity and application. Curr Opin Environ Sci Health Elsevier 31:100432. https://doi.org/10.1016/j.coesh.2022.100432

Malhotra A, Dolui SK (2019) Nanotechnology for precision agriculture: present status and future perspectives. J Nanosci Nanotechnol 19(1):1–22

Mariyam S, Upadhyay SK, Chakraborty K, Verma KK, Duhan JS, Muneer S, Meena M, Sharma RK, Ghodake G, Seth CS (2024) Nanotechnology, a frontier in agricultural science, a novel approach in abiotic stress management and convergence with new age medicine-a review. Sci Total Environ 912:169097. https://doi.org/10.1016/j.scitotenv.2023.169097

Mditshwa A, Khaliq G, Hussein Z, Ejaz S (2023) Editorial: Sustainable post-harvest management practices for fresh produce. Front Sustain Food Syst 7. https://doi.org/10.3389/fsufs.2023.1143759

Mishra P, Singh R, Kumar A (2022) Microclimatic modifications and their role in climate change adaptation for sustainable agriculture. Environmental Research Communications 4(3):031002. https://doi.org/10.1088/2515-7620/ac5f2b

Mishra P, Muralidaran Y, Vaishnavi G, Dhoni S, Jyotsana S, Singh M, Kumar V (2023) Microbial nanotechnology and its indispensable role in sustainable agriculture. In: Kumar V, Singh M (eds) Microbial nanotechnology: advances in agriculture, industry and health sectors, vol 2022. De Gruyter, Berlin, Boston, pp 53–72. https://doi.org/10.1515/9783110754476-003

Mondal S (2022) A review: nanofertilizers for the biofortification and sustainable agriculture. Vidyodaya J Sci 25(02). https://doi.org/10.31357/vjs.v25i02.6172

Mude L, Kuppusamy G, Karri VVSR, Arun R, Kumar N, Thanzim M (2022) Regulatory aspects of herbal nanoformulations. In: Nanophytomedicine, pp 255–264. https://doi.org/10.1201/9781003231745-23

Muthukumar J, Kumar S, Arunachalam N, Kumar L, Chidambaram R (2023) Chapter 18: Quality assurance of packaged food using nanotechnology. In: Nanotechnology applications for food safety and quality monitoring. Academic, pp 341–372. https://doi.org/10.1016/b978-0-323-85791-8.00019-7

Nandini B, Geetha NP (2021) 15: Smart delivery mechanisms of nanofertilizers and nanocides in crop biotechnology. In: Advances in nano-fertilizers and nano-pesticides in agriculture. A smart delivery system for crop improvement, Woodhead publishing series in food science, technology and nutrition. Woodhead Publishing, pp 385–414. https://doi.org/10.1016/B978-0-12-820092-6.00015-X

Nawaz S, Maqsood I, Batool F, Sandhu ZY, Hassan S, Akram F, Rashid B (2023) Improvement of abiotic stress tolerance in plants with the application of nanoparticles. IntechOpen. https://doi.org/10.5772/intechopen.110201

Padmakumar A, Eswar K, Kong L, Yang W, Gopalakrishnan S, Cahill DM, Rengan AK (2023) Bacteria-premised nanobiopesticides for the management of phytopathogens and pests. ACS Agric Sci Technol 3(5):370–388. https://doi.org/10.1021/acsagscitech.3c00025

Patil ND, Tiwari RK, Kumar A (2022) Pre and post harvesting using deep learning techniques: a comprehensive study. In: 4th international conference on recent trends in computer science and technology (ICRTCST), Jamshedpur, India, pp 207–211. https://doi.org/10.1109/ICRTCST54752.2022.9781959

Pavoni L, Pavela R, Cespi M, Bonacucina G, Maggi F, Zeni V, Canale A, Lucchi A, Bruschi F, Benelli G (2019) Green micro- and nanoemulsions for managing parasites, vectors and pests. Nano 9(9):1285. https://doi.org/10.3390/nano9091285

Prakash A, Quazi S, TiwariP. (2023) The nanotechnological approaches in agriculture and its allied fields, Advances in environmental engineering and green technologies book series, pp 134–13s. https://doi.org/10.4018/978-1-6684-5533-3.ch007

Prasad K, Singh SK, Bhattocharjee P, Rudrapaul J, Kumar U, Yadav SK, Maurya S, Mani A (2023) Impact of climate change on postharvest quality of vegetables. In: Solankey SS, Kumari M (eds) Advances in research on vegetable production under a changing climate, Advances in olericulture, vol 2. Springer, Cham. https://doi.org/10.1007/978-3-031-20840-9_16

Qayyum A, Nadeem F, Bibi Y, Ullah R, Bajwa HR, Jawad H, Sher A (2022) Role of nanoparticles in improving stress tolerance in crop plants. In: Biostimulants for crop production and sustainable agriculture, pp 386–394. https://doi.org/10.1079/9781789248098.0024

Rajput VD, Minkina T, Feizi M, Kumari A, Khan M, Mandzhieva S, Sushkova S, El-Ramady H, Verma KK, Singh A, van Hullebusch ED, Singh RK, Jatav HS, Choudhary R (2021) Effects of silicon and silicon-based nanoparticles on rhizosphere microbiome, plant stress and growth. Biology 10:791. https://doi.org/10.3390/biology10080791

Rajput VD, Kumari A, Upadhyay SK, Minkina T, Mandzhieva S, Ranjan A, Sushkova S, Burachevskaya M, Rajput P, Konstantinova E, Singh J, Verma KK (2023) Can nanomaterials improve the soil microbiome and crop productivity? Agriculture 13:231. https://doi.org/10.3390/agriculture13020231

Rani K (2022) Role of nanotechnology in improving quality of edible crop plants. IP Int J Compr Adv Pharmacol 7(3):112–114. https://doi.org/10.18231/j.ijcaap.2022.023

Sabrine S, Tarek B (2023) Nano green pesticide, an innovative approach. All Sci Abstracts 1(3):4–4. https://doi.org/10.59287/as-abstracts.925

Santo Pereira EA, Oliveira CH, Fraceto FL, Santaella C (2021) Nanotechnology potential in seed priming for sustainable agriculture. Nano 11:267. https://doi.org/10.3390/nano1102026

Sciarelli F (2023) Advances in pre-harvest management of avocado. In: BurleighDodds series in agricultural science, pp 135–204. https://doi.org/10.19103/as.2023.0144.04

Sharma M, Singh PK (2023) Newly engineered nanoparticles as potential therapeutic agents for plants to ameliorate abiotic and biotic stress. J Appl Nat Sci 15(2):720–731. https://doi.org/10.31018/jans.v15i2.4603

Shende S, Rajput V, Minkina T, Sushkova S, Mandzhieva S (2023) Strategic role of nanotechnology in plant growth improvement and crop production. In: Nanotechnology in agriculture and agroecosystems, Micro and Nano Technologies. Elsevier, pp 25–49. https://doi.org/10.1016/b978-0-323-99446-0.00015-5

Singh P, Sharma R, Kumar V (2021) Organic and inorganic fertilizers in sustainable agriculture: Impacts on soil health and crop productivity. Agronomy 11(7):1325. https://doi.org/10.3390/agronomy11071325

Singh S, Singh VK, Singh PK, Modi AR, Kumar A (2022) Chapter 13: Microbial antagonists in post-harvest management of fruit. In: Research and technological advances in food science. Academic, pp 333–346. https://doi.org/10.1016/b978-0-12-824369-5.00005-1

Srivastava S (2023) Ethical issues regarding the use of nanobiotechnology-based products. In: Nanobiotechnology for the livestock industry. Elsevier, pp 435–473. https://doi.org/10.1016/b978-0-323-98387-7.00005-7

Srivastava N, Priyanka P, Chattopadhyay J (2023) Chapter 22: Role of engineered nanomaterial in food safety of agricultural products. In: Nanotechnology applications for food safety and quality monitoring. Academic, pp 495–512. https://doi.org/10.1016/b978-0-323-91933-3.00013-1

Sumita P, Kachhwaha K (2023) Enhancement of seed germination mediated by OH-functionalised multiwalled carbon nanotubes (OH-MWCNTs) and their effect on morphological parameters of *Pennisetum glaucum* (L.) R.Br. genotypes. Res J Biotechnol 18(3):57–63. https://doi.org/10.25303/1803rjbt57063

Sun Y, Zhu G, Zhao W, Jiang Y, Wang Q, Wang Q, Rui Y, Zhang F, Gao L (2022) Engineered nanomaterials for improving the nutritional quality of agricultural products: a review. Nano 12(23):4219–4219. https://doi.org/10.3390/nano12234219

Tighe-Neira R, Recio G, Rengel Z, Inostroza-Blancheteau C, Reyes-Díaz M (2023) Chapter 11: Role of some nanomaterials in management of nutrients stress in crop plants. In: Engineered nanomaterials for sustainable agricultural production, soil improvement and stress management, Plant biology, sustainability and climate change. Academic, pp 241–255. https://doi.org/10.1016/b978-0-323-91933-3.00004-0

Tripathi S, Verma A, Singh N (2023) Adoption of sustainable agricultural practices for climate resilience: Cover cropping and improved seed varieties. Sustainability 15(2):1120. https://doi.org/10.3390/su15021120

Tsantili E, Bai Y (2022) Improving crop resilience to climate change through genetic and agronomic innovations. Frontiers in Plant Science 13:841234. https://doi.org/10.3389/fpls.2022.841234

Vasudevan V, Arumugam V (2023) Chapter 14: Role of metal oxide nanomaterials in the preservation of harvested crops. In: Nonomaterial oxides in horticulture and agronomy. Academic, pp 275–289. https://doi.org/10.1016/b978-0-323-91809-1.00013-5

Venkateswaran SP, Palanisamy VK, Divya P, Geethamani P, Periakaruppan R (2023) Chapter 6: Impact of metal oxide nanomaterials on seed germination, plant growth and development. In: Nanometal oxides in horticulture and agronomy. Academic, pp 81–99. https://doi.org/10.1016/b978-0-323-91809-1.00008-1

Verma KK, Singh P, Song X-P, Malviya MK, Singh RK, Chen G-L, Solomon S, Li YR (2020) Mitigating climate change for sugarcane improvement: role of silicon in alleviating abiotic stresses. Sugar Tech 22(5):741–749. https://doi.org/10.1007/s12355-020-00831-0

Verma KK, Song XP, Joshi A, Rajput VD, Singh M, Sharma A, Singh RK, Li DM, Arora J, Minkina T, Li YR (2022a) Nanofertilizer possibilities for healthy soil, water and food in future: an overview. Front Plant Sci 13:865048. https://doi.org/10.3389/fpls.2022.865048

Verma KK, Xiu-Peng S, Abhishek J, Dan-Dan T, Vishnu DR, Munna S, Jaya A, Tatiana M, Yang-Rui L (2022b) Recent trends in nano-fertilizers for sustainable agriculture under climate change for global food security. Nano 12(1):173. https://doi.org/10.3390/nano12010173

Verma KK, Song X, Degu HD, Guo D, Joshi A, Huang H, Xu L, Singh M, Huang D, Rajput VD, Li Y (2023) Recent advances in nitrogen and nano-nitrogen fertilizers for sustainable crop production: a mini-review. Chem Biol Technol Agric 10:1–14. https://doi.org/10.1186/s40538-023-00488-3

Verma KK, Song XP, Kumari A, Jagadesh M, Singh SK, Bhatt R, Singh M, Seth CS, Li YR (2024a) Climate change adaptation: challenges for agricultural sustainability. Plant Cell Environ:1–12. https://doi.org/10.1111/pce.15078

Verma KK, Song X-P, Singh S, Kumari A, Arora J, Singh SK, Solanki MK, Seth CS, Li Y-R (2024b) Synergistic interactions of nanoparticles and plant growth promoting rhizobacteria enhancing soil-plant systems: a multigenerational perspective. Front Plant Sci 15:1376214. https://doi.org/10.3389/fpls.2024.1376214

Vibala BV, Anooj ES, Jani Subha T (2021) Recent studies on agricultural nanotechnology research and environmental sustainability. In: Nakkella AK, Yadav SK (eds) Global environmental justice institutions policies and implementations. Bharti Publication, New Delhi. isbn:978-93-90818-44-0. Accessed 13 Mar 2024

Vinay JU, Kurjogi M, Uppar A, Vanti G (2023) Engineered nanomaterials in crop plants temperature and or heat stress management. In: Engineered nanomaterials for sustainable agricultural production, soil improvement and stress management, Plant biology, sustainability and climate change. Academic, pp 227–239. https://doi.org/10.1016/b978-0-323-91933-3.00006-4

Violante HGM, Pérez MVA (2022) Pre-harvest and post-harvest application of microbial inoculants influencing post-harvest storage technology in horticultural crops. In: Sustainable horticulture. Microbial inoculants and stress interaction, Developments in applied microbiology and biotechnology. Academic, pp 399–436. https://doi.org/10.1016/b978-0-323-91861-9.00004-5

Weisany W, Khosropour E (2023) Engineered nanomaterials in crop plants drought stress management. In: Engineered nanomaterials for sustainable agricultural production, soil improvement and stress management, Plant biology, sustainability and climate change. Academic, pp 183–204. https://doi.org/10.1016/b978-0-323-91933-3.00005-2

Wu T, Zou G, Zhang G, Shan Y, Liu B, Ding Z, Assis T, Nogueira R, Nawaz M, Zhao F, Abideen Z, He Z (2023) The efficiency of nanoparticles on improving seed germination and mitigating ammonium stress of water spinach (*Ipomoea aquaticaForssk.*) and Hami Melon (*Cucumismelo L.*). Sustain For 15(13):10083–10083. https://doi.org/10.3390/su151310083

Yasmeen F, Mustafa G, Jhanzab HM, Komatsu S (2023) Chapter 16: Current scenario and future trends of plant nano-interaction to mitigate abiotic stresses: a review. In: Impact of engineered nanomaterials in genomics and epigenomics. Wiley Online Library, pp 355–396. https://doi.org/10.1002/9781119896258.ch16

Yimer AH (2022) Post-harvest management practices of maize in Ethiopia: a review. Agric Sci Dig. https://doi.org/10.18805/ag.df-414

Zehra A, Rai A, Singh SK, Aamir M, Ansari WA, Upadhyay RS (2021) Chapter 10: An overview of nanotechnology in plant disease management, food safety, and sustainable agriculture. In: Food security and plant disease management. Woodhead Publishing, pp 193–219. https://doi.org/10.1016/B978-0-12-821843-3.00009-X

Ziogas V (2023) Principles and practices in fruit tree production and post-harvest management. Agronomy 13(2):408. https://doi.org/10.3390/agronomy13020408

Chapter 19
Role of Nanobioformulations on Preharvest and Postharvest Crop Management

Rachid Ezzouggari, Ilham Dehbi, Mohammed Khadiri, Oumaima Benaissa, and Rachid Lahlali

19.1 Introduction

Recent decades have been marked by exponential growth in the world's population. According to (UN DESA 2022) predictions, there will be 10.87 billion people on Earth by 2100, up from 9.7 billion in 2050. This is a major and worrisome growth tendency, especially in less developed countries. This sharp rise in population highlights how urgently food supply must improve to keep up with the growing demand. However, using traditional agricultural methods to produce enough food agricultural goods to fulfil the demands of the world's population is severely hampered by

R. Ezzouggari · O. Benaissa
Phytopathology Unit, Department of Plant Protection, Ecole Nationale d'Agriculture de Meknès, Menkès, Morocco

Laboratory of Biotechnology, Conservation and Valorization of Natural Resources (LBCVNR), Faculty of Sciences Dhar El Mehraz, Sidi Mohamed Ben Abdallah University, Fez, Morocco

I. Dehbi
Phytopathology Unit, Department of Plant Protection, Ecole Nationale d'Agriculture de Meknès, Menkès, Morocco

Laboratory of Plant Biotechnology and Molecular Biology, Faculty of Sciences, Moulay Ismail University, Zitoune, Meknes, Morocco

M. Khadiri
Microbial Biotechnology and Plant Protection Laboratory, Faculty of Science, Ibn Zhor University, Agadir, Morocco

R. Lahlali (✉)
Phytopathology Unit, Department of Plant Protection, Ecole Nationale d'Agriculture de Meknès, Menkès, Morocco
e-mail: rlahlali@enameknes.ac.ma

C. S. Seth et al. (eds.), *Nano-formulations for Sustainable Agriculture*,
Nanotechnology in the Life Sciences, https://doi.org/10.1007/978-3-031-90617-6_19

the degradation of arable land, nutrient depletion, and nutrient unavailability to plants (Verma et al. 2024a).

The importance of agriculture in supplying the world's food demands despite a rapidly expanding population and changing dietary habits is one of the report's main points of emphasis. Food security, poverty reduction, and economic development especially in rural areas are all made possible by sustainable farming methods (Verma et al. 2022a; FAO 2022b). A scenario like this accelerated the search for substitute tactics that would guarantee the least amount of environmental damage and the highest level of output. Many strategies, including integrated pest management (IPM), were put into practice to guarantee the prudent use of pesticides with the least amount of dangers to the ecosystem (Miller et al. 2012). However, since the green revolution, excessive use of chemical fertilizers especially those high in nitrogen, phosphorus, and potassium (NPK) has been a standard practice in contemporary agriculture to solve soil micronutrient deficiencies. Still, because of the air, soil, and water pollution, this has raised environmental issues, worldwide eutrophication and water pollution are caused by the buildup of fertilizers in soil and plants, as well as by microbiological and leaching processes (Verma et al. 2022b). Furthermore, adding fertilizers and pesticides to the soil changes its composition and can interfere with the dynamics of the soil microflora, which can impact soil function (Deshmukh et al. 2022; Choudhary et al. 2024).

Alternatively, in response to the environmental concerns and challenges posed by the excessive use of chemical fertilizers and pesticides in agriculture, there has been a growing emphasis on utilizing biological control agents to manage plant diseases (VanderGheynst et al. 2006; Verma et al. 2023). Naturally occurring biofertilizers including *Bacillus subtilis*, *Paenibacillus elgii*, and *Pseudomonas fluorescens* have been shown to boost crop fertilization efficacy when NPs like silver and gold are present. By successfully lowering metal pollutants in the rhizosphere and enhancing plant development and production, bacterial-based nanobioformulations can significantly contribute to environmental remediation efforts (Baker et al. 2019).

A different approach to managing weeds, fungi, and insects that constitute plant pests has been suggested recently: the use of nanomaterials. Nanoparticles are used as antibacterial agents in food packaging and agriculture, as reported in several research. Apart from their antimicrobial properties, some of these nanoparticles, i.e., Ag, Fe, TiO_2, CeO_2, Al_2 O_3, Cu, Si, Al, Zn, ZnO, and carbon nanotubes have been demonstrated to have certain detrimental effects on plant growth. Out of all of these nanomaterials, silver nanoparticles have shown the greatest promise (Kumar et al. 2024b). Food manufacturers increasingly employ nanoparticles to enhance food quality and maintain its nutritional value (Deshmukh et al. 2022). The fast breakthroughs in nanotechnology have raised concerns due to the accumulation of manufactured nanomaterial and its potential to penetrate the food chain (Priester et al. 2012; Rajput et al. 2021a, b).

The use of nanoparticles is not inherently dangerous. For example, when employed in agriculture, water, and food, many nanoscale elements contained in conventional meals (e.g., proteins in milk, fat globules in mayonnaise, carbs, DNA, etc.) may be hazardous to human health and the environment (Gruère et al. 2011).

High dosages of fibrous and tubular nanostructured materials, such as single- and multiple-walled carbon nanotubes, have been linked to fibrotic lung responses, which raise the risk of carcinogenesis and cause inflammation, according to several toxicological investigations (Deshmukh et al. 2022). In nanobiotechnology, nanobioformulation is the most recent development. Given the state of nanotechnology and the state of agricultural development today, nanobioformulation is a highly inventive field that needs greater attention from scientists worldwide. Plant extract may be used to create nanoparticles using green technologies that are easy to use, quick to react, easy on the environment, and straightforward (Harish et al. 2023). The green method to creating nanobioformulations expands agriculture's potential to enhance plant growth regulators, fertilization processes, wastewater treatment, pesticide delivery of an active component at a specific spot, and nutrient absorption in plants (Verma et al. 2024b). Nanotechnology is a fast-growing science that has drawn interest because of its many uses in a variety of industries, including environmental protection, agriculture, and health (Jampílek and Králová 2020).

With its many uses as biofertilizers, biostimulants, biopesticides, and bioinoculants, nanobioformulations have become more beneficial in agricultural operations, especially in the preharvest stages. Because the nanoparticles in these formulations allow for improved absorption, penetration, and targeted distribution of active substances to plants, they are more effective. This is because nanotechnology is being used in their creation. They strengthen plant defenses against diseases and pests, increase nutrient absorption, promote plant growth, and improve stress tolerance. Reducing chemical inputs, limiting environmental effects, and raising crop yields are some of the ways that using nanobioformulations may contribute to sustainable agriculture practices (DeRosa et al. 2010; Khodakovskaya et al. 2011).

In this chapter, we will discuss in detail crop management before and after harvest through the use of nanobioformulations, we will examine how these innovative technologies can be applied to protect crops against pathogenic infections, enhance their growth and yield, as well as prolong their shelf-life postharvest,we will also explore the challenges and opportunities associated with adopting these approaches in modern agriculture, highlighting their potential for more effective and sustainable crop management.

19.2 Nanotechnology in Agriculture

19.2.1 What Is the Nanotechnology?

The first global issue that needs to be addressed is providing food security for a world population that is expanding quickly. Based on projections, demand for food could potentially increase by 59–98% by the year 2050, when there would be 9.7 billion people on the planet (Deshmukh et al. 2022; Duro et al. 2020). Global food supplies are threatened by using biological resources for energy production, manufacture of chemicals, large postharvest losses, decreased added value, inadequate distribution, and other issues, even in spite of the world's population growth, especially in emerging nations (Barrett 2021; Neme et al. 2021; Jagadesh et al. 2024). Farmers everywhere will concentrate on utilizing cutting-edge innovations and technologies to increase agricultural yields through vast and intensive farming. Precision farming and the usage of nanomodified stimulants have significantly enhanced the existing efforts. Basic aspects of food security that could be enhanced by developments in nanotechnology research include agricultural productivity, soil enhancement, safe water usage, food distribution in shops, and food quality (Ashraf et al. 2021; El-Beltagy and Madkour 2012; Marchiol 2018; El-Ramady et al. 2022) (Fig. 19.1).

The field of crafting devices with atomic precision, where each atom and chemical bond is meticulously identified, is referred to as nanotechnology (Shea 2008). Nanotechnology entails the examination of materials at scales ranging from 1 to 100 nanometers, leveraging unique physical characteristics to spearhead innovative uses. This definition is a little tight in terms of measurements of size. The materials' ability to solve problems may have been emphasized further. "Particulate between 10 and 1,000 nm in size dimensions that are simultaneously colloidal particulate" is one definition of nanoparticles that has been proposed in an attempt to characterize them from an agricultural perspective (Mukhopadhyay 2014; Sajid and Płotka-Wasylka 2020).

Excessive use of synthetic chemicals, fertilizers, and herbicides can permanently harm soil ecosystems and microflora, rendering them unfit for long-term agricultural use. The international community is turning to biological sources of biofertilizers and biopesticides due to a lack of sustainable and environmentally acceptable

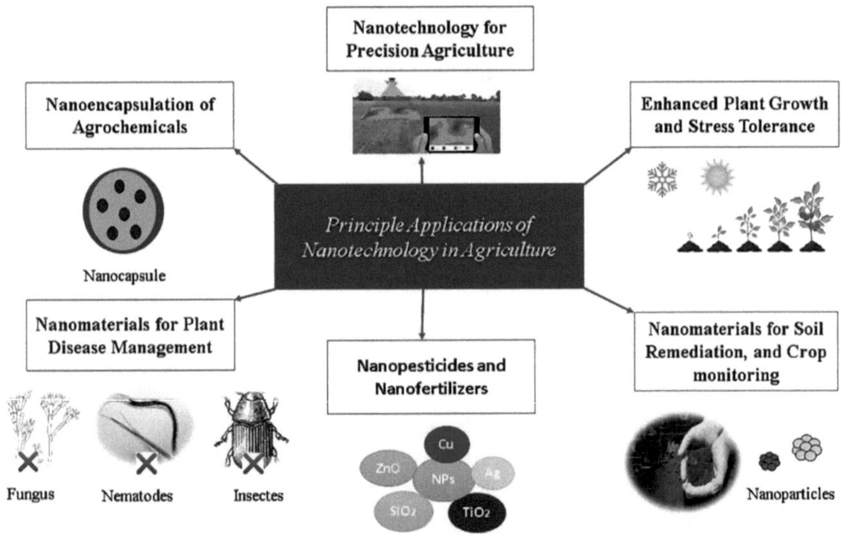

Fig. 19.1 The fundamentals of nanotechnology application in contemporary agriculture

technologies and increased knowledge of agriculture; yet, these products still have issues with shelf life, field stability, active dosage, and coverage area (Prashar and Shah 2016). Applying nanomaterials is essential to increase fertilization use efficiency, yields, and reduce the need for pesticides (An et al. 2022). They are also needed to detect pathogens and toxic chemicals in food items quickly and accurately, to develop smart pesticide and fertilizer delivery systems, to develop intelligent food packaging, processing systems, and to regulate food security in agriculture (Neme et al. 2021).

Additional areas of use for providing support in food processing include the device for producing food being deposited (by biofilm coating), nano-filters, sieves, and nanomembranes, as well as catalytic agents and adsorbents with nanoscale and nanocomposite structures (Bartolucci 2017; Neme et al. 2021). Food packaging using nanoparticles is intended to reduce the difficulty encountered when packing items and potentially reduce the creation of waste and the utilization of precious raw materials (Ashfaq et al. 2022; Neme et al. 2021). Numerous advancements in nanotechnology for food production and processing are underway, encompassing areas like nano-sensors, nanoencapsulation, food additives, nano-packaging, smart delivery systems utilizing nanoparticles, as well as applications in pharmaceuticals and healthcare (Rashidi and Khosravi-Darani 2011; Thiruvengadam et al. 2018; Kumar et al. 2024c; Mariyam et al. 2024). Applications also include, biopolymer matrices, encapsulation formation, simple solutions, emulsions, and associated colloids (micelles) providing efficient delivery systems (John et al. 2011; Zabot et al. 2022).

19.2.2 Plant Pathology and Nanobioformulations

Every year, a colossal 4.6 million tonnes of pesticides are deployed in agriculture, with almost 90% of this quantity ending up in the ecosystem. This phenomenon causes severe ecological imbalances, pollutes soils and water sources, promotes bioaccumulation in the food chain and contributes to a decline in biodiversity. However, recent approaches to modern agriculture, based on nanobioformulation, have provided a solution for replacing pesticides with a bioactive ingredient. Nanotechnology offers innovative solutions to preharvest and postharvest crop management generally focuses on controlling antimicrobial, insecticidal and nematicidal activities based on various bioformulations, while maintaining fruit and vegetable quality, improving the efficiency of agricultural inputs, resistance to environmental stresses and promoting crop growth (Dinesh et al. 2023; Jampílek and Kráľová 2019). It is possible to develop targeted treatments to control crop pests and diseases, while extending the shelf life of agricultural products thanks to smart packaging and nanotechnological coatings (Rana et al. 2021; Kumar et al. 2024a).

Nanobioformulations based on bacteria (BBCA) hold significant potential for plant protection efforts by efficiently decreasing infection diseases (Padmakumar et al. 2023), thereby enhancing plant growth and productivity (Bidyarani et al. 2022). Microbial biopesticides are developed taking into account their pathogenic capacity towards infectious organisms. Many crops are vulnerable to attack by various fungal and bacterial pathogens that cause remarkable production losses, such as *Bipolaris sorokiniana*, *Penicillium digitatum*, *Fusarium* sp., *Alternaria* sp., *Aspergillus* sp., *Xanthomonas* sp., etc. (Deshmukh et al. 2022; Hassan and Chang 2017; Taheri et al. 2023). The use of antimicrobial chemicals (fungicides, bactericides) is widespread to control these infections, and compromises the sustainability of normal soil flora and environmental safety (Lamichhane et al. 2018). Various NPs based on metals, such as Ag, CNT, Cu, S,Ti, Zn, ZnO have been documented to be more effective at very low concentrations against certain pathogenic fungi instead of using synthetic chemicals (Ameen et al. 2021).

However, these NPs likewise exert bioaccumulation and genetic mutation problems in the biosphere (González-Morales et al. 2020). Therefore, advances in nanobiologics seem to be a better solution to all safety issues and concerns. Nanoformulations that use biologically active substances from micro-organism-based products (antibiotics, enzymes, etc.) and their marketable metabolites as biostimulants, bio-fertilisers and biopesticides to stimulate plant growth, nutrient acquisition and infectious disease control while respecting the environment are known as nanobioformulations (Arit et al. 2021; Saharan et al. 2023). Bioformulation preparations based on several microbial strains are frequently used, including *Bacillus*, *Lactobacillus*, *Paenibacillus*, *Pseudomonas*, *Actinobacteria*, *Azospirillum*, *Acetobacter*, *Herbaspirillum*, *Serratia*, *Burkholderia* and *Rhodococcus* (Backer et al. 2018; Bidyarani et al. 2022; Haldar and Sengupta 2023), and fungi, as *Trichoderma* spp. and mycorrhizal fungi (Afzal et al. 2020; Bidyarani et al. 2022). Nutrients can be loaded ontonano-biological products prepared using a variety of methods, such as encapsulation/trapping, uptake and ligand binding (Deshmukh et al. 2022; Pirzadah et al. 2019). Chitosan nanoparticles are commonly used as biocompatible nanotechnology carriers for delivering biopesticides.

These chitosan nanoparticles are used to deliver various active agents such as etofenprox (Hwang et al. 2011), *Lippia sidoides* (Paula et al. 2010), and oregano (Hosseini et al. 2013). A recent study demonstrated the acquisition of chitosan nanoparticles incorporating α-pinene (P-CSNP) through nano-precipitation, wherein an edible coating (EC-P-CSNP) effectively controlled *Alternaria alternata* growth during the cold storage of bell peppers, thereby preserving their physicochemical quality (Hernández-López et al. 2020). Nanoencapsulation facilitates the gradual release of host-targeted chemicals to control insect pests, through various mechanisms such as biodegradation, dissolution, osmotic pressure and diffusion at a specific pH. Garlic oil-containing nanoparticles prove effective in controlling *Tribolium castaneum* (Rana et al. 2021). Some organisms treated with nanoparticles show a significant improvement in their efficacy against infectious pests.

19.3 Nanobioformulations in Preharvest Crop Management

According to the Food and Agriculture Organization of the United Nations (FAO) report from 2011, global food production losses were estimated at approximately one-third, or around 1.3 billion tonnes per year (Redlingshöfer 2017). It is noted that preharvest crop losses amount to about 35%, exceeding 1 billion tonnes (Mt). Concurrently, losses during harvesting and storage are estimated at around 690 million tonnes (Mesterházy et al. 2020), due to various factors including yield losses induced by biotic stresses, which vary depending on crops and regions. In developing countries, cereal crop losses before and after harvesting vary considerably, ranging from 15–100% and 10–60%, respectively. Insects, responsible for about 20% of global annual crop losses, can also transmit pathogens, causing plant diseases. Damage caused by pests varies between 5% and 50% depending on the crop, while pathogenic diseases can lead to losses of up to 30% (Rathee and Dalal 2018; Van Esse et al. 2020).

However, abiotic stresses such as drought, salinity, heavy metals, and mineral toxicity induce morphological, physiological, biochemical, and molecular alterations in plants, negatively impacting their growth, development, and productivity (Choudhary et al. 2025). These stresses can result in significant losses, ranging from 50% to 70% in terms of growth and yield (Zahra et al. 2021). The widespread use of synthetic pesticides to counter abiotic and biotic stresses in agriculture has led to concerning environmental and health consequences, such as water and soil pollution, biodiversity decline, and increased health risks including cancer and mental disorders. These challenges underscore the urgent need to find sustainable alternative solutions to meet global food needs while preserving the environment and human health (Zhang 2018). Generally, pesticides, fertilizers, and antibiotics are sprayed in agriculture and can be efficiently eliminated. Moreover, the production cost of these pesticides and fertilizers is very high and must be controlled. The use of nanomaterials in agriculture aims to minimize nutrient losses to increase yields and reduce the amount of products needed to protect plants (Usman et al. 2020).

Bioformulation is seen as a promising alternative to chemicals and fertilizers in sustainable agriculture (Mishra et al. 2015), although it faces challenges in terms of conservation, stability, and efficacy, particularly on a large scale. At the same time, agro-nanobiotechnology is emerging as a promising solution to address global challenges of food security, sustainability, and environmental preservation, while avoiding additional obstacles related to biosafety regulation (Sangeetha et al. 2017a,b). Nanobioformulations resulting from the integration of nanotechnology with biologically active substances from microbial-based products (green products) and their metabolites have the potential to optimize the efficiency of nutrients applied in agriculture. Nanobioformulation is an approach that uses biostimulants, bio-inoculants, biofertilizers, and biopesticides to promote plant growth, enhance crop protection, optimize the efficiency of agricultural inputs, deliver plant stimulants, and increase crop productivity, while allowing for increased production and acquisition of nutrients and disease control without compromising the environment (Bidyarani et al. 2022; Patel et al. 2020) (Fig. 19.2).

Nanotechnologies and nanobioformulations offer new perspectives by harnessing the characteristics of ordinary flowers and microbes in an innovative way. This approach involves encapsulating these elements and their derivatives in nanoparticles, ensuring prolonged and regulated action with minimal side effects. Through these nanoparticles, active substances can be released in a controlled manner, thus limiting soil contamination and promoting precision agriculture.

19.4 Utilization of Nanobioformulations for Biofertilizer Deliverys

The increase in demand for agricultural products has led to an intensification of production, partly achieved through increased use of fertilizers. The consumption of inorganic fertilizers has risen by approximately 23% for nitrogen, 13% for phosphorus, and 56% for potassium. These fertilizers are beneficial for increasing agricultural yields, improving the quality of food for human or animal consumption, and reducing both the necessary arable land and the need to convert new land for agricultural purposes (UN DESA 2022). In 2018, agriculture used nearly 190 million tons of inorganic fertilizers, with a projected demand expected to exceed 197 million tons by 2024. Global revenue generated by sales of inorganic fertilizers was

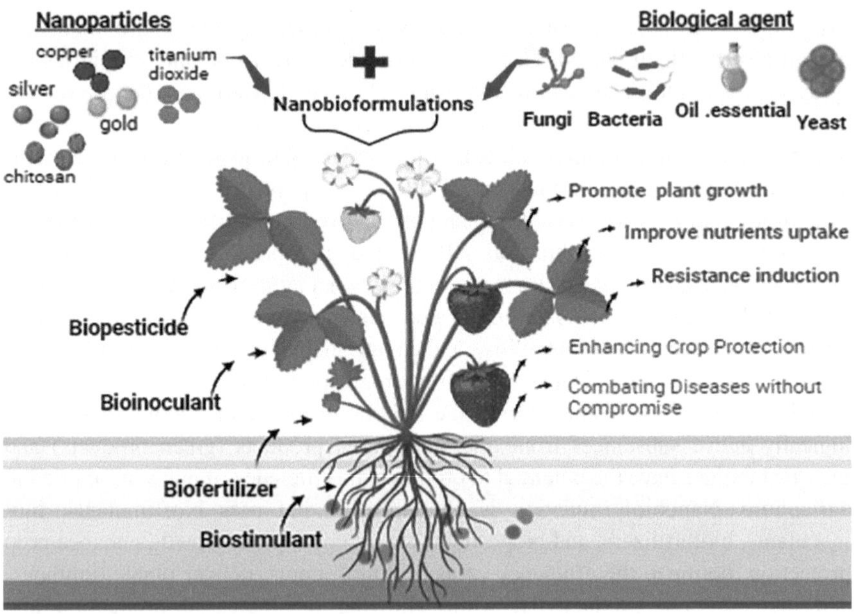

Fig. 19.2 Nanobioformulation: Boosting agricultural efficiency and sustainability

about $151 billion in 2018. According to projections for the period 2020–2025, the market is expected to achieve a Compound Annual Growth Rate (CAGR) of 3.8%. Despite the use of significant quantities of organic fertilizers, the consumed volumes and monetary value of these products are not systematically recorded (FAO 2022a).

The use of biofertilizers provides an effective alternative to minimize the use of synthetic fertilizers, allowing microbes to convert organic matter into simpler compounds. These rhizosphere microbes produce extracellular molecules that promote plant growth and help them better respond to environmental stress. However, biofertilizers are sensitive to temperature, prone to desiccation during storage, and have a reduced shelf life. Biofertilizers containing various strains of microorganisms, such as *Bacillus*, *Pseudomonas, Glomus,* etc., can be used alone or in combination in commercial agriculture (Bidyarani et al. 2022).

Nanobioformulations based on bacteria can play a significant role in environmental decontamination by effectively reducing metallic contaminants present in the rhizosphere, resulting in improved plant growth and productivity (Dikshit et al. 2013). The combined use of nano-fertilizers with bacteria like *Azotobacter* and *Pseudomonas* on wheat crops has shown positive results, increasing spike length, spike, and seed numbers. These nanobiofertilizers enhance plant growth by creating a favorable soil environment in the rhizosphere. Despite some challenges such as performance under variable climatic conditions and the need for high dosages for optimal coverage, the unique characteristics of nanobioformulations, including a high surface-to-volume ratio and improved physicochemical properties, enhance their effectiveness (Bidyarani et al. 2022).

19.5 Use of Nanobioformulations Against Plant Diseases

The widespread use of antimicrobial chemicals to control plant infections compromises the sustainability of normal soil flora and environmental safety. To reduce the use of these chemicals, nanotechnologies offer promising alternatives, including nanoparticles based on metals, carbon, and metal oxides, which have proven effective at very low concentrations. However, these nanobioformulations may also pose challenges such as bioaccumulation and genetic mutations in the biosphere. Nevertheless, advancements in nanobioformulations appear to offer a safer and more sustainable solution for plant disease control and increasing agricultural productivity.

19.5.1 Use of Bacteria and Fungi Based Nanobioformulations

Several studies have shown the interest of use of nanobioformulations based on bacteria or their metabolites for the control of plant pathogens. Indeed, many bacterial species have been included in the development of nanobioformulations with high efficiency against plant diseases. These bacterial species belong to different genera such as Bacillus, Pseudomonas and Streptomyces. In this context, *Bacillus siamensis* silver nanoparticles were demonstrated to have a potential effect in preventing brown stripe infections and bacterial leaf blight on rice plants. Moreover, these nanobioformulations have also showed a good ability in promoting plant growth (Ibrahim et al. 2019). Besides, *Bacillus cereus RNT6* zinc oxide nanobioformulation has been revealed effective in the management of rice panicle blight bacterial disease (Ahmed et al. 2021).

Regarding the genus Pseudomonas, *Pseudomonas poae* based silver nanoparticles have been found to suppress spore germination and germ tube elongation, in addition to prevent mycotoxin production and to cause membrane damage of *Fusarium graminearum*; the causal agent of wheat head bligh (Ibrahim et al. 2020). Furthermore, (Shabani et al. 2024), have developed a nanobioformulation utilizing one of two combinations of polymer monomers (alginate, starch, and chitosan), incorporating Talaromyces flavus as a biocontrol agent. This formulation has shown effectiveness against the pathogenic fungus *Verticillium dahliae*, the causative agent of cotton wilt disease. The researchers have successfully engineered nanobioformulations in the forms of nanocapsules, nanoemulsions, and nanopowders. Another study aimed to produce nanoformulations based on the antagonistic fungus *Talaromyces flavus,* focusing on their ease of application and effectiveness in biological control of plant diseases. These nanoformulations included two types of nanocapsules, a nanoemulsion, and a powdered form. The objective was then to assess the effect of these nanoformulations on sporulation, active population, and inhibition of the growth of certain important terrestrial pathogenic agents, such as *Verticillium dahliae* and *Fusarium oxysporum* (Naraghi et al. 2018).

Moreover, *Pseudomonas rhodesiae* silver nanoparticles have been shown to be effective in the control of soft rot diseases in sweet potato plants through inducing biofilm formation inhibition and protecting potato tubers (Hossain et al. 2019). Furthermore, it's important also to draw exemples on nanobioformulations based on bacteria of the genus Streptomyces. Indeed, *Streptomyces griseus* copper nanoparticles have exhibited a potential effect against the red rot disease in Tea plants (Ponmurugan et al. 2016). Moreover, *Streptomyces pseudogriseolus* ACV-11 silver nanoparticles have resulted in an important activity against different phytopathogenic fungi and insect pests including *Alternaria alternata, Culex pipiens, Fusarium oxysporum, Aspergillus niger Pythium ultimum,* and *Musca domestica* (Fouda et al. 2020). In addition to the application of Streptomyces bacteria in the elaboration of nanobioformulations efficient against plant pathogens. An interesting macrocyclic lactone produced by *Streptomyces avermitilis* has also been used for this purpose. Indeed, for exemple, abamectin viruses nanoparticles have exhibited an important toxicity against nematodes (Cao et al. 2015). Similarly, the lactone avermectin used for the synthesis of chitosan nanoparticles has been demonstrated efficient against wood nematodes (Liang et al. 2018).

Fungi were also used successfully in the development of nanobioformulations controlling plant diseases. Thus, for instance, Trichoderma based silica and copper nanoparticles have been revealed a good solution for the management of collar canker and red root rot disease of tea (Natesan et al. 2021). Moreover, Talaromyces Flavus based nanobioformulations have been shown to exhibit a good effect against different plant diseases. It's the case for exemple of the control of cotton Verticillium wilt and Tomato Fusarium Wilt Diseases using nanobioformulations including this fungus '*Talaromyces flavus*' (Naraghi et al. 2018; Naraghi and Negahban 2019).

19.5.2 Use of Plant Extracts and Essential Oils Based Nanobioformulations

Plant extracts and essential oils represent an important mean for the control of different plant diseases, thanks to their antimicrobial and insecticidal activities. Thus, recently, several studies have been conducted in order to develop nanobioformulations based on these plants' secondary metabolites. In this context, silver nanoparticles including extracts of different plants species, namely *Moringa oleifera*, *Conyza dioscoridis* and *Melia azedarach* have been found to exhibit an important insecticidal activity against juveniles and eggs of the root-knot nematode *Meloidogyne incognita* (Abbassy et al. 2017). Moreover, silver nanoparticles synthesized with extracts from Urtica urens have also displayed an important effect against this nematode (Nassar 2016). Furthermore, plants secondary metabolites based nanobioformulations have also exhibited a potential activity against a wide range of microorganisms. It is the case of *Artemisia absinthium* essential oil based copper nanoparticles which have exerted a potential effect against several bacteria including *Pseudomonas aeruginosa* (Celebi et al. 2021).

Essential oils derived from ginger have been extensively utilized in the form of nanoemulsions, as well as solid-lipid nanoparticles, to encapsulate natural bioactive compounds, thus enhancing their antimicrobial activity. Nano-droplets, with a size less than or equal to 100 nm, are widely employed to combat tropical plant diseases. Various polymer-based nanoparticles utilizing garlic essential oil, such as poly (DL-lactide-co-glycolide), chitosan, and zein, are widely employed to encapsulate active ingredients, namely cinnamaldehyde, eugenol, and carvacrol. Various emulsion-based nanobioformulations have been extensively reported to control plant pathogens both in vitro and in field trials. Targeted living entities include *Pseudomonas aeruginosa,* Salmonella sp., *Escherichia coli, Fusarium oxysporum*, etc. (Abdullahi et al. 2020). moreover, it has been also targeted successfully by cadmium oxide nanoparticles including leaves extracts of *Leucaena leucocephala.* Indeed, this nanobioformulation has induced a potent effect against *Pseudomonas aeruginosa* (Ghotekar 2017). Besides, recently, silver nanoparticles containing extracts of Hedera nepalensis have also showed an important activity against two bacterial phytopathogens, namely, *Ralstonia solanacearum* and *Erwinia carotovora* (Bibi et al. 2023) Overall, nanobioformulations constitute a green and a good solution for the management of plants diseases caused by different pathogenic agents. Therefore, their application in the agricultural sector represent a natural, promising and efficient mean replacing chemical pesticides.

19.6 Use of Nanobioformulation for Enhancing Crop Productivity

A multitude of products utilizing nanotechnologies are currently employed in the agricultural sector to address various challenges and enhance productivity. These products are designed to operate at nanoscale levels, thus providing significant advantages. By reducing the use of chemicals, they contribute to a more sustainable and environmentally friendly agriculture. Additionally, these products improve soil quality and fertility, thereby promoting optimal growth conditions for crops. By decreasing soil and water contamination from potentially harmful chemicals, nanotechnology-based products also contribute to environmental preservation and ecosystem health. They also help minimize health risks for consumers by limiting the presence of pesticide residues in food products. Similarly, they offer safer working conditions for farmers by reducing their exposure to potentially hazardous substances (FAO/OMS 2011). Due to these benefits, nanobioformulation emerges as the optimal solution for ensuring prolonged release, administering targeted active ingredients, reducing toxicity, enhancing bioavailability, and safeguarding ecosystem balance (Duhan et al. 2017).

Recently, Pereira et al. (2019) evaluated the effects of priming seeds using nanocarrier systems containing gibberellic acid (GA_3) on the growth and productivity of *Solanum lycopersicum,* cultivated under field conditions. This innovative study aims to address a gap in the field, focusing specifically on the use of nanosystems for plant growth regulators (PGRs). The results revealed that the use of these nanocarrier systems could significantly enhance fruit production, with productivity increasing nearly fourfold with the use of nanoALG/CS-GA3. Thus, this research highlights the promising potential of nanocarrier systems containing PGRs for agricultural applications, thereby contributing to promoting sustainable agriculture. Other studies have also explored the use of newly synthesized polymer nanocapsules (NC) as biocompatible vectors for delivering bioactive compounds to plants, particularly by loading the NC with gibberellic acid (GA). These research efforts aim to develop innovative plant cultivation methods using sustainable nanoparticles, by harnessing lignin as a raw material, a by-product of wood transformation. They seek to evaluate the effects of these nanocapsules on the germination, growth, and health of model plants *Eruca vesicaria* and *Solanum lycopersicum*, both in vitro and in vivo. (Falsini et al. 2019).

19.7 Nanobioformulations in Post Harvest Crop Management

The diversity of the world's agroclimatic zones favors a wide variety of plant crops, requiring special attention to ensure optimum yield and quality. However, physical damage and microbial diseases are major factors leading to crop losses. Fruit, with its delicate texture, is particularly sensitive to these risks (Jayasekhar Babu et al. 2022). Bioformulation is emerging as a promising alternative to fertilizers and agricultural chemicals, offering benefits for sustainable agriculture (Mishra et al. 2015). Nevertheless, despite their potential, bioformulations have limitations such as limited stability and efficacy when applied in large quantities in the field. To overcome these obstacles, nanotechnology has become a solution of choice in the agricultural field, paving the way for agro-nanobiotechnology (Mishra et al. 2014.

Nanoparticles are seen as efficient vectors for transporting nutrients to target plants, contributing to their growth and yield. This convergence between nanotechnology and biologically active products derived from micro-organisms has given rise to nanobioformulation. Nanobioformulation is widely used to provide biofertilizers, biostimulants and biopesticides to stimulate plant growth, control disease and increase agricultural productivity (Palmqvist et al. 2015; Naraghi et al. 2018; Patel et al. 2020). Several microorganisms have been extensively employed in the production of nanoparticles (NPs) utilizing various inorganic metals like Ag, Al, Au, MnO, ZnO, and TiO_2 in recent decades. Employing these NPs has notably enhanced agricultural output by mitigating losses linked to crop processing, transportation, and storage (Kumari et al. 2019). Microorganisms frequently employed in the preparation of these formulations include *Bacillus*, *Pseudomonas*, *Lactobacillus*, *Paenibacillus*, *Serratia*, *Burkholderia*, and fungi such as *Trichoderma* spp. and mycorrhizal fungi (Backer et al. 2018; Aamir et al. 2020).

A variety of techniques are employed to load nutrients into nanobioformulations, including encapsulation, uptake and binding via ligands (Pirzadah et al. 2019). Nanobioformulation has many advantages (Fig. 19.3) such as lower toxicity, increased durability, versatility, improved effectiveness, and enhanced solubility (Kumar et al. 2021) Nanomaterials can be synthesized using two main approaches: top-down and bottom-up. On a commercial scale, nanomaterial production is primarily achieved through the bottom-up method, where bulk precursors are reduced to a nanoscale size using techniques such as nanolithography, precision engineering, or grinding (Fadiji et al. 2022). Grinding is often employed to obtain fine powder with small particles and high water retention capacity. For instance, a study demonstrated that green tea powder with 1000 nm particles exhibits enhanced

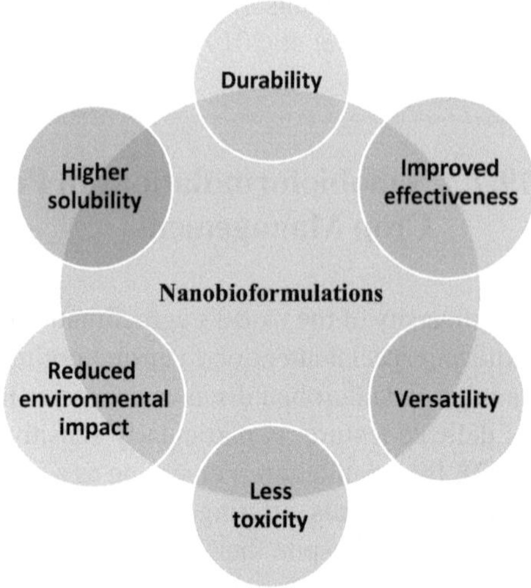

Fig. 19.3 Benefits of nanobioformulations in sustainable agriculture

nutrient digestion capability, resulting in increased antioxidant activity due to elevated dismutase enzyme activity (Sanguansri and Augustin 2006).

The bottom-up technique is associated with nanotechnology-based devices, which are often controlled by external forces to produce the necessary parameters and initiate particle formation from higher levels through packing and fragmentation, to achieve the required size (Sangeetha et al. 2017a,b). In the top-down approach, atoms are transformed into nanoscale materials, which can involve complex processes. The matching of attractive and repulsive interactions among paired molecules used as components to achieve efficient supramolecular assemblies is essential for self-assembly (Sanguansri and Augustin 2006). Self-assembly or clustering of casein micelles, for example, results in the creation of protein-carbohydrate complexes. The arrangement of casein micelles, structures generated within protein-polysaccharide liposomes, and their aggregates are examples of self-assembled nanostructures. According to research, the top-down strategy allows for the creation of nanostructures with fewer defects and a more uniform chemical arrangement (Pathakoti et al. 2017).

Advancements in the field of nanobioformulation appear to offer a more effective solution to safety and health concerns (Asadishad et al. 2018). Formulations such as nanocapsules, nanoemulsions, and nanopowders have been developed by researchers, demonstrating effective control over plant pathogenic fungi (Naraghi et al. 2018). Essential oils from ginger, converted into nanoemulsions, as well as solid lipid nanoparticles, are widely used to encapsulate natural bioactive compounds, thereby enhancing their antimicrobial activity. Surfactants, utilized to form nanodroplets smaller than 100 nm, are also employed in combating tropical plant diseases. Various polymer nanoparticles based on garlic essential oil, such as poly (DL-lactide coglycolide), chitosan, and zein, are commonly used to encapsulate active ingredients such as cinnamaldehyde, eugenol, and carvacrol. It is well-documented that different emulsion-based nanobioformulations are effective in controlling plant pathogens (Bidyarani et al. 2022). The synthesis of nanoparticles through green nanotechnology, utilizing plants and microorganisms, offers more attractive alternatives due to their eco-friendly nature and simplified production processes (Babu et al. 2018). The unique properties of nanoparticles are adeptly utilized to transform agricultural practices into sustainable ones. This results in minimizing economic losses caused by postharvest diseases in fruits and vegetables (Youssef and Roberto 2014; Feregrino-Perez et al. 2018).

In nanotechnology, the use of edible coatings offers multiple benefits, including the potential reduction of water loss, slowing down the ripening process, protection against mechanical damage and cooling, as well as prevention of rot and addition of shine to treated products. These coatings can also function as antimicrobial, antioxidant, or anti-maturation composites They can be made from various substances such as proteins, lipids, or polysaccharides, or combinations of these biomacromolecules (Upadhyay et al. 2022). Different edible coatings can be utilized on vegetables and fruits (Table 19.1). Furthermore, chitosan enhances tomatoes' resistance to gray mold caused by *Botrytis cinerea* and shows promise as a bio-based composite material to partially replace synthetic fungicides (Liu et al. 2007). Additionally, the application of a chitosan coating, combined with 1% arabic gum on bananas, reduced fruit ripening and increased shelf life by several days compared to untreated bananas (Maqbool et al. 2011).

A novel nanotechnology-based coating composed of pullulan and polymer nanocapsules, incorporating thyme essential oil, was applied to extend the shelf life of table grapes. Grapes treated with these nanocapsules maintained their color, firmness, and total soluble solids content for a prolonged period compared to untreated grapes. Additionally, the coated grapes showed a reduced incidence of gray mold in contrast to untreated ones. The nanocapsule coating effectively slowed down the evaporation of volatile compounds, leading to their prolonged presence on the grape surfaces. This mechanism contributed to delaying the ripening process and ultimately extending the shelf life of the grapes (Piña-Barrera et al. 2019).

Utilizing green nanosilver synthesis (Ag NP) through black tea extract as a nanopreservative has emerged as a potent, cost-effective, and eco-friendly approach. This method has extended the shelf life of apples and mitigated postharvest damage induced by *Monilinia fructigena*, all without posing any detrimental effects on human health (Madbouly 2021). The incorporation of pomegranate bark polyphenols into silica nanoparticles could be a reasonable method for controlling the fungi responsible for postharvest blue rot, replacing chemical fungicides with environmental and human health risks (Balooch et al. 2017). Various sizes of chitosan-based nanoparticles were examined in freshly cut apples regarding their microbiological profile, color, polyphenol oxidase and peroxidase activity, and hardness. Apples coated with chitosan nanoparticles measuring 110 nm demonstrated improved antibacterial characteristics and an overall reduction in microbial contamination compared to other treatments. Slices of apples treated with plain or untreated chitosan displayed slightly elevated levels of browning compared to those coated with nano-chitosan (Pilon et al. 2014).

Table 19.1 Examples of edible coatings applied to vegetables and fruits

Edible coatings	Vegetables/ fruits	Response	References
Ag nanoparticles (NPs)	Apple	Extending the shelf life of apples and reducing postharvest damage caused by *Monilinia fructigena*	Madbouly (2021)
Clay-chitosan nanocomposite	Orange	Antifungal activity against *Penicillium digitatum*	Youssef and Hashim (2020)
Chitosan NPs	Grapes	Improving moisture retention, slowing down the ripening process	Melo et al. (2018)
Gelatin/Kafirin	Vegetables	Reducing water loss	Khafar et al. (2018)
Gelatin/cellulose nanocrystal	Strawberries	Improving the shelf life of strawberries	Fakhouri et al. (2014)
Chitosan/ Tripolyphosphate	Banana	Slow down the ripening process	Esyanti et al. (2019)
Chitosan-silver NPs	Mango	Effectiveness in controlling *Colletotrichum gloeosporioides*	Dean et al. (2012)
Se NPs	Tomato	Preservation of tomato quality	Hernández-Hernández et al. (2019)
Gelatin-fiber/Titanium dioxide (TiO_2)	Melon	Effectiveness for mold control	Qiao et al. (2019)
Aloe vera-nanocomposite	Mango	Enhancing titratable acidity.	Dubey et al. (2019)
ZnO NPs	Cucumber	Increased carotene, zinc and iron content.	Seray et al. (2021)
Gelatin-Chitosan/ (Ag/ ZnO)	Okra	Antimicrobial polyethylene packaging for maintaining vegetable quality	Al-Naamani et al. (2018)
Microencapsulated ascorbic acid	Guava	Decreased the respiration rate	Martínez-Ortiz et al. (2019)

In fresh-cut apples stored at 4 °C for 2 weeks, ZnO NPs combined with polylactic acid were effective in preventing the growth of pathogens such as bacteria, molds, and yeasts (Li et al. 2017). Moreover, the 10–15 nm silver nanoparticle composite inhibited fungal growth, particularly the germination of *Colletotrichum gloeosporioides* conidia affecting various postharvest fruits (Dean et al. 2012). Similarly, a clay-chitosan nanocomposite was developed to counteract the effects of *Penicillium digitatum*, the pathogen affecting the sweet orange 'Valencia Late'. It exhibited total inhibition of the pathogen at a concentration of 20 $\mu g.mL^{-1}$ (Youssef and Hashim 2020). In another study, gelatin-based edible coatings incorporating cellulose nanocrystals notably prolonged the strawberries shelf life (Fakhouri et al. 2014). The utilization of chitosan-assisted nano-silica coating enhanced the physicochemical and physiological attributes of longan fruits at room temperature in comparison to alternative treatments, effectively producing an outstanding semi-permeable film (Shi et al. 2013). Likewise, nano-ZnO coating decreased microbial deterioration and maintained the quality of specific fruits after harvest while in storage (Sogvar et al. 2016).

On the other hand, nanotechnology provides the opportunity to create fruit packaging at the nanoscale, characterized by reduced relative humidity, increased oxygen permeability, and superior strength. These packages have been manufactured by blending polyethylene with nanopowders such as nano-silver, kaolin, anatase titanium dioxide, and mineral titanium dioxide, demonstrating a significant improvement in fruit quality preservation throughout storage (Sujithra and Manikkandan 2019). For instance, packaging using nanotechnology can help preserve the sensory attributes and chemical as well as physiological characteristics of strawberries more effectively than traditional packaging (Yang et al. 2010). Nanoemulsion is one of the flagship applications of nanotechnology. It comprises tiny emulsion droplets at the nanoscale (oil/water system) with sizes below 100 nm (Anton and Vandamme 2011). Forming these nanoemulsions requires high mechanical energy, typically achieved through steps such as high-pressure homogenization, the use of a microfluidizer, or sonication protocols to disperse the phases. Nanoemulsions reduce the need for stabilizers as they protect against food breakage and fragmentation, significantly reducing the required amount of fat. The final products of nanoemulsions closely resemble conventional food products, with no notable alteration in taste (Dasgupta et al. 2015). Nanoemulsions act by protecting the surface of fruits or as controlled release systems for lipophilic compounds such as antioxidants and other additives (Silva et al. 2012).

A previous study reported that nanoemulsions prepared with α-tocopherol and nopal mucilage had a significant impact on the activity of polyphenol oxidase and pectin methylesterase in freshly cut apples. Submicron-sized systems helped maintain color and firmness characteristics while controlling weight loss, suggesting that particle size plays a crucial role in regulating alterations that may affect product quality during storage (Zambrano-Zaragoza et al. 2014). A nanoemulsion coating made from carnauba wax with a diameter of 42 nm is a widely accepted approach. The wax, derived from carnauba palm leaves at a concentration of 2.4%, was applied to papaya fruits, which were then stored for 9 days at 22 °C and 60–70% relative humidity. Fruit rot is shown after reduction (Rhim 2004).

Chitosan in the form of nanoemulsions could be highly effective as a biofungicide to control anthracnose in tropical fruits while preserving their quality during cold storage (Zahid et al. 2012). In conclusion, bioformulations utilizing nanotechnology prove to be effective and sustainable in postharvest crop management. The future is heading towards an environmentally friendly approach with nanobioformulation in agriculture, aiming to control plant diseases and enhance crop productivity.

19.8 Challenges and Considerations

Nanomaterials NMs can have unique physico-chemical properties that could make them toxic to living organisms or cause undesirable effects on ecosystems, and so they present one of the main concerns in agriculture with their effects on the environment and human health. The challenges associated with the development of nanoparticles for agricultural applications are still under study, and issues of safety and standards assessment, public concerns, registration policies, manufacturing expenses, and other concerns need to be resolved. A rigorous evaluation of the effects of nanoparticles is needed prior to determining whether their use is safe (Singh and Gurjar 2022).

The safety threats posed by NMs to human health and other living things are a source of worry. According to a study by (Juárez-Maldonado et al. 2021), is a risk that the direct interaction between NMs and microbial cells will be less diverse and abundant of particular microbial groups in the soil (Yang et al. 2021). Any type of cell can proliferate as a result of this interaction's potential to cause particular modifications in gene expression and physiology of microbial cells (Juárez-Maldonado et al. 2021). However, the assessment of these risks remains mainly linked to the design and prediction of the destiny of nanoscale agrochemicals in real applications (field conditions) (Grillo et al. 2021; Okeke et al. 2022). Combining current models of agricultural yield and nutrient cycling with nano-informatics approaches to optimize the targeting, transport, capture of nutrients on soil microbial communities in order to devise designs that combine optimal safety and functional attributes of nanoscale agrochemicals (Zhang et al. 2021).

Although nanotechnology-based agrochemicals can offer many benefits, they have not yet reached their full potential. Multiple businesses have filed patent applications, including for many nanotechnology production and application protocols, but the number of commercial products based on nanotechnology is still relatively small and requires more advancement (Singh et al. 2021). According to the nanotechnology products database provided by the statistical database (StatNano), 9420 nanotechnology-based products have been marketed, of which only 229 nanoproducts are used for agricultural applications (An et al. 2022). The majority of these products are nano-fertilisers (43%), while nanoformulations designed to increase plant survival against diseases, pests and other stresses account for 28% of these products. However, utilizing NMs in agriculture represents a sector with narrow profit margins, making it challenging to yield substantial economic gains relative to the initial investment in manufacturing. This difficulty arises from the elevated production expenses and intricate manufacturing processes involved (An et al. 2022; Mariyam et al. 2024).

In addition, there are no unified methods and standards for NMs from government authorities (Kah et al. 2019). The government's current procedures for regulating nanopesticides (NPs) are either inapplicable or inadequate for evaluating the dangers to public health. It is imperative to do research and establish relevant assessment techniques and guidelines applied widely for NPs in agrochemicals (de

Oliveira Pereira et al. 2020). Due to the intricate nature of nano-structures, size, reactivity, shape and electric charge, characterization of the biological safety of NPs is a challenging task. This is because lipid preparations, such as liposomes, liquid crystals, and micelles, are used to deliver agrochemical ingredients (AcIs) that are poorly soluble in water, requiring the use of surfactants and potentially increasing pesticide toxicity (Chaud et al. 2021).

Future research on nano-agrochemicals will focus on the competition with conventional formulations regarding functionality, expense and large-scale production technology, particularly from a marketing point of view, creation of novel criteria to evaluate the safety and the quality of NMs products, along with the establishment of a meticulously regulated release mechanism capable of adapting to variations in both biotic and abiotic stimuli, such as enzymatic activity, temperature, light, and pH, etc., and developing specific usage standards and efficient technologies for the extensive field implementation of NMs to promote their advancement in agriculture (An et al. 2022).

19.9 Future Directions and Implications

In order to improve agricultural practices and guarantee global food security, the use of nanobioformulations to pre- and post-harvest crop management holds great potential. As technology develops, chances to address major issues facing the agriculture sector arise from the combination of nanotechnology and biological agents. Applications of nanotechnology in agriculture are still limited to lab-scale research, despite the potential benefits and the expanding trends in publications and patents. An investigation of the World Intellectual Property Nearly 4000 patent applications have been filed for the food and nano industries in the last 10 years, according to an organization database (Deshmukh et al. 2022). Different strategies are being used by OECD member nations and non-member (non-OECD) countries to regulate nanotechnology in the agri/feed/food sectors (Mishra et al. 2017).

When applying NPs to plants, it is essential to determine how their physicochemical characteristics affect phytotoxicity. To comprehend the impacts of modified NPs (MNPs) on agricultural applications compared to pristine NPs, long-term research carried out utilizing standard guidelines and GLPs are crucial. To understand MNP behavior, it is essential to perform research on their properties in a variety of media (such as soil, hydroponics, and synthetic medium). It is vital to optimize the dose at which nanoagri inputs are applied to certain plant species. It is also necessary to pay attention to studies on biogenic MNPs, which are thought to be less harmful to the environment and human health than their chemically manufactured equivalents. In-depth research is required to fully comprehend the significance of these MNPs for various trophic levels in the food chain, as well as their impacts on edible portions of plants and biomagnification. Moreover, a thorough database for in silico modeling and comparable nanomaterial hazard prediction would improve the utilization of nano-based agro-products and supplement assessment processes (Deshmukh et al. 2022). Insufficient research has been done on BCA nanobioformulations, nevertheless. To fully understand how nano delivery systems that are appropriate for host-pathogen interactions interact and to forecast their eventual destiny in the agricultural environment, comprehensive research must be carried out. An understanding of the mutual interaction of nanocarriers and BCAs is necessary to regulate BCA distribution and release at the desired location. Nanoformulated BCAs that are produced also need to be economical for farmers or end users and technologically sophisticated (Chaitanya et al. 2023).

Successful applications of nanotechnology in agriculture mightaccomplished by utilizing sophisticated scientific models of natural processes, with nature acting as an essential guide. For example, in order to support environmental sustainability and production, efforts may concentrate on improving soil nutrient efficiency. Future research should focus on reducing the dangers connected with the use of nanoparticles (NPs) in order to move this field forward and make sure that NPs used in laboratory settings are accepted more widely in the agricultural community. This means working together to improve NP synthesis techniques so they can be used in practical settings. Furthermore, it is crucial to comprehend the proper dosage levels of NPs within safety limits by conducting thorough dose-dependent investigations in natural soil systems. Furthermore, it is essential to look into the physio-chemical characteristics of agricultural areas where nanobioformulations might be used to reduce dangers to soil biomes and plants. It is advisable to do comprehensive study on biosynthesized nanoparticles (NPs) due to their reduced toxicity, in order to adequately investigate their potential applications for sustainable agriculture practices (Mishra et al. 2017).

19.10 Conclusions and Future Prosepcts

In this summary, we observe that nanobioformulations offer a promising outlook in the field of agriculture, providing innovative solutions to address the growing challenges related to food security, environmental sustainability, and climate resilience. By leveraging the benefits of nanotechnologies, these formulations enable more efficient use of agricultural resources, enhance crop protection against diseases and pests, and improve agricultural productivity in a sustainable manner. However, it is crucial to acknowledge that the use of nanomaterials also poses potential risks to human health and the environment. Therefore, ongoing research is necessary to assess their safety and long-term impacts, and to develop appropriate guidelines and regulations for their responsible use. Despite these challenges, the potential benefits of nanobioformulations in terms of sustainability, efficiency, and resilience justify their continued exploration and gradual adoption in modern agriculture. By combining the latest technological advancements with traditional farming practices and integrated crop management approaches, we can shape a future where agriculture is both productive, environmentally friendly, and capable of meeting the growing global food needs.

Acknowledgements This research received financial support from the Phytopathology Unit within the Department of Plant Protection and Environment at The Ecole Nationale d'Agriculture de Meknès-Morocco.

References

Aamir M, Rai KK, Zehra A, Dubey MK, Kumar S, Shukla V, Upadhyay RS (2020) Microbial bioformulation-based plant biostimulants: a plausible approach toward next generation of sustainable agriculture. In Microbial Endophytes (pp. 195–225). Elsevier. https://doi.org/10.1016/B978-0-12-819654-0.00008-9

Abdullahi A, Ahmad K, Ismail IS, Asib N, Ahmed OH, Abubakar AI, Siddiqui Y, Ismail MR (2020) Potential of using ginger essential oils-based nanotechnology to control tropical plant diseases. Plant Pathol J 36(6):515–535. https://doi.org/10.5423/PPJ.RW.05.2020.0077

Abbassy MA, Abdel-Rasoul MA, Nassar AMK, Soliman BSM (2017) Nematicidal activity of silver nanoparticles of botanical products against root-knot nematode, Meloidogyne incognita. Archives of Phytopathology and Plant Protection 50(17–18):909–926. https://doi.org/10.1080/03235408.2017.1405608

Afzal A, Ijaz M, Shah SRA (2020) Determination of suitable growth stage for application of fungicide against stripe rust in wheat. Pak J Agric Res 33(4):714–719. https://doi.org/10.17582/journal.pjar/2020/33.4.714.719

Ahmed T, Wu Z, Jiang H, Luo J, Noman M, Shahid M, Manzoor I, Allemailem KS, Alrumaihi F, Li B (2021) Bioinspired green synthesis of zinc oxide nanoparticles from a native Bacillus cereus strain RNT6: characterization and antibacterial activity against rice panicle blight pathogens Burkholderia glumae and B. gladioli. Nanomaterials 11(4):884. https://doi.org/10.3390/nano11040884

Al-Naamani L, Dutta J, Dobretsov S (2018) Nanocomposite zinc oxide-chitosan coatings on polyethylene films for extending storage life of okra (Abelmoschus esculentus). Nanomaterials 8(7):479

Ameen F, Alsamhary K, Alabdullatif JA, ALNadhari, S. (2021) A review on metal-based nanoparticles and their toxicity to beneficial soil bacteria and fungi. Ecotoxicol Environ Saf 213:112027. https://doi.org/10.1016/j.ecoenv.2021.112027

An C, Sun C, Li N, Huang B, Jiang J, Shen Y, Wang C, Zhao X, Cui B, Wang C, Li X, Zhan S, Gao F, Zeng Z, Cui H, Wang Y (2022) Nanomaterials and nanotechnology for the delivery of agrochemicals: strategies towards sustainable agriculture. J Nanobiotechnol 20(1):1–19. https://doi.org/10.1186/s12951-021-01214-7

Anton N, Vandamme TF (2011) Nano-emulsions and micro-emulsions: clarifications of the critical differences. Pharm Res 28:978–985

Arit MD, Rahman MA, Milu MMH, Siddik AB, Hogue ME (2021) Green nanomaterials: synthesis, properties and spectroscopic applications. In: Nanomaterials for spectroscopic applications, pp 213–272. https://doi.org/10.1201/9781003160335-8/GREEN-NANOMATERIALS-SYNTHESIS-PROPERTIES-SPECTROSCOPIC-APPLICATIONS-MUAMMER-DIN-ARIT-MD-ASADUR-RAHMAN-MD-MAHMUDUL-HAGUE-MILU-ABU-BAKAR-SIDDIK-MD-ENAMUL-HOGUE

Asadishad B, Chahal S, Akbari A, Cianciarelli V, Azodi M, Ghoshal S, Tufenkji N (2018) Amendment of agricultural soil with metal nanoparticles: effects on soil enzyme activity and microbial community composition. Environ Sci Technol 52(4):1908–1918

Ashfaq A, Khursheed N, Fatima S, Anjum Z, Younis K (2022) Application of nanotechnology in food packaging: pros and cons. J Agric Food Res 7(September 2021):100270. https://doi.org/10.1016/j.jafr.2022.100270

Ashraf SA, Siddiqui AJ, Elkhalifa AEO, Khan MI, Patel M, Alreshidi M, Moin A, Singh R, Snoussi M, Adnan M (2021) Innovations in nanoscience for the sustainable development of food and agriculture with implications on health and environment. Sci Total Environ 768:144990. https://doi.org/10.1016/j.scitotenv.2021.144990

Babu PJ, Raichur AM, Doble M (2018) Synthesis and characterization of biocompatible carbon-gold (C-Au) nanocomposites and their biomedical applications as an optical sensor for creatinine detection and cellular imaging. Sensors Actuators B Chem 258:1267–1278

Backer R, Rokem JS, Ilangumaran G, Lamont J, Praslickova D, Ricci E, Subramanian S, Smith DL (2018) Plant growth-promoting rhizobacteria: context, mechanisms of action, and roadmap to commercialization of biostimulants for sustainable agriculture. Front Plant Sci 871. https://doi.org/10.3389/FPLS.2018.01473

Baker S, Satish S, Prasad N, Chouhan RS (2019) Nano-agromaterials: influence on plant growth and crop protection. In: Industrial applications of nanomaterials. Elsevier, pp 341–363. https://doi.org/10.1016/B978-0-12-815749-7.00012-8

Balooch MR, Sabahi H, Aminian H, Nourinejhad S (2017) Antifungal effects of pomegranate extract intercalated into silicate nanoparticles against blue mold of apple. Iranian J Plant Pathol 53(2):151–164

Barrett CB (2021) Overcoming global food security challenges through science and solidarity. Am J Agric Econ 103(2):422–447. https://doi.org/10.1111/ajae.12160

Bartolucci C (2017) Nanotechnologies for agriculture and foods: past and future. In: Nanotechnology in agriculture and food science. Wiley, pp 1–14. https://doi.org/10.1002/9783527697724.ch1

Bibi Z, Ali M, Abohashrh M, Ahmad I, Khan H, Ali M, Akbar F, Ahmad N, Iqbal A, Ullah F, Ullah Z, Muhmood T (2023) Biologically Synthesized Silver Nanoparticles Efficiently Control Plant Pathogenic Bacteria-Erwinia carotovora and Ralstonia solanacearum. Inorganics 11(7):309. https://doi.org/10.3390/inorganics11070309

Bidyarani N, Vyas G, Jaiswal J, Deshmukh SK, Kumar U (2022) Recent approaches in nanobioformulation for sustainable agriculture. In: Nanotechnology in agriculture and environmental science. CRC Press, pp 166–182. https://doi.org/10.1201/9781003323945-14

Cao J, Guenther RH, Sit TL, Lommel SA, Opperman CH, Willoughby JA (2015) Development of abamectin loaded plant virus nanoparticles for efficacious plant parasitic nematode control. ACS Appl Mater Interfaces 7(18):9546–9553. https://doi.org/10.1021/acsami.5b00940

Celebi O, Tolga Cinisli K, Celebi D (2021) Antimicrobial activity of the combination (Nano-Bio) of Artemisia absinthium with copper nanoparticles. Materials Today: Proceedings 45:3809–3813. https://doi.org/10.1016/j.matpr.2021.01.824

Chaitanya AK, Jamedar HVR, Shanmugam A, Kaniganti S, Devi YL, Kumar PG, Mekala R, Pitha CC, Wani SH (2023) Advances in QTL mapping for biotic stress tolerance in wheat. In: QTL mapping in crop improvement, pp 119–148. https://doi.org/10.1016/B978-0-323-85243-2.00025-8

Chaud M, Souto EB, Zielinska A, Severino P, Batain F, Oliveira-Junior J, Alves T (2021) Nanopesticides in agriculture: benefits and challenge in agricultural productivity, toxicological risks to human health and environment. Toxics 9(6):131. https://doi.org/10.3390/toxics9060131

Choudhary R, Rajput VD, Ghodake G, Ahmad F, Meena M, Rehman RU, Prasad R, Sharma RK, Singh R, Seth CS (2024) Comprehensive journey from past to present to future about seed priming with hydrogen peroxide and hydrogen sulfide concerning drought, temperature, UV and ozone stresses- a review. Plant Soil 500(1–2):351–373. https://doi.org/10.1007/s11104-024-06499-9

Choudhary R, Ahmad F, Kaya C, Upadhyay SK, Muneer S, Kumar V, Meena M, Liu H, Upadhyay H, Seth CS (2025) Decrypting proteomics, transcriptomics, genomics, and integrated omics for augmenting the abiotic, biotic, and climate change stress resilience in plants. J Plant Physiol. https://doi.org/10.1016/j.jplph.2025.154430

Dasgupta N, Ranjan S, Mundekkad D, Ramalingam C, Shanker R, Kumar A (2015) Nanotechnology in agro-food: from field to plate. Food Res Int 69:381–400 de Oliveira Pereira FS, Agarrayua DA, Quines CB, Ávila D (2020) Risk assessment of nanofertilizers and nanopesticides. In: Nanopesticides. Springer International Publishing, pp 299–316. https://doi.org/10.1007/978-3-030-44873-8_10

Dean R, Van Kan JAL, Pretorius ZA, Hammond-Kosack KE, Di Pietro A, Spanu PD, Rudd JJ, Dickman M, Kahmann R, Ellis J (2012) The Top 10 fungal pathogens in molecular plant pathology. Mol Plant Pathol 13(4):414–430

DeRosa MC, Monreal C, Schnitzer M, Walsh R, Sultan Y (2010) Nanotechnology in fertilizers. Nat Nanotechnol 5(2):91–91. https://doi.org/10.1038/nnano.2010.2

Deshmukh SK, Kochar M, Kaur P, Singh PP (2022) Nanotechnology in agriculture and environmental science. In: Nanotechnology in agriculture and environmental science. https://doi.org/10.1201/9781003323945

Dikshit A, Shukla SK, Mishra RK (2013) Exploring nanomaterials with PGPR in current agriculture scenario

Dinesh GK, Karthika S, Ragul S, Sinduja M, Sathya V, Sivashankari L, Divyabharathi P, Elangovan A, Prasad S (2023) Effect of nano-formulated agrochemicals on rhizospheric communities in millets, pp 293–330. https://doi.org/10.1007/978-981-99-2166-9_15

Dubey PK, Shukla RN, Srivastava G, Mishra AA, Pandey A (2019) Study on quality parameters and storage stability of mango coated with developed nanocomposite edible film. Int J Curr Microbiol App Sci 8(4):2899–2935

Duhan JS, Kumar R, Kumar N, Kaur P, Nehra K, Duhan S (2017) Nanotechnology: the new perspective in precision agriculture. Biotechnol Rep (Amst) 15:11–23. https://doi.org/10.1016/j.btre.2017.03.002

Duro JA, Lauk C, Kastner T, Erb KH, Haberl H (2020) Global inequalities in food consumption, cropland demand and land-use efficiency: a decomposition analysis. Glob Environ Chang 64(May):102124. https://doi.org/10.1016/j.gloenvcha.2020.102124

El-Beltagy A, Madkour M (2012) Impact of climate change on arid lands agriculture. Agric Food Secur 1(1):1–12. https://doi.org/10.1186/2048-7010-1-3

El-Ramady HR, Verma KK, Rajput VD, Minkina T, Elbehiry F, Elbasiouny HY, Elsakhawy T, Omara AED, Amer MM (2022) Sources of silicon and nano-silicon in soil and plants. In: Silicon and nano-silicon in environmental stress management and crop quality improvement: progress and prospects. Academic Press, Elsevier. https://doi.org/10.1016/B978-0-323-91225-9.00003-0

Esyanti RR, Zaskia H, Amalia A, Nugrahapraja, dan H. (2019) Chitosan nanoparticle-based coating as post-harvest technology in banana. J Phys Conf Ser 1204:12109

Fadiji AE, Mthiyane DMN, Onwudiwe DC, Babalola OO (2022) Harnessing the known and unknown impact of nanotechnology on enhancing food security and reducing postharvest losses: constraints and future prospects. Agronomy 12(7):1657

Fakhouri FM, Casari ACA, Mariano M, Yamashita F, Mei LHI, Soldi V, Martelli SM (2014) Effect of a gelatin-based edible coating containing cellulose nanocrystals (CNC) on the quality and nutrient retention of fresh strawberries during storage. IOP Conf Ser Mater Sci Eng 64(1):12024

Falsini S, Clemente I, Papini A, Tani C, Schiff S, Salvatici MC, Petruccelli R, Benelli C, Giordano C, Gonnelli C, Ristori S (2019) When sustainable nanochemistry meets agriculture: lignin nanocapsules for bioactive compound delivery to plantlets. ACS Sustain Chem Eng 7(24):19935–19942. https://doi.org/10.1021/acssuschemeng.9b05462

FAO (2022a) Marché mondial des engrais: bilan du resserrement actuel du marché. https://www.fao.org/3/ni280fr/ni280fr.pdf

FAO (2022b) Food and Agriculture Organization of the United Nations. http://www.fao.org/statistics/en/

FAO/OMS (2011) 5pXQLRQ_G·H[SHUWV_)$2_206_ VXU_O·DSSOLFDWLRQ_GHV_ QDQRWHFKQRORJLHV_ GDQV_OHV_VHFWHXUV_GH_O·DOLPHQWDWLRQ_ HW_ GH_O·DJULFXOWXUH__LQFLGHQFHV_SRVVLEOHV_ VXU_OD_VpFXULWp_DOLPHQWDLUH 5DSSRUW_GH_OD_UpXQLRQ. Экономика Региона, 32

Feregrino-Perez AA, Magaña-López E, Guzmán C, Esquivel K (2018) A general overview of the benefits and possible negative effects of the nanotechnology in horticulture. Sci Hortic 238:126–137

Fouda A, Hassan SE-D, Abdo AM, El-Gamal MS (2020) Antimicrobial, antioxidant and larvicidal activities of spherical silver nanoparticles synthesized by endophytic Streptomyces spp. Biol Trace Elem Res 195(2):707–724. https://doi.org/10.1007/s12011-019-01883-4

Ghotekar S (2017) Green Synthesis of Fluorescent CdO Nanoparticles using Leucaena leucocephala L. Extract and their Biological Activities. Journal of Bacteriology & Mycology: Open Access 5(5). https://doi.org/10.15406/jbmoa.2017.05.00148

González-Morales S, Parera CA, Juárez-Maldonado A, De la Fuente MC, Benavides-Mendoza A (2020) The ecology of nanomaterials in agroecosystems. In: Nanomaterials for agriculture and forestry applications. Elsevier, pp 313–355. https://doi.org/10.1016/B978-0-12-817852-2.00013-5

Grillo R, Fraceto LF, Amorim MJB, Scott-Fordsmand JJ, Schoonjans R, Chaudhry Q (2021) Ecotoxicological and regulatory aspects of environmental sustainability of nanopesticides. J Hazard Mater 404:124148. https://doi.org/10.1016/j.jhazmat.2020.124148

Gruère G, Narrod C, Abbott L (2011) Agricultural, food, and water nanotechnologies for the poor opportunities, constraints, and role of the consultative group on international agricultural research. Communications, February. http://www.ifpri.org/sites/default/files/publications/ifpridp01064.pdf

Haldar S, Sengupta S (2023) An overview of the multifaceted role of plant growth-promoting microorganisms and endophytes in sustainable agriculture: developments and prospects. In: Microbial symbionts and plant health: trends and applications for changing climate, pp 179–208. https://doi.org/10.1007/978-981-99-0030-5_8

Harish S, Sendhilvel V, Rajendran L, Parthasarathy S, Raguchander T (2023) Plant associated endophytes as potential agents for the protection of crops from phytopathogens. In: Microbial biocontrol: molecular perspective in plant disease management, pp 89–115. https://doi.org/10.1007/978-981-99-3947-3_4

Hassan O, Chang T (2017) Chitosan for eco-friendly control of plant disease. Asian J Plant Pathol 11(2):53–70. https://doi.org/10.3923/ajppaj.2017.53.70

Hernández-Hernández H, Quiterio-Gutiérrez T, Cadenas-Pliego G, Ortega-Ortiz H, Hernández-Fuentes AD, Cabrera de la Fuente M, Valdés-Reyna J, Juárez-Maldonado A (2019) Impact of selenium and copper nanoparticles on yield, antioxidant system, and fruit quality of tomato plants. Plan Theory 8(10):355

Hernández-López G, Ventura-Aguilar RI, Correa-Pacheco ZN, Bautista-Baños S, Barrera-Necha LL (2020) Nanostructured chitosan edible coating loaded with α-pinene for the preservation of the postharvest quality of *Capsicum annuum* L. and Alternaria alternata control. Int J Biol Macromol 165:1881–1888. https://doi.org/10.1016/j.ijbiomac.2020.10.094

Hossain A, Hong X, Ibrahim E, Li B, Sun G, Meng Y, Wang Y, An Q (2019) Green synthesis of silver nanoparticles with culture supernatant of a bacterium Pseudomonas rhodesiae and their antibacterial activity against soft rot pathogen Dickeya dadantii. Molecules 24(12):2303. https://doi.org/10.3390/molecules24122303

Hosseini SF, Zandi M, Rezaei M, Farahmandghavi F (2013) Two-step method for encapsulation of oregano essential oil in chitosan nanoparticles: preparation, characterization and in vitro release study. Carbohydr Polym 95(1):50–56. https://doi.org/10.1016/j.carbpol.2013.02.031

Hwang IC, Kim TH, Bang SH, Kim KS, Kwon HR, Seo MJ, Youn YN, Park HJ, Yasunaga-Aoki C, Yu YM (2011) Insecticidal effect of controlled release formulations of etofenprox based on nano-bio technique. J Fac Agric Kyushu Univ 56(1):33–40. https://doi.org/10.5109/19633

Ibrahim E, Fouad H, Zhang M, Zhang Y, Qiu W, Yan C, Li B, Mo J, Chen J (2019) Biosynthesis of silver nanoparticles using endophytic bacteria and their role in inhibition of rice pathogenic bacteria and plant growth promotion. RSC Adv 9(50):29293–29299. https://doi.org/10.1039/C9RA04246F

Ibrahim E, Zhang M, Zhang Y, Hossain A, Qiu W, Chen Y, Wang Y, Wu W, Sun G, Li B (2020) Green-synthesization of silver nanoparticles using endophytic bacteria isolated from garlic and its antifungal activity against wheat fusarium head blight pathogen fusarium graminearum. Nanomaterials 10(2):219. https://doi.org/10.3390/nano10020219

Jagadesh M, Dash M, Kumari A, Singh SK, Verma KK, Kumar P, Bhatt R, Sharma SK (2024) Revealing the hidden world of soil microbes: metagenomic insights into plant, bacteria, and fungi interactions for sustainable agriculture and ecosystem restoration. Microbiol Res 285:127764. https://doi.org/10.1016/j.micres.2024.127764

Jampílek J, Kráľová K (2019) Nanobiopesticides in agriculture: state of the art and future opportunities. In: Nano-biopesticides today and future perspectives, pp 397–447. https://doi.org/10.1016/B978-0-12-815829-6.00018-8

Jampílek J, Králová K (2020) Impact of nanoparticles on toxigenic fungi. In: Nanomycotoxicology. Elsevier, pp 309–348. https://doi.org/10.1016/B978-0-12-817998-7.00014-8

Jayasekhar Babu P, Saranya S, Longchar B, Rajasekhar A (2022) Nanobiotechnology-mediated sustainable agriculture and post-harvest management. Curr Res Biotechnol 4(July):326–336. https://doi.org/10.1016/j.crbiot.2022.07.004

John RP, Tyagi RD, Brar SK, Surampalli RY, Prévost D (2011) Bio-encapsulation of microbial cells for targeted agricultural delivery. Crit Rev Biotechnol 31(3):211–226. https://doi.org/10.3109/07388551.2010.513327

Juárez-Maldonado A, Tortella G, Rubilar O, Fincheira P, Benavides-Mendoza A (2021) Biostimulation and toxicity: the magnitude of the impact of nanomaterials in microorganisms and plants. J Adv Res 31:113–126. https://doi.org/10.1016/j.jare.2020.12.011

Kah M, Tufenkji N, White JC (2019) Nano-enabled strategies to enhance crop nutrition and protection. Nat Nanotechnol 14(6):532–540. https://doi.org/10.1038/s41565-019-0439-5

Khafar EAA, Zidan NS, Aboul-Anean HED (2018) The effect of nano materials on edible coating and films' improvement. Int J Pharm Res Allied Sci 7(3–2018):20–41

Khodakovskaya MV, de Silva K, Nedosekin DA, Dervishi E, Biris AS, Shashkov EV, Galanzha EI, Zharov VP (2011) Complex genetic, photothermal, and photoacoustic analysis of nanoparticle-plant interactions. Proc Natl Acad Sci 108(3):1028–1033. https://doi.org/10.1073/pnas.1008856108

Kumar P, Pandhi S, Mahato DK, Kamle M, Mishra A (2021) Bacillus-based nano-bioformulations for phytopathogens and insect–pest management. Egypt J Biol Pest Control 31:1–11

Kumar D, Mariyam S, Gupta KJ, Thiruvengadam M, Ghodake GS, Xing B, Seth CS (2024a) Comparative investigation on chemical and green synthesized titanium dioxide nanoparticles against chromium (VI) stress eliciting differential physiological, biochemical, and cellular attributes in Helianthus annuus L. Sci Total Environ 930:172413. https://doi.org/10.1016/j.scitotenv.2024.172413

Kumar D, Singh R, Upadhyay SK, Verma KK, Tripathi RM, Liu H, Dhankher OP, Tripathi RD, Sahi SV, Seth CS (2024b) Review on interactions between nanomaterials and phytohormones: novel perspectives and opportunities for mitigating environmental challenges. Plant Sci 340:111964. https://doi.org/10.1016/j.plantsci.2023.111964

Kumar D, Biswas JK, Mulla SI, Singh R, Shukla R, Ahanger MA, Shekhawat GS, Verma KK, Siddiqui MW, Seth CS (2024c) Micro and nanoplastics pollution: sources, distribution, uptake in plants, toxicological effects, and innovative remediation strategies for environmental sustainability. Plant Physiol Biochem 213:108795. https://doi.org/10.1016/j.plaphy.2024.108795

Kumari B, Mallick MA, Solanki MK, Hora A, Mani M (2019) Applying nanotechnology to bacteria: an emerging technology for sustainable agriculture. In: Role of plant growth promoting microorganisms in sustainable agriculture and nanotechnology. Elsevier, pp 121–143

Lamichhane JR, Osdaghi E, Behlau F, Köhl J, Jones JB, Aubertot JN (2018) Thirteen decades of antimicrobial copper compounds applied in agriculture. A review. Agron Sustain Dev 38(3). https://doi.org/10.1007/s13593-018-0503-9

Li W, Li L, Cao Y, Lan T, Chen H, Qin Y (2017) Effects of PLA film incorporated with ZnO nanoparticle on the quality attributes of fresh-cut apple. Nanomaterials 7:207

Liang W, Yu A, Wang G, Zheng F, Jia J, Xu H (2018) Chitosan-based nanoparticles of avermectin to control pine wood nematodes. Int J Biol Macromol 112:258–263. https://doi.org/10.1016/j.ijbiomac.2018.01.174

Liu J, Tian S, Meng X, Xu Y (2007) Effects of chitosan on control of postharvest diseases and physiological responses of tomato fruit. Postharvest Biol Technol 44(3):300–306

Madbouly AK (2021) The efficacy of green synthesized nanosilver in reducing the incidence of post-harvest apple fruit brown rot. J Fungi 7(6). https://doi.org/10.3390/jof7060473

Maqbool M, Ali A, Alderson PG, Zahid N, Siddiqui Y (2011) Effect of a novel edible composite coating based on gum arabic and chitosan on biochemical and physiological responses of banana fruits during cold storage. J Agric Food Chem 59(10):5474–5482

Marchiol L (2018) Nanotechnology in agriculture: new opportunities and perspectives. In: New visions in plant science, vol i. https://doi.org/10.5772/intechopen.74425

Mariyam S, Upadhyay SK, Chakraborty K, Verma KK, Duhan JS, Muneer S, Meena M, Sharma RK, Ghodake G, Seth CS (2024) Nanotechnology, a frontier in agricultural science, a novel approach in abiotic stress management and convergence with new age medicine-a review. Sci Total Environ 912:169097. https://doi.org/10.1016/j.scitotenv.2023.169097

Martínez-Ortiz MA, Palma-Rodríguez HM, Montalvo-González E, Sáyago-Ayerdi SG, Utrilla-Coello R, Vargas-Torres A (2019) Effect of using microencapsulated ascorbic acid in coatings based on resistant starch chayotextle on the quality of guava fruit. Sci Hortic 256:108604

Melo NFCB, de MendonçaSoares BL, Diniz KM, Leal CF, Canto D, Flores MAP, da Costa Tavares-Filho JH, Galembeck A, Stamford TLM, Stamford-Arnaud TM (2018) Effects of fungal chitosan nanoparticles as eco-friendly edible coatings on the quality of postharvest table grapes. Postharvest Biol Technol 139:56–66

Mesterházy Á, Oláh J, Popp J (2020) Losses in the grain supply chain: causes and solutions. Sustain For 12(6):2342

Miller RJ, Bennett S, Keller AA, Pease S, Lenihan HS (2012) TiO_2 nanoparticles are phototoxic to marine phytoplankton. PLoS One 7(1):e30321. https://doi.org/10.1371/journal.pone.0030321

Mishra S, Singh A, Keswani C, Singh HB (2014) Nanotechnology: exploring potential application in agriculture and its opportunities and constraints. Biotech Today 4(1):9–14

Mishra S, Singh A, Keswani C, Saxena A, Sarma BK, Singh HB (2015) Harnessing plant-microbe interactions for enhanced protection against phytopathogens. In: Plant microbes symbiosis: applied facets. Springer India, pp 111–125. https://doi.org/10.1007/978-81-322-2068-8_5

Mishra S, Keswani C, Abhilash PC, Fraceto LF, Singh HB (2017) Integrated approach of agri-nanotechnology: challenges and future trends. Front Plant Sci 8. https://doi.org/10.3389/fpls.2017.00471

Mukhopadhyay SS (2014) Nanotechnology in agriculture: prospects and constraints. Nanotechnol Sci Appl 7(2):63–71. https://doi.org/10.2147/NSA.S39409

Naraghi L, Negahban M (2019) Biological control of cotton Verticillium wilt by nanoformulations containing Talaromyces flavus. Eurasia J Biosci 13(July):1177–1185

Nassar AMK (2016) Effectiveness of Silver Nano-particles of Extracts of Urtica urens (Urticaceae) Against Root-knot Nematode Meloidogyne incognita. Asian Journal of Nematology 5(1): 14–19. https://doi.org/10.3923/ajn.2016.14.19

Naraghi L, Negahban M, Heydari A, Razavi M, Afshari-Azad H (2018) Growth inhibition of fusarium oxysporum f. Sp. lycopercisi, the causal agent of tomato fusarium wilt disease by nanoformulations containing talaromyces flavus. Ekoloji 27:103–112

Natesan K, Ponmurugan P, Gnanamangai BM, Manigandan V, Joy SPJ, Jayakumar C, Amsaveni G (2021) Biosynthesis of silica and copper nanoparticles from Trichoderma, Streptomyces and Pseudomonas spp. evaluated against collar canker and red root-rot disease of tea plants. Arch Phytopathol Plant Protect 54(1–2):56–85. https://doi.org/10.1080/03235408.2020.1817258

Neme K, Nafady A, Uddin S, Tola YB (2021) Application of nanotechnology in agriculture, post-harvest loss reduction and food processing: food security implication and challenges. Heliyon 7(12):e08539. https://doi.org/10.1016/j.heliyon.2021.e08539

Okeke ES, Ezeorba TPC, Mao G, Chen Y, Feng W, Wu X (2022) Nano-enabled agrochemicals/materials: potential human health impact, risk assessment, management strategies and future prospects. Environ Pollut 295:118722. https://doi.org/10.1016/j.envpol.2021.118722

Padmakumar A, Pavani C, Eswar K, Kong L, Yang W, Gopalakrishnan S, Cahill DM, Rengan AK (2023) Bacteria-premised nanobiopesticides for the management of phytopathogens and pests. ACS Agric Sci Technol 3(5):370–388. https://doi.org/10.1021/acsagscitech.3c00025

Palmqvist NGM, Bejai S, Meijer J, Seisenbaeva GA, Kessler VG (2015) Nano titania aided clustering and adhesion of beneficial bacteria to plant roots to enhance crop growth and stress management. Sci Rep 5(1):10146. https://doi.org/10.1038/srep10146

Patel KV, Nath M, Bhatt MD, Dobriyal AK, Bhatt D (2020) Nanofomulation of zinc oxide and chitosan zinc sustain oxidative stress and alter secondary metabolite profile in tobacco. 3 Biotech 10(11):477. https://doi.org/10.1007/s13205-020-02469-x

Pathakoti K, Manubolu M, Hwang H-M (2017) Nanostructures: current uses and future applications in food science. J Food Drug Anal 25(2):245–253

Paula HCB, Sombra FM, Abreu FOMS, de Paul RCM (2010) Lippia sidoides essential oil encapsulation by angico gum/chitosan nanoparticles. J Braz Chem Soc 21(12):2359–2366. https://doi.org/10.1590/S0103-50532010001200025

Pereira A d ES, Oliveira HC, Fraceto LF (2019) Polymeric nanoparticles as an alternative for application of gibberellic acid in sustainable agriculture: a field study. Sci Rep 9(1):7135. https://doi.org/10.1038/s41598-019-43494-y

Pilon L, Spricigo PC, Miranda M, Assis OBG, Mattoso LHC, Ferreira MD (2014) Original article Chitosan nanoparticle coatings reduce microbial growth on fresh-cut apples while not affecting quality attributes. 2013:1–9. https://doi.org/10.1111/ijfs.12616

Piña-Barrera AM, Álvarez-Román R, Báez-González JG, Amaya-Guerra CA, Rivas-Morales C, Gallardo-Rivera CT, Galindo-Rodríguez SA (2019) Application of a multisystem coating based on polymeric nanocapsules containing essential oil of Thymus vulgaris L. to increase the shelf life of table grapes (Vitis vinifera L.). IEEE Trans Nanobioscience 18(4):549–557

Pirzadah TB, Malik B, Maqbool T, Rehman RU (2019) Development of nano-bioformulations of nutrients for sustainable agriculture. In: Nanotechnology in the life sciences., July, pp 381–394. https://doi.org/10.1007/978-3-030-17061-5_16

Ponmurugan P, Manjukarunambika K, Elango V, Gnanamangai BM (2016) Antifungal activity of biosynthesised copper nanoparticles evaluated against red root-rot disease in tea plants. J Exp Nanosci 11(13):1019–1031. https://doi.org/10.1080/17458080.2016.1184766

Prashar P, Shah S (2016) Impact of fertilizers and pesticides on soil microflora in agriculture. (Issue April). https://doi.org/10.1007/978-3-319-26777-7_8

Priester JH, Ge Y, Mielke RE, Horst AM, Moritz SC, Espinosa K, Gelb J, Walker SL, Nisbet RM, An Y-J, Schimel JP, Palmer RG, Hernandez-Viezcas JA, Zhao L, Gardea-Torresdey JL, Holden PA (2012) Soybean susceptibility to manufactured nanomaterials with evidence for food quality and soil fertility interruption. Proc Natl Acad Sci 109(37). https://doi.org/10.1073/pnas.1205431109

Qiao G, Xiao Z, Ding W, Rok A (2019) Effect of chitosan/nano-titanium dioxide/thymol and tween films on ready-to-eat cantaloupe fruit quality. Coatings 9(12):828

Rajput, V. D., Singh, A., Minkina, T. M., Shende, S. S., Kumar, P., Verma, K. K., Bauer, T., Gorobtsova, O., Deneva, S., & Sindireva, A. (2021a). Potential Applications of Nanobiotechnology in Plant Nutrition and Protection for Sustainable Agriculture. In Nanotechnology in Plant Growth Promotion and Protection (pp. 79–92). Wiley, Hoboken, NJ, USA. https://doi.org/10.1002/9781119745884.ch5

Rajput VD, Singh A, Minkina T, Rawat S, Mandzhieva S, Sushkova S, Shuvaeva V, Nazarenko O, Rajput P, Komariah VKK, Singh AK, Rao M, Upadhyay SK (2021b) Nano-enabled products: challenges and opportunities for sustainable agriculture. Plan Theory 10:2727. https://doi.org/10.3390/plants10122727

Rana RA, Siddiqui MN, Skalicky M, Brestic M, Hossain A, Kayesh E, Popov M, Hejnak V, Gupta DR, Mahmud NU, Islam T (2021) Prospects of nanotechnology in improving the productivity and quality of horticultural crops. Horticulturae 7(10):1–19. https://doi.org/10.3390/horticulturae7100332

Rashidi L, Khosravi-Darani K (2011) The applications of nanotechnology in food industry. Crit Rev Food Sci Nutr 51(8):723–730. https://doi.org/10.1080/10408391003785417

Rathee M, Dalal P (2018) Emerging insect pests in Indian agriculture. Indian J Entomol 80(2):267–281

Redlingshöfer B (2017) 3.9. Pertes et gaspillages alimentaires. In: L'alimentation à découvert. https://doi.org/10.4000/books.editionscnrs.10336

Rhim JW (2004) Increase in water vapor barrier property of biopolymer-based edible films and coatings by compositing with lipid materials. Food Sci Biotechnol 13(4):528–535

Saharan P, Kumar V, Kaushal I, Mittal A, Shukla SK, Kumar D, Sharma AK, Om H (2023) A comprehensive review on the metal-based green valorized nanocomposite for the remediation of emerging colored organic waste. Environ Sci Pollut Res 30(16):45677–45700. https://doi.org/10.1007/S11356-023-25998-3

Sajid M, Płotka-Wasylka J (2020) Nanoparticles: synthesis, characteristics, and applications in analytical and other sciences. Microchem J 154. https://doi.org/10.1016/j.microc.2020.104623

Sangeetha J, Thangadurai D, Hospet R, Harish ER, Purushotham P, Mujeeb MA, Shrinivas J, David M, Mundaragi AC, Thimmappa SC, Arakera SB, Prasad R (2017a) Nanoagrotechnology for soil quality, crop performance and environmental management. In: Nanotechnology. Springer Singapore, pp 73–97. https://doi.org/10.1007/978-981-10-4573-8_5

Sangeetha J, Thangadurai D, Hospet R, Purushotham P, Karekalammanavar G, Mundaragi AC, David M, Shinge MR, Thimmappa SC, Prasad R (2017b) Agricultural nanotechnology: concepts, benefits, and risks. In: Nanotechnology: an agricultural paradigm, pp 1–17

Sanguansri P, Augustin MA (2006) Nanoscale materials development–a food industry perspective. Trends Food Sci Technol 17(10):547–556

Seray M, Skender A, Hadj-Hamou AS (2021) Kinetics and mechanisms of Zn 2+ release from antimicrobial food packaging based on poly (butylene adipate-co-terephthalate) and zinc oxide nanoparticles. Polym Bull 78:1021–1040

Shabani MH, Naraghi L, Maleki M, Negahban M (2024) Evaluation of the efficacy of different concentrations of nano-capsules containing Talaromyces flavus with two forms of powder and suspension in reducing the incidence of cotton Verticillium wilt. Braz J Biol 84. https://doi.org/10.1590/1519-6984.262480

Shea J (2008) Nanotechnology – basic calculations for engineers and scientists [Book Review]. IEEE Electr Insul Mag 22(6). https://doi.org/10.1109/mei.2006.253431

Shi S, Wang W, Liu L, Wu S, Wei Y, Li W (2013) Effect of chitosan/nano-silica coating on the physicochemical characteristics of longan fruit under ambient temperature. J Food Eng 118(1):125–131

Silva HD, Cerqueira MÂ, Vicente AA (2012) Nanoemulsions for food applications: development and characterization. Food Bioprocess Technol 5:854–867

Singh D, Gurjar BR (2022) Nanotechnology for agricultural applications: facts, issues, knowledge gaps, and challenges in environmental risk assessment. J Environ Manag 322:116033. https://doi.org/10.1016/J.JENVMAN.2022.116033

Singh H, Sharma A, Bhardwaj SK, Arya SK, Bhardwaj N, Khatri M (2021) Recent advances in the applications of nano-agrochemicals for sustainable agricultural development. Environ Sci: Processes Impacts 23(2):213–239. https://doi.org/10.1039/D0EM00404A

Sogvar OB, Koushesh Saba M, Emamifar A, Hallaj R (2016) Influence of nano-ZnO on microbial growth, bioactive content and postharvest quality of strawberries during storage. Innovative Food Sci Emerg Technol 35:168–176. https://doi.org/10.1016/J.IFSET.2016.05.005

Sujithra S, Manikkandan TR (2019) Application of nanotechnology in packaging of foods: a review. Int J ChemTech Res 12(4):7–14

Taheri P, Soweizy M, Tarighi S (2023) Application of essential oils to control some important fungi and bacteria pathogenic on cereals. J Nat Pest Res 6(March):100052. https://doi.org/10.1016/j.napere.2023.100052

Thiruvengadam M, Rajakumar G, Chung IM (2018) Nanotechnology: current uses and future applications in the food industry. 3 Biotech 8(1):1–13. https://doi.org/10.1007/s13205-018-1104-7
UN DESA (2022) World population prospects 2022
Upadhyay TK, Varun Kumar VS, Sharangi AB, Upadhye VJ, Khan F, Pandey P, Kamal MA, Baba AY, Hakeem KR (2022) Nanotechnology-based advancements in postharvest management of horticultural crops. Phyton-Int J Exp Bot 91(3):471–487. https://doi.org/10.32604/phyton.2022.017258
Usman M, Farooq M, Wakeel A, Nawaz A, Cheema SA, Rehman H u, Ashraf I, Sanaullah M (2020) Nanotechnology in agriculture: current status, challenges and future opportunities. Sci Total Environ 721:137778. https://doi.org/10.1016/j.scitotenv.2020.137778
Van Esse HP, Reuber TL, van der Does D (2020) Genetic modification to improve disease resistance in crops. New Phytol 225(1):70–86
VanderGheynst JS, Scher H, Guo H-Y (2006) Design of formulations for improved biological control agent viability and sequestration during storage. Ind Biotechnol 2(3):213–219
Verma KK, Song XP, Joshi A, Rajput VD, Singh M, Sharma A, Singh RK, Li DM, Arora J, Minkina T, Li YR (2022a) Nanofertilizer possibilities for healthy soil, water and food in future: an overview. Front Plant Sci 13:865048. https://doi.org/10.3389/fpls.2022.865048
Verma KK, Song XP, Joshi A, Tian DD, Rajput VD, Singh M, Arora J, Minkina TM, Li YR (2022b) Recent trends in nano-fertilizer for advancing sustainable agriculture under the era of climate change to ensure food security for future generations. Nanomaterials 12:173. https://doi.org/10.3390/nano12010173
Verma KK, Song X, Degu HD, Guo D, Joshi A, Huang H, Xu L, Singh M, Huang D, Rajput VD, Li Y (2023) Recent advances in nitrogen and nano-nitrogen fertilizers for sustainable crop production: a mini-review. Chem Biol Technol Agric 10:1–14. https://doi.org/10.1186/s40538-023-00488-3
Verma KK, Song XP, Kumari A, Jagadesh M, Singh SK, Bhatt R, Singh M, Seth CS, Li YR (2024a) Climate change adaptation: challenges for agricultural sustainability. Plant Cell Environ:1–12. https://doi.org/10.1111/pce.15078
Verma KK, Song X-P, Singh S, Kumari A, Arora J, Singh SK, Solanki MK, Seth CS, Li Y-R (2024b) Synergistic interactions of nanoparticles and plant growth promoting rhizobacteria enhancing soil-plant systems: a multigenerational perspective. Front Plant Sci 15:1376214. https://doi.org/10.3389/fpls.2024.1376214
Yang FM, Li HM, Li F, Xin ZH, Zhao LY, Zheng YH, Hu QH (2010) Effect of nano-packing on preservation quality of fresh strawberry (Fragaria ananassa Duch. cv Fengxiang) during storage at 4 C. J Food Sci 75(3):C236–C240
Yang X, He Q, Guo F, Sun X, Zhang J, Chen Y (2021) Impacts of carbon-based nanomaterials on nutrient removal in constructed wetlands: microbial community structure, enzyme activities, and metabolism process. J Hazard Mater 401:123270. https://doi.org/10.1016/j.jhazmat.2020.123270
Youssef K, Hashim AF (2020) Inhibitory effect of clay/chitosan nanocomposite against Penicillium digitatum on citrus and its possible mode of action. Jordan J Biol Sci 13(3):349–355
Youssef K, Roberto SR (2014) Applications of salt solutions before and after harvest affect the quality and incidence of postharvest gray mold of 'Italia' table grapes. Postharvest Biol Technol 87:95–102
Zabot GL, Schaefer Rodrigues F, Polano Ody L, Vinícius Tres M, Herrera E, Palacin H, Córdova-Ramos JS, Best I, Olivera-Montenegro L (2022) Encapsulation of bioactive compounds for food and agricultural applications. Polymers 14(19). https://doi.org/10.3390/polym14194194
Zahid N, Ali A, Manickam S, Siddiqui Y, Maqbool M (2012) Potential of chitosan-loaded nanoemulsions to control different Colletotrichum spp. and maintain quality of tropical fruits during cold storage. J Appl Microbiol 113(4):925–939
Zahra N, Hafeez MB, Shaukat K, Wahid A, Hussain S, Naseer R, Raza A, Iqbal S, Farooq M (2021) Hypoxia and anoxia stress: plant responses and tolerance mechanisms. J Agron Crop Sci 207(2):249–284

Zambrano-Zaragoza ML, Gutiérrez-Cortez E, Del Real A, González-Reza RM, Galindo-Pérez MJ, Quintanar-Guerrero D (2014) Fresh-cut Red Delicious apples coating using tocopherol/mucilage nanoemulsion: effect of coating on polyphenol oxidase and pectin methylesterase activities. Food Res Int 62:974–983

Zhang W (2018) Global pesticide use: profile, trend, cost/benefit and more. Proc Int Acad Ecol Environ Sci 8(1):1

Zhang P, Guo Z, Ullah S, Melagraki G, Afantitis A, Lynch I (2021) Nanotechnology and artificial intelligence to enable sustainable and precision agriculture. Nat Plants 7(7):864–876. https://doi.org/10.1038/s41477-021-00946-6

Chapter 20
Opportunities and Challenges of Nano-engineering

Venkatratna Shrinivas Rao Bhiksham Perike

20.1 Introduction

The Sire of Nanotechnology is premeditated to be American physicist Richard Feynman. He instigatethe postulation in 1959, during his orate, "There's Plenty of Room at the Bottom". Succeeding fifteen years, Norio Taniguchi, a Japanese scientist was the first to manoeuvre and define the locution "nanotechnology" in 1974 as: "nanotechnology mainly consists of the processing of separation, consolidation, and deformation of materials by one atom or one molecule". 1 to 100 nanometers within the orbit of waged with the matter and anatomy at the molecular and atomic levels. This automation is termed as "Nanotechnology". Pursuits of design, develop and implement innovative solutions at Nanoscale in application of nanotechnology and nanoscience. Inventors which comply this quality is called as "Nano Engineers". Monopolies like electronics, materials science, medicine, energy and manufacturing etc vital their proems. Steering and administer matter at the nanoscale to invent new materials with idiosyncratic properties and functionalities as well as to revamp existing high tech.

20.2 Opportunities and Types of Engineers

As a nanoengineers your premier part will be thrive latest products, policies and affairs. Copious sphere and trades are fetes you like biomedical technologies, DNA Computers, targeted drug sytems, carbon-based electronics etc. Master's degree in Nanotechology unlatches a threshold and metier elbow room which are baggy and

V. S. R. B. Perike (✉)
Department of Agriculture, Rocky Ponds Produce, Gumlu, Queensland, Australia

C. S. Seth et al. (eds.), *Nano-formulations for Sustainable Agriculture*,
Nanotechnology in the Life Sciences, https://doi.org/10.1007/978-3-031-90617-6_20

ginormous. Wherefore appends Nanomaterial engineers, Biomedical Nanotechnologists, nanoelectronicsResearchers, Nanoparticle Toxicologists, Nanosytems Architects, Nanomedicine scientists, Quantum Dot Technologists, Nanomanufacturing engineers, Nanocomposite Researchers, Nanorobiotics Engineers whatnot. Conservatoires, Bureaus, Capitalistic sytems etc. are the plant for nanoengineers. Research facilities, legwork, Maneuver with rebbes, white collars.

Shepherd colloquys, virtual meetings, ateliers etc are bet for sculpt the edicts of Erudition, Transmogrifications and Burgeonings. Furthest up 10 realms for orb nanotechnology are USA, China, Germany, Japan, Korea, France, UK, India, Italy, and Spain. Best known and funded program in Nanotechnology is The National Nanotechnology Initiative (NNI) is a USA Government research and development (R&D) initiative. Nanotechnology by National Aeronautics and Space Administration (NASA) USA Directorate bureau, neoteric acme include: nano-aluminum/ice rocket propellant, ultrasensitive chemical sensors for monitoring trace gases and vapors, nano-structured composites for thermal isolation applications, and smart electroactive materials.

20.3 Duties and Responsibilities

Nanotechnology Engineers have various fidelity and burdens to enact. Indagation and Evolution in Nanomaterials are been assembled and thrived by nanoengineers with error-free chattels via scrutiny. Covet facets of mediums are been wangled like strength, conductivity or reactivity by manoeuvre the materials at nanoscale for copious enactments like electronics, sensors, drug delivery and energy storage appertain to the studying of nanoscale physics and engineering. Quality control and testing in nanomaterials caliber and uniformity is epoch-making Nanoengineers are liable for nanomaterials' persusal. Appraise, chattels and gig of nanomaterials are done using cutting-edge technology. As Electron Microscopy, Atomic force microscopy, Spectroscopy. Rackets exigencies are stumble on by pennant quality control by indsutry merits and managerial exigencies. Innovation and Snag fathoming in Nanoengineers are trailblazer of upheavals, new odds, inscribing trails at nanoscales. Problem deciphers fortes, technical snags, hone of nanotechnology are the salient slants. Promulgation and Evidences plays Complex concept, Sundry patrons etc. are nanoengineers prime facets to bequeath their notions as Gen and Facts is the key with other duties includes methodologies of research, results, and project progression. Obedience of Safety Concordat & Learning's role by Nanoengineers slate welfare as premier facet as per the idiosyncratic properties of nanomaterials like apt handling, de regale disposal stratagems, palliate potential perils for souls and air. Nanotechnology is a pronto transmogrifying field, and engineers must linger contemporaneous with the modern advancements and emerging technologies. Continual swotting through professional development activities, conferences, and collaboration with peers is requisite.

20.4 Challenges

Gubbins at the nanoscale have odd optical, magnetic, and electronic facets. The dare for nanoengineers and nano scientists is to forge trails to a limpid ken of what exactly is affairs in the nanoworld, so they can more prudent and potently yoke the thew of nanotechnology. Varsities and Guilds across the globe are rigorously clocking how atoms fit in cahoots to form goodly structures. Nanotechnology can even be used in the time ahead to dose- life baleful maladys like cancer. Anyhow, it does have some snags, for example, virulency, environmental torment and organ mar caused by nanoparticles. There are some moral affairs perturbed with the use of nanotechnology too. The utmost dispute facing nanotechnology is inventing finer modeling tacks. Erecting the large calibers needed is still cadence consuming and exorbitant. Erst in the blood stream, nanomaterials can circulate throughout the body and can lodge in organs and tissues including the brain, liver, heart, kidneys, spleen, bone marrow and nervous system. Whilom inside cells, they may interfere with normal cellular function, cause oxidative damage and even cell death.

20.5 Conclusion

Nanotechnology has the flair to rejig our lives. This is because it presents almost untold latent to make wondrous changes in essentially all domains ranging from medicine, computer technology, construction, environmental remediation, food industry, to new energy sources. Nanotechnology is a sear pristine hi-tech that has just embarked on, it is a Belshevik science that will barter all what we knew before. The doom that we were gazing just in science will soon be unfeigned.

References

https://www.careerexplorer.com/careers/nanotechnology-engineer/
https://www.uts.edu.au/study/science/course-experience/careers-and-employment/chemistry-physics-and-nanotechnology-career-options
https://vlsifirst.com/blog/nanotechnology-careers
https://www.linkedin.com/advice/1/what-some-common-nanotechnology-career-paths-sectors

[illegible]

Correction to: Nanomaterials Strategies for Enhancing Plant Adaptation in the Adverse Agroclimatic Variables

Yogesh Dashrath Naik, Anil Namdev Kale, Viabhav Kumar Upadhayay, Adita Sharma, and Ajay Veer Singh

Correction to:
Chapter 7 in: C. S. Seth et al. (eds.), *Nano-formulations for Sustainable Agriculture*, Nanotechnology in the Life Sciences, https://doi.org/10.1007/978-3-031-90617-6_7

The chapter was inadvertently published with the incorrect author name. The name has been updated as Ajay Veer Singh.

The updated version of this chapter can be found at
https://doi.org/10.1007/978-3-031-90617-6_7

C. S. Seth et al. (eds.), *Nano-formulations for Sustainable Agriculture*,
Nanotechnology in the Life Sciences, https://doi.org/10.1007/978-3-031-90617-6_21

Index

C. S. Seth et al. (eds.), *Nano-formulations for Sustainable Agriculture*, Nanotechnology in the Life Sciences,
https://doi.org/10.1007/978-3-031-90617-6

MIX
Papier aus verantwortungsvollen Quellen
Paper from responsible sources
FSC® C105338

If you have any concerns about our products,
you can contact us on
ProductSafety@springernature.com

In case Publisher is established outside the EU,
the EU authorized representative is:
Springer Nature Customer Service Center GmbH
Europaplatz 3, 69115 Heidelberg, Germany

Printed by Libri Plureos GmbH
in Hamburg, Germany